ARCTIC AND ALPINE MYCOLOGY II

ENVIRONMENTAL SCIENCE RESEARCH

Recent Volumes in this Series

Volume 27 – SHORT-TERM BIOASSAYS IN THE ANALYSIS OF COMPLEX ENVIRONMENTAL MIXTURES III
Edited by Michael D. Waters, Shahbeg S. Sandhu, Joellen Lewtas, Larry Claxton, Neil Chernoff, and Stephen Nesnow

Volume 28 – UTILIZATION OF MAMMALIAN SPECIFIC LOCUS STUDIES IN HAZARD EVALUATION AND ESTIMATION OF GENETIC RISK
Edited by Frederick J. de Serres and William Sheridan

Volume 29 – APPLICATION OF BIOLOGICAL MARKERS TO CARCINOGEN TESTING
Edited by Harry A. Milman and Stewart Sell

Volume 30 – INDIVIDUAL SUSCEPTIBILITY TO GENOTOXIC AGENTS IN THE HUMAN POPULATION
Edited by Frederick J. de Serres and Ronald W. Pero

Volume 31 – MUTATION, CANCER, AND MALFORMATION
Edited by Ernest H. Y. Chu and Walderico M. Generoso

Volume 32 – SHORT-TERM BIOASSAYS IN THE ANALYSIS OF COMPLEX ENVIRONMENTAL MIXTURES IV
Edited by Michael D. Waters, Shahbeg S. Sandhu, Joellen Lewtas, Larry Claxton, Gary Strauss, and Stephen Nesnow

Volume 33 – ACID RAIN: Economic Assessment
Edited by Paulette Mandelbaum

Volume 34 – ARCTIC AND ALPINE MYCOLOGY II
Edited by Gary A. Laursen, Joseph F. Ammirati, and Scott A. Redhead

---

A Continuation Order Plan is available for this series. A continuation order will bring delivery of each new volume immediately upon publication. Volumes are billed only upon actual shipment. For further information please contact the publisher.

ARCTIC AND ALPINE MYCOLOGY II

Edited by
Gary A. Laursen
School of Agriculture and Land Resources Management
University of Alaska
Fairbanks, Alaska

Joseph F. Ammirati
Department of Botany
University of Washington
Seattle, Washington

and

Scott A. Redhead
Biosystematics Research Centre
Ottawa, Ontario, Canada

Springer Science+Business Media, LLC

Library of Congress Cataloging in Publication Data

International Symposium on Arcto-Alpine Mycology (2nd: 1984: Fetan, Switzerland)
Arctic and alpine mycology II.

(Environmental science research; v. 34)
"Proceedings of the Second International Symposium on Arctic and Alpine Mycology, held August 26-September 2, 1984, in Fetan, Switzerland"—T.p. verso.
Bibliography: p.
Includes index.
1. Fungi—Arctic regions—Congresses. 2. Mycology—Congresses. 3. Alpine flora—Congresses. I. Laursen, Gary A. II. Ammirati, Joseph F. III. Redhead, Scott A. IV. Title. V. Title: Arctic and alpine mycology. 2. Vi. Series.
QK615.I57 1984 589.2'0998 87-7814

ISBN 978-1-4757-1941-3 ISBN 978-1-4757-1939-0 (eBook)
DOI 10.1007/978-1-4757-1939-0

Proceedings of the Second International Symposium on Arctic and Alpine Mycology, held August 26–September 2, 1984, in Fetan, Switzerland

Originally published by Plenum Press, New York in 1987.

PREFACE

During the summer of 1980, the First International Symposium on Arctic and Alpine Mycology (ISAM-I) was held at the then extant Naval Arctic Research Laboratory near Barrow, Alaska, U.S.A., well within the Arctic Circle (Laursen and Ammirati, Arctic and Alpine Mycology. The First International Symposium on Arcto-Alpine Mycology. Univ. Wash. Press, 1982). The facility is currently owned and operated by the Utkeagvik Inupiat Community and is named the National Academic and Research Laboratory, thus retaining its acronym NARL.

Twenty-five scientists participated in that historic first meeting. Their interests in the fungi spanned a vast geographic area of cold dominated habitats in both the northern and southern hemispheres that included four continents (N. and S. America, Eurasia,and Antarctica), nine countries, and numerous islands ranging from Greenland to Jan Mayen in the Svalbard group.

ISAM-I helped to develop ongoing interests and initiate others. This is what ISAM-I founders hoped would happen. As a result, the organizing committee for ISAM-II was formed. Its mandate was to: involve a maximum of one third new participants in future ISAM meetings: divide the responsibility for organizing future meetings at sites located in areas of interest to research thrusts in Arctic and alpine environments: keep the number of participants small enough to ensure manageability, taking full advantage of field collecting opportunities with minimal complications and cost.

ISAM-II was held in eastern Switzerland in the Unter-Engadin at the Hochalpines Töchter-Institut at Fetan/Ftan near the Swiss National Park, from August 26 to September 2, 1984. The National Park was a favorite collecting area of a Swiss pioneer in alpine mycology, Jules Favre. Twenty-eight participants, all pictured in Figure 1, represented twelve countries as follows: Austria (1); Canada (1); Denmark (3); Finland (2); France (2); Germany (1); Japan (1); Norway (3); Scotland (2); Sweden (2); Switzerland (7); and the U. S. A. (3). They visited a variety of alpine sites, including type localities, in a carefully planned and executed itinerary of field trips organized by Prof. E. Müller and Dr. E. Horak and their students. The group was fortunate enough to have good weather for the duration of the symposium, as the following week the entire area became snowed under with an early storm. Evening sessions were devoted to the presentation of papers and examination of specimens which often led to animated discussions between seasoned older researchers and younger scientists, between North Americans and Europeans, and between taxonomists and ecologists. There was an important balance between senior, middle-level, younger and graduate-level scientists, insuring continued growth, an influx of new research ideas, and an infusion of individuals researching mycological problems in Arctic and alpine habitats.

Growing popularity in Artic and alpine mycology requires that organizers of future ISAM meetings restrict the number in forays to about 25-30 individuals. This is mainly due to the difficulties involved in supporting and transporting large groups to remote sites. Holding formal symposia in conjunction with other international meetings in more central metropolitan areas may be necessary to facilitate the interaction of a larger group. This could be coupled with a 7-10 day foray to a remote site with a smaller group of participants. This problem needs resolving if groups of more than 30 participants are going to meet at future symposium gatherings.

The establishment of Arctic and Alpine Mycology Mycology as a branch of study was addressed at ISAM-II. A preamble for "Arranging the ISAM" was drafted by M. Lange (Copenhagen) and adopted at the ISAM-II business meeting. It sets guidelines for organizing future ISAM gatherings in a meaningfull way and allows for the development of all aspects of Arctic and alpine mycology.

The preamble is as follows:

1. Invitations for the next assembly are considered and determined at each ISAM. The session should accept one invitation and indicate a second priority.

2. A representative of the inviting nation is selected as President for the next ISAM period. He/she has the responsibility for making all arrangements.

3. The ISAM is open only by invitation. Persons invited shall constitute a qualified group with broad expertise in arcto-alpine mycology. No past ISAM members may claim the privilage of an invitation.

4. It shall be the right of members of the last ISAM to suggest participants to the next assembly.

5. The President will draw up a list of potential participants from nations participating in arcto-alpine research in mycology, keeping in mind geographical distribution. He/she shall consult with previous presidents on matters of principle or of major importance.

6. Definite invitations shall be issued not later than nine months before the ISAM assembly.

7. If the President must cancel the arrangement, then he/she shall inform the two previous Presidents before defintive action is taken. He/she shall inform a representative of the nation that extended the second priority. The country in question shall then appoint a President who shall assume the above responsibilities.

8. ISAM members shall submit a relevant paper to be printed in a special ISAM publication in accordance with decisions and arrangements made by the President. To facilitate this, the President may appoint an editorial committee.

The organizational structure and membership, albeit small, exists. Previously unresolved and long range objectives may be developed for adoption. These and other professional activities have resulted in our making positive strides toward meeting and correcting criticisms of the ISAM-I volumne.

A third ISAM is scheduled to take place in 1988 at Svalbard (Norway). The site for the meeting is positioned above 78° N latitude. The following committees have been established:

Executive Committee:
Sigmund Sivertsen (Norway-Trondheim) President
Egon Horak (Switzerland-Zürich) Past President
Gary Laursen (U. S. A.-Alaska) Secretary

Organizing Committee (Norway):
Sigmund Sivertsen, Chairperson
Gro Gulden
Trond Schumacher

ACKNOWLEDGEMENTS

No meeting of this magnitude is ever conducted without support given from several sources. From Vice Chancellor for Research and Advanced Study at the University of Alaska came a travel grant: from the Mikrobiologisches Institut and the Geobotanisches Institut, Eidegenossische Technische Hochschule, Zürich, came all logistical support and arrangements, and from President H. Ursprung's office came financial support; the Swiss Army provided vehicles for local foray transportation; the Hochalpines Töchter-Institut and its summer staff provided gracious living accommodations; the Swiss National Park provided collecting permits for several forays into various sections of the park; and Mr. G. A. Mulligan, Director of the Biosystematics Research Centre, Ottawa, provided the resources necessary to prepare and index all manuscripts in a camera-ready format.

G. A. Laursen
J. F. Ammirati
S. A. Redhead

CONTENTS

INTRODUCTION

Research topics from the following papers which constitute the second International Symposium on Arctic and Alpine Mycology (ISAM) are largely biased in favour of taxonomy, perhaps because of the pioneering work being conducted in the field of Arctic and alpine mycology. The present volumne contains a collection of 20 papers dealing with fungi from Arctic, Arctic-like, or alpine habitats (c.f. Watling's paper); eight of which are devoted wholly or inpart to the Ascomycotina and eleven that are dedicated to investigations mainly on the Basidiomycotina. Abstracts for most these works were published separately (Laursen and Ammirati, Univ. Alaska, Agric. Exp. Stn., Misc. Pub. 84-2, 1984).

Leading papers of these proceedings are two biogeographically oriented papers which give overviews on the historically linked origins of alpine fungi, illucidated by examples of parasitic ascomycetes from the Alps (Müller and Magnuson), and contemporary climatic factors influencing the distribution of Arctic-alpine fungi, with relatively low altitude areas of Scotland serving as an unusual example (Watling). These are followed by three papers detailing specific factors related to the ecological niches and fruiting periods of mycorrhizal and saprophytic agarics in alpine *Dryas* mounds, subarctic birch and spruce forests, and heaths, and endophytic ascomycetes in *Loiseleuria* (Debaud, Metsänheimo, Petrini). Finally there are two sections largely devoted to taxonomy of first the Ascomycetes, and secondly the Basidiomycetes. Two of these address problems in the large fleshy terrestrial or coprophilous discomycetes in the genera *Saccobolus* and *Sarcoleotia* (Schumacher and Sivertsen), another documents microascomycetes specifically parasitizing a characteristic Arctic-alpine moss, *Polytrichum sexangulare* (Döbbeler), and a fourth summarizes data on Nordic juncicolous species in the parasitic/saprophytic genus *Mycosphaerella* (Holm and Holm). A key to the Arctic and alpine *Phaeosphaeria* is given in another (Leuchtmann). Finally, this section leads into the Basidiomycete section with a paper covering a variety of Ascomycetes and Basidiomycetes from Svalbard (Huhtinen).

In the Basidiomycete section one contribution is an in depth examination of the historical facts concerning the study of the omphaloid, mainly Arctic and alpine basidiolichens, and the repercussions of these studies with novel interpretations (Redhead and Kuyper). Two are studies on mycorrhizal species in *Astrosporina/Inocybe*; one historically linked to the studies and collecting sites of J. Favre (Horak), the other an exploration of new territory (Miller). Two contributions concentrate on the genus *Cortinarius* - the subgenus *Telamonia* from European alpine sites (Lamoure), and notable taxa from alpine sites in western North

America (Moser and McKnight). Another summerizes the current state of knowledge for six families of agarics and boletes in Greenland (Knudsen & Borgen). Hygrophoraceae from Arctic and subarctic Alaska are treated (Laursen, Ammirati and Farr), as are *Galerina* species from Svalbard (Gulden). Gasteromycetes form a final topic in the paper on Greenland species of *Bovista* (Lange).

ON THE ORIGIN AND ECOLOGY OF ALPINE PLANT PARASITIC FUNGI

Emil Müller

and

J.A. Magnuson

Department of Microbiology
Swiss Federal Institute of Technology
CH-8092 Zürich, Switzerland

Key words: Alpine Fungi, parasites, rusts, Ascomycetes, Basidiomycetes

ABSTRACT

A summary is given on the origin of Alpine plants after a short discussion of the term "Alpine" and a comparison of Alpine and Arctic climate.

The ice periods of the past one million years nearly destroyed the original Tertiary flora of the Alps; however, during the last 10,000 years many of the preglacial plants have re-invaded those sites being gradually freed of ice cover. These plants were accompanied by plant parasitic fungi. The present geographic distribution of some plant species may indicate their location during the glaciation. A small portion of Alpine plants, with their fungi, survived at ice-free sites within the glaciated Alps. These facts are compared with the situation in the Arctic, and examples are cited.

Strategies of survival under alpine conditions are discussed with reference to Savile's (1972) paper on the adaptation of plants to the Arctic climate. Frequently, the adaptation of fungi to the short Alpine summers has been effected by a simplification of the life cycle; e.g., the suppression of conidial states or the development of systemic infections. Adaptation to winter conditions of deep snow cover is manifested by the ability of some fungi to grow at temperatures at or even below the freezing point. Many Alpine fungi have adapted to the high amount of UV radiation by developing thick, darkly pigmented cell walls and gelatinous sheaths around the spores.

INTRODUCTION

Plant parasitic fungi cannot be discussed without a consideration of the hosts which offer, although perhaps not voluntarily, the essential nutrients. Fungal penetration and spread within plant tissues is hindered by defense mechanisms, so that successful invasion requires certain fungal specializations. The permanent interaction between host and parasite, an arrangement which has existed for millions of years, is considered to be an important stimulant for evolution of the involved organisms (Parleviet 1979). For alpine parasitic fungi, survival depends on their capacity to resist the harsh climate, especially the short summers, and to maintain a functioning reproductive ability despite unfavorable conditions. The principles of the Arctic relationships between host, parasite, and climate were masterfully discussed by Savile (1972).

The term "alpine" is unfortunately ambiguous. Used in a geographical sense it indicates the Alps, but used in a climatological or ecological sense it indicates a certain zone of any lofty mountains. In an attempt to reduce the resultant confusion, the terms used in this paper will be defined, somewhat arbitrarily, for the purpose of clarification. The terms "alpine" and "arctic" will be used here to describe ecological zones. In both sites the tree line represents the best boundary between subarctic and arctic, and between subalpine and alpine zones (Savile 1972, Landolt 1983). The term "Alpine" will be used to indicate the alpine zone of the Alps, while "Arctic", of course, will indicate that geographic region. Subdivision of both the arctic and the alpine zones is difficult, and will not be attempted here. Terms such as low arctic and high arctic, alpine and nival, are not easily defined since "we inevitably have a continuous gradation of terrestrial climate from tree line to the limit of land" (= permanent ice; Savile 1972).

Alpine climate corresponds in many respects to the Arctic climate. The Alpine climate is similar to the Arctic climate in its low temperatures and short summer periods of plant growth. In general, the average summer temperature decreases with increasing altitude, as it does with increasing northern latitude. Local conditions may, however, raise or lower the temperature so that a given altitude does not necessarily indicate a certain average summer temperature. Alpine climate does differ from Arctic in certain respects. Alpine summer daylength is considerably shorter than Arctic, although light intensity is higher and therefore sunlight may warm the ground and rocks appreciably. Rainfall tends to increase with increasing altitude, but the central mountain chains of the Alps are distinctly dryer than the peripheral chains to the north of the south. Precipitation is considerably higher in the Alps than in the Arctic.

Table 1. July temperature means and yearly rainfall at some Swiss localities.

Locality	Altitude m.s.m.	July temp. mean, °C	Yearly rain-fall, mm	Number of plant spp.
St. Gallen	670	15.8	1300	
Davos	1540	12.0	1000	
Grand-St. Bernard	2470	6.6	2500	238[1)]
Säntis	2500	5.0	3000	
Weissfluhjoch(Davos)	2640	5.1	1600	224[2)]
Jungfraujoch	3450	-1.8	2500	

[1)]Tissière 1868 [2)]Schibler 1929, vascular plants above 2600 m

It is apparent from Tables 1 and 2 that July temperature mean is not totally dependent on either altitude (in the Alps) or northern latitude

Table 2. Effect of summer temperature on Artic flora (Savile 1972).

Locality	latitude °N	July temp. mean, °C	number of plant spp.
Ella (Greenland)	72°50'	9.0	184
Coral Harbour	64°12'	7.8	169
Central Peary Land	82°80'	6.4	106
Hazen Camp	81°49'	6.0	105
Resolute	74°41'	4.3	70
Isachsen	79°47'	3.6	48

(in the Arctic). However, it seems that the July temperature mean correlates well with the number of plant species in the area, and that this number in the Alps is double that in the Arctic. In the Alps, the number of plants decreases rapidly with increasing altitude (and, therefore, decreasing temperature). Within the Davos region, which includes the Weissfluhjoch and Silvretta chains, Schibler (1929) recorded the data given in Table 3.

The number of plant species in the Alpine zone tends to be higher than in the Arctic, due mainly to two reasons: the Alpine zone contains a greater number of favorable or tolerable habitats, and is in proximity to lower sites containing numerous plants, some of which are capable of spreading to the alpine zone. Most of the Alpine plants and fungi are not restricted to the Alpine zone but may also be found under more favorable conditions.

Table 3. Altitude and number of plant species at Davos.

Altitude m	Number of plant species
2600-2762	253
2762-2925	155
2925-3087	85
3087-3250	38
3250-3512	21

ORIGIN OF ALPINE PLANTS AND FUNGI

Hess and Landolt (1968, 1970, 1972) listed about 3000 indigenous plants (pteridophytes and phanerogams) for Switzerland and some neighboring regions, covering a total of about 80,000 km; about 22% (670) of these plant species were able to grow under alpine conditions. Speculations on the origin of Alpine plants must involve a consideration of the ice periods during which large portions of the Alps and the lower land to the north were covered with ice. During the past million years several ice periods occurred, alternating with warmer periods which were similar to the present climate. During the last glaciation (Würm), arctic ice advanced southwards to northern Germany. The original flora of the Alps was almost destroyed at this time, but some plant survived at sites similar to the small rocky islands existing within current glaciers. A certain number of originally Alpine plants emigrated to the ice-free zones of lower lands, north and south of the Alps. After the last ice period, plants from the lower neighboring regions again invaded the ice-free zones within the Alps, gradually progressing to the present alpine zone. A considerable number of plants from southern European mountains also may have colonized the Alps. Recolonization is summarized by Hess and Landolt (1968).

Table 4. Alpine plants of the central Alps: present geographic distribution

Groups	Arctic-alpine (above treeline)	Central and southern European mountains	Alps only	Others	Total
Pteridophyta	10	0	0	10	20
Conifers	1	0	0	1	2
Monocotyledons	33	20	10	55	118
Dicotyledons	68	206	158	99	531
	112	226	168	165	671

The present geographic range of alpine species is represented in Table 4. About 16% are arctic-alpine, occurring in Eurasiatic and American mountains as well as in the Arctic. The largest group, 34%, plants from central and southern European mountains, represents, to a great extent, those plants living in the Alps during the Tertiary and colonizing these sites again after the ice period. Recolonization may have occurred directly from the south as well as from the east and the west. A considerable segment, 25% of present alpine plants are endemic to the Alps, either having survived at favored localities within the glaciated zone; e.g., *Androsace helvetica* with *Pleospora phyllophila* (Crivelli 1983), *Primula glutinosa*, *Rumex nivalis*, and *Callianthemum coriandrifolium*, or having recolonized their present geographic range from the northern, or more often, the southern foothills of the Alps. Certain species still inhabit places thought to be their initial refuge during glaciation, continuously inhabited during the ice period. A further portion of present Alpine plants are either Eurasian or Eurasian-North American mountain plants.

Since the Alps were largely populated, as was the present Arctic region, by repeated invasions of plants from various regions (Savile 1972, Hess and Landolt 1968), parasitic fungi may have also reached the Alps and the alpine regions with their hosts. However, in certain cases the present host range may be misleading, as infection may have spread from the original introductory host to related plant species. It is even possible that certain parasitic fungi evolved within the Alps only after the recent ice period. Some observations even indicate that speciation is continuing today. The ascomycetous *Nodulosphaeria cirsii*, cultured from different species of *Cirsium*, showed uniform esterase pherograms in disc-electrophoresis. These clearly differed from pherograms of cultures isolated from *Carduus defloratus*, even though the fructification found on the different hosts were morphologically identical (Bucher 1974, Bresinsky 1977). Unfortunately, and due to the lack of investigations and incomplete information concerning the presence of plant parasitic fungi in

Table 5. Present geographic distribution of some alpine parasitic fungi typical for the Alps.

Host species	Parasitic fungi	Sites outside the Alps
Polygonum viviparum	*Bostrichonema polygoni* *Venturia polygoni-vivipari* *Wettsteinina eucarpa*	central European and North American mountains, Arctic
Dryas octopetala and other *Dryas* species	*Chaetapiospora islandica* *Isothea rhytismoides* *Leptosphaerulina dryadis* *Synchytrium cupulatum* *Wettsteinina dryadis*	Eurasiatic and North American mountains, Arctic
Salix reticulata	*Venturia subcutanea*	Arctic
Kobresia (Elyna) myosuroides	*Anthracoidea elynae*	Arctic
Saxifraga aizoides and other *Saxifraga* species	*Puccinia jueliana* *Exobasidium warmingii*	Arctic, Subarctic Carpathians
Astragalus frigidus	*Polystigma volkartianum* *Uromyces phacae-frigidae*	Arctic
Carex sempervirens	*Anthracoidea sempervirentis* *Micropeziza verrucosa*	southern European mountains
Carex curvula	*Anthracoidea curvulae*	southern European mountains
Saxifraga rotundifolia	*Exobasidium schinzianum*	Carpathians
Callianthemum coriandrifolium	*Puccinia sardonensis* *Puccinia Kochiana* *Puccinia calliantheni*	Tatra Tatra restricted to the Alps
Daphne striata	*Dothidea muelleri*	restricted to the Alps
Epilobium fleischeri	*Nodulosphaeria epilobii* *Puccinia epilobii-Fleischeri*	restricted to the Alps
Peucedanum ostruthium	*Puccinia imperatoriae*	restricted to the Alps
Laserpitium halleri	*Nodulosphaeria ladina*	restricted to the Alps
Bupleurum stellatum	*Puccinia bupleuri stellati* *Venturia braunii*	restricted to the Alps

other mountains of central and southern Europe, it is not yet possible to reasonably determine the origin of most Alpine parasitic fungi.

In some cases, however, it seems possible to fix the age, at least the postglacial age, and origin of some examples. Most arctic-alpine plants and their attendent parasites populated mountainous regions with cool climates long before the glacial epoch in the Northern Hemisphere. These plants survived in the ice-free zones during that epoch, and followed the retreating ice when it ended. It seems, therefore, quite logical for such abundant arctic-alpine plants as *Polygonum viviparum* or *Dryas octopetala*, together with related *Dryas* species, to have the same characteristic fungal parasites both in the Arctic and in the Alps (Table 5). This geographical similarity of parasites also occurs in such species as *Venturia subcutanea* on *Salix reticulata*, (Nüesch 1960); *Polystigma volkartiana* and *Uromyces phacae-frigidae* on *Astragalus frigidus*); *Anthracoidea elynae* on *Kobresia myosuroides*); and *Puccinia jueliana* and *Exobasidium (Arcticomyces) warmingii* on *Saxifraga* spp., (Savile 1959, Müller 1977 a).

An interesting evolutionary theory is connected with *Exobasidium warmingii*, a surculicolous parasite (Nannfeldt 1981) on *Saxifraga aizoides*, *S. aizoon*, and *S. oppositifolia* in the Arctic. According to Engler (1916), these three host species originally were European-alpine (Savile 1963). Engler assumed a quick post glacial spread from central Europe to the north, made possible by the distributory capacity of the small seeds by wind. It seems likely that during that spread the parasite accompanied the host. During the Tertiary division the fungus was restricted to the Alps, living on the three above mentioned host species and also on *S. aspera* and *S. bryoides*. On *S. rotundifolia*, a second *Exobasidium* causing leaf spots is sometimes present. It is quite probable that *E. warmingii* originates from the leaf spot fungus, *Exobasidium schinzianum*, which moved gradually to the Alps from southeastern Europe during the Tertiary division. *Exobasidium schinzianum* is also still present in the Carpathians, according to a sample collected by M. and K. Vanky (which was not available for inclusion in Müller 1977 a).

A considerable number of Alpine plants also currently live in the mountains of central and southern Europe: the Pyrenees, the Apennines, the Yugoslavian and Albanian mountains, and the Carpathians including the Tatra, Bulgarian, and Greek mountains. A number of fungi have been found: *Anthracoidea sempervirentis* on *Carex sempervirens* and a number of related *Carex* species; *Micropeziza verrucosa* on *Carex sempervirens*; and *Anthracoidea curvulae* on *Carex curvula*. All have similar distributions which may have originated in the southern European mountains.

One of the Alpine plant species considered to have survived glaciation at its presently populated localities is *Callianthemum coriandrifolium* (Ranunculaceae). Three rust species are known for this host: the heterocyclic *Puccinia sardonensis*, with the dikaryon on *Anthoxantum alpinum* (Poaceae), and the distantly related microcyclic *Puccinia kochiana*, both of which occur in the Alps and in the Czechoslvoakian Tatra, and the microcyclic relative of *Puccinia sardonensis*, *Puccinia calliantheml*, which is restricted to small, isolated sites in the Alps. It seems probable that the last evolved only after isolation of the area in the Alps during the ice period.

A similar situation occurs with *Dothidea muelleri* on *Daphne striata* (Thymelaeaceae), an exlcusive inhabitant of the eastern Alps and some small, isolated sites in the western Alps. The fungus is similar to, and may be derived from, *Dothidea* (*Plowrightia*) *mezereum* living on plants such as *Daphne mezereum*, a widespread species with a range including the subalpine zone. *Dothidea muelleri* differs considerably from *D. mezereum* in that all parts of its fructification, the stroma, asci, and ascospores, are smaller and the ascospores are only unicellular (Loeffler 1957).

ECOLOGY OF ALPINE PARASITIC FUNGI

Savile (1972) discussed the ability of parasitic fungi to survive under arctic conditions such as short, cool summrs. One of the most basic requirements for survival is effective spore production to insure a regular infection capacity. One method of guaranteeing regular sporulation is to simplify the life cycle. Savile's observations indicate the several possible survival strategies listed below, which also may apply to Alpine conditions.

Table 6. Cultured Ascomycetes: occurring in Switzerland and including alpine species.

Genus	non-alpine species		alpine species		total	
	cultured	+ anamorph	cultured	+ anamorph	cultured	+ anamorph
Clilioplea *	0	0	2	0	2	0
Cistella *	1	1	1	0	2	1
Clathrospora *	0	0	3	0	3	0
Diplonaevia *	0	0	2	0	2	0
Dothidea *	5	5	1	1	6	6
Entodesmium *	3	0	2	0	5	0
Gnomonia[1)]	24	7	6	0	30	7
Hyaloscypha *	2	2	1	1	3	3
Hysteropezizella *	1	0	4	0	5	0
Laetinaevia[2)]	2	2	3	0	5	2
Leptosphaeria *	13	11	3	3	16	14
Leptosphaerulina *	3	0	13	0	16	0
Massariosphaeria *	7	3	1	1	8	4
Montagnula *	2	1	3	1	5	2
*Mycosphaerella**[3)]	23	20	6	2	29	22
Naevala[2)]	1	1	1	1	2	2
Nimbomollisia *	0	0	1	0	1	0
Nodulosphaeria *	13	0	7	0	20	0
Phaeosphaeria *	28	17	12	3	40	20
Pleospora *	14	8	20	1	34	9
Pyrenopeziza *	9	6	4	1	13	7
Venturia *[4)]	28	13	12	0	40	13
total	179	97	108	15	287	112

* cultural experiments performed at the Swiss Federal Institute of technology by several authors: [1)]Monod 1983; [2)]Hein 1976; [3)]von Arx 1949; Klebahn 1918; Brefeld and von Tavel 1891; [4)]von Arx 1952; Bachmann 1963; Müller 1958; Nüesch 1960.

Suppression of conidial states

A summary of cultural studies, performed in part at the Swiss Federal Institute of Technology, on ascomycetous genera containing some alpine species is given in Table 6. Anamorph formation is compared between alpine and non-alpine fungi, those growing at lower altitudes below treeline. It is obvious that the percentage of ascomycetes producing anamorphs is considerably lower in alpine than in non-alpine fungi (11% vs. 55%). This tendency, however, cannot be viewed too simplistically since certain genera do not follow the general rule while in others there is no species, either alpine or non-alpine, with a conidial state. Information presented in Table 6 strongly indicates fewer anamorph-producing ascomycetes occurring in the alpine zone, although it is not absolutely certain that species never forming anamorphs in pure culture will not form anamorphs under other conditions. Certain species; e.g., the polyphagous *Phaeosphaeria alpina* (Leuchtmann 1984), do not behave uniformly in pure culture.

Some strains form ascomata and conidiomata (belonging to the genus *Stagonospora*), while others form only ascomata or only conidiomata.

Although teleomorphs are very uniform, the conidia produced may vary considerably in size. It is possible that these cases may exhibit the early stages of differentiation, probably connected with host specialization (Leuchtmann 1984).

The number of microcyclic rust fungi is considerably higher in the alpine zone than in zones of lower altitudes (Table 7). Heteroecious rusts, with their need for two close hosts and requirement of a much longer developmental period, are suppressed; autoecious rusts, in many cases, may omit formation of either the aecia or the uredinia (Gäumann 1959; compare also Fischer 1904).

Table 7. Life cycles of non-alpine and alpine Swiss rust species.

Species	heteroecious	autoecious: aecia and/or uredinia only	autoecious: microcyclic	total
non-alpine	246	191	57	494
alpine	39	34	62	135
total	285	225	119	629

Suppression of the ascomatal state

Although this strategy may be logical, it is much more difficult to prove since the occurrence of teleomorphs in laboratory cultures is rare, with teleomorphs mostly forming within a narrow ecological range (Müller 1977 c). Savile (1972) cites *Mycosphaerella tassiana* with its *Cladosporium* anamorph as a possbile example for this strategy. That fungus was mentioned by several authors to be the most common ascomycete in the Arctic, as it is in the Alps. However, cultural experiments with *Mycosphaerella tassiana* collected in the Alps from alpine and non-alpine sites (von Arx 1949) do not fully confirm the reduction of either anamorph or teleomorph. The only example of an alpine fungus with a clear relationship to a certain ascomycete genus seems to be *Ascochyta pedicularidis* (Fuck.) v. Arx (= *Phoma pedicularidis* Fuck., von Arx 1957, 1964), an imperfect fungus causing systemic infections on Scrophulariaceae, mainly *Pedicularis* species. Savile (1968) listed the arctic records. The fungus is also found in the Alps. *Ascochyta* species correlate to *Didymella*, but *Didymella* was never found in our collections; the cultures continued to yield the *Ascochyta* state which significantly differs from the anamorph of *Didymella pedicularidis* (Corbaz 1958).

Complicated life cycles with longer intervals between states

Convincing examples of this type are difficult to find. The only real observations concern asocmycetes producing ascomata not in the first but in the second or third year after the stem or leaf died. An example with this kind of behaviour is the inoperculate discomycete *Cenangiopsis oxyparaphysata* (Müller 1977 b).

Simultaneous occurrence of the anamorph and teleomorph

Regular simulataneous occurrence of conidial and ascogenous states is known for some species of *Leptosphaeria*; e.g., *L. acuta* on *Urtica*, *L. macrospora* on Asteraceae, and *L. anemones* on *Anemone* and *Pulsatilla*, all of which are occasionally alpine. This type of behavior is not distinctly different from that of species growing in more favorable climates, however.

Simplification of the breeding system

Savile (1972) suggested that the breeding system tends to be simpler under arctic conditions. In the case of rust fungi, he mentions that all species known to him are self-compatible. However, from results of cultural studies with mono-ascosporic cultures, it seems that non-alpine as well as alpine ascomycetes are mainly self-compatible, the majority of non-alpine species following that breeding system. In general, self-incompatibility seems to be rare within ascomycetes.

Additional measures may be taken to guarantee fructification by an optimal utilization of alpine conditions. Even the winter with its permanent snow cover may be utilized to some advantage. The strictly alpine *Pyrenophora ephemera*, on *Luzula lutea* and *L. spadicea*, survives under low temperatures (below 18°C) for long periods. At 3°C that fungus requires six months on malt agar for formation of ascomata, asci, and young ascospores, and another six months for maturation of the spores. That behavior suggests a slow development of the ascoma under the winter snow cover, which regularly reaches one to several meters at altitudes above 2000 m, and for which ground temperatures at the freezing point have been demonstrated. After the snow has gone, ascomata soon mature and the ascospores infect new leaves, thus leaving most of the short summer available for the endophytic colonization of the host. Since many alpine ascomycetes form their ascomata in early summer, that type of behavior would seem to be widespread.

Hypoxylon diathrauston, on *Pinus montana* var. *prostrata*, shows an even more highly developed adaptation to low temperatures. The host, and therefore the parasite, is not strictly alpine, although it grows up to 2400 m even on slopes with northern exposures and deep winter snow cover. The fungal spore will germinate only after treatment for several weeks with temperatures just below the freezing point, followed by incubation at -3°C in a humidity chamber. Germination begins after several days incubation, and mycelial growth becomes continuous (Ouellette and Ward 1970). The fungus forms a conidial state and the teleomorph on malt agar. The ascigerous state requires 18 months at 3°C for maturation. *Herpotrichia juniperi*, the black snow mold of conifers, also grows at temperatures below the freezing point and is therefore able to develop under the snow cover (Gäumann et al. 1934).

The ability of spores to germinate at low temperatures may have a curious effect on the range of two rust species living on plant species belonging to the *Solidago virga-aurea* complex. Under natural conditions these rusts, *Puccinia virgaureae* and *Uromyces solidaginis*, are found only at high elevations within the subalpine and alpine zones, where *Solidago alpestris* represents the only species belonging to that complex. In the laboratory, however, infection of other species is effected as easily as with *S. alpestris*. It is suggested that the long snow cover at higher altitudes prevents an early germination. At lower sites, however, the commonly short-lived snow cover ends long before the host plants begin to develop, allowing the teleospores to germinate and produce basidiospores which are fated to die by starvaton before suitable hosts can develop.

Species of the genera *Polystigma* and *Diachora* (Polystigmatles) that are restricted within temperate zones to the host families Rosaceae and Fabaceae, are able to block transport of assimilates from the infected leaves, causing a considerable storage build-up in the infected tissues. Most of the species form mature perithecia toward the end of the growing period within which the ascospores may overwinter. Host starch reserves may be necessary for use in further developmental processes of the fungus, which proceeds during winter at high altitudes. Regular development during winter, with further development in spring, is typical for *Polystigma volkartianum* (on *Astragalus frigidus*), a segregate of *Polystigma astragali* occurring on other *Astragalus* species found usually only in lower zones. In the case of *P. volkartianum*, adaptation to deep snow cover has resulted in a capacity for growth under the snow allowing the utilization of the stored starch.

Venturia subcutanea, when compared to other *Venturia* species growing on leaves of *Salix* spp., has the highest temperature limit for growth (33°C compared to 27°C of other species). *Venturia subcutanea* lives on *Salix reticulata*, *S. herbacea*, *S. retusa*, and *S. serpyllifolia* in the Alps, and on *S. reticulata* in the Arctic. All these species are espalier willows, creeping over the ground. Sunshine may raise the temperature of the ground significantly above the air temperature so that even the well protected fungal thalli must be adapted to higher temperatures (Nüesch 1960).

As in the Arctic (Savile 1972), systemic and perennial infections are also more than usually widespread in alpine zones. Many of them are visibly manifested by changes in growth intensity or habit of the host plant. In other cases, systemic colonization is not easily substantiated. Research on endophytic fungi has shed some light on the range of systemic infections for alpine plants. Widler (1982) compared endophytes of *Arctostaphylos uva-ursi* from two sites at 1220 m and 2320 m altitude. As expected, the number of fungal species was lower at the higher altitude, but reached as many as 105 species isolated from living plant tissues from 2320 m, compared to 157 from the lower altitude (Table 8).

Table 8. Endophytic fungi from *Arctostaphylos uva-ursi*, isolated species (Widler 1982).

	Parsenn (Davos) 2310 m	Alvaneu 1200 m	Occurring at both sites
Basidiomycetes	1	3	1
Ascomycetes	25	28	15
Fungi imperfecti	79	126	51
Total	105	157	67

Widler took samples every 6 weeks during two years and stated that many of these fungi survived winter conditions within the green leaves. Therefore, plants with perennial leaves, such as Ericaceae, Polygonaceae, Thymelaeaceae, and Conifers, may help many fungi to persist thrugh the winter under the protecting snow cover. Fructification of fungi occurs mainly on dead plant material.

Systemic infections with fructifications on living plant tissues are rare, except in the basidiomycetous Exobasidiales, Uredinales, Ustilaginales, and Tilletiales. The ascomycetous *Muellerites juniperi* (on *Juniperus nana*) forms ascomata continuously. These are situated at the needle base inside the leaf furrow, and all developmental stages may be observed on infected shoots. The ascospores mature throughout the season so that ascospores are always available for new infections. In spite of this, infected plants are rare. It seems that infections are not easily successful. *Euphorbia cyparissias* is the host for a number of heteroecious and microcyclic rust fungi belonging to the genus *Uromyces*. These rusts, including some alpine species, are mainly systemic and remain within the buds during winter. Symptoms vary with the parasite species concerned.

Ecological adaptation has also affected fungal cell walls. Savile (1972) noted that deeply pigmented walls of mycelia ascomata, and spores are predominant in the arctic species of fungi that attack the aerial parts of their hosts. Pigmentation is thought to aid heat absorption, therefore better utilizing sunlight; to shield protoplasm from harmful ultraviolet radiation; and to somehow aid in the prevention of water loss. The increase of temperature and reduction of desiccation are probably the chief functions of wall pigments in the Arctic.

Thick, deeply pigmented walls of ascomata and spores are also typical for many fungi gorwing in higher altitudes of the Alps. Such characters are genetically fixed, and are also found in the mycelia, ascomata, and spores of laboratory cultures from such fungi, even when kept in darkness for several months. Wall thickening, pigmentation, and ornamentation is common within genera which include alpine and subalpine species; e.g. *Leptosphaerulina*, *Massariosphaeria*, *Montagnula*, *Phaeosphaeria*, and *Pleospora* (Crivelli 1983, Leuchtmann 1984). Such characters may even provide a clearer differentiation of and within certain groups of species. Compared to the other species of the genus, impressively thicker ascomata walls were found for *Phaeosphaeria alpina* (on different Poaceae), *P. oreochloa* (on *Sesleria disticha*), and in the arctic *P. caricinella* (Leuchtmann 1984). The ascoma walls of *Phaeosphaeria nardi* (on *Nardus stricta*) are conspicuously thicker at boreal sites and high altitudes in the Alps than at lower localities (Holm 1957, Leuchtmann 1984). The *Phaeosphaeria herpotrichoides* complex may be differentiated into groups on the basis of ascospore shapes, but these groups often cannot be considered to be separate species (Eriksson 1967). However, some alpine host specialized forms of *P. herpotrichoides* differ distinctly from lower altitude forms by producing dark, rough spore walls (Leuchtmann 1984). Within the genus *Pleospora* (Crivelli 1983), several species groups may be characterized by their ascospore wall shapes and ornamentation; e.g., a small group with *P. paronychiae*, *P. androsaces*, *P. brachyspora* (all on Caryophyllaceae), and *P. phaeospora* (polyphagus), in which younger ascospores are coated with a very dark, verrucose outer layer which later breaks into smaller portions. Most species of the *P. discors* group have a distinctive striate ascospore ornamentation, with longitudinal stripes at the end cells and transverse stripes at the central cells of the ascospores. One of the most conspicuous of all Alpine fungi is *Phaeosphaeria pleurospora* on Poaceae and Cyperaceae. The long, slender, multiseptate ascospores are dark with regular, longitudinal, light-colored stripes (Leuchtmann 1984).

Teleospores of alpine Uredinales may also show thicker cell walls and darker pigmentation than related species of lower altitudes; such is the case with the microcyclic *Puccinia alpina* and *P. ruebeli*, on *Viola* spp. as compared to the non-alpine heteroecious *Puccinia violae*.

The ascospores of many alpine ascomycetes are typically coated with a gelatinous outer layer, as in *Pleospora* (Crivelli 1983) and *Phaeosphaeria* (Leuchtmann 1984). Unfortunately, the gelatinous sheaths disappear when the fungi are kept dry for long periods, and therefore cannot be observed in herbarium samples. The gelatinous layers may perform functions similar to those performed by thicker and darker cell walls.

CONCLUSIONS

During the glaciation of the Alps most plants emigrated to more favorable climates, re-invading the Alps between and after the ice periods. In a few cases, an evolution within the Alps during or after the glaciation may be suggested. Alpine plant parasitic fungi invaded their present habitats along with their hosts, so that the origin of these parasites is closely tied to that of Alpine plants. Development of such plant-host connections could have occurred during the Tertiary division when the Alps were formed.

The examples given in this paper of Alpine fungi and their adaptation to alpine conditions represent only a small portion of the total Alpine fungi. Unfortunately, no summary of these fungi currently exists. Any species list compiled would of necessity be incomplete, since there are new records made and new species discovered every year.

Alpine parasitic fungi often agree in many respects, such as behavior, morphology, and even identity, with parasitic fungi living in the Arctic. Nevertheless, certain differences cannot be ignored; e.g., the Alpine predominance of fungi with a one-year cycle, while two and three year cycles are more common in Arctic species. The number of plant parasitic fungi present in the Alpine area tends to be considerably larger than in the Arctic. The reasons for these larger Alpine numbers include the proximity to lower, more favorable habitats, the inclusion of a greater number of favorable habitats, and, possibly, the frequent occurrence of locally high humidity due to high precipitation.

REFERENCES

von Arx, J. A., 1949, Beiträge zur Kenntnis der Gattung Mycosphaerella, *Sydowia,* 3: 28-100.

______, 1952, Studies on *Venturia* and related genera, *Tijdschr. Plantenziekten,* 58: 260-266.

______, 1957, Die Arten der Gattung *Colletotrichum* Cda., *Phytopath. Z.,* 29: 413-468.

______, 1964, Revision der zu *Gloeosporium* gestellten Pilze, Nachtrage und Berichtigungen, *Koninkl. Nederl. Akademie van Wetensch., Amsterdam, Proc. C,* 66: 172-182.

Bachmann, C., 1963, Untersuchungen au Geraulaceen bewohnenden *Venturia* - Arten, *Phytopath. Z.,* 47: 197-206.

Brefeld, O., and von Tavel, F., 1891, "Untersuchungen aus dem Gesaultgebiel gebiel der mykologie," 10: 157-378.

Bresinsky, A., 1977, Chemotaxonomie der Pilze, *in:* "Beiträge zur Biologie niederer Pflanzen," W. Frey, H. Hurka, F. Oberwinkler, eds., Gustav Fischer Verl., Stuttgart, S. 25-42.

Bucher, J., 1974, Anwendung der diskontinuierlichen Polyacrylamidgel - Elektrophorese in der Taxonomie der Gattung *Nodulosphaeria* Rbh. (Ascomycetes), *Viertelj. Schr. Naturf. Ges. in Zurich,* 119: 125-164.

Corbaz, R., 1958, Recherches sur le genre *Didymella* Sacc. *Phytopath. Z.,* 28: 375-414.

Crivelli, P., 1983, "Über die heterogene Ascomycetengattung *Pleospora* Rbh., Vorschlag für eine Aufteilung," Diss. ETH Nr. 7318, 213 p. Kommissionsverlag F. Flück, CH-9053 Teufen, Switzerland.
Engler, A., 1916, Beiträge zur Entwicklung der Hochgebirgsfloren, *Abh. Königl. Preuss. Akad. Wiss., Physik. Math. Kl.*, 1: 1-113.
Eriksson, O., 1967, On graminicolous pyrenomycetes from Fennoscandia. 2. Phragmosporous and scolecosporous species, *Arkiv for Botanik*, 6(9): 381-440.
Fischer, E., 1904, Die Uredineen der Schweiz, *Beitr. Kyypt. Fl. Schweiz*, 2(2): 1-590.
Gäumann, E., 1959, Die Rostpilze Mitteleuropas, *Beitr. Krypt. Fl. Schweiz*, 12: 1-1407.
Gäumann, E., Roth, C., and Anliker, J., 1934., Über die Biologie der *Herpotrichia nigra*, *Z. Pflanzenkrh.*, 44: 97-116.
Hein, B., 1976, Revision der Gattung *Laetinaevia* Nannf. (Ascomycetes) und Neurodnung der Naevioideae, *Willdenowia, Beih.*, 9: 1-136.
Hess, H., and Landolt, E., 1968, "Flora der Schweiz," Birkhäuser, Basel. Vol. 1: 1-858.
______, 1970, "Flora der Schweiz," Birkhauser, Basel. Vol. 2: 1-956.
______, 1972, "Flora der Schweiz," Birkhauser, Basel. Vol. 3: 1-876.
Holm, L., 1957, Etudes taxonomiques sur les Pléosporacées, *Symb. Bot. Upsal.*, 14(3): 1-188.
Klebahn, H., 1918, "Haupt-und Nebenfruchtformen der Ascomyceten," Verlag Borntraeger, Leipzig.
Landolt, E., 1983, Probleme der Höhenstufen in den Alpen, *Botanica Helvetica*, 93: 255-268.
Leuchtmann, A., 1984, "Über *Phaeosphaeria* Miyake und ander bitunicate Ascomyceten mit mehrfach querseptierten Ascosporen," Diss. ETH 7545, 157 p.
Loeffler, W., 1957, Untersuchungen über die Ascomyceten-Gattung *Dothidea* Fr., *Phytopath. Z.*, 30: 349-386.
Monod, M., 1983, Monographie taxonomique des Gnomoniacées, Ascomycètes de l'ordre des Diaporthales, *Sydowia, Beih.*, 9: 1-314.
Müller, E., 1958, Systematische Beuerkungen uber einige *Venturia*-Arten, *Sydowia*, 11: 79-92.
______, 1977a, Reflections on the geographical distribution of *Exobasidium warmingii*, *Kew Bulletin*, 31: 545-551.
______, 1977b, Zur Pilzflora des Aletschwaldreservats (Kt. Wallis, Schweiz), *Beitr. Krypt. Fl. Schweiz*, 15(1): 1-126 (Kommissionsverl, F. Flück, CH-9053 Teufen, Switzerland).
______, 1977c, Factors inducing asexual and sexual sporulation in Fungi (mainly Ascomycetes), *in*: "The whole fungus," B. Kendrick ed., National Museum of Natural Sciences, Ottawa (Can.) p. 265-282.
Nannfeldt, J. A., 1981, *Exobasidium*, a taxonomic reassessment applied to the European species, *Symb. Bot. Upsal.*, 23(2): 1-72.
Nüesch, J., 1960, Beiträge zur Kenntnis der Weiden-bewohnenden Venturiaceae, *Phytopath. Z.*, 39: 329-360.
Ouellette, G. B., and Ward, E. W. B., 1971, Low temperature requirements for ascospore germination and growth of *Hypoxylon diathrauston*, *Can. J. Bot.*, 48: 2223-2225.
Parleviet, J. E., 1979, The co-evolution of host-parasite systems, *Symb. Bot. Upsal.*, 22(4): 39-45.
Savile, D. B. O., 1959, The botany of Somerset Island, District of Frankling, *Can. J. Bot.*, 37: 959-1002.
______, 1963, Mycology in the Canadian Arctic, *Arctic*, 16: 17-25.
______, 1968, Some fungal parasites of Scrophulariaceae, *Can. J. Bot.*, 46: 461-472.
______, 1972, Arctic adaptations in plants, *Can. Dept. of Agric. Monogr.*, No. 6: 1-81.

Schibler, W., 1929, "Die Flora des Davoser Landwassertales über 2600 m. Festschr.," Jahresvers. SNG in Davos. Naturf. Ges. Davos, 93-118.
Tissière, P. G., 1868, "Guide de Botanique sur le Grand St. Bernard," Imp. Dulex-Ansermoz, Aigle, 117 p.
Widler, B., 1982, "Untersuchungen über endophytische Pilze von *Arctostaphylos uva-ursi* (L.) Sprengel (Ericaceae)," Diss. ETH-Zurich, 133 p.

LARGER ARCTIC-ALPINE FUNGI IN SCOTLAND

Roy Watling

Royal Botanic Garden
Edinburgh EH3 5LR
Scotland

Key words: Alpine flora, Scotland, *Amanita*, *Russula*, *Laccaria*, *Omphalina*, *Inocybe*, *Cortinarius*, Basidiomycetes, Ascomycetes

ABSTRACT

The larger Arctic-alpine fungi found in the islands and highlands of Scotland are considered and their distribution even at sea-level discussed in relation to climate, geology, soil and vegetation. Particular attention is paid to the distribution of *Amanita nivalis*, *Russula* spp. and *Laccaria* spp., and *Omphalina alpina* and *hudsoniana*; *Inocybe* spp., and *Cortinarius* spp. Larger ascomycetes and other basidiomycetes are briefly considered.

INTRODUCTION

The geology, spanning rocks of all epoche, and the strongly oceanic climate (Atlantic) of the British Isles makes these west European islands unique. The British Isles enjoys a climate which is considerably ameliorated by the effects of the Gulf Stream, and so creates conditions not in keeping with their latitude (50-63°N); indeed Shetland is further north than the southernmost tip of Greenland and more northerly than parts of Alaska (Figure 1). The British Isles have a geology that can be linked in many cases directly to that of continental Europe and their vegetation is considered a rather depauperate fragment of the W. European flora. Some specialized elements are found in the south (Mediterranean and Lusitanian), and in Scotland the range of vascular plant species, although less diverse, possesses many interesting taxa, some representing northern elements (Arctic-alpine). It is the larger fungi associated with these latter elements that are presented in this paper.

MATERIALS AND METHODS

Except for a limited survey of the fungi associated with *Salix herbacea* (Watling, 1981), and then in only one main collecting site, little has been published on Scottish Arctic-alpine fungi, although Robert Kaye Greville as early as 1822 described the montane *Amanita nivalis*. No one has inquired as to whether or not the distribution of the larger fungi in communities containing Arctic-alpine vascular plants: 1) have slavishly followed their associates and hosts; 2) show a similar pattern to those plants because of overall climatic factors; or 3) unlike the

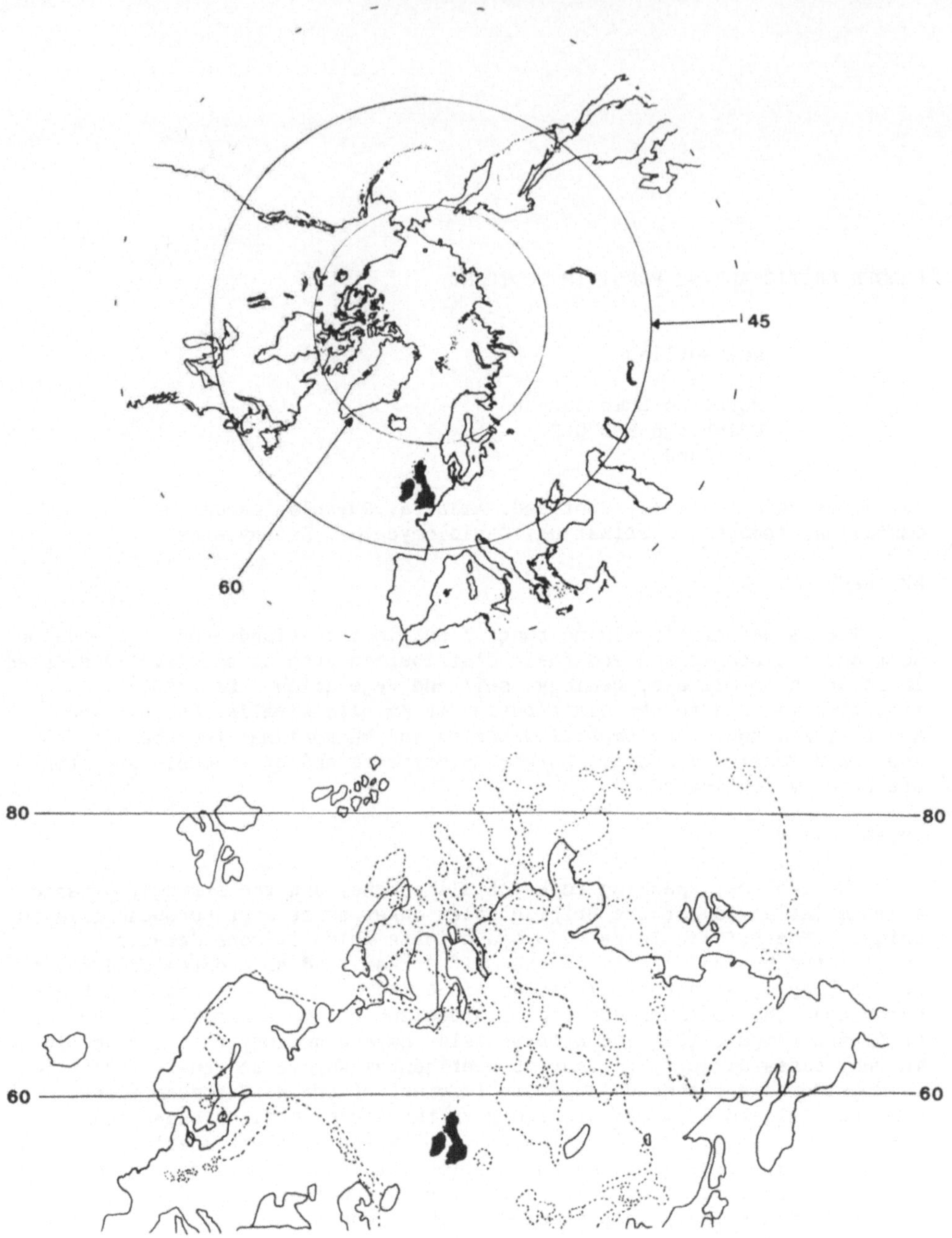

Fig. 1. British Isles in relation to the North temperate world. A. Gnomonic projection showing polar region. B. Superimposed Mercator projections of Europe and Asia, and North America, on the British Isles. Degrees north indicated.

vascular plants are restricted to mountain tops. Based on Rev. J. Stevenson's *Mycologia Scotica* (1879), Foister (unpubl.) tabulated the mountain fungi known to occur between 1000 and 4000 ft. in Scotland, i.e. 305-1219 m. He listed 89 Hymenomycetes and five Gasteromycetes, and since that compilation further records have been slowly accumulating over the yeaars, e.g., Dennis (1955) and Henderson (1958).

The Outer and Inner Hebrides are comparatively well-known with several studies being reviewed by Dennis & Watling (1983) to which can now be added further records (Watling, 1983a&b); Kirk & Spooner, 1984; Dennis, (pers. comm.). In the far north, areas of Sutherland have been studied (Dennis, 1955; Watling in Kenworthy, 1976, and unpubl.); more recently, a survey of Shetland has been undertaken and preliminary results are available (Watling, unpubl.). All this material has been called upon for this presentation. Localities personally studied are indicated in Figure 2.

SCOTLAND: SETTING THE SCENE

(a) Physical features: Geology & Geomorphology

Although a full spectrum of rock types is found in Scotland, igneous (acidic and basic) and metamorphic rocks dolomite the country both in the north and south. The Central Massif (Grampian Mountains) with mountains reaching over 1300 m is a mixture of igneous and highly twisted metamorphic rocks with sandstones (Devonian) composing the northeastern 'shores' and the closest of the northern islands (Orkneys). To the west are a series of Pre-Cambrian sandstones and limestones, many contorted and metamorphised to some degree. The central area is a dissected plateau separated from the Southern Uplands, which act as a boundary with England, by the Central Valley; the valley formed by the drainage courses of the River Forth to the east and River Clyde to the west is composed mainly of Carboniferous marine and fluviatile deposits. Scattered amongst these deposits are the stumps of former Ordovician and Silurian mountains (Figure 3). Metamorphic rocks make up the greatest part of the south, some folds giving high land reaching upwards to 838 m. All areas north, central and south are intruded with igneous rocks whilst a little Devonian (Old Red Sandstone) can be found on each side of the Central Valley.

In the West, the rocks are also metamorphic with the Inner and Outer Hebrides having some of the oldest rocks in the Britain Isles, indeed in the Northern Hemisphere. Some of these rocks constitute mountains rising to 1109 m in the Cullins of Skye (Inner Hebrides) and 620 m at Beinn Mhór in S. Uist (Outer Hebrides). The island group of St. Kilda is farthest west of the Outer Hebrides a little over 139 km from mainland Scotland; the main island Hirta rises to 430 m at Conochair. The islands are a mixture of dolerite and granophyre in the main with some ultrabasic rocks. In the extreme north Shetland is 160 km from mainland Scotland. The islands making up this archipelago are again mixtures of igneous rocks, many ultrabasic, and a few sedimentary deposits. The highest point on Shetland in Ronas Hill, 450 m on Mainland.

(b) Physical features: Climate

Generally the east of Scotland is drier than the west. Indeed, small areas in the Firth of Forth and Moray Firth receive an annual rainfall as low as many of the drier parts of England, i.e., under 63.5 cm, Table 1. In contrast, over 155 cm falls on the northwest of Scotland, particularly in the higher areas, and a third of the precipitation falls from December to February. Such figures are equivalent to the English Lake District and Snowdonia in North Wales, both of which are characterized by mountainous terrain.

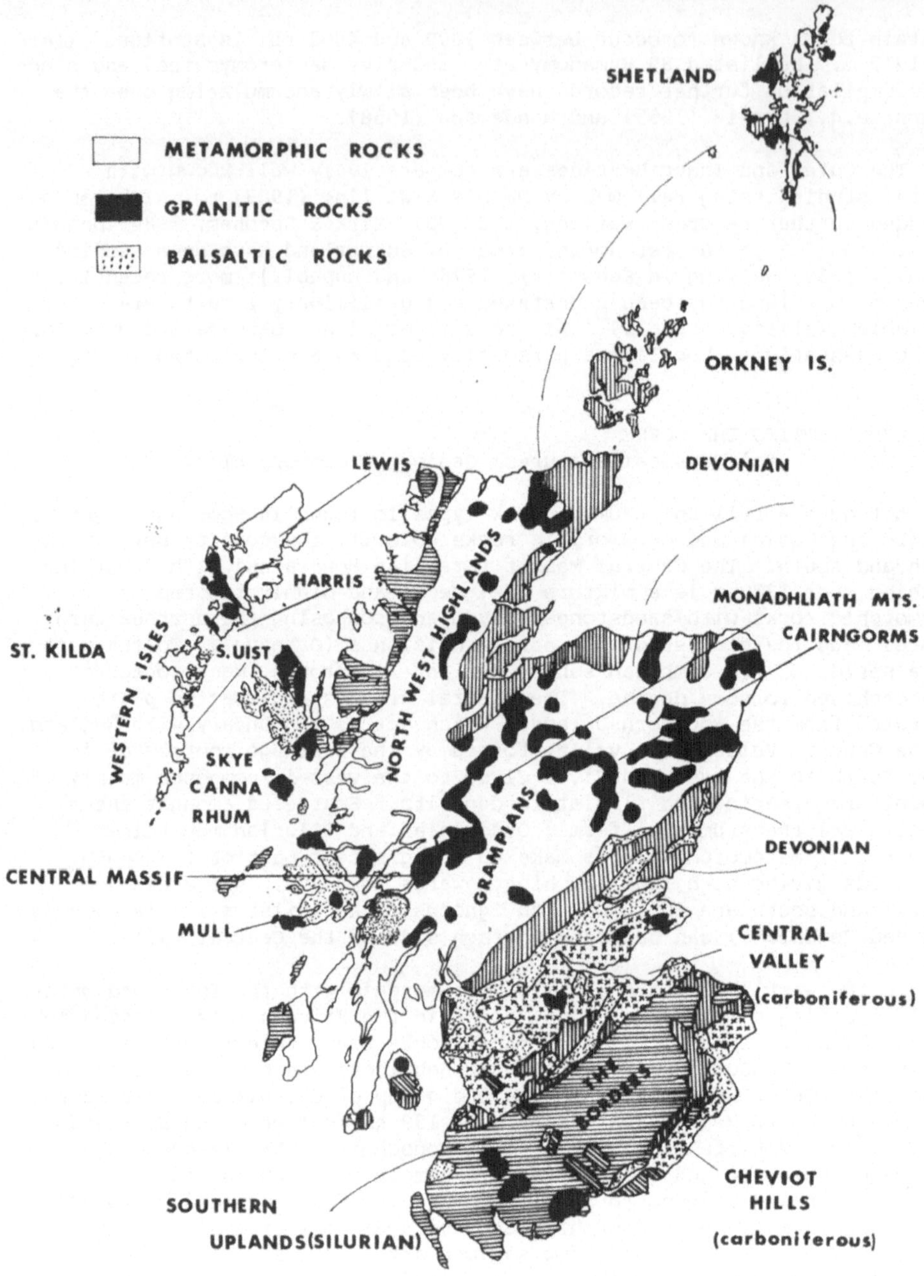

Fig. 2. Map showing major geological and topographical features of Scotland.

Relative humidities of < 70 percent in eastern Scotland are equivalent to areas of central England but measurements of 80-85 percent characterize the Orkneys, Shetland, Outer Hebrides and the northernmost tip of Ireland. Coastal areas of Scotland generally have high relative humidity (75-80 percent) but in the east it is accompanied by low precipitation.

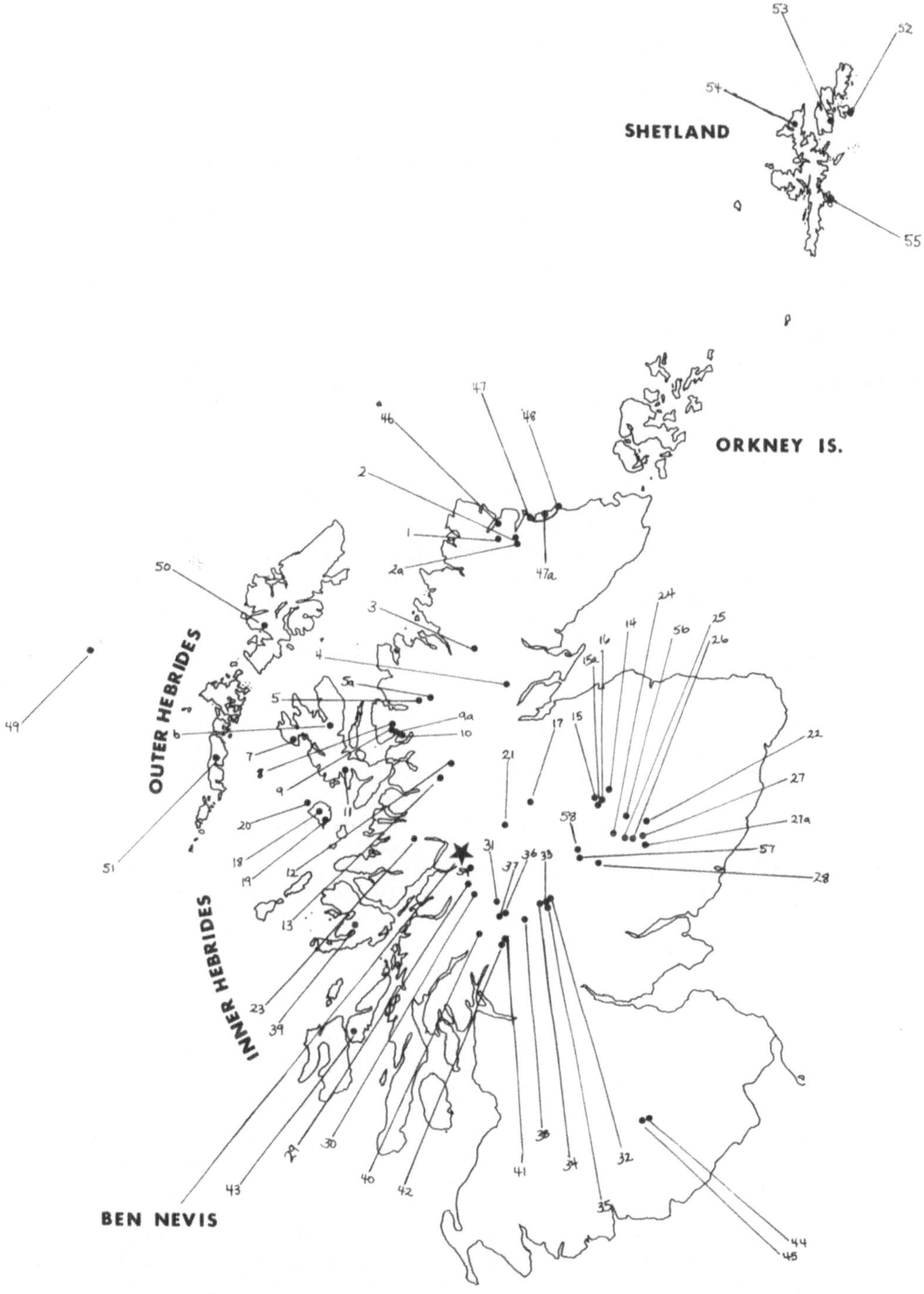

Fig. 3. Map showing collecting sites and localities mentioned in text. Also see Table 2.

In annual accumulated temperature (number of degrees above 6°C totalled over 12 months) the Central Massif is equivalent to Shetland and the Orkneys; whereas northwest Scotland is, in general, equivalent to the northwest part of England, particularly the Lake District.

The minimum average February temperatures in Shetland and the Orkneys are 0.5-1°C, parallel to Edinburgh in the Central Valley; whereas the Outer Hebrides are 1.5-2°C, and the Inner Islands, except for Skye with the influence of the high mountain range, are similar (2.5°C or more) to southern and western Ireland and to the Scilly Isles off Southern England. The pattern of July maximum temperature for the British Isles is completely different to that of the winter minimum. Instead of a basically west to east relationship, there is in addition to a south to north pattern superimposed on the area. When the London basin is experiencing 22 to greater than 22.5°C, Shetland is experiencing temperatures 8°C lower, and the Orkneys 7-7.5°C lower. The Outer and Inner Hebrides are slightly higher at 16-17°C and parallel those temperatures experienced in Ireland. Oceanity can be measured by plotting the range of average monthly temperature (Page, 1982). For Scotland these range from less than -8.5°C in the west, Shetland and the Orkneys, to -6.5 to -6°C in the southeast Southern Uplands and Central Massif; the lower the figure the more continental the climate.

Following mid-summer the first air frosts are recorded by August 15 in the Central Massif in contrast to December 1 for Long Island (Outer Hebrides) and November 1 in Shetland. The last air frost may be as late as June 1 in the Central Massif, April 1 in Long Island and late April or even May in Shetland. Over 100 days of frost are recorded in the Central Massif, Southern Uplands and in England in the northern Pennines and the Lake District.

The wind as well as bringing rain is both a drying and cooling agent. The British Isles, and Scotland in particular, are exposed. Winds blow predominantly throughout the year generally from the west (Atlantic). The general effect of the vegetation and terrain is to decrease the winds' effects west to east. The northerly and westerly coastal regions receive the highest wind speeds. Thus the annual average number of days exceeding the figure of 32 kmph at the standard elevation of 10 m for the Outer Hebrides and Shetland is over 30 days. Generally the west of Scotland is also cloudier than the east with readings expressing an easterly skew which take in the Cairn Gorm mountains and is only equalled in dullnes in England by the eastern Pennines.
The low summer temperatures, because of cloud cover and high winds on the Scottish mountains, contrast markedly with the Alps. Thus, Manley (1952) has shown that a small change in altitude in the Scottish mountains gives a very great reduction in the length of the growing season for plants in contrast to a similar change in the Alps.

On Ben Nevis (1340 m) there are eight months of temperatures at or below freezing. Six degrees centigrade would appear to be the threshold of plant growth and such a fall in air temperature would ensure permanent ice-fields in many mountain regions. Thus only a slight general worsening of the climate would lead to permanent snow. Five degrees is equivalent to the difference between the average summer temperatures for London and Shetland.

Although probably based on underestimates of potential evaporation one method of successfully comparing the climate of different areas in Scotland has been to consider the potential water deficit (PWD) of an area (Green, 1964). This water deficit increases either where rainfall is low

Table 1. General climatic data for selected areas of Scotland. Modified from Page (1982), and Green and Harding (1983).

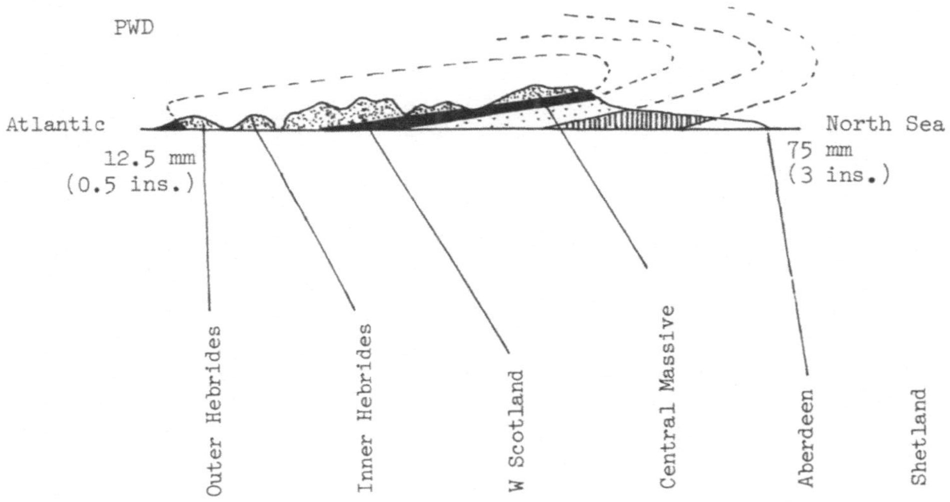

Area	Outer Hebrides	Inner Hebrides	W. Scotland	Central Massif	Aberdeen	Shetland	Orkney	N. Scotland (eg. Betty Hill, Sutherland)	Glasgow	Edinburgh
Cloudiness Days %	70-75	70-75	70-75	75-80	65-70	75-80	70-75	75-80	70-75	65-70
July Max. Temp.°C	14.5-16	16-17	16-17	19-19.5	< 19	14.5-15	15-15.5	16-16.5	19-19.5	< 19
Feb. Min. Temp.°C	1-1.5	1.5-2.5	0-1	< -1	0-0.5	0.5-1	0.5-1	0-0.5	1-1.5	0-0.5
Frost Days	20-40	20-40	40-60	> 100	40-60	40-60	40-60	40-60	40-60	40-60
Snow Cover (Days)	< 10-20	< 10	10-20	> 100	20-30	20-60	20-60	40-60	10-20	10-20
Average Rainfall (mm)	1200-1600	1600-3200	1600-3200 (> 3200)	1600-3200	600-800	800-1600	800-1200	800-1600	1200-1600	600-800
Potential Water Deficit	< 0.5-1	< 0.5	< 0.5	< 0.5	1-3	< 0.5-1	1-3	1-3	0.5-1	1-3

or potential evaporation is high, or both. Thus, southeast of Edinburgh where an overall low annual precipitation is experienced a high PWD is recorded because of rather high potential evaporation. A considerably higher potential evaporation in the southwest allows for a nearly as high a PWD even though the rainfall is much higher. Green (1964) attempted to map the PWD for Scotland and showed that a unique climatic pattern is presented in areas with potential water deficits ranging between 12.5 and 25 mm. In a transect taken from sea level in the west to eastern Scotland the PWD increases because of the increase in elevation as one travels eastwards (Table 1). Thus the sea level areas in the Western Isles are climatically similar in many ways to the highlands of Aberdeenshire and Inverness-shire in the Central Massif. This climatic data becomes significant when one considers the distribution of Arctic-alpine plants in Scotland. Indeed Green's (1964) data has successfully been applied to an understanding of the vegetation of the Outer Hebrides (Boyd, 1979) and the Inner Hebrides (Boyd & Bowes, 1983). In Shetland the PWD is parallel to western Scotland although the rainfall is more in keeping with that of eastern Scotland, as it is with the Orkney Islands, yet the number of cloudy wet days is high because of a coastal fog parallel to that experienced in the Faeroes.

(c) Physical features: Soils

The sum of the climatic and geological characters of Scotland lead, on the whole, to rather poor soils ranging from skeletal soils in the mountainous areas of both east and west and in the north and south, through various podsols on the various acidic igneous, metamorphic and sedimentary rock systems, to podsoilised mulls. Only where calcareous rocks occur, and then those not covered in deposits of glacial drift, are there base rich soils to be found. In Scotland, such soils are few and rather scattered. When present, they are an important factor in the distribution of some of Scotland's rarer plant communities and plant species, including Arctic-alpine taxa referred to below.

SCOTLAND: VEGETATION

(a) Vascular Plants

A full account of Scottish Vegetation has been given by McVean & Ratcliffe (1962) and in this work they discuss, amongst others, the montane plant communities. However, the appearance of Arctic-alpine vascular plants in Scotland at low altitude in the west was only fleetingly dealt with. In fact, although this phenomenon has been known for many generations, it is almost botanical folklore (Matthews, 1937). Little, except in general terms, has been published on this distinct pattern of distribution in Britain; the phenomenon is parallel in the Faeroes and in western Norway.

The absence of similar vascular plants in lowland Scotland and elsewhere in the British Isles, except for very small pockets, has been explained as possibly a result of competition from aggressive incomers after the last retreat of the ice at the termination of the Pleistocene. Parallel competition, reduced or absent in Northwest Scotland, allowed the Arctic-alpine species to persist. This may not be the complete story and one must turn to work elsewhere for an explanation of such a pattern of distribution.

Dahl (1951) discussed this phenomenon as applicable to Scandinavia and suggested that one limiting factor for the distribution of Arctic-alpine plants was high summer temperatures. A map of selected isotherms reduced to sea level was offered as an indicator as to whether Arctic-alpine plants might be found at lower levels. Dahl (1951) supplied a table giving taxa which are either absent or very rare outside a specific isotherm; whilst inside that same isotherm those same species are more or less frequent, at least in some localities. Those species in his table of importance to this discussion are: *Alchemilla alpina* (27°C isotherm); *Betula nana* (27°C); *Dryas octopetala* (27°C); *Oxytropis lapponica* (23°C); *Salix arbuscula* (25°C); *S. herbacea* (26°C); *S. myrsinites* (28-29°C) and *S. reticulata* (26°C). *Salix glauca* (29°C) and *S. polaris* (25°C) are included in the table but are not found in the British Isles. There is also some debate as to whether true *Betula tortuosa* grows in Scotland; its isotherm is 27°C.

A plant species inhabits a district provided there are localities with a microclimatic which satisfies the demands of that species, and provided that viable diaspores have reached it in order to become established. Dahl (1951) calculated the figure at the highest localities of the terrain in all parts of Fennoscandia and Denmark, assuming that there is a decrease of 0.6°C per 100 m rise in altitude. He supplied data for maximum summer temperature on mountain peaks in all parts and showed the great significance of the 25°C isotherm from Lindesness northwards through Norway, Sweden and Finland.

Application of both Dahl's (1951) information on Scandinavian plants and Green's (1964) climatological potential water deficit data to the distribution of Scottish Arctic-alpine plants offers admirable correlations. Thus, in conjunction with degrees of competition from more widespread plants, a working hypothesis can account for Scottish plant communities.

In Scotland, the vascular plants which show the descending pattern to sea level in a north westerly direction have been listed by McVean & Ratcliffe (1962). Of these, those important to this study are *Arctuous alpina* (610-152.5 m); *Betula nana* (457-91.5 m); *Dryas octopetala* (457-0 m); *Loiseleuria procumbens* (610-305 m); *Salix herbacea* (457-91.5 m) and *S. myrsinites* (305-61 m). Other characteristic plants of this group include *Carex bigelowii*, *Juniperus nana*, *Juncus trifidus*, *Saxifraga oppositifolia* and *Tofieldia pusilla*.

Although Arctic-alpine plants are to be found south of the Central Valley, it is the Central Massif to which one looks for the characteristic montane floras and montane agarics. The Lake District and Wales, although with montane areas, are poor in Arctic-alpine plant species.

(b) Vegetational communities

The Cairn Gorms are part of the Grampian Mountains and form a dissected plateau offering the largest area of continuous high ground over 850 m in the British Isles. They cover 390 square kilometres of mountainous country and include Ben Macdhui, Braeriach, Cairn Toul and Cairn Gorm (see Figure 2 and Table 2) itself, all exceeding 1200 m; several other mountains approach very near their heights. This high plateau composed of rather coarsely weathered granite produces a freely drained soils which in the east with its relatively low rainfall supports predominantly moorland vegetation, surprisingly with little development of peat-bog. Only in very restricted areas does the vegetation change dramatically and then is associated with limestone outcrops and the like (Raven & Walters, 1956). In geology, Lochnagar is related to the Cairn Gorm system.

These mountains contrast with the volcanic hills of the Western Highlands and Islands where rock outcrops are less than half the height of Cairn Gorm yet support rich mountain flora. However, two areas of Scotland stand out as phenomenal: the mica schistose rocks of the Central Highlands, particularly the Breadalbane range (Figure 2); and the limestones of the northwest where Arctic-alpines are found at a very low altitude.

The Central Highlands contrast sharply with the moorland dominated Cairn Gorms with their Ericaceae and Empetraceae, as they are generally clothed in sheep-grazed grasslands especially within the original tree-zone which has long since disappeared through man's activities. Where Ericaceae occur the underlying rock is probably of a more acidic metamorphic nature, e.g., quartzite.

The mountain tops, however, because of their exposure and persistently colder climate are very similar in their overall plant communities. Indeed the agaricologist finds more in the *Salix herbacea* communities, so widespread on the Scottish mountain tops than the vascular plant collector who is rewarded more by an examination of the ledges below the summits.

The floras of Shetland and South Uist (Outer Hebrides) are considered to be equivalent to subarctic shrub communities of mainland Europe (Spence, 1979). The southernmost islands of the Hebrides are equivalent to the south and west of Ireland where more southerly elements are to be found. Although some parallels might be expected between the Hebrides and W. Ireland this is masked by these additional southern elements to the flora (see Scannell, 1982). It is not really possible to make meaningful comparison as the fungi of the western areas of Ireland are poorly known (Ramsbottom, 1938; Muskett & Malone, 1980).

Table 2. Localities mentioned in text with spot heights. Numbers refer to those on Fig. 3. Sites of particular interest and visited as part of this survey are marked with an asterisk.

a. MONTANE COMMUNITIES

9a	Bealach na Ba, Meall Gorm, Ross & Cromarty. 640 m.
31	Beinn Achaladair, Perthshire/Aberdeenshire. 1037 m.
*8	Beinn Bhan, Ross & Cromarty. 396 m.
*5	Beinn Eighe, Ross & Cromarty. 1010 m.
35	Beinn Ghlas, Perthshire. 1115 m.
36	Beinn Heasgarnich, Perthshire. 1078 m.
*2a	Ben Hiel, Sutherland.
1	Ben Hope, Sutherland. 927 m.
*33	Ben Lawers, Perthshire. 1214 m (incl. Coire Odhar)
*2	Ben Loyal, Sutherland. 764 m.
*40	Ben Lui, Perthshire/Argyllshire. 1130 m.
16	Ben Macdui, S.W. Aberdeenshire. 1309 m.
51	Ben Mhór, South Uist, 620 m.
*39	Ben More, Mull. 966 m.
	Ben Nevis, Inverness-shire. 1340 m (see text)
43	Beinn an Orr, Jura. 784 m.
*6	Ben Tote, Skye. 112 m.
*28	Ben Vrackie, Perthshire. 841 m.
*4	Ben Wyvis, Ross & Cromarty. 1046 m.
11	Blaven (Bla Bheinn), Skye. 927 m.
15	Braeriach, Inverness-shire/Aberdeenshire. 1296 m.

cont'd on next page

29 Buchaille, Etive Mor (Stob Dearg), Argyllshire. 1022 m.
26a Caenlochan Druim Mor, Perthshire. 961 m.
*14 Cairn Gorm, Inverness-shire/Banffshire. 1245 m.
15a Cairn Toul, Inverness-shire. 1293 m.
*25 'The' Cairnwell, Perthshire/Aberdeenshire. 933 m.
*10 Cearcall Dubh, Kishorn. 410 m.
13 Ciste Dubh, Inverness-shire. 982 m.
30 Clach Leathad, Argyllshire. 1098 m.
*49a Conochair, Hirta, St. Kilda. 304 m.
24 Creag an Lochain, Pershire. 836 m.
*21 Creag Meagaidh, Inverness-shire. 1127 m.
38 Creag Mhór, Killin, Perthshire. 719 m.
37 Creag Mhór, Perthshire. 1048 m.
42 Cruach Ardrain, Perthshire. 1045 m.
23 Fraoch Bhenn, Glen Finnian, Inverness-shire. 856 m.
*18 Fionchra, Rhum. 609 m (also Bloodstone Hill)
17 Geal Charn, Monadhleath, Inverness-shire. 915 m.
*27 Glas Maol, Perthshire. 1068 m.
59 Glencoe (457 m), see text under *Gyromitra*.
*27a Glen Isla, Angus. Rising to 950 m - see Glas Maol
7 Healaval Beg, Skye. 480 m.
44 Head of Kerrcleuch, Selkirkshire. 518 m.
22 Lochnager, Perthshire. 1155 m.
32 Meall Gharb, Perthshire. 1116 m.
*34 Meall nan Tarmachan, Perthshire. 1043 m.
12 Móm Sodhail, Inverness-shire. 1181 m.
*54 Ronas Hill, North Collafirth, Shetland. 450 m (incl. Midfield., 388 m).
3 Seana Bhrough, Ross & Cromarty. 926 m.
*9 Sgurr an Chaorachain, Ross & Cromarty. 1053 m.
*41 Stob Garbh, Perthshire. 960 m.
*19 Trallaval, Rhum, Inner Hebrides. 772 m.
*45 Whitecombe, Moffat, Dumfries & Galloway. 820 m.

b. COASTAL COMMUNITIES

*47-48 Bettyhill (47a), Sutherland. 30 m - 100 m (transect from Skerray to Strathy including Strathnaver Reserve).
*55 Bressay, Shetland. 20 m.
*53 Burravoe, Shetland. 40 m.
20 Canna; rising to Carn a' Ghaill (210 m) Inner Hebrides.
*52 Fetlar, Shetland. 50 m.
50 Glen Meavaig, North Harris (rising to 90 m).
*49 Hirta, St. Kilda Group, Outer Hebrides (see also Conachair above).
*46 Kempie, Loch Enboll, Sutherland. ±30 m.

c. SUBARCTIC BIRCHWOODS: GRASSLANDS

*5a Kinlochewe, Ross & Cromarty. 50 m.
*5 Morrone Birkwood, Braemar, Aberdeenshire. 450 m.
*57 Struan Wood, Calvine, Perthshire. 200 m.
*58 Tulach Hill, near Blair Atholl. 470 m.

(c) Historical and anthropogenic factors:

The vegetation one now sees in Scotland is a result of the colonization after the retreat of the ice at the close of the Pleistocene and an expansion outwards from the refugia known to have occurred. During the ice-age, ice moved from two major areas; a central one stretching south to West Scotland giving a glacier with a western skew, and a second narrow arc with western and southern movement across the Southern Uplands. Ice moved from the first over the Outer and Inner Hebrides, and north and eastwards scouring across what are now the plains of Fife, Aberdeenshire and Sutherland and the Central Valley. These barren areas were then serially colonized and subsequently matured with the restrictions imposed by the insular nature of the British Isles; e.g., the lack of *Picea*, *Larix*, etc. in the flora (see Huntley & Birks, 1983).

There is evidence also to substantiate that there have been fairly drastic changes in the vegetation of Scotland since the Pleistocene. Even as late as the Bronze age (c. 3500-2500 BP), closed native pine forests (*Pinus sylvestris*) of the Central Massif ascended to about 610 m, although the precise height is in debate. Oscillations in tree-line since the Bronze Age demonstrated in Northern Europe cannot be seen in Scotland but this is probably related to Britain always having had a low natural tree-line with trees, never as is Scandinavia, ascending above 720 m. In fact no natural tree-line occurs now in Scotland because of man's activities. Trees have long since disappeared from felling for charcoal and establishment of grazing property. Kenworthy (1976) has discussed these factors for the northernmost parts of mainland Scotland, an area (Bettyhill) which will be frequently referred to in the text below.

(d) Fungal flora

The fungal flora is a reflection of the vegetation which is itself a reflection of the degree and speed of colonization and subsequent stabilization at a given locality. One of the pivotal species in the studies of Dahl (1951) was the Lesser Willow, *Salix herbacea*, an important Scottish mountain ectomycorrhizal plant. Unlike Scotland's other native dwarf willows it grows in relatively large closed communities; *S. reticulata*, *S. myrsinites*, etc. are locally restricted to mountain ledges which because of drying out are inhospitable to the extensive fruiting of larger fungi (Watling, 1981). *S. herbacea* is therefore paramount to the studies of larger fungi in the Scottish mountains because of its range of habitats on both acid and basic rocks, and its high frequency in suitable sites in the mountains. Connelly & Dahl (1970) give a useful review to accompany a map of the distribution of this willow in Europe and this can indicate areas with which the fungal floras of Scotland might be usefully compared.

Generally the fungus flora of the Scottish mountains, is parallel in many ways to those of Scandinavia and the Alps. One inexplicable exception is the absence of *Lactarius* spp. at higher altitudes. However, the vegetation of Scotland differs from other key areas in one major respect, that of a long period of high grazing pressure experienced from sheep, red deer and the mountain hare. These perturbations open up communities to colonization by a wide range of Hygrophoraceae and Entolomataceae. Thus, there is an important anthropogenic factor which must be incorporated into any understanding of the Scottish montane flora of larger fungi. Watling (1981) has suggested that field data indicates that some agarics can switch hosts and that some montane communities are relics of a more mesophytic woodland cover. Thus *Collybia dryophila*, *Cystoderma amianthinum* and *C. carcharias* are as equally at home at 900 m in the Central Massif, as in lowland woods: *Nolanea cetrata* is found amongst *Luzula sylvatica* in the north at high altitudes and to the west in exposed lowland areas. However, the majority of species to be discussed below are biotrophs associated with *Salix herbacea*.

NOTES ON INDIVIDUAL TAXA

All material is in the Royal Botanic Garden, Edinburgh (E) unless otherwise stated. Herbaria abbreviations follow Lanjow & Stafleu (1964). All localities and their spot heights appear in Table 2 and localities visited during the present survey indicated in Fig. 3.

1. Amanitaceae

Amanita nivalis Greville, Scottish Cryptogamic Flora 1(4) pl. 18, 1822.

Greville's original material came from the Cairn Gorms (Breariach; Ben-ne Bourd) in the Central Massif. Unfortunately no type material exists in Edinburgh (E) where Greville's fungal material is housed. Although it has been found on Breariach since, insufficient numbers of basidiomata have been collected to distribute as neotype material. Indeed, although it is consistently found on the schistose rock systems a little further south in Scotland than the Cairn Gorms, in the type locality it is a rather unreliable fruiter. Twelve collections are housed in E and one in Kew (K) and the fungus is also known from at least four additional sites. It ranges from the Central Massif of the Cairn Gorms and Breadalbanes, where it occurs at over 900 m, to the northwest at Sgurr a' Chaorachaan and Beinn Bhan (Ross & Cromarty). It is exclusively associated with *Salix herbacea*, and although this willow may be found at lower altitudes, no records of *A. nivalis* are available below 305 m. It may be located on bare peaty soil or in *Rhacomitrium* heath with *S. herbacea*, or in mountain turf with *S. herbacea* on podsolic or skeletal soils having a pH of 4.2-5.9 (McVean & Ratcliffe, 1962). It is a snow-bed agaric associated with such plants as *Carex bigelowii*, *Galium* and *Polytrichum*, a character from which Greville coined the epithet.

A. nivalis is a relatively small whitish to pale greyish buff member of the 'vaginata' group. It is characterized by the glabrous to faintly pubescent stipe lacking any remnants of an annulus or velar floccules, prominent membranous sulcate volva and smooth pileus with shortly sulcate margin, and few or no velar plaques; the basidiospores are non-amyloid, subglobose (rarely globose) measuring 10-12.5 x 9.5-12 µm.

The intermediates observed by Lange (1955) Kühner (1972) have not been seen in Scotland; although, a collection from the SW summit of Lochnagar was pale drab cinnamon buff (*Wat.* 11767). It commences white but soon discolours brownish or ochraceous especially towards the centre and particularly when sun scorched. It, however, more frequently seen rather water soaked from autumnal storms and it then appears drab greyish.

Greville's fungus is the same as that described by Kühner (1972) as *A. hyperborea* (Karst.) Fayod and by Favre (1955) as *A. vaginata* f. *oreina* (Watling, 1985). Karsten's *A. hyperborea* has been shown to be quite a different fungus related to *A. friabilis* (Karst.) Bas; for further details one should refer to Bas (1982). *A. nivalis* is the only truly montane *Amanita* in Scotland although two other white taxa are known, one from the northern coast of mainland Scotland and a second from the upland birch-woods. The first has been found, but rarely, and always associated with *Salix repens* in cliff turf. This taxon may possess a dark cinnamon buff volva but is generally much taller and has a more silvery white pileus. It remains to be seen if this and *A. nivalis* are conspecific. The second white *Amanita* is *A. vaginata* var. *alba* Gillet (= var. *fungites* (Batsch) J. Lge.) with which many continental authors have erroneously synonymized Greville's agaric. Indeed Stevenson (1879) also places this mountain agaric under '12. *A. vaginatus** Bull. var. *albida**', and indicates the fungus has been found on 'the bleak summits of the loftiest Grampians' and 'except in colour which is snow white does not differ from *A. vaginata* more than the common variety *fulvus*'. Stevenson indicates that he had found it at 121.5 m at Hunters Hill, Glamis but this is probably a record of the true var. *alba*. *A. vaginata* var. *alba*, however, is a tall, elegant fungus and much the same in stature as *A. fulva*.

It could in fact be mistaken in the field by the unaware for *A. virosa* Secr. or the N. American *A. bisporigera* Atk. but these are of course totally different.

A. vaginata var. *alba* has been found in upland birchwoods (*Wat.* 11518) although it is less common than *A. fulva* (Glen Tilt, White 1879). *A. rubescens* (Fr.) Gray has been found in maritime communities with *Salix repens* in Sutherland on mainland Scotland, but as yet has not been recorded for similar communities on Shetland. *A. rubescens* and *A. vaginata* are of course widespread in the British Isles, but have not been found with *S. herbacea*.

Material examined and additional records: Breadalbanes (Ben Lawers, *Henderson* 4308; Carn Creag, *Wat.* 11819; Meall nan Tarmachan, *Wat.* 14178; Coire Odhar); Cairn Gorms (Braeriach, *Reid*; Ben Macdhui, *Henderson* 2266); Lochnagar, *Wat.* 11767; The Cairnwell, *Wat.* 17489 and VIP: *OK MILLER*; Western Highlands (Sgurr a' Chaorachaan,, *Henderson* 7145; Bheinn Bhan, *Henderson* 9464); Glen Affric Forest (Ciste Dhubh and Móm Sodhail; *Orton* 418 and 2542) and Beinn Achaladair, *Wat.* 13466. Also A' Chailleach, nr. Dundonnell, legit *A. Bennell* and Druir Reich. The Scottish collections have been compared with material from Finmark (north of Kevo, *Wat.* 9631), Greenland (JB/F/41; see Watling, 1977), and Switzerland (*Wat.* 8254, 8292 and 8763).

Russulaceae

Russula spp.

At least four taxa must be considered under this heading and there is evidence that further species will be involved when more records are accumulated.

R. nana Killerman is relatively common on The Cairnwell amongst *Salix herbacea* (*Wat.* 17507), although, collections have been found at lower altitudes in the same mountain complex, possibly having been transported there by erosion from higher slopes. Records are also available from Ben Achaladair, Ben Hope (legit Alexander), and in the Western Isles on Canna (*Dennis*; mat in K, det. R. Singer), possibly with *S. repens*, *R. aquosa* Lecl. differs in the less strongly coloured pileus and absence of septate dermatocystidia; it is frequent in alpine birch-woods.

Reid (1972) has described a second alpine *Russula*, *R. norvegica* Reid, which differs in its very dark blackish purple pileus, which fades to pinkish purple and finally becomes almost white. Collections of *R. norvegica* from Cairn Gorm are in *K* and *E* (*Wat.* 17432) and material from the Cairnwell is in *E*, all with *S. herbacea* (*Wat.* 17432). Judging from the spore size given by Henderson (1958), some of his collections may have been *R. norvegica*.

R. pascua (Favre) Romagn. has been found with *Salix herbacea* and *Alchemilla alpina*, also on The Cairnwell. Only a single basidioma on two separate occasions were taken (*Wat.* 17529), but it is also known from Cairn Gorm (*Reid* in *K*). On Shetland it is apparently widespread both with *S. herbacea* on Ronas Hill, the highest land in the archipelago, (*Wat.* 16834), and with *S. repens* at 25 m on Bressay, and at 90 m at Burravoe and Fetlar in cliff turf, (*Wat.* 16666-8; 16672-4; 16677; & 16679 & 16718; 16853). The Shetland collections resemble very closely Møller's illustrations (1945), whereas the Cairnwell collections are smaller and more compact and resemble Favre's illustration (1955).

In the Ronas Hill locality *R. pascus* was associated with *Entoloma ameides*,* *Boletus edulis*, etc. On Bressay it is with *Lactarius lacunarum*, *Inocybe fastigiata* and *I. lacera*, whilst at Burravoe it is found intermixed with larger taxa in the Viridantinae, i.e., *R. xerampelina* var. *graveolens* J. Lange (*Wat.* 16699; 16718 & 16755); also found in sand-dunes communities in other parts of Scotland.

R. persicina Krombh., although common in alpine birch-woods also colonizes upland communities. Thus it is found with *Salix herbacea* on The Cairnwell (*Wat.* 17486), St. Kilda (Watling & Richardson, 1971), and the northernmost maritime communities in mainland Scotland (Bettyhill, Sutherland). It has been found accompanying *R. fragilis* (Pers.: Fr.) Fr., i.e., *R. fallax* auct. pl. This latter species is similar in many ways macroscopically to Reid's *R. norvegica* having a purple pileus, with a darker, almost black centre, and an off-white spore-print. It differs in the broader units in the pileus and spore-ornamentation. It too has been found on St. Kilda (Watling & Richardson, 1971) and Shetland (*Wat.* 16683 & -85) as well as in Sutherland.

One as yet unnamed taxon has been found near the summit of Cairn Gorm (*Wat.* 17444), with *Salix herbacea*. It has a pale yellowish buff pileus, concolorous furfuraceous stipe, white spore-print and slightly acrid taste; the basidiospores are ornamented with isolated warts in much the same way as *R. farinipes* Romell but this is well outside that species' habitat range. *R. raoultii* Quél. which it also resembles possesses reticulate ornamented spores.

Until now no species of *Lactarius* has been found at high altitude. *L. lacunarum* Romagn. ex Hora has been found associated with *Salix repens* and accompanied by *Russula pascua* on Bressay in Shetland from cliff turf, and with *R. persicina* on St. Kilda with the same dwarf willow (*Wat.* 16684). This same *Lactarius* has been found on other Hebridean Isles, but is not specific to such communities. On St. Kilda *R. persicina* is also found with *Lactarius controversus* (Fr.: Fr.) Fr., a rather uncommon fungus in Scotland, although not restricted to Arctic-alpine communities. *L. controversus* has been found with *S. repens* in maritime communities in eastern Scotland (Barry Links, near Dundee; *Wat.* 13903) and in island communities with the same host and wind blown *Populus tremula* on Mingulay (Ball & Watling, 1976). *L. aspideoides* Kühn. is recorded from *Dryas* turf overlooking Loch Eriboll (*Wat.* 17553) and *L. helvus* (Fr.) Fr. and *L. rufus* (Scop.: Fr.) Fr. from amongst *Betula nana* accompanying an as yet unidentified *Russula* sp. of Sect. *Emeticinae*.

Boletaceae

Boletes are rare in the Scottish mountains. In subarctic shrub they are often dominant, especially *Leccinum* spp. which are associated with *Betula*. Even at heights of 457.50+m isolated groups of trees have their

---

*Authorities if not given will be offered under appropriate family headings.

associated boletes, e.g. *L. versipelle* (Fr. & Hok) Snell, in Coill a' Choire, Creag Meagandh.

Three taxa have been located above 850 m in the Central Massif, i.e. *Boletus* (*Xerocomus*) *spadiceus* Rostk. (*Wat.* 17526) on Ben Vrackie, *Lecicnum salicola* Watling on Cairn Gorm (*Wat.* 17397) and *B. edulis*. The first has been regularly found on Ben Vrackie but nowhere else at high altitudes; it is of course common in lowland wooded communities. The second has only once been found in a montane environment, it being previously only known from wind and salt tolerant, cliff turf communities of *Salix repens* in Sutherland (Watling, 1976; 1979) and in the Western Isles (Noble, pers. comm.). *L. salicola* may in fact be a permanent member of the montane fungal flora but it might be suggested that the single locality for *B. spadiceus* on Ben Vrackie is a result of a single colonization from adjacent birch-woods, as might be suspected for the single record of *B. edulis* L.: Fr. at 450 m on Midfield, Shetland (*Wat.* 16740). *B. edulis* is common in Arctic birch-woods and other lowland communities. All the montane boletus were associated with *Salix herbacea*. *Leccinum holopus* (Rostk.) Watling is recorded from a *Betula nana* community on the slopes of Ben Hiel, *Wat.* 17546 and 17547.

CORTINARIACEAE

Cortinarius spp. are frequent members within montane floras of the Scottish mountains, although only a few have been critically studied and perhaps as many as a third await elucidation.

Subgenus *Seriocybe*: *C. anomalus* (Fr.: Fr.) Fr. (*Wat.* 16960,16961), and *C. lepidopus* Cooke (*Wat.* 17539 & *Miller* 17801 in VPI) may be found associated with a range of trees in lowland communities yet both grow with *S. herbacea* at or above 336 m in the Central Massif, e.g., The Cairnwell (Watling, 1981). The closely related *C. spilomeus* (Fr.: Fr.) Fr. has not been found in montane communities but occurs with *Dryas* in coastal turf near Kempie Loch Eriboll, Sutherland (*Wat.* 17548 & 17549) associated with *Lactarius aspideoides*. These coastal turfs in the north are noted for their alpine plant components. *C. lepidopus* is known from St. Kilda (Watling & Richardson, 1971).

Subgenus *Myxacium*: *C. favrei* Henderson is rare in Scotland. Although the original collection was described from the Beinn Eighe Nature Reserve (Henderson, 1958) at 335 m, it has also been collected in the Central Massif. This is Favre's interpretation of *C. alpinus* Boudier (Favre, 1955) which differs by having smaller basidiospores. Scottish material has been compared with material from Finnmark (*Wat.* 9774), Iceland (*Wat.* 13514), Greenland (Watling, 1977 & 1983c), Svalbard (Watling, 1983c), and more recently from the Swiss Alps. A member of the *C. pseudosalor* group is also found at high altitude in the Central Massif with *S. herbacea* and on the exposed Western Isles and northern shores of the mainland with *S. repens*. *C. pseudosalor* is known from St. Kilda (Watling & Richardson, 1971).

Subgenus *Dermocybe*: The recently described *C. norvegicus* Høiland (1983) a member of the *C. cinnamomeus* complex, has been recognized on several summits in the Central Massif. Material from The Cairnwell was confirmed by Høiland (Watling, 1984a). *C. cinnamomeo-badius* R. Henry (= *C. croceus* (Schaeff.: Fr.) Høiland (*Wat.* 17510), and *C.* cf. *pratensis* (Bon & Gaugé) Høiland (*Wat.* 17516) have also been collected on The Cairnwell. Cairnwell in many ways resembles certain communities in the Alps, in fact one recent collection from Cairnwell (*Wat.* 17485) is similar to collections close to *C. polaris* Hoiland found in the Alps.

Subgenera *Telamonia/Hydrocybe*: *C. pertrisis* Favre is known from bare summit peat on Ben Lawers associated with *S. herbacea* (Henderson, 1958) and *C.* cf. *rufostriatus* Favre in a similar habitat on Cairnwell (Watling, 1984a) and Cairn Gorm (*Wat.* 17392 & 17393). On Cairn Gorm both *C. gausapatus* Favre (*Wat.* 17396) and *C. scotoides* Favre (*Wat.* 17394 & 17395) have also been found, both associated with *S. herbacea* but in a more mineral soil derived from granite. Although the latter may have been found elsewhere in Scotland (*Henderson*, 14 ix 1966) generally the species so-far recorded are known only from single collections. More work in both laboratory and field is required. One group, however, is widespread, viz. the richly fulvous coloured, striate capped Hydrocybes of the *C. obtusus* group.

C. striatuloides Henry (*Wat.* 17371, 17372 & 173733) from Midfield, Shetland with *Salix herbacea* is similar in all ways, except for size, to collections from St. Kilda with *S. repens*. As pointed out by Watling & Richardson (1971), it is yet to be seen whether Henry's taxa prove to be autonomous; they agree with *C. pseudoscandens* Orton, nom. prov.

The genus *Inocybe* is very well represented in the Scottish mountains, but members have only recently been studied. All species apparently grow with *S. herbacea* at the montane sites, but in more coastal regions in the north and west taxa are associated with *S. repens* and *Dryas octopetala*. Because of the relatively recent documentation, the true distribution is not known. The flora is undoubtedly rich; thus taxa are recorded for Cairn Gorm, *I. decipientoides* var. *taxocystis* Favre (= *lanuginella* forma; probably on autonomous taxon: *Henderson*, 14.ix.1966), *I. dulcamara* (A.& S.: Pers.) Kummer (*Wat.* 17376; as var. *squamosoannulata* described by Favre), *I. giacomi* Favre (*Wat.* 17379), *I. lacera* (Fr.) Kummer (*Wat.* 17380) and *f. heterosporique* noted by Favre (*Wat.* 17377), and *I. umbrina* Bres. (*Wat.* 17389). Favre's taxa were all described from the Alps (Favre, 1955).

I. lacera is relatively widespread being found with maritime heath (Mingulay, Western Isles, legit Pankhurst, *Wat.* 17099), with *Salix repens* (Straloch, Perthshire, *Wat.* 12792 and Shetland, *Wat.* 16676), with *S. herbacea* (Cairnwell, *Henderson* 7179), and with subarctic *Betula* (Beinn Eighe Reserve, *Henderson* 2348). A close taxon, differing by having smaller spores, 9.5-11 x 4.5-5.5 µm, swollen, dark yellow cystidia, some very prominently thick-walled (in NH_4OH), has been found. In these respects the fungus resembles *I. obscura* (Pers.: Pers.) Gill. found in association with *S. herbacea* (Ben Lawers, *Wat.* 17341).

I. fastigiata, common in lowland communities is known from The Cairnwell (as var. *alpina* Heim; *Wat.* 17502) as well as with *S. repens* in Shetland (*Wat.* 16788, 16732), often associated with *I. lacera* (*Wat.* 16675). *I. pyriodora* (Pers.: Fr.) Kummer has been found with *Dryas* near Kempie, Loch Eriboll, Sutherland (*Wat.* 17555).

I. scabella var. *fulvella* Heim has been collected on Ben Lawers (*Wat.* 17338 & *Henderson* 4604) on relatively base rich rocks and on Cairn Gorm amongst granite rubble (*Wat.* 17378); both habitats supported *S. herbacea*. *I. mixtilis* (Britz.) Sacc., which Dennis, Orton and Hora (1960) consider the correct name for *I. scabella* Fr. s. Heim, has also been found on Ben Lawers (*Wat.* 17340) and Cairnwell (*Wat.* 11770 *I. rhacodes* Favre has been found once in the British Isles at Straloch, Perthshire with *Salix repens* (Watling, 1984a): mat. in L.

Hebeloma spp. are relatively rare in the Scottish mountains although they play a very important role in the associated subarctic birch-woods. However, *H. marginatulum* (Favre) Bruchet a rather common species in the Alps (Favre, 1955; Bruchet, 1970) and Greenland (Watling, 1977) has been found with *S. herbacea* on The Cairnwall (*Wat.* 17475); *H. marginatulum* has also been recorded for Svalbard (Watling, 1983c). Watling & Richardson's (1971) record of *H.* aff. *hiemale* Bres. from St. Kilda is probably the same species.

The montane *Galerina* spp. still pose a problem in Scotland in that collections have not been systematically made. *G. paludosa* (Fr.) Kühn., *G. tibiicystis* (Atk.) Kühn. and *G. sphagnorum* (Fr.) Kühn. are common and widespread in *Sphagnum* bogs throughout the Highlands and Islands. *G. clavata* (= *heterocystis*), *G. hypnorum* (Schrank: Fr.) Kühn, *G. mycenopsis* (Fr.: Fr.) Kühn. *G. mniophila* (Lasch) Kühn. and *G. vittaeformis* (Fr.) Moser are common on moss cushions or in mossy grasslands. Other taxa, which undoubtedly occur, are under consideration.

For example *G. subcerina* is known from Rannoch Moor (*Wat.* 11756). A closely related taxon is recorded from Ben Vrackie, as is *G. pseudocerina* from The Cairnwell (*Wat.* 11446). *G. atkinsoniana* Smith & Sing. is apparently fairly widespread and occurs in mossy carpets amongst *Polytrichum, Empetrum, Oxycoccus, Calluna* and *Carex bigelowii*; mat. in O. It is distinguished by 2-spored basidia, broad basidiospores and the presence of prominent pileocystidia. *G. luteofulva* Orton, originally described from England, also has been found in montane regions (The Cairnwell). Another feature of Scottish mountains is the occurrence of *G. unicolor*; however, so much confusion exists with this and closely related taxa that all records must be re-considered. *G. moelleri* Bas (= *G. pseudopumila* Orton) is also found in both montane and island communities, e.g., Cairn Gorm accompanying *Inocybe* spp. *G. tundrae* Smith & Singer has been recorded from Hirta in St. Kilda (Watling & Richardson, 1971).

Flammulaster (Phaeomarasmius) harrisonii (Dennis) Watling was described from amongst *Nardia scalaris* (Hepaticae) and *Oligotrichum hercyinum* (Musci) on Rhum at 600 m. Two recent collections one from a scree-slope on Creag Meagadh (Coille a' Choire) in the Monadhliath mountains, and one from Cairn Gorm have been made, differing only in the smaller basidiospores and narrower cheilocystidia.

The *Nardia/Oligotrichum* consortium is common in the Scottish mountains. It remains to be seen whether or not the differences represented indicate a distinct species or whether the Monadhliath material (*Wat.* 17332) comes within the limits of a single species. I think not! Generally, members of the genus *Flammulaster* are found in base rich lowland woodlands. *Naucoria bohemica* Vel. is also encountered in lowland communities with *Alnus* and/or *Salix*. It is therefore surprising to have one collection from 1176 m on Cairn Gorm (*Wat.* 17388), with *S. herbacea*.

Gymnopilus fulgens (Favre & Mre.) Singer is not uncommon on peaty soils in the mountains. It is sometimes found in association with members of the genus *Hypholoma*; e.g., Braeriach 762 m (*Henderson* 3583), in the Western Isles (Kintra, *Wat.* 6087) or the Orkneys. It is also known from Devon and Dorset (*Orton* 272 & 3; 549; 1537).

Members of the genus *Phaeogalera* are apparently common northwards in Europe or are to be found in montane habitats. Thus *P. stagninoides* (P.D. Orton) Pegler & Young has as yet only been found from the type area, viz. localities in Shetland (*Orton* 1648 & 1649), *P. stagnina* (Fr.) Pegler & Young either above 305 m (Orton, 1960, Pg. 311 mat. in O), nearer sea-level in the north, e.g., Sutherland, (*Dennis* 29); Shetland (*Wat.* 16705). *P. zetlandica* (P.D. Orton) Kühn. was also described from Shetland (*Orton* 1650 & 4421), and subsequently refound there (*Wat.* 16765; 16813); it is now known not to be confined to the north. Thus, in E material is available from Glen Esk, Angus (*Orton 84); Abernethy Forest, Inverness-shire (*Orton *4421); Rannoch, Perthshire (*Orton 4422) and the Island of Mull (*Orton* 3360); indeed it is known from Herefordshire in the west Midlands of England. *P. zetlandica* has in addition been collected on The Cairnwell (*Wat.* 17466).

HYGROPHORACEAE

No member of the genus *Hygrophorus* Fr. s. stricto has been found in Scottish Arctic-alpine localities. Members of the segregate genus *Camarophyllus* (Fr.) Wünsche (= *Hygrocybe* subgenus *Cuphophyllus* Donk), however, have been found e.g. *H. niveus* Scop.: Fr. (in O) and *H. pratensis* (Pers.: Fr.) Fr. (*Wat.* 17516). Both have been collected on The Cairnwell and are common in the northern and western parts of mainland and island Scotland. In addition, members of the segregate genus *Hygrocybe* (Fr.) Wünsche, e.g., *H. cantharella* (Schw.) Murr. and its allies, are frequent in montane situations (*Wat.* 17498) but are really members of the Callunetum and associated plant communities. Like *H. pratensis* and its allies they are not strictly Arctic-alpine. *H. conica* (Scop.: Fr.) Kummer, *H. langei* Kühn. (*Wat.* 17501), *H. nitrata* (Pers.: Pers.) Wünsche (*Wat.* 17468) and *H. quieta* (Kühner) Singer have all been found on The Cairnwell in association with the *Salix herbacea* community there, and with *Amanita nivalis*, *Russula nana*, etc. in true alpine communities. In contrast, *H. flavescens* (Kauffman) Singer has been found on Cairn Gorm (*Wat.* 17505). All five, however, are widespread in the British Isles including Shetland, and the Inner and Outer Hebrides. Thus, although (unlike the genus *Hygrophorus*) the genus *Hygrocybe* is widespread in Arctic-alpine areas only one, even of this genus, is confined to such localities, i.e., *H. lilacina*.

Hygrocybe lilacina (Laestadius) Moser in Gams Kryptogamenflora von Mitteleuropa IIb/2, 64 (1967).

Because of its stature this species was originally described as an '*Omphalia*'. Lange (1955) recognized it as a *Hygrophorus* under the name *Hygrophorus violeipes*, which he described from Greenland. This agaric is rather rare in Scotland; only three collections exist at E, whilst none exists in K. In all three cases, the material was found above 760 m and usually in *Rhacomitrium* heath communities. It is not found at lower altitudes even in the Inner Hebrides, where on Ben More it was high near the summit. *H. lilacina* is found in the Alps, Newfoundland (Redhead, DAOM 189779 pers. comm.) and Greenland. Fries (1874) knew of it through Laestadius.

Material examined: Central Massif (Ghlas Maol, *Henderson*, 28.ix.1962; Cairn Gorm, legit Wallace, *Orton* 2143): Inner Hebridean Islands, Mull (Ben More, *Watling & Henderson* 9018).

TRICHOLOMATACEAE

Laccaria

Some attention has been paid in the last few years to montane collections of *Laccaria* spp. Two species appear to dominate the montane plateau, often in association with *Salix herbacea*. *Laccaria proximella* Sing. and *L. laccata* (Scop.: Fr.) Cooke are such species.

Indeed, in respect to *L. laccata* and all those collections whose spores have been subjected to statistical analysis the taxon involved appears to be var. *subalpina* Singer. *L. proximella* appears to be the alpine equivalent to *L. proxima* and differs in its smaller stature, distinct lilaceous tinge to the gills, and ellipsoid spores (Q1.13-1.28(-1.35)) with low ornamentation. The distinct smell of radish found in lowland collections of *L. proxima* is lacking in *L. proximella* and neither are the stipe and pileus of *L. proximella* as distinctly (coarse) fibrillose. *L. laccata* var. *subalpina* is also of small stature and is recorded from the Central Massif and Northwest Highlands. The 'Q' value of this variety ranges from 1.0-1.07 (-1.15). *L. proximella* is probably the commonest of the pair, being recorded from many localities in the Central Massif and western mountains. Both species have been found at Collafirth (Ronas Hill) on Shetland.

Laccaria bicolor (Maire) P.D. Orton also has been found at high altitudes both in the Central Massif (The Cairnwell) of mainland Scotland, on the inner Hebrides, e.g., Trallival, Rhum 580 m (*Wat.* 1047C) and in Shetland (Ronas Hill). All were with *Salix herbacea*. The basidiospores in these collections were minutely ornamented as in *L. proximella*, and with a similar Q (1.06-1.30). In comparison, a collection of *L. bicolor* from a conifer plantation in the Strathardle valley, gave basidiospores up to 9 um and with a Q value of 1.25-1.36.

Collections from Engalnd and lowland woodlands in Scotland probably refer to the recently recognized *L. farinacea* (Huds.) Singer with the amethyst stipe. On making *L. bicolor* a species Orton (1960) indicated he had not found this species below 275 m, which suggests he had a more northern to highland fungus in mind. His collections are from Tomich, Glen Affric; Loch Rannoch etc. Obviously, more work is required.

Material examined: *Laccaria laccata* var. *subalpina* Singer in Plant Syst. Evol. 126: 365 (1977). Ghlas Maol, *Henderson* 833; 4152; Bealach na Ba, *Henderson* 4588, 4590; The Cairnwell, *Wat.* 12880, 13281, Coire Odhar, Ben Lawers, *Wat.* 17334, 17335 & 17336.
Laccaria proximella Singer apud Singer & Moser in Mycopath. Mycol. appl. 26: 146 (1965) Ben Lawers, *Henderson* 3511; Braeriach, *Henderson* 3575, 3577;; Beinn Bhan, *Henderson* 7153; The Cairnwell, *Henderson* 7171, *Wat.* 12900, & 16957; Ciste Dubh (above Cluane), legit A.P. Bennell, *Wat.* 13574; Buchaille, Etive Mor, *A.P. Bennell*; Ronas Hill, Collafirth, Shetland, *Wat.* 17739; Coire Odhar, Ben Lawers, *Wat.* 17337; Coire Raibeirt, ledges overlooking Coire an Sneachda and Coire Domhain, Cairn Gorm, *Wat.* 17384, 17385 & 17386 respectively.

Omphalina

At least six species of *Omphalina* are found in the Scottish mountains of which members of the *O. ericetorum* s. lato - are most widespread. Other species are *O. oniscus* (Fr.: Fr.) Quél., which is not infrequent in *Sphagnum* beds in the mountains (19 collections in E), the recently recorded *O. parvivelutina* Clémencon & Irelt, *O. alpina* (Britz.) Bresinsky & Stangl and *O. hudsoniana* (Jenn.) Bigelow.

The distribution of *O. parvivelutina* in Scotland is unknown although it will probably be found to be widespread. During a single week in the autumn of 1983, it was found on two mountain tops in unconnected ranges, on Ben Vrackie a relatively low outlying mass towards the southern fringe of the Highland mixture of acidic and strongly calcareous Dalradian schists, and on The Cairnwell (Watling, 1984a; mat. in O). *O. ericetorum* (Fr.: Fr.) M. Lange ex Bigelow (= *O. umbellifera* (L.: Fr.)

Quélet) until recently has been uncritically studied in the British Isles. The distribution of it and its segregates is really unknown. Certainly, *O. ericetorum* s. lato is widespread and common in both lowland and montane communities. The newly recognized *O. fulvopallens* P.D. Orton, and what is considered to be Møller's interpretation of *O. pseudoandrosacea* (Bull.: Fr.) Møller (1945), have been found in both the Western Isles and the Scottish mountains.

About *O. obscurata* (Kühner) Reid, Lange (1955) writes "it is truly belonging to the most typical elements in the arcto-alpine flora." The material on which Reid (1958) validated the name came from Loch Loy (Nairn), Inverness-shire. Apparently it is rare in Scotland having been collected twice since the 1955 collection (Kinlochewe, Ross-shire, *Orton* 1310, and Glen Esk, Angus, *Henderson* 1643). These and a record from Dawlish, Devon (*Orton* 807) are all non-Arctic-alpine communities. The alpine *O. obatra* Favre is recorded in the New Checklist (Dennis, Orton & Hora, 1960).

Thus only two well-documented British alpine species are found in the Scottish mountains, viz. *O. alpina* and *O. hudsoniana* and demand greater attention.

O. alpina (Britzelmayr) Bresinsky & Stangl in Zeits. f. Pilzk. 40: 73 (1974).

This species apparently fruits more frequently in the British Isles than does *O. hudsoniana*. Records are available from the Inner Hebrides (Eigg, Rhum, Mull and Skye), Outer Hebrides (South Uist), and St. Kilda (Hirta), and from Shetland, as well as the Southern Uplands (Moffat) and the Central and adjacent Massifs (Cairn Gorms, Breadalbane range, Ben Loyal, Beinn Eighe, Cearcall Dubh) of mainland Scotland. Elsewhere in the British Isles, material from E & K indicate it to be present in Snowdonia, N. Wales. It should also be looked for in the English Lake District.

O. alpina has been known in Britain as *O. luteovitellina* (Pilát & Nannfeldt) M. Lange, originally described from Finnmark (Pilát & Nannfeldt, 1954) and it is also Cooke's *Agaricus* (*Omphalia*) *umbelliferus* var. *flavus*. It was known to Stevenson (1879) from Scotland under this last name. Henderson (1958) gives a modern description based on material from the Scottish mountains. He writes, "It seems to favour peat covered with an algal scum as a substrate." Indeed, the "algal scum" is often called '*Botrydina vulgaris*', a lichen which can be observed in a large number of the herbarium collections examined or in fact reference is made to it in the collection data. The author has found it with *Salix herbacea*, and *S. arbuscula*, *Empetrum hermaphorditum*, *Polygonum viviparum*, *Alchemilla alpina*, *Lycopodium alpinum*; however, these plants reflect the habitat not biotrophic associates. The single record of *O. alpina* from Clare Is., Ireland (Rea & Hawley, 1912) may represent the only example known to date of a phenomenon parallel to the occurrence of Arctic-alpine vascular plants in Ireland (Raven & Walters, 1956).

O. alpina grows under overhangs in summit peat, on steep rocks in gorges or screes and margins of peat banks. One collection made by Gilbert (E) from St. Kilda (Hirta) is associated with the foliose lichen *Coriscium*. However, this association is usually considered to be confined to *O. hudsoniana* (see below). Unfortunately, field data did not accompany the material but if such an association does occur it should be searched for in other circumboreal areas. Singer (1969) records it from the Andes.

Table 3. Members of the Trichotomataceae found in the mountains, but recorded for exposed island groups and northern communities (Watling & Richardson, 1971; Henderson & Watling, 1978; Watling 1976 and unpubl.).

	West		North	
	Inner Heb.	St. Kilda	Shet-land	Suther-land
Calocybe carnea (Bull.: Fr.) Kühn.	x	x	x	x
Clitocybe catinus (Pers.: Fr.) Kummer (Wat. 16949 and mat. in O)	–	–	–	–
C. clavipes (Pers.: Fr.) Kummer	x	–	–	–
C. infundibuliformis (Schaef.) Quél.	x	–	–	x
C. sericella Kühn. & Romagn.	–	–	–	–
C. subdryadicola Harmaja (mat. in L)	–	–	–	–
Collybia confluens (Pers.: Fr.) Kummer	x	–	–	–
C. dryophila (Bull.: Fr.) Kummer (incl. dark variety)	x	x	x	x
Hygrophoropsis aurantiaca (Fr.) Mre.	x	x	x	x
+ var. *pallida*	x	–	x	–
Lepista rickenii Singer (Wat. 17053)	–	–	x	–
Marasmius androsaceus (L.: Fr.) Fr.	x	–	x	x
Mycena amicta (Fr.) Quél.	x	x	x	x
M. epipterygia (Scop.: Fr.) S.F. Gray	x	x	–	x
M. galopus (Pers.: Fr.) Kummer	x	–	x	x
M. leptocephala (Pers.: Fr.) Gillet	x	x	x	x
M. megaspora Kauffman; 2-spored (Wat. 17487)	–	–	–	–
M. latifolia Peck (mat. in C)	–	–	–	–
M. pura (Pers.: Fr.) Kummer	x	–	x	x
M. sanguinolenta (A.& S.: Fr.) Kummer	x	x	x	x
M. uracea Pearson	x	x	–	x
Omphaliaster asterospora (J. Lge.) Lamoure	x	–	–	–
Tephrocybe palustris (Peck) Donk	x	–	x	x
T. tesquorum (Fr.) Moser	–	x	x	–

In the Central Massif, *O. alpina* grows from 457 m to 1175 m, usually at the upper limit. In the west it may be found even as low as 305 m, e.g., on slopes of Ben More (Mull) or 425 m (Beinn Eighe) and 487 m above Kishorn. On the slopes of Conochair (St. Kilda) it is found at 275 m and on the cliffs of Bloodstone Hills and Fionchra (Rhum) at 305 m (Inner Hebrides). It also occurs at 426.5 m on Shetland (Midfield) and on Beinn Mhór (S. Uist, Outer Hebrides, legit J. Canon, *Wat.* 17092).

Henderson (1958) gives a selection of material which has been examined along with other collections. A full list of exsiccata with localities and dates is therefore not given.

O. alpina (as *O. luteovitellina* in E)

Material examined: *Bennell* 594. *Henderson* 1512; 2768; 2787; 2779; 2851; 3601; 3915; 3972; 4088; 4589; 4895; 7029; 7097; 7209; 7260; 9135; 9136. *Watling* 517C; 518C; 644C; 957C; 11151; 11739, 13311, 13312, 13313; 16848; 17092; 17333.

Omphalina hudsoniana (Jenn.) Bigelow, Mycologia 62: 15 (1970).

Twenty-eight collections of *O. hudsoniana* (two in K) indicate that this agaric is probably potentially more widespread than *O. alpina* as collections of it are known from much lower altitudes, e.g., Pools of Dee, Braemar, 31 vii 1975, (*Reid* in K). However, *O. hudsoniana* has been found on Braeriach in the Cairn Gorms close to 1219 m as well as 182-305 m on St. Kilda (Hirta), 960 m on Ben More (Mull), and 426.5 m at Midfield on Shetland. It is found in the New Checklist of British Agarics and Boletes as *O. luteolilacina* (Favre) Henderson. When Henderson (1958) validated this name, he offered a description based on Scottish material. In addition to the Central Massif and the localities given above, it has been found in the following localities from which *O. alpina* is not known; Blaven (Skye); Kerrcleuch (Selkirkshire); Glen Isla (Angus); Beinn an Orr (Jura); and Fraoch Bhenn (Glen Finnan). A full list of exsiccata examined during the study is given, but field data is omitted; records of *Coriscium viride* (Ach.) Vain., the thallose state of this agaric, are not included.

Material examined: *Henderson* 2788; 330a; 3589; 4104; 4578; 6784; 7138; 9134; 9225; Orton 2179, 2180, 21881; 5013; Watling 358C; 959C; 2697C; 5697; 5717; 6110; 9492; 11155; 12556; 13314; 14279; 16687; 17094; 17375.

A wide range of tricholomataceous fungi have been located in Arctic-alpine communities many of which are familiar members of either lowland woodlands or grasslands, or both. Some species accompany *Hypholoma* spp. on more peaty soils. Table 3 summaries the most important finds and compares them with their occurrence on St. Kilda, Sutherland and Shetland and in the Inner Hebrides. The genus *Cystoderma* perhaps should be mentioned independently as half the number of British species have been found on The Cairnwell alone. Thus *C. amianthinum* (Scop.: Fr.) Fayod, *C. carcharias* (Pers.) Fayod and *C. granulosum* (Batsch.: Fr.) Fayod (*Wat.* 17493) are all known there in grassland on limestone. All three are not strictly montane agarics and they may be found at sea level.

LEPIOTACEAE

A single record of *Lepiota cristata* (Fr.) Kummer is known from The Cairnwell, but it is locally common and widespread in the *Dryas* lawns overlooking Loch Eriboll, Sutherland (*Wat.* 17552).

AGARICACEAE

Members of the genus *Agaricus* are unexpected in the Scottish mountains, although they characterize the maritime pastures on mainland Scotland, the unique grassland on shell-sand (machair) of the Western Isles, and the wind swept grasslands of Shetland and St. Kilda. Of the several taxa found in these coastal localities, originally formed and now relying on sheep grazing for their maintenance, only *Agaricus campestris* L.: Fr. has been found in alpine communities. This agaric has been found in grasslands on the limestone pavement with *Calocybe gambosa* (Fr.) Singer above Blair Atholl (Tulach Hill), Perthshire, on grassy ledges of Seana Bhraigh, Ross & Cromarty at 550 m (legit Bennell), and amongst *Salix herbacea* on The Cairnwell (*Wat.* 17504 & 17474).

STROPHARIACEAE & COPRINACEAE

Many of the familiar coprophilous fungi belong to one or the other of these families (see below) but no member of the Coprinaceae typifies Scottish mountains. The genus *Hypholoma* is well represented in the montane flora but none are considered strictly Arctic-alpine elements. *H. elongatum* (Pers.: Fr.) Ricken, *H. subericaeum* (Fr.) Kühn., *H. udum* (Pers.: Fr.) Kühn. and *H. myosotis* (Fr.) Moser are very common in moorland communities on peaty soil, boggy areas and in *Sphagnum*, especially *H. elongatum*. *H. ericaeum* (Pers.: Fr.) Singer and *H. ericaeoides* P.D. Orton are less frequent, but are found in the same communities. As one proceeds north and/or west in the British Isles these agarics dominate the agaric flora over vast areas of countryside. In the Outer Isles and in Shetland they are exceedingly common below 30 m and on mainland Scotland they may rise to over 1000 m.

The genus *Psathyrella* has been found occasionally, in the Scottish mountains but collections await critical appraisal. Members of this genus are not major components of the flora. *Panaeolus rickenii* Hora is not uncommon in base rich montane grasslands.

BOLBITIACEAE

Several members of the genus *Conocybe* have been found in montane areas particularly where base rich flushes appear or where limestone or similar calcareous outcrops are found. These include *C. macrocephala* Kühn. ex Watl., *C. piloselloides* Watl., including a collection with some lecythiform cells on the stipe, *C. pseudopilosella* Kühn. ex Watl., *C. siennophylla* (B. & Br.) Singer and *C. subpubescens* P.D. Orton. None are characteristically montane and may in some cases be a direct result of dung having been incorporated into otherwise rather poor podsolic soils. *C. coprophila* Kühn. described originally from the French Alps has a wide distribution in the British Isles, but often only in areas with the most extreme climatic conditions; i.e., Dartmoor; Malham, Yorkshire (*Wat.* G106-108; G112-114), northern mainland Scotland (*Wat.* 17545), and the windblown Western Isles (Dennis, 1955). Although not strictly montane in Britain, it has been found in Scotland (The Cairnwell) on sheep dung in calcareous turf. It is interesting to note that in the Italian Alps it was collected once in otherwise acidic areas on dung closely associated with the concrete foundations for ski lifts from which calcareous material had leached.

Agrocybe arvalis (Fr.) Singer, although again not a characteristic montane agaric, has been found in sheep and deer grazed areas on The Cairnwell. It is not infrequent in subarctic Scottish birchwoods. *A. paludosa* (J. Lge.) Kühn. & Romagn. has been found in montane base rich flushes early in the year (June).

ENTOLOMATACEAE

Entoloma ameides (Berk. & Br.) Sacc. (*Wat.* 16773) and *E. fuscomarginatum* P.D. Orton (*Wat.* 17093) are common montane taxa. *Leptonia sericella* (Fr.: Fr.) Barbier and *Nolanea staurospora* Bres. (= *N. conferenda* (Britz.) Sacc.) are also common and widespread. Although very rich in species assignable to *Leptonia* and *Nolanea*, the entolomataceous flora contains no species indicative of a particular distribution pattern.

It would appear that many species are more indicative of a change in land management where grazing has eliminated what would otherwise have been birch woodland even above 365 m. Thus *N. cetrata* (Fr.: Fr.) Kummer has been found on Ben Loyal (Dennis, 1955), and at 275 m with *Luzula sylvatica* close to a community of *S. herbacea* on Hirta (St. Kilda), and at 932 m on The Cairnwell. But are such species members of a montane community or are they remnants of former communities? Certainly the entolomataceous flora of subarctic birchwood and associated hill pastures in Scotland is very rich (see Watling, 1984b). *Leptonia catalaunica* Singer and *L. pyrospila* Romagn. ex P.D. Orton are particularly beautiful members of this flora and are known from as far north as Shetland. At least two of Machiel Noordeloos' provisionally defined taxa have been found on The Cairnwell (Watling, 1984a).

OTHER BASIDIOMYCOTINA

Clavulina cinerea (Fr.) Schröt. (*Wat.* 17381, 17382, 17383) is the only clavarioid fungus so far seen on the mountain summits. More westwardly *Clavaria fumosa* Fr. and *C. vermicularis* Fr. are frequent (Watling, 1983b), along with *Clavulinopsis corniculata* (Fr.) Corner and *C. fusiformis* (Fr.) Corner. *Clavaria argillacea* Fr. is common on peat and var. *sphagnicola* Corner has been found on Fetlar, Shetland (*Wat.* 16883).

Bovista nigrescens Pers., *Calvatia utriformis* (Bull.: Pers.) Jaap, *Lycoperdon foetidum* Bon. and *L. molle* Pers. are four montane puffballs. Thus far the *Calvatia* and *L. molle* (*Wat.* 17458) are restricted in their mountain habitats to the mica-schist area of the central range. The other two species are very common in a whole range of grassland communities.

Many coprophiles, e.g., *Coprinus cordisporus* Gibbs, *Stropharia semiglobata* (Batsch: Fr.) Quél., *Panaeolus sphinctrinus* (Fr.) Quél., *Psilocybe subcoprophila* Britz. (*Wat.* 16964), as might be expected, are found up to 1066 m on both sheep and deer dung which is strewn over calcareous, siliceous and cumulose soils in the mountains. Morphologically the only differences between these and the lowland basidiomata are the generally smaller size and if on south facing slopes they often feature a cracked pileipellis; often the basidiospores are slightly larger in these montane collections.

LARGER ASCOMYCETES

The more conspicuous members of the Acomycotina, except for coprophiles, are rarely encountered in the Scottish mountains. Three interesting Pezizales, however, are worthy of mention, i.e., *Helvella acetabulum* (L.) Quél. (= *Paxinia*), a *Gyromitra* sp. and a *Geopora* sp.

The alpine form of the first as documented by Dissing (1966) has been collected once on The Cairnwell (*Miller* 17820 in VPI; Dupl. in E); the more normal form is not uncommon in lowland woods. A single ascoma of a species of *Gyromitra* in general agreement with Velenovsky's *G. bubaci* Vel. was located in Glencoe (near Aonach Dubh, around *Sorbus aucuparia*; *Coppins* 1759). This interesting false morel (lorel) has not been found since. Finally, a group of ascomata of a taxa close to *Geopora cervina* (Vel.) Schum. (*Coppins* 10283) has been found below the summit of Ben Loyal (*Coppins*, pers. comm.). The material agrees with that documented by Schumacher (1979); its partially buried apothecia, in this instance in a moss cushion of *Amphidium mougeotia*, can easily be overlooked and this suggests a possible reason why it has not previously been seen.

Elaphomyces asperulus Vitt. with small scruffy warts on the peridium, pink buff cortex, dark chocolate gleba, and ascospores (18.5-22 um diam.) with only slightly roughened walls with ornamentation of blunt spines coalescing to give an irregular surface pattern, has been found twice in widely separated areas. One collection from *Agrostis/Salix herbacea* community on Bealach na Ba near Kishorn, Northwest Scotland (*Henderson* 4910) and with *Salix herbacea* on Ronas Hill, Shetland (*Wat.* 17013). The first collection consisted of some ascomata parasitized by *Corydceps capitata* (Holmsk.: Fr.) Link, with part spores 19-22 x 2-2.8 µm (*Henderson* 4911).

SUBARCTIC BIRCHWOODS

Spence (1974) considers that the birches found in the Western Isles and the remnants on islands in the lochs of Shetland are examples of a former scrub birch copse found presently in the subarctic. Characteristic 'birch' fungi are located in these communities and follow the same tree to outliers high up in corries etc., of montane areas.

Only recently has attention been paid to to the British Highland birchwoods. Those in the northwest are particularly interesting in that they are the nearest birchwoods in Europe to those of Iceland. Some of the Scottish birchwoods have now been studied in detail, and their fungal flora documented (Watling, 1984b). They include Struan, near Calvine, a Roe, Red deer and sheep grazed Highland birchwoods on Dalradian schist overlaid by glacial deposits that are locally calcareous. The herbaceous cover includes *Ajuga, Primula* and *Anemone. Myrica* occurs in the wetter areas. *Trollius, Cirsium heterophyllum* and *Listera* are less common locally. Struan is an example of the Rannoch/Tummel birchwoods, which are internationally known. Morrone Birkwood is one of the few subalpine woods (350-500 m) on base rich soils remaining in Britain and consists of *Betula pubescens* subsp. *odorata* with a clear understorey of *Juniperus communis*. The wood is on calcareous schist and limestone with brown podzolics and brown earths derived from base rich glacial drift. *Potentilla crantzii, Polygonum viviparum, Rubus saxatilis* and *Galium boreale*, all montane herbs, are present along with northern species such as *Trientalis* and *Linnaea*, which are common locally. Reference should be made to Watling (1984b) for larger fungi of these communities.

Betula nana occurs in the British Isles on mountain moors from 244 m to 853.5 m. It is found in Northumberland (England) and Peebleshire (Southern Uplands of Scotland) and northwards from Perthshire and Argyllshire to Sutherland although extremely local. Only one site has been investigated to any large extent; it is a peaty area on the slopes of Ben Loyal and Ben Hiel. Here *B. nana* grows in Callunetum, wet peat bog and plant communities between the two extremes. On the drier peats *Lactarius helvus* (*Wat.* 17550) has been collected accompanying *L. rufus* (*Wat.* 17551) and an unnamed *Russula* probably of the *Emeticinae*. *Leccinum holopus*, as indicated above, was found in wetter depressions in two separate areas. *Mycena epipterygia* and other Tricholomataceae although found with *B. nana*, can be found in adjacent communities.

CONCLUSIONS

The species of flowering plants of an area can indicate the nature of the vegetation. The lack of endemism in the Scottish mountain plant communities is considered to indicate a relatively recent flora. It is impossible at this stage to say which agarics, if any, are endemic, although evidence so far shows a continuity with other European cold-climate fungus floras. The naturally low tree-line, the considerable modification of the vegetation by man and the descending nature of many Arctic-alpine plant species gives to workers in Scotland a unique opportunity to study the inter-relations between plants and fungus communities. Recent studies in asco-lichens show that here too the Arctic-alpine elements exhibit a northern and western descent in their distribution in much the same way as vascular plants (Coppins, pers. comm.); agarics at least follow this trend. The close proximity of typical mountain vegetation and lowland forest offers many interesting facets to be critically studied.

Much has still to be done in Scotland. The vagrancy of the weather and the often isolated nature of suitable plant communities makes collecting often unrewarding. It is only after many seasons of collecting that a picture unfolds. However, the macro-fungal flora of Scotland like the Arctic-alpine plants is very rich, albeit isolated in small often very restricted localities.

ACKNOWLEDGEMENTS

It is with great pleasure that I acknowledge the assistance of Dorothy Brunton in the preparation of the maps, Norma Gregory in the checking of locality data and collectors' numbers, and Dr. B.J. Coppins in offering his expert knowledge on the ascomycetes, and for reading through the completed manuscript. I would also wish to show my gratitude to the Royal Society of Edinburgh for allowing me to incorporate the late F.H.W. Green's figure on potential water deficit into my text.

REFERENCES

Ball, T., and Watling, R., 1976, The Agarics of Mingulay, *Bull. Brit. Mycol. Soc.,* 10(2): 67-69.

Bas, C., 1982, Studies in *Amanita*-II. Miscellaneous notes, *Persoonia,* 11(4): 429-442.

Boyd, M. J., 1979, The natural environment of the Outer Hebrides, *Proc. Roy. Soc. Edinb.* Sec. B, 77: 1-561.

Boyd, M. J., and Bowes, D. R., 1983, The natural environment of the Inner Hebrides, *Proc. Roy. Soc. Edinb.* Sect. B., 83: 1-648.

Bruchet, G., 1970.,Contribution à l'étude du genre *Hebeloma* (Fr.) Kummer., *Bull. Mem. Soc. Linn. Lyon* Suppl., 39: 1-131.

Conolly, A. P., and Dahl, E., 1970, Maximum summer temperature in relation to the modern Quaternary distributions of certain Arctic-montane species in the British Isles. Part I. The modern relationships, *in:* "Studies in the Vegetational History of the British Isles", Cambridge, D. Walker, and R. G. West, eds.

Dahl, E., 1951, On the relation between summer temperature and the distribution of alpine vascular plants in the lowlands of Fennoscandia, *Oikos,* 3: 22-52.

Dennis, R. W. G., 1955, The larger fungi in the northwest highlands of Scotland, *Kew Bull.,* 1955: 111-126.

Dennis, R. W. G., Orton, P. D., and Hora, F. B., 1960, New checklist of British agarics and boleti, *Trans. Brit. Mycol. Soc.,* Suppl.

Dennis, R. W. G., and Watling, R., 1983, Fungi in the Inner Hebrides, *in:* The natural environment of the Inner Hebrides, Boyd, M. J., and Bowes, D. R., eds., *Proc. Roy. Soc. Edinb.,* Sect. B, 83: 415-432.

Dissing, H., 1966, The genus *Helvella* in Europe with special emphasis on the species found in Norden, *Dansk. Bot. Arkiv.,* 25: 1-172.

Favre, J., 1955, "Les champignons supérieurs de la zone alpine du Parc National Suisse," Liestal.

Fries, E. M., 1874, "Hymenomycetes Europeae," Upsala.

Green, F. H. W., 1964, The climate of Scotland, *in:* "The vegetation of Scotland," Burnett, J. H., ed., Oliver & Boyd, Edinburgh.

Green, F. H. W., and Harding, R. J., 1983, Climate of the Inner Hebrides, *in:* "The natural environment of the Inner Hebrides," Boyd, M. J., and Bowes, D. R., eds., *Proc. Roy. Soc. Edinb.* Sect. B 83: 121-140.

Greville, R. K., 1882, "Scottish Cryptogamic Flora," Edinburgh.

Henderson, D. M., 1958, New and interesting Scottish fungi 1, *Notes Roy. Bot. Gdn., Edinb.,* 22: 593-597.

Henderson, D. M., and Watling, R., 1978, Fungi, *in:* "The Island of Mull," A. C. Jermy, and J. A. Crabbe, eds., 15.1-15.74.

Høiland, K., 1983, *Cortinarius* subgenus *Dermocybe*, *Opera Botanica*, 71: 1-113.
Huntley, B., and Birks, H. J. B., 1983, "An atlas of past and present pollen maps for Europe: 0-13000 years ago," Cambridge.
Kenworthy, J. B., 1976, "John Anthony's Flora of Sutherland," Botanical Society of Edinburgh.
Kirk, P. M., and Spooner, B. M., 1984, An account of the fungi of Arran, Gigha and Kintyre, *Kew Bull.*, 38: 503-597.
Kühner, R., 1972, Agaricales de la zone alpine. Amanitacees, *Ann. Sci. Univ. Besançon* Sér. 3, Bot. fasc. 12: 31-38.
Lange, M. , 1955, Den Botaniske ekspedition til Vestgrønland 1946. Macromycetes. Part II, Greenland agaricales, *Meddelelser om Grønland*, 147(11): 1-69.
Lanjouw, J., and Stafleu, F. A., 1964, Index Herbariorum. Utrecht, *Regnum Vegetabile*, 31.
McVean, D., and Ratcliffe, D., 1962, "Plant Communities of the Scottish Highlands," HMSO, London.
Manley, G., 1952, "Climate and the British Scene," London.
Matthews, J. R., 1937. Geographical relationships of the British flora, *J. Ecology, London*, 25: 1-90.
Møller, F. H., 1945, "Fungi of the Faeroes," Copenhagen.
Muskett, A. E., and Malone, J. P., 1980, Catalogue of Irish Fungi - II. Hymenomycetes, *Proc. Roy. Ir. Acad.*, 80B: 197-276.
Orton, P. D., 1960, New checklist of British agarics and boleti. Part III. Notes on genera and species in the list, *Trans. Brit. Mycol. Soc.*, 43: 159-439.
Page, C. N., 1982, "The Ferns of Britain & Ireland," C.U.P., Cambridge.
Pearsall, W. H., 1950, "Mountains and Moorlands," London.
Pilát, A., and Nannfeldt, J. A., 1954, Notulae ad cognitionem hymenomycetum Lapponiae Tornensis (Sueciae), *Friesia*, 5: 6-38.
Ramsbottom, J., 1958, The Killarney Foray (20-25 ix 36), *Trans. Brit. Mycol. Soc.*, 22: 5-11.
Raven, J., and Walters, M. R., 1956, "Mountain Flowers," London.
Rea, C., and Hawley, H. C., 1912, Clare Island Survey. Part 13. Fungi, *Proc. Roy. Ir. Acad.*, 31B: 1-26.
Reid, D. A., 1958, New or interesting records of British Hymenomycetes. II, *Trans. Brit. Mycol. Soc.*, 41: 419-445.
Reid, D. A., 1972, Coloured illustrations of rare and interesting fungi. V, *Fungorum Rariorum Icones Coloratae*.
Scannell, M. J. P., 1982, Southern elements in the cryptogamic flora of Ireland, *J. Life Sci. Roy. Dubl. Soc.*, 3: 267-276.
Schumacher, T., 1979, Notes on taxonomy, ecology and distributions of operculate discomycetes (Pezizales) from river banks in Norway, *Norw. J. Bot.*, 26: 53-83.
Singer, R., 1969, Mycoflora Australis, *Beih. Nova Hedwigia*, 29: 1-405.
Spence, D. H. N., 1974, Subarctic debris and scrub vegetation of Shetland, *in:* "The Natural Environment of Shetland," R. Goodier, ed., Nature Conservancy Council, Edinburgh.
Spence, D. H. N., 1979, "Shetlands Living Landscape: A study in island plant ecology," Sandwick.
Stevenson, J., 1879, "Mycologia Scotica," Edinburgh.
Watling, R., 1976, Notes on the fungal flora of Sutherland, *in:* "John Anthony's Flora of Sutherland," J. B. Kenworthy, ed., Botanical Society of Edinburgh.
Watling, R., 1977, Larger fungi of Greenland, *Astarte*, 10(2): 61-72.
Watling, R., 1979, A new British Bolete, *Notes Roy. Bot. Gdn., Edinb.*, 31: 139-142.
Watling, R., 1981, Relationships between macromycetes and the development of higher plant communities, *in:* "The Fungal Community," D. T. Wicklow, and G. C. Carroll, eds., New York.

Watling, R., 1983a, Additions to the fungus of the Hebrides, *Trans. Bot. Soc. Edinb.*, 44: 127-138.
Watling, R., 1983b, Fungi of Skye, *The Glasgow Naturalist*, 20: 269-311.
Watling, R., 1983c, Larger cold-climate fungi, *Sydowia, Ann. Mycol.* Ser. II, 36: 308-325.
Watling, R., 1984a, Larger fungi around Kindrogan, Perthshire, *Trans. Bot. Soc. Edinb.*, 44(3): 237-259.
Watling, R., 1984b, Macrofungi of birchwoods, *in:* "Birches," D. M. Henderson and D. Mann, eds., *Proc. Roy. Soc. Edinb.*, 85: 129-140
Watling, R., 1985, Observations on *Amanita nivalis* Greville, *Agarica*, 6: 327-335.
Watling, R., and Richardson, M. D., 1971, The agarics of St. Kilda, *Trans. Bot. Soc. Edinb.*, 41: 165-187.
White, F. B., 1879, Glen Tilt: its fauna and flora, *The Scottish Naturalist*, 5: 85-93.

ECOPHYSIOLOGICAL STUDIES ON ALPINE MACROMYCETES: SAPROPHYTIC *CLITOCYBE* AND MYCORRHIZAL *HEBELOMA* ASSOCIATED WITH *DRYAS OCTOPETALA*

J.C. Debaud

Université Claude Bernard, Lyon I
Laboratoire de Mycologie associé au C.N.R.S.
43 Bd du 11 Novembre
F. 69 622 Villeurbanne Cedex France

Key words: Alpine, saprophyte, *Clitocybe*, mycorrhizae, *Hebeloma*, *Dryas octopetala*, Rosaceae, fruiting, temperature, ecophysiology

ABSTRACT

In alpine areas numerous saprophytic or mycorrhizal Agaricales are associated with a pioneer Rosaceae: *Dryas octopetala*. Ecophysiological studies of the relation between mycorrhizal fungi and *Dryas* revealed influences of some biological and/or physical factors on fungal fruiting and their probable distribution.

Results concerning soil temperature influence on *in situ* fruiting of *Clitocybe* spp. show that:

- fruiting is induced by the drop of maximal litter temperature if the drop is greater than or equal to 4°C.
- full sporocarp development occurs only after this drop when the maximal temperature is between 13° and 20° and the minimal temperature is between 4° and 9°C. If temperature drops are below these limits, the development is delayed.
- mycelia survive low winter temperature.

It seems that temperature is probably the most important factor for the distribution of these *Clitocybe* species in alpine areas. Their occurrence in the *Dryas octopetala* community is probably best explained by other influencing factors; i.e., litter chemical composition.

Synthetic Mycorrhizae synthesized between alpine *Hebeloma* species and *Dryas octopetala* permitted laboratory fruiting of these fungi, which was not the case with other attempts at synthesis with *Pinus* spp., a non-host of these fungi. The ecological significance of mycorrhizal synthesis that allows fruiting is discussed as is the importance of nutrients and storage in the mantle and Hartig net for rapid sporocarp formation.

INTRODUCTION

The association of a mycorrhizal fungus with a plant can at least be explained partially by the presence of a symbiotic relationship. For the saprophytic fungus, the problem seems to be more complex. This association may be explained by demonstrating close dependence of the fungus on the microenvironment of the plant; i.e., litter and humus layers, or by identifying identical chemical and/or physical requirements to those of the plant and microclimatic factors, in particular.

In the alpine area, some Agaricales are associated with *Dryas octopetala*. This is the case for many saprophyte fungi; e.g., species of the genus *Clitocybe* (Lamoure, 1972), as well as mycorrhizal fungi such as *Hebeloma* spp. (Bruchet, 1973 and 1974). The purpose of this work is to precisely define the influence of certain biological and/or physical factors on the development of the fungi from these two genera, and in so doing, determine limits of their distribution in alpine area, and in particular *Dryas octopetala* communities. Among numerous factors we studied, microclimatic factors such as temperature and the ability to form mycorrhizae were important considerations.

Little is known or reported on the influence of microclimatic conditions on the development of Agaricales. Important characteristics resulting from the influence of temperature and humidity on higher fungi fruiting in the temperate zones were shown by Wilkins and Patrick (1940), Wilkins and Harris (1946), Lange (1948) and Hering (1966). Precise data concerning some Arctic macromycetes were provided by Petersen (1977).

Ectomycorrhizal symbiosis of Arctic or alpine fungi has received less studied than those of temperate zone fungi, except for the works of Miller and Laursen (1978), Antibus and Linkins (1978), Linkins and Antibus (1980) and Antibus, Croxdale, Miller and Linkins (1981) on the *Salix rotundifolia* mycorrhizae. For numerous alpine Agaricales, the ability to form mycorrhizae remains to be clearly demonstrated, particularly synthesis in pure culture.

MATERIALS AND METHODS

Investigation area: (Plate I, fig. 1, 2 and 3) The study area is in Vanoise National Park, 2100 m, on the right moraine of Epena glacier, under the Grande Casse (3852 m). Because of the melting of the glacier, pioneer vegetation has colonized the base of the moraine for only about 100 years. This vegetation is dominated by *Dryas octopetala* (Plate I, fig. 5 and 6). Growth of *Dryas* on rocky calcareous soil of the moraine (Plate II) makes a rather thin litter (1 to 2 cm) and a 2 to 5 cm humus layer allowing the growth of Agaricales (Plate III). These Agaricales are not found on bare soil.

Field data concerning *Clitocybe* spp.: Three areas of investigation represent 1000 m^2. Plant cover is about 50 percent. The available surface area is 500 m^2. Numerous *Clitocybe* spp. growing on *Dryas octopetala* litter were determined and cultured by D. Lamoure. The data herein presented concern all species of the genus *Clitocybe* present in

---

PLATE I: Fig. 1. General view of Grande Casse (3852 m) and Epena Glacier. Fig. 2. Right side moraine ending at Lac de la Glière (2070 m). Fig. 3. Locations of investigation areas (->). Fig. 4. One investigation area with pioneer flora dominated by *Dryas octopetala*. Fig. 5. *Dryas octopetala* flower. (Rosaceae) (x 1.8) Flowering from mid-July to mid-August. Fig. 6. *Dryas* fruits (akènes) (x 1.8) Fruit formation and seed maturation from mid-August to end of September.

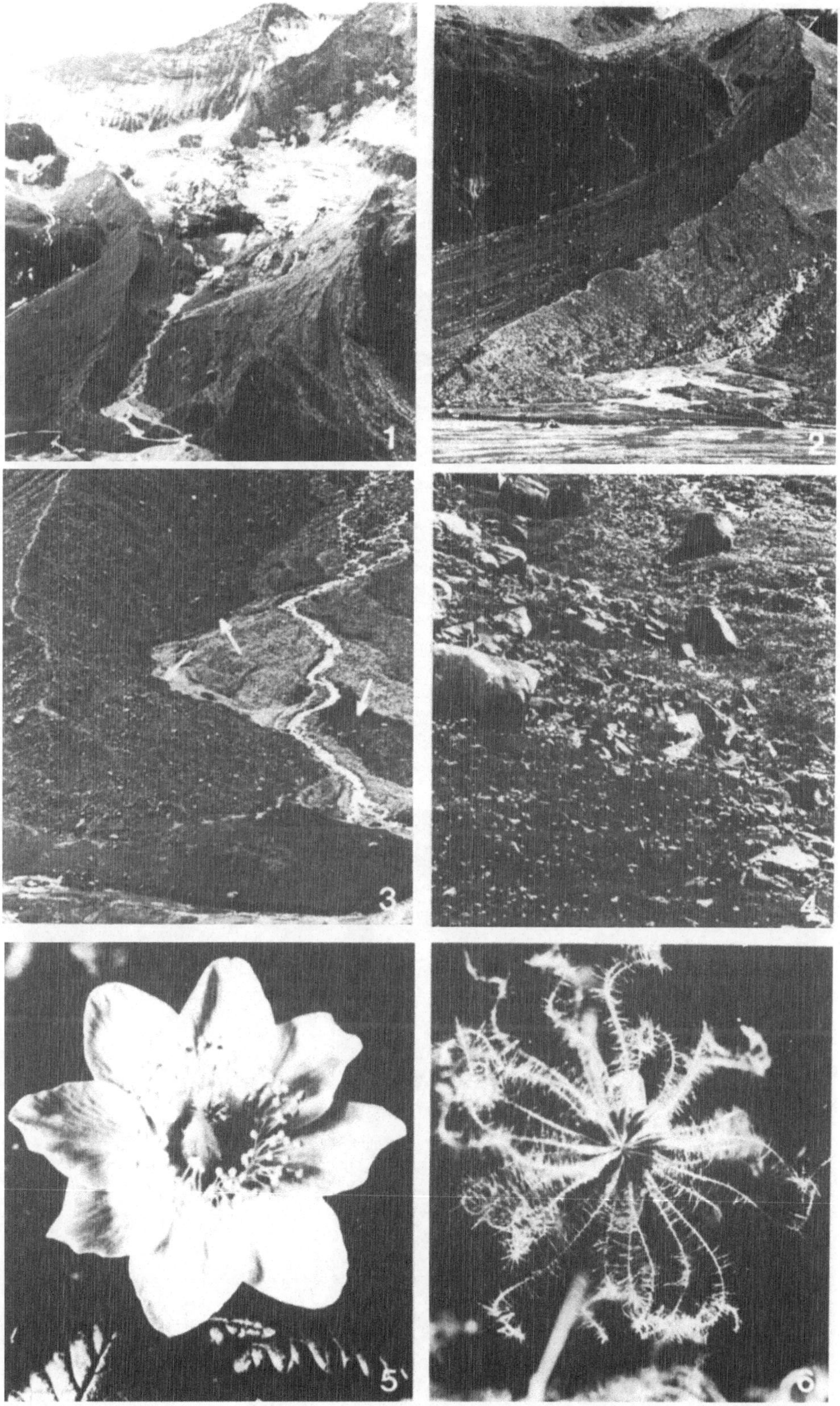
1
2
3
4
5
6

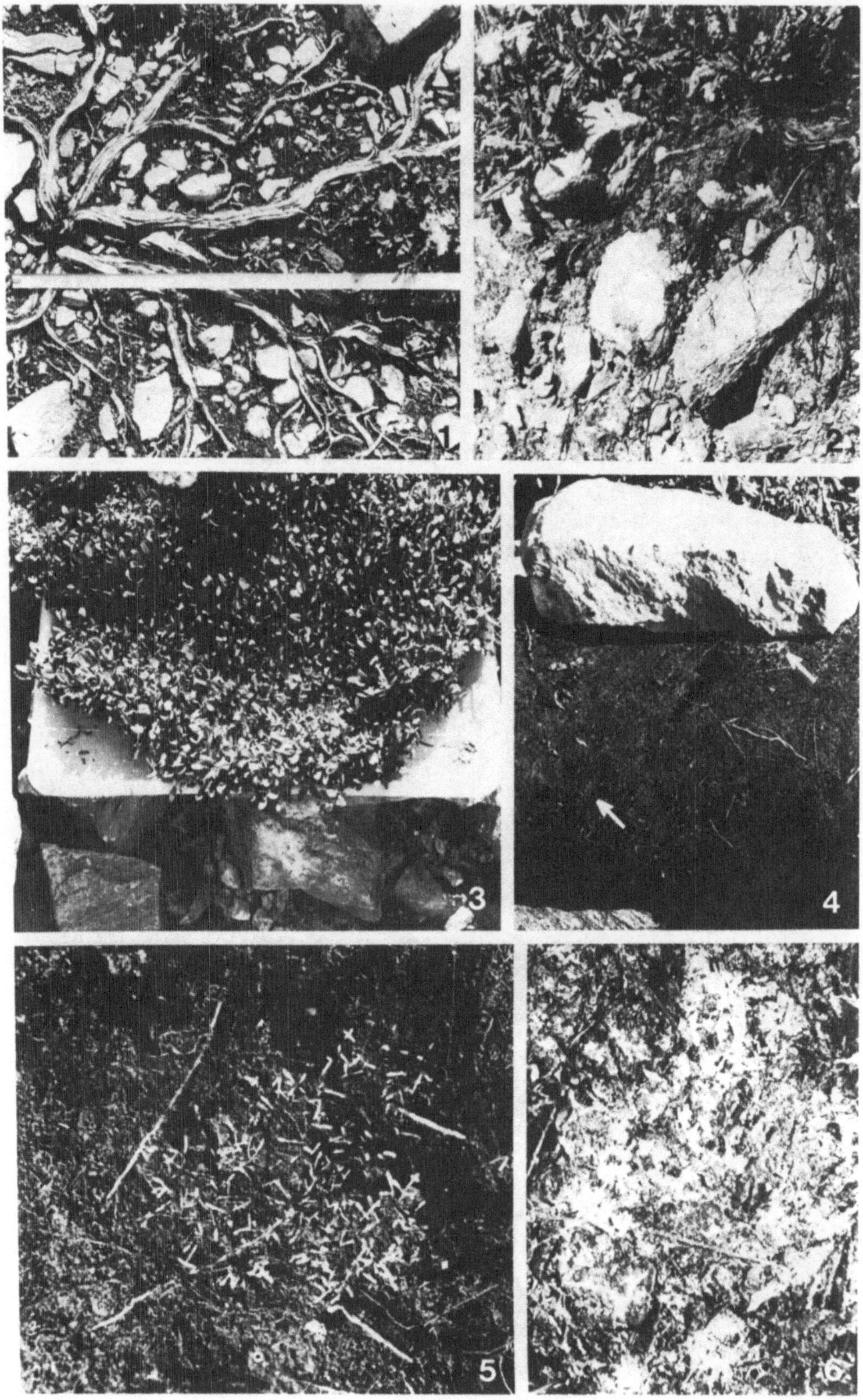
1
2
3
4
5
6

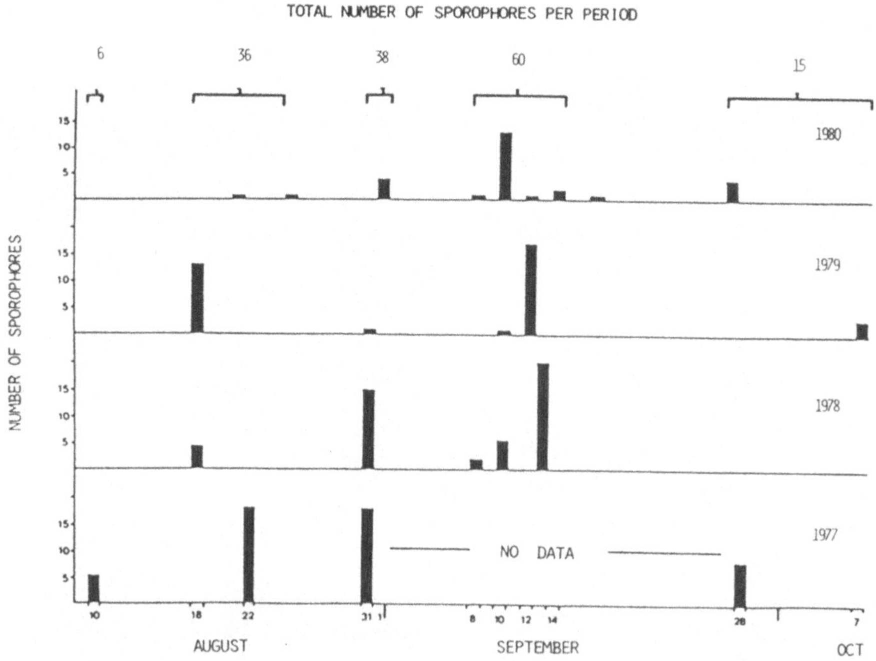

Fig. 1. Seasonal variation in *Clitocybe* spp. sporocarp production.

this 500 m^2 area during four successive years. In 1977, data were gathered about every fortnight. During 1978, 1979 and 1980, daily data were gathered throughout the fruiting season (mid-August to mid-September). Other collections were made at the beginning and at the end of the season.

Microclimatic condition measurements: Soil temperature was measured with max-min thermometers placed at the surface of the soil and at depths of 2, 5, 10 and 20 cm. These thermometers remained in place for four years. In addition, temperatures at depths of 2 and 5 cm in litter and humus were recorded with thermographs during the fruiting period.

Mycorrhizal syntheses: Culture conditions on artificial substrates and mineral medium were defined so as to induce mycorrhizal syntheses between *Dryas octopetala* and *Hebeloma* spp. in the laboratory (Debaud, Pepin and Bruchet, 1981 a). Conditions were defined, thus allowing an ultrastructural study of the mycorrhizae harvested during fungal fruiting (Debaud, Pepin and Bruchet, 1981 b).

PLATE II: Fig. 1. *Dryas octopetala* shrub (about 100 years old) Surface: $2m^2$. Leaves and litter layers taken off, stems developed in dark humus. Fig. 2. Soil under *Dryas*. Alpine ranker with thin litter and humus layers and calcareous rocky soil. Fig. 3. Basin under *Dryas* carpet for throughfall analysis. Fig. 4. Inferior part of *Dryas* humus layer with numerous mycorrhizae (-> see detail Fig. 5) and mycelial colonization (-> see detail Fig. 6). Fig. 5. *Dryas octopetala* mycorrhizae in humus layer. Unidentified fungus (x 1.8). Fig. 6. Unidentified mycelium - probably saprophytic - in *Dryas* humus (x 1.8).

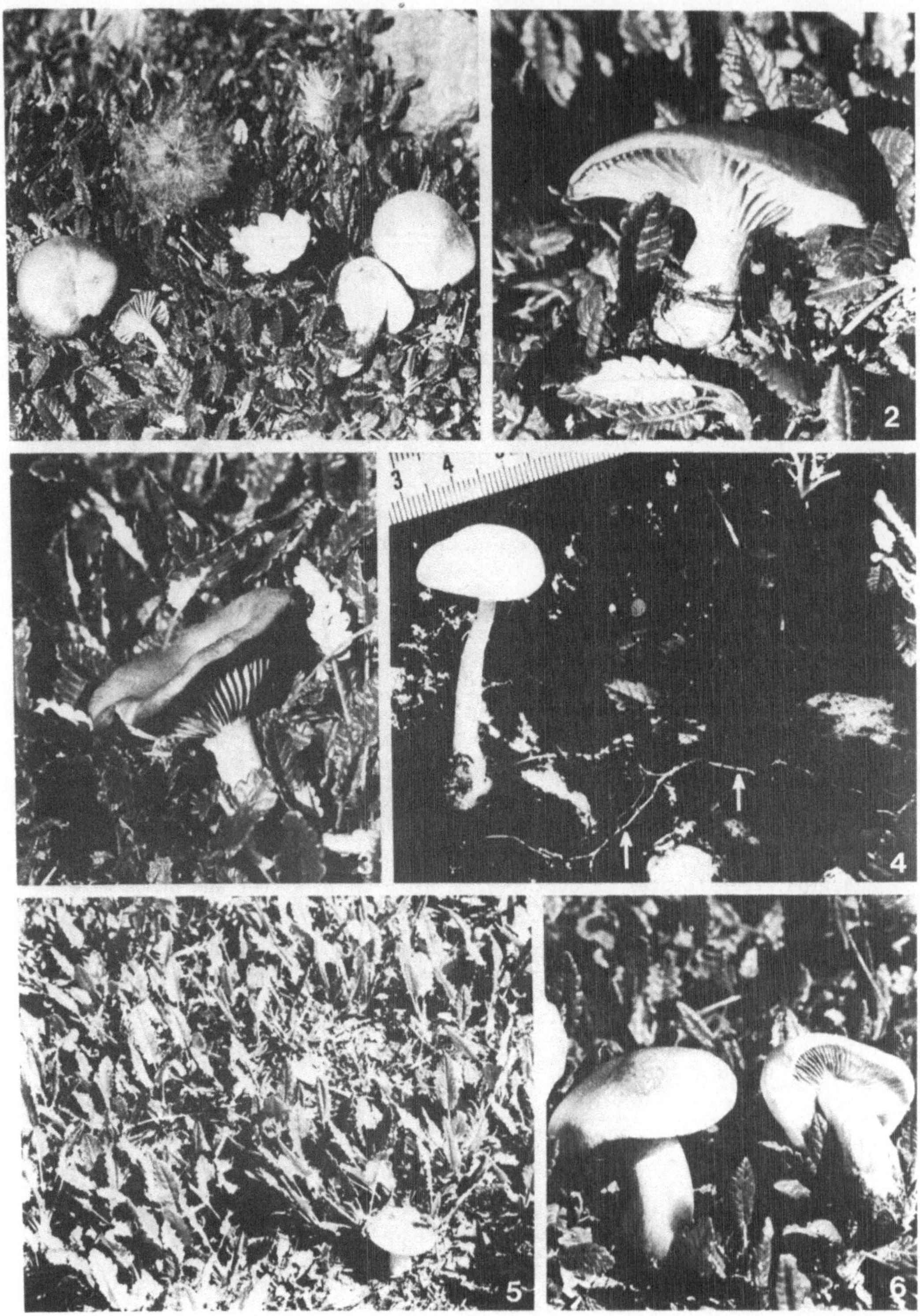
2
3
4
5
6

Table 1. Sporocarps of *Clitocybe* spp./ season

		Growing season 1977	Winter 1977-78	Growing season 1978	Winter 1978-79	Growing season 1979	Winter 1979-80	Growing season 1980
Minimal soil temperature at the depth of :	2cm		-0°5C		-8°C		-2°C	
	5cm		0°C		-7°C		-1°C	
	10cm		0°C		-6°C		0°C	
	20cm		0°C		-5°C			
Time of snow disappearance		June 15th		July 10th		June 20th		July 15th
Total number of sporocarps		50		44		35		28

RESULTS/CONCLUSIONS/DISCUSSION

ALPINE CLITOCYBE SPP. ECOPHYSIOLOGY: EFFECT OF TEMPERATURE ON GROWTH AND FRUITING.

Clitocybe spp. fruiting period: Four years data (Fig. 1) show that fruiting time ranges from August 10 to October 8. During this time three periods can be distinguished as:

- an initial period before mid-August (probably underestimated because of the small number of observations)
- and a main period from mid-August to mid-September. During this period there are 3 fruiting flushes at 8 to 10 day intervals, for the three years with daily data
  - the first between August 10 and 22.
  - the second August 31 and September 1,
  - the third between September 8 and 14.

  This distribution was noted every year, but cannot be explained.
- a final period at the end of September or the beginning of October. During this final period the number of sporocarps is suspected to be important because of climatic conditions (frost and snowfall).

Comparison of the different species *Clitocybe*: Certain species appear early in the season, for instance *Clitocybe lateritia* (from August 10) whereas others appear later; i.e., *Clitocybe serotina* (fruiting on October 7). The duration of the fruiting period varies according to

---

PLATE III: Fig. 1. Alpine Agaricales associated with *Dryas octopetala*. Left: *Clitocybe lateritia*. Right: *Hebeloma alpinum* (x 1.65). Fig. 2. *Clitocybe* sp. Stirps *Infundibuliformes* (x 0.8). Fig. 3. *Clitocybe* sp. (x 0.8). Fig. 4. Sporocarp of *Clitocybe candicans* var. *dryadicola* with mycelial strands (->) in humus layer. Fig. 5. Young sporocarp of *Hebeloma alpinum* in *Dryas* carpet (x 1.65). Fig. 6. Mature sporocarps of *Hebeloma alpinum* (x 0.8).

Table 2. Maximum/minimun aestival temperatures

	At the depth of	Days with minimal temperature inferior to 4°C	Days with maximal temperature	
			inferior to 12°C	superior or equal to 24°C
1977	2 cm	6	9	3
	5 cm	6	16	0
1978	2 cm	6	2	10
	5 cm	2	4	0
1979	2 cm	1	0	2
	5 cm	0	0	0
1980	2 cm	4	6	3
	5 cm	1	5	0

(1) In vitro Clitocybe spp. mycelial growth shows that temperature could be considered as unfavourable when it is inferior to 12°C or superior to 24°C

species. For example, that of *C. lateritia* is important, from August 10 to September 14; whereas, that of *C. serotina* is shorter (end of September to the beginning of October). Despite these differences, and partly because no precise taxonomic determinations were made of all sporocarps, all species belonging to the genus *Clitocybe* were considered as a whole. This approximation does not prevent us from pointing out the main characteristics demonstrated by the fungus in response to temperature.

Total number of sporocarps per year: In the same investigation area the annual number of *Clitocybe* spp. sporocarps varied from 20 to 50 (Table 1). Three hypotheses could be made to explain this variation:

- the duration of the snow cover, or more precisely the time of snow disappearance
- the influence of winter temperatures
- the temperature influence during the growing and fruiting periods.

Influence of time of snow disappearance: Eynard (1977) suggested that the time of snow disappearance could have an effect on the fungal fruiting period. According to this author, the interval between snow disappearance and fruiting should be constant. Our observations do not confirm this hypothesis, however, the main *Clitocybe* fruiting flushes of 1978 (Table 1) (- snow disappearance July 10-) are similar to the main flushes of 1979 (snow disappearance on June 20). As for species appearing late in the season, no rule can be drawn (see for example the final flushes of 1977 and 1980 with fruiting at the same period, but with snow disappearance June 15 and July 15 respectively. The effect of a late snow disappearance seems to be reduced during the growing season. A correlation could be better found between fungal fruiting and the *Dryas octopetala* physiological state.

Effect of temperature: Previous work on *in vitro* mycelial growth of 24 alpine *Clitocybe* spp. isolates show that most of these fungi have an optimal temperature of around 18°C. They grow rather well at 12°C. One third of these fungi grow slowly at 4°C and a temperature of 30°C is lethal after 3 weeks (Debaud, 1975).

PLATE IV: Laboratory fruiting of *Hebeloma* spp. after eight months of culture with *Dryas octopetala*, and the formation of ectomycorrhizae: Fig. 1. *Hebeloma alpinum* (x 1.65). Fig. 2. *Hebeloma marginatulum* (x 1.23).

Influence of winter temperatures: In the field, over an eight to nine month period, thermal conditions do not allow mycelia to grow; (i.e., temperatures are close to or lower than 0°C). These *Clitocybe* spp. show substantial resistance to subfreezing temperatures. This fact is confirmed by the *in vitro* study of Eynard (1977) as well as the fruiting of the same fungus during several years at the same place, and therefore most probably from the same mycelium. For instance, each year from 1975 to 1980 *C. lateritia* fruited in the same place. Fungal biomass measurements also showed that mycelial concentrations were the same at the end of a season as they were at the beginning of the next season (Debaud, 1983). Only the reaching of lowest minimal temperatures, despite their supposed short duration, could explain any decrease of the total number of sporocarps seen the following year. Low minimal temperatures can destroy mycelium of some of the more cryosenstive species. Despite the existence of little data concerning winter temperatures, the very low temperature of the 1978-79 winter (-7°C at the depth of 5 cm) could explain the observed one-third decrease in the total number of sporocarps for two successive years.

Influence of aestival temperature: Litter and humus temperatures during the growing season almost always fall between 0° and 25°C (Table 2). The minimal temperature is rarely below 4° and most of the time is found to be between 4° and 10°C. Maximal temperature varies from 10° to 25°C and it is generally above 12° during more than half of each day, which allows for good mycelial growth. The duration of the fruiting season, the high number of temperature variations during the growing season, and the interference of other factors do not allow the drawing of conclusions as a general rule for predicting the abundance of the annual fruiting. On the other hand, there is no correlation between the total number of sporocarps that appeared during a season, the number of fruiting flushes, and the number of thermal stimuli (for instance 50 sporocarps distributed into 4 flushes in 1977 and 28 into 9 flushes in 1980).

Variations of litter temperature and induction of *Clitocybe* spp. fruiting: *In vitro* studies on Agaricales showed that temperature variations can induce or increase fruitbody formation. Minimal and maximal temperature variations were analyzed at the depth of 2 cm in litter and humus layers. Temperature drops with an amplitude greater or equal to 4°C were especially studied because the number of temperature drops in each growing season is approximately the same as the number of fruiting flushes. Variations with an amplitude below 4°C are frequent, but they seem to have no effect on *Clitocybe* spp., fruiting. Among the 18 flushes and 135 sporocarps studied during four consecutive years, it was determined that:

- 11 flushes (102 sporocarps) appear 24 or 48 hours after a temperature drop with an amplitude greater than or equal to 4°C at the 2 cm litter depth;
- 3 flushes (1 sporocarp each) appear 72 to 96 hours after the drop; and
- 4 flushes (30 sporocarps among which all were late species) appear between 6 and 9 days after the drop.

For most of these alpine *Clitocybe* spp. a sudden drop of temperature seems necessary to induce fruitbody formation. These frequent drops have an effect only within a precise time duration during the growing season.

PLATE V: Ultrastructure of *Hebeloma alpinum* x *Dryas octopetala* ectomycorrhizae at the time of fungal fruiting.: Fig. 1. Mantle, epidermal cells, Hartig net and cortical cells. Internal part of mantle and Hartig net hyphae with great amounts of glycogen and some polyphosphate granules (x 2640). Fig. 2. Uni-or multiseriate Hartig net and important starch reserves in cortical cells (x 3712). Fig. 3. Hyphal detail in Hartig net showing great amounts of glycogen and lipid reserves (x 7920).

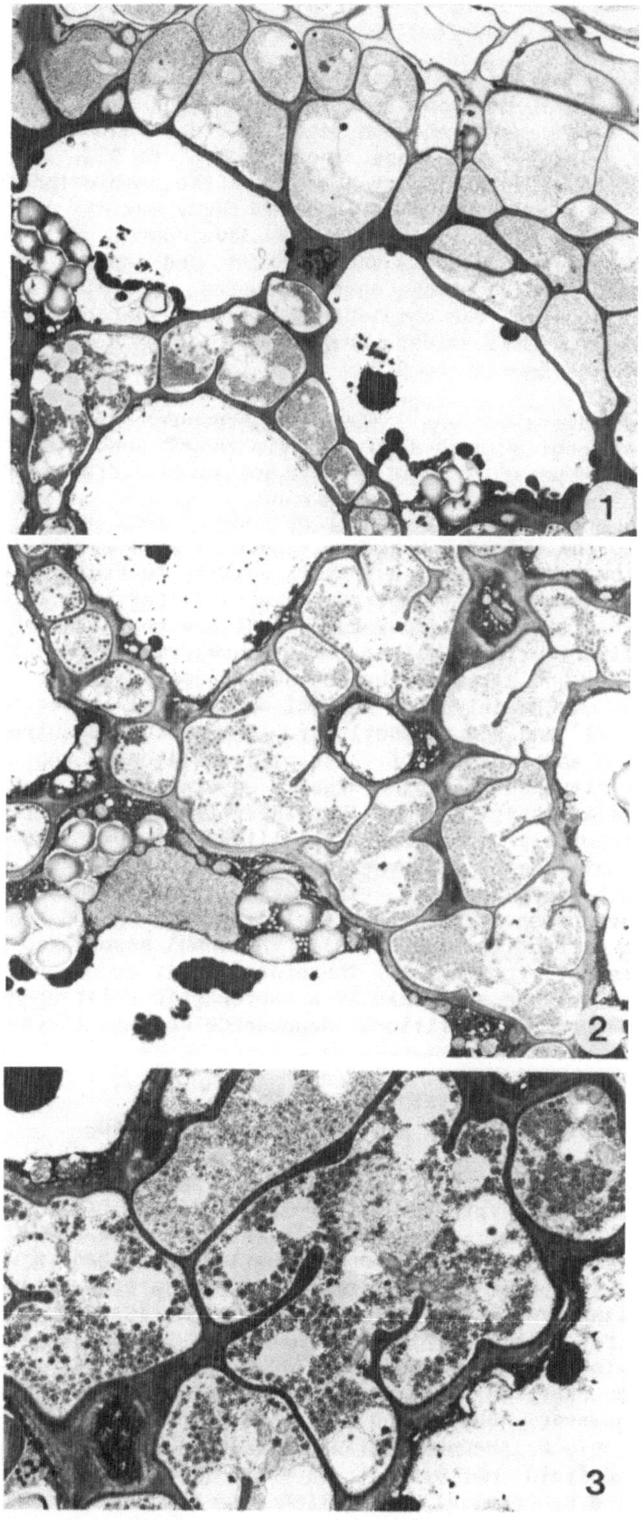
1
2
3

Beyond these fruiting limits temperature drops have no effect. This is probably due to the fact that mycelia are not receptive to the fruiting response.

Temperature required for the completion of sporocarp-development: After the inducing temperature drop, in 15 flushes among the 18 studied, the daily maximal temperature, at 2 cm in the litter, ranges from 10° to 15°C, and the daily minimal temperature ranges from 6° to 9°C. The level of these temperatures play an important role in the completion of sporocarp development. In a certain number of cases these maximal and/or minimal temperatures become too low to allow full development in mature sporocarps. The development is thus delayed, and can end when temperatures increase to within certain limits. For all alpine *Clitocybe* spp. studied, the completion of mature sporocarp development is possible when maximal temperature ranges from 13° to 21°C and minimal temperature ranges from 4° to 9°C.

Temperature and distribution: Temperature strongly influences the life cycle of these fungi at each step in their growth and development. Alpine *Clitocybe* spp. temperature requirements are quite different from those of temperate zone Agaricales. Temperatures that are too low certainly define the upper distribution limits of fungi. This upper limit corresponds to the effects of high altitude because mycelial resistance is very low to low temperatures, and their ability to fruit is reduced when maximal and minimal litter temperatures are greater than 13°C and 4°C respectively. Similarly, temperatures that are too high define the lower altitude of distribution. With a temperature greater than 25°C, mycelial growth decreases and a temperature of 30°C becomes lethal. On the other hand, fruiting occurs only when maximal and minimal litter temperatures are less than 21° and 9°C respectively. Temperature requirements for the mycelial growth and fruiting of these alpine *Clitocybe* spp. can probably explain their presence in alpine areas. However, their distribution in *Dryas octopetala* litter is probably related to other factors that may include the chemical compostion of the litter and humus layers. Another work showed that aqueous litter extract strongly increases the *in vitro* mycelial growth of these fungi. This stimulation is due in large part to greater amounts of seven vitamins (including thiamin and biotin) present in the litter extract (Debaud, 1980). The fungi associated with *Dryas octopetala* are not mycorrhizal. Therefore, their occurrence in *Dryas octopetala* can only be explained by a saprophytic relationship and probably by a specific nutritional dependence on some litter component(s).

ALPINE HEBELOMA SPP. ECOPHYSIOLOGY: MYCORRHIZAE SIGNIFICANCE FOR DISTRIBUTION AND FRUITING.

The two *Hebeloma* species studied; i.e., *Hebeloma alpinum* and *H. marginatulum*, formed synthetic ectomycorrhizae with *Dryas octopetala* (Debaud et al., 1981 a). The symbiotic relationships formed are probably obligate associations and such an explanation is needed to understand the distribution of *Hebeloma* spp. in *Dryas octopetala* communities. But does mycorrhizal syntheses have ecological significance? It is probably not the case for all mycorrhizal syntheses. For example, the two *Hebeloma* species studied formed *in vitro* mycorrhizae with *Pinus* and *Picea* spp. (Bruchet, 1972; Hacskaylo and Bruchet, 1972). These two fungi were, however, never found in the field under *Pinus* or *Picea* spp. It can only be assumed that an artificial symbiosis cannot always survive in the field, nor will it allow the fungus to fruit. It is possible that a mycorrhizal association does not explain entirely the distribution of these fungi under natural conditions. The distribution of

H. alpinum and *H. marginatulum* could be the result of other influencing factors; i.e., symbiosis, microclimatic conditions, and nutritional requirements, which must allow full development of the fungus and its host. The mycorrhizal syntheses obtained with *H. alpinum* and *H. marginatulum* was a necessary precursor to the fruiting of these two fungi (Plate IV) and to the full development of *Dryas octopetala* (up to seed formation). Therefore, the ecological significance of these syntheses is probably as good as the significance of natural mycorrhizae. On the other hand, ultrastructural studies of these synthetic ectomycorrhizae, harvested at the time of fungal fruiting, showed important nutrients to be stored in hyphal cells (Debaud et al., 1981 b). Except for the external hyphae of the mantle, most of the fungal cells contain a great amount of glycogen (Plate V, fig. 1, 2 and 3). There are also polyphosphate grànules (dark bodies, fig. 1) and lipid droplets (fig. 3) present. In the same way, cortical cells showed abundant amounts of starch (fig. 1 and 2). This nutrient reserve, particularly polysaccharides, is probably related to fungal fruiting. Alpine *Hebeloma* spp. fruitbody formation is rapid, and it seems to be influenced by a thermal stimulus as well as *Clitocybe* spp. despite a lack of precise data concerning temperature effect on *Hebeloma* spp. fruiting under natural conditions. The important and rapid synthesis of fungal biomass during fruiting is probably due mainly to the nutrients stored in the mycorrhizal fungal cells. In our laboratory and field experiments, *Hebeloma* spp. fruiting is related to a precise step of *Dryas octopetala* development. Sporocarp formation occurs when *Dryas* seeds are forming and when the first signs of senescence is indicated from the leaves. Fungal fruiting seems to correspond to the moment when nutrients return to the roots. Mycorrhiza metabolism probably changes and at that time and exchanges between fungus and host probably become favorable to the fungus. For example, after fruiting of *Hebeloma alpinum* following ectomycorrhizae synthesis, phosphorus rates in *Dryas octopetala* leaves were 30 percent lower than the control. The decrease in phosphorus in the host could be related to fungal requirements for fruiting. In these experiments, the decrease in phosphorus was certainly important for field studies because laboratory conditions for *Hebeloma alpinum* fruitbody weight was about 10 percent that of *Dryas* weight, which is simply not the case under natural conditions.

ACKNOWLEDGEMENTS

The author wishes to extend his appreciation to the Vanoise National Park for financial support of this research.

REFERENCES

Antibus, R. K., Croxdale, J. G., Miller, O. K., and Linkins, A. E., 1981, Ectomycorrhizal fungi of *Salix rotundifolia* III, Resynthesized mycorrhizal complexes and their surface phosphatase activities, *Can. J. Bot.*, 59: 2458-2465.

Antibus, R. K., and Linkins, A. E., 1978, Ectomycorrhizal fungi of *Salix rotundifolia* Trautv. I, Impact of surface applied Prudhoe Bay crude oil on mycorrhizal structure and composition, *Arctic*, 31: 366-380.

Bruchet, G, 1973, "Contribution à l'étude du genre *Hebeloma* (Fr.) Kumm.: essai taxonomique et écologique," Thèse Doct. d'Etat, Univ. Lyon, 114 p.

Bruchet, G., 1974, Recherches sur l'écologie des *Hebeloma* arctico-alpins, étude spéciale de l'aptitude ectomycorhizogène des espèces. Travaux mycologiques dédiés à R. Kühner, *n° spécial. Bull. Soc. Linn. Lyon.*, 85-96.

Debaud, J. C., 1975, Influence de la température sur la croissance des mycéliums de Clitocybes alpins (Basidiomycètes-Agaricales), *Trav. Scient. Parc National Vanoise*, 6: 167-174.
Debaud, J. C., 1980, Ecophysiologie des champignons alpins associés à *Dryas octopetala* (Rosacées), *Bull. Soc. Ecophysiol*, 5: 139-144.
Debaud, J. C., 1983, "Recherches écophysiologiques sur des espèces alpines des genres *Clitocybe* et *Hébeloma* (Agaricales) associées à *Dryas octopetala* (Rosacées)," Thèse Doct. Etat Univ. Lyon. tome I: 206 p., tome II: 124 p.
Debaud, J. C., Pepin, R., and Bruchet, G., 1981a, Etude des ectomycorhizes de *Dryas octopetala*. Obtention de synthèses mycorhiziennes et de carpophores d'*Hebeloma alpinum* et *H. marginatulum*, *Can. J. Bot.*, 59: 1014-1020.
Debaud, J. C., Pepin R., and Bruchet, G., 1981b, Ultrastructure des ectomycorhizes synthétiques à *Hebeloma alpinum* et *Hebeloma marginatulum* de *Dryas octopetala*, *Can. J. Bot.*, 59: 2160-2166.
Eynard, M., 1977, "Contribution à l'étude écologique des Agaricales des groupements à *Salix herbacea*," Thèse Doct. Spéc. Univ. Lyon. 202 p.
Hacskaylo, E., and Bruchet, G., 1972, *Hebelomas* as mycorrhizal fungi, *Bull. Torrey Bot. Club*, 99: 17-20.
Hering, T. F., 1966, The terricolous higher fungi of Four Lake District woodlands, *Trans. Brit. Mycol. Soc.*, 49: 369-383.
Lamoure, D., 1972, Agaricales de la zone alpine. Genre *Clitocybe*, *Trav. Sci. Parc National Vanoise*, 2: 107-152.
Lange, M., 1948, The agarics of Maglemose, *Dansk. Bot. Arkiv.*, 13: 1-141.
Linkins, A. E., and Antibus, R. K., 1980, Mycorrhizae of *Salix rotundifolia* in coastal arctic tundra, *in:* "Arctic and Alpine Mycology, the First International Symposium on Arcto-Alpine Mycology," G. A. Laursen and J. F. Ammirati, eds., 509-525.
Miller, O. K., and Laursen, G. A., 1978, "Ecto and endomycorrhizae of arctic plants at Barrow, Alaska, *in:* "Vegetation and production ecology of an Alaskan Arctic tundra," L. L. Tieszen, ed., Springer Verlag, Berlin, 229-237.
Petersen, P. M., 1977, Investigations on the ecology and phenology of the macromycetes in the Arctic, *Meddelelser on Grønland*, 199: 72 p.
Wilkins, W. H., and Harris, G. C. M., 1946, The ecology of the larger fungi. V. An investigation into the influence of rainfall and temperature on the seasonal production of fungi in a beech wood, *Ann. Appl. Biol.*, 33: 179-188.
Wilkins, W. H., and Patrick, S. H. M., 1940, The ecology of the larger fungi. IV. The seasonal frequency of grassland fungi with special reference to the influence of environmental factors, *Ann. Appl. Biol.*, 27: 17-34.

SOCIOLOGY AND ECOLOGY OF LARGER FUNGI IN THE SUBARCTIC AND OROARCTIC ZONES IN NORTHWEST FINNISH LAPLAND

Katriina Metsänheimo

Botanical Museum, University of Oulu
SF-90570 Oulu, Finland

Key words: birch forest, oroarctic heath, snow bed, larger fungi, fruiting body production, Boletaceae, Cortinariaceae, Russulaceae

ABSTRACT

The ecology, sociology and fruiting body production of larger fungi was studied at Kilpisjärvi in northwestern Finnish Lapland (69°05'N lat. 22°15'E long.) from 1976-83. The investigated habitats were subarctic (orohemiarctic) birch (*Betula pubescens* subsp. *tortuosa*) forests and low oroarctic heaths and snow beds just above the tree line and in the middle oroarctic zone. The middle oroarctic habitats were only observed from 1980-83. There were 17 permanent study plots of 100 m^2 in birch forests and 12 in the oroarctic zone. Information is also given on the vegetation and soil conditions of the plots.

The identified genera in the habitats studied are presented. The total number of taxa identified in all the study plots was 219 (206 in birch forests and 58 in oroarctic habitats). The number of agaric genera was 46 (46 and 21 respectively). The greatest number of identified taxa (148) was found in herb-rich forests and the smallest (9) in middle oroarctic heath. The number of identified species will increase along with the progress of the determination work.

In the birch forests the most productive genera were *Leccinum*, *Cortinarius* subgenus *Myxacium*, *Lactarius* and *Russula*. In the low oroarctic heaths the most productive genera were *Leccinum*, *Cortinarius* subgenus *Myxacium* and *Rozites*, and in snow beds *Entoloma* and *Amanita*.

INTRODUCTION

The present investigation takes place at Kilpisjärvi, the municipality of Enontelkiö, in northwestern Finnish Lapland (69°05'N lat. 22°15'E long., see map in Ohenoja and Metsänheimo (1982). This northwestern corner is the only part of Finland which includes an edge of the Scandinavian mountains. Typical of the area are maritime subarctic birch forests and wide oroarctic regions.

The fungus flora of subarctic Fennoscandia has been studied and described by several authors (Ohenoja and Metsänheimo 1982), but detailed sociological and quantitative studies are few in number. Mycosociology, in particular, is a young research field in the Nordic countries (Gulden 1982).

Until the beginning of the present investigation (1976), NW Lapland was mycologically rather poorly known (e.g. distribution maps in Heikkilä, 1982) and there is still much to do in several fields of mycology.

Some results of the present investigation have already been published (Ohenoja 1980, Metsänheimo 1981, Ohenoja and Metsänheimo 1982). The research began as a quantitative study, but from the very beginning there was also a special interest in studying the whole agaric fungus flora of the area as well. The period of quantative field work lasted for eight seasons (1976-83). Plans are made to concentrate on the determination of species during 1984-86. The quantitative work has been very laborious and intensive during the field seasons, and that is one reason for the incomplete descriptions and determinations of many difficult groups.

The mycosociological questions roused the author's interest during the first few research years, and special observations on the sociology of fungus species were made from 1979 onward. The methods used in the mycosociological investigation will be published later as will more detailed information on fungal species and associations in the research area.

In this paper, the fungus flora of the studied habitats is presented at the genus level. The fruiting body production of different taxonomical groups, mainly of the families, is also reported and information on the vegetation and soil conditions of the study plots is given.

MATERIALS AND METHODS

The fruiting bodies of larger fungi were collected from stable study plots of 100 m^2 (2 x 50 m) mostly once a week during the autumn season (Ohenoja and Metsänheimo 1982). The species were determined macroscopically primarily from fresh material at Kilpisjärvi and partly microscopically from dried material at Oulu. Some Aphyllophorales, *Cortinarius*, *Dermocybe*, *Inocybe* and *Russula* species have been determined by specialists familiar with these groups.

The main part of the fungus material is preserved at the Botanical Museum of Oulu University. Yield data have been treated with a UNIVAC-1100 computer at the University of Oulu, and are preserved on magnetic tape.

RESEARCH HABITATS

The habitats studied are the following. The birch forests have been typified according to Hämet-Ahti (1963).

The vegetation of each plot has been analyzed on ten (on five in the low oroarctic heath) systematically selected 1 m^2 squares (field and bottom layer). The tree and shrub layer has been analyzed over the whole 100 m^2 plot (Table 1).

Birch forests (more detailed information in Ohenoja and Metsänheimo 1982)	Number of plots	Observation time
1. sET, dryish	5	1976-83
2. CoEMT, dryish	5	1976-83
3. CoMT, mesic	3	1976-83
4. herb-rich forest	4	1976-83

Oroarctic heaths and snow beds		
Low oroarctic zone: 620-640 m, just above the birch line		
5. Heath on N-slope, characterized by *Betula nana* and *Empetrum nigrum* subsp. *hermaphroditum*	4	1977-83
6. Snow bed on N-slope, meadow along a brook	2	1979-83
Middle oroarctic zone: 830-870 m		
7. Heath on SE-slope, characterized by *Betula nana*, *Empetrum nigrum* subsp. *hermaphroditum* and *Cassiope tetragona*	3	1980-83
8. Snow bed on SE-slope, characterized by *Carex bigelowii* and *Salix herbacea*	3	1980-83

TABLE 1. Vegetation of the habitats studied at Kilpisjärvi. (Vegetation of habitats 6-8 has not yet been analyzed.)

	birch forest				low oroarctic heath
Habitat	1.	2.	3.	4.	5.
Trees					
height, m	3	4	5	5.5	
percentage cover	15	30	35	40	
Shrubs					
height, cm	70	120	120	100	40
percentage cover	8	3	9	15	9
number of species	6	4	6	6	3
Field layer					
height, cm	9	10	10	14	6
percentage cover	65	65	75	75	55
number of vascular plants	34	24	34	64	26
Bottom layer					
height, cm	3	2	3	2	2
percentage cover	30	30	20	15	50
number of mosses	13	11	13	4	10
number of lichens	11	14	5	8	18

Climatological factors: the precipitation and temperature conditions at Kilpisjärvi are presented by Ohenoja and Metsänheimo (1982). Some ecological measurements of soil conditions were also made. The pH and conductivity of the humus layer are presented in table 2.

RESULTS AND CONCLUSIONS

The number of identified taxa found from the study plots is presented in table 3. There were 219 identified fungal taxa on 29 study plots during 1976-83.

If the herbarium specimens, and the species which have been observed outside the plots, are included, there are records of about 300 species from the Kilpisjärvi area. However, the area is far from being thoroughly studied and new finds are being made continuously.

TABLE 2. pH and conductivity of the humus layer in the habitats studied at Kilpisjärvi.

Habitat		pH	conductivity uS	number of soil samples	number of sample years
	1.	4.8	63	12	5
subarctic	2.	4.4	129	12	5
birch	3.	4.6	148	12	5
forest	4.	6.2	158	12	5
	5.	4.5	38	3	3
oroarctic	6.	5.0	93	2	2
zone	7.	4.9	24	2	2
	8.	4.9	17	2	2

Some species of genera characteristic of the area are still undetermined, and this makes any detailed comparison with other studies difficult. Common and typical genera of arcto-alpine and subalpine regions have been reported by many authors, namely *Cortinarius*, *Dermocybe*, *Entoloma*, *Galerina*, *Hebeloma*, *Inocybe*, *Laccaria*, *Omphalina*, and excluding the most arctic regions, *Mycena* (Lange 1946, 1957, Kallio 1960, Lange and Skifte 1967, Gulden and Lange 1971, Ohenoja 1971, Petersen 1977). The number of identified species of these genera which are also typical for Kilpisjärvi area, will increase with the progress of the determination work.

A comparison of the number of species showed that the difference between the subarctic and oroarctic zones at Kilpisjärvi was great, obviously too great, because the observation time for most of the plots of the low oroarctic region was much shorter and did not consist of equally productive seasons as in the birch forest. The number of fruiting species decreases significantly when the shelter of the forest is lacking. This is especially true of the saprophytic species, as Moser (1982) reports. The number of fruiting mycorrhizal species decreases, partly for the same reason; although, the lack of a suitable mycorrhizal partner is of more concern for them. The influence of increased transpiration on windswept heaths (habitat 7) in the middle oroarctic zone is seen when their total number of identified taxa is compared with that of the snow bed (habitat 8) at about the same altitude. The ratio, heath to snow bed, was almost 1:2.

All of the deficiencies of the genera in oroarctic habitats shown in the table 3 are not real. For instance the "lacking" *Omphaliaster* and *Calvatia* (Lycoperdaceae) were seen near but not in the study plots during the years of observation.

In birch forest habitats, the greatest number (148) of identified taxa was found in the herb-rich forest. The number of vascular plants (table 1) was also much higher there than in the other habitats, and the soil conductivity value (table 2) was highest there, indicating good nutrient conditions.

Genera which include the greatest numbers of identified species in all birch forest habitats were *Leccinum*, *Clitocybe*, *Collybia*, *Mycena*, *Cortinarius* subgenus *Myxacium* and *Telamonia*, *Lactarius* and *Russula*. *Leccinum*, *Lactarius* and *Russula* species are typical and abundant in subarctic and low oroarctic regions elsewhere in Fennoscandia and in Greenland, where oroarctic *Salix* and *Betula nana* thickets are common (see literature cited above). The herb-rich forest

TABLE 3. Number of identified taxa found from the study plots in different habitats at Kilpisjärvi from 1976-83. Taxonomic order according to Ainsworth et al. (1965-73), for Agaricales according to Moser (1983).
(Not all taxa are determined as to species. The genera including several undetermined species are marked by a broken line under the number of identified taxa.)

	birch forest				oroarctic zone				
Habitat	1.	2.	3.	4.	5.	6.	7.	8.	total
MYXOMYCOTA	1	1	2	1					2
BASIDIOMYCOTINA									209
Tremellales	1	3	3	4					4
Aphyllophorales									13
Stereaceae	1			1					1
Thelephoraceae	1			1					1
Auriscalpiaceae	1	1	1	1					1
Clavariaceae s. lat.				1	1				2
Hydnaceae	1	1							1
Polyporaceae s. lat.		2	6	5					7
Agaricales									188
Boletaceae									8
Chalciporus	1	1							1
Leccinum	5	4	4	3	1				6
Xerocomus	1	1		1					1
Paxillaceae									1
Paxillus	1	1	1						1
Hygrophoraceae									6
Camarophyllus				1		1			2
Hygrocybe	1		1	1		1			3
Hygrophorus		1							1
Tricholomataceae									52
Armillaria			1	1					1
Clitocybe	6	6	7	7	2		1		7
Collybia	5	5	5	5	3				6
Gerronema				2					2
Laccaria	1	1	1	2	2	2	1	2	3
Lepista			2	1					2
Marasmius	1	1	1	1	1				1
Mycena	12	12	15	18	3	1			19
Omphaliaster	1	1							1
Omphalina	2	1	1	1	2				2
Panellus			1						1
Ripartites			1						1
Tephrocybe	1	1	1						1
Tricholoma	1	3	1	2					4
Xeromphalina	1		1	1					1
Entolomataceae									7
Entoloma	2	3	3	4	2	4		1	5
Rhodocybe	2	1	1	1	1	1		1	2
Pluteaceae									2
Pluteus	1		2	1					2
Amanitaceae									1
Amanita	1		1	1		1		1	1

Table 3 continues...

Table 3 continued....

Agaricaceae									4
Agaricus				1					1
Cystoderma	1	1	2	1	1			2	3
Coprinaceae									2
Coprinus				1					1
Psathyrella	1	1	1						1
Strophariaceae									9
Kuehneromyces		1							1
Phaeomarasmius		1	1	1					1
Pholiota	1	2	2	1					3
Psilocybe	1		1			1		1	1
Stropharia			1						1
Tubaria				2					2
Crepidotaceae									1
Crepidotus				1					1
Cortinariaceae									64
Cortinarius									38
Leprocybe	1	1	1	1					1
Phlegmacium	4	3	1	4					5
Sericeocybe	3	4	3	1	1				4
Myxacium	6	7	6	7	3	1		1	8
Telamonia	9	12	10	11	5	2	2	2	19
Dermocybe	2	2	2		2	1	1		3
Galerina	1	1	1	2	1	1	1	1	2
Gymnopilus	1	1	1						1
Hebeloma	3	2	2	4					4
Inocybe	4	2	4	13	5	1	1	1	16
Rozites	1	1	1		1				1
Russulaceae									31
Lactarius	5	6	5	9	5		1	1	12
Russula	9	9	11	13	5	4	1	1	19
Gasteromycetes									4
Lycoperdaceae	1	1	3	3					4
ASCOMYCOTINA									8
Discomycetes									6
Pezizales				2	1				3
Helotiales		1	3	1					3
Pyrenomycetes									1
Sphaeriales			1	1					1
Plectomycetes									1
Eurotiales								1	1
total	106	110	126	148	49	22	9	16	219
	birch forest total 206				oroarctic total 58				

further contained several *Inocybe* species, some of which, are characteristic of limestone areas; i.e., *I. leucoblema* Kühn. The genus *Inocybe* was also typical of oroarctic study plots, as were also *Entoloma* and *Laccaria*.

In birch forest habitats only a few species were found of the genera *Tricholoma* and *Amanita*, which are well represented in coniferous forests. The family Boletaceae, excluding the genus *Leccinum*, and the family Agaricaceae were also poorly represented. Lange's (1946) report from the Abisko region agree with the present results from Kilipsjarvi.

Quantitative fruiting body production of the taxonomic groups in different habitats (Fig. 1 and table 4) is one more aspect to

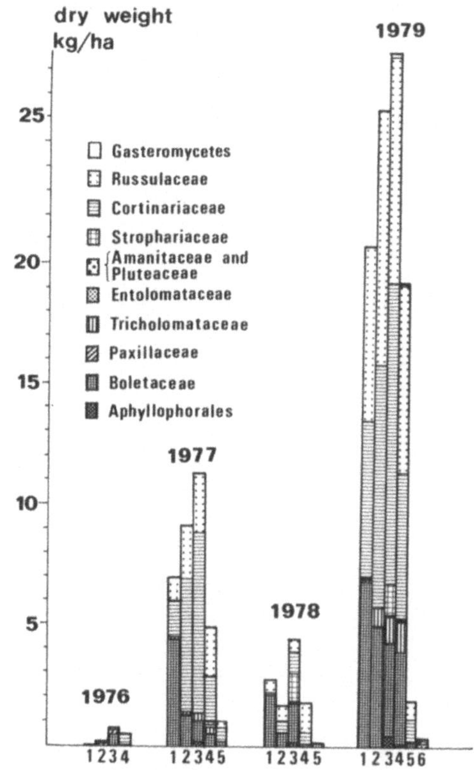

Fig. 1. Mean proportions of taxonomic groups of the total yields (dry weight kg/ha) in different habitats (1-6) at Kilpisjärvi in 1976-79.

mycosociological research. It can be assumed that the mycorrhizal larger fungi, which are ecologically most significant for a habitat, are those which are able to produce large quantities of fruiting bodies. Kallio (1982) mentions that it is well known that fruiting body formation can rarely occur at the limit of the distribution of the species. In extremely poor years, however, it is quite unpredictable which species happen to produce the few fruiting bodies that can be found. Such a year was 1976 at Kilpisjärvi. Variation in the phenology of the fungus seasons also gives rise to differences between the taxonomical groups contributing to total yields. A season with an early fruiting peak induces a species component quite different from a season with a retarded peak as in the years 1979 and 1978 at Kilpisjärvi (Ohenoja and Metsänheimo 1982). Also the above mentioned aspects indicate a need for several years of observations in mycosociological research.

The most productive genera in birch forests were *Leccinum*, *Cortinarius* subgenus *Myxacium*, *Lactarius* and *Russula*. The genus *Leccinum* was approximately the most productive in habitat 1.; *Cortinarius* subgenus *Myxacium* in the habitats 2. and 3.; *Russula* in habitat 4. In the low oroarctic heath (habitat 5.), the most productive genera were *Leccinum*, *Cortinarius* subgenus *Myxacium* and *Rozites*, and in the snow beds (habitat 6.) *Entoloma* and *Amanita*. The percentage proportion of *Leccinum*, *Cortinarius* subgenus *Myxacium*, *Lactarius* and *Russula* of the total dry weight yields in 1977-79 are shown in table 5.

Table 4. Mean proportions of taxonomic groups of the total yields (dry weight g/ha) in different habitats at Kilpisjärvi in 1976-79.

	subarctic birch forest																low oroarctic zone			
Habitat	1.				2.				3.				4.				5.			6.
	1976	1977	1978	1979	1976	1977	1978	1979	1976	1977	1978	1979	1976	1977	1978	1979	1977	1978	1979	1979
Tremellales	-	-	-	-	-	-	-	<1	-	-	-	<1	-	1	-	1	-	-	-	-
Aphyllophorales	7	-	-	-	-	8	-	-	28	219	182	439	1	51	3	83	-	-	-	-
Agaricales																				
Boletaceae	-	4422	2157	6785	-	1268	585	4924	443	769	1473	3815	-	381	-	3725	212	86	151	-
Leccinum	-	4422	2066	6785	-	1095	585	4842	443	769	1473	3815	-	381	-	3725	212	86	151	-
Paxillaceae	-	-	-	-	69	44	-	37	216	59	-	-	-	-	-	-	-	-	-	-
Hygrophoraceae	-	-	-	1	-	-	-	4	-	-	-	1	-	-	-	52	-	-	-	36
Tricholomataceae	1	85	37	90	-	75	38	745	5	282	85	1104	38	219	33	1377	2	3	69	2
Entolomataceae	-	1	-	12	-	7	-	5	-	30	-	10	-	6	4	12	27	-	1	111
Amanitaceae+																				
Pluteaceae	-	7	-	3	-	-	-	-	-	9	-	25	-	106	6	67	-	-	-	96
Agaricaceae	-	-	-	<1	-	-	-	1	-	4	3	14	-	-	-	-	-	-	4	-
Coprinaceae	-	24	-	5	-	-	-	1	-	13	-	18	-	-	4	2	-	-	-	-
Strophariaceae	1	-	-	84	14	14	-	13	2	3	1204	1311	-	156	135	102	-	-	-	1
Cortinariaceae	71	1405	76	6450	48	5452	415	10060	33	7476	875	12484	441	1813	425	5977	488	40	886	78
Cortinarius incl.																				
Dermocybe	71	1402	73	6330	-	5080	386	9056	33	7106	855	11387	436	1256	410	5134	320	36	282	64
Myxacium	71	1001	34	3033	-	4136	27	5538	33	6145	329	6911	420	255	-	2006	251	36	199	-
Russulaceae	-	999	421	7236	16	2192	551	10593	17	2356	540	9276	3	2012	1120	7664	231	-	718	38
Russula	-	349	195	4557	-	1162	70	5727	17	138	83	3704	3	1632	1080	5873	6	-	330	38
Lactarius	-	650	226	2679	16	1030	481	4866	-	2218	457	5572	-	380	40	1791	225	-	388	-
Gasteromycetes	-	-	-	6	-	-	-	5	-	-	-	186	-	-	-	46	-	-	-	-
Ascomycetes	-	-	-	-	-	-	-	1	-	<1	-	7	-	-	-	88	-	-	1	-
total	79	6943	2691	20672	147	9060	1589	26389	744	11220	4362	28690	483	4745	1730	19196	960	129	1830	362

Table 5. The percentage proportions of *Leccinum*, *Cortinarius* subgenus *Myxacium*, *Lactarius* and *Russula* of the total dry weight yields in 1977-79.

Habitat	1.			2.			3.			4.			5.			6.
years	77	78	79	77	78	79	77	78	79	77	78	79	77	78	79	79
% of the total dry weight yield	92	94	82	82	73	79	83	54	70	55	65	70	72	95	58	10

An examination of the yields confirms the status of the above mentioned genera as the most charactierstic groups of mycorrhizal fungi in birch forests and the low oroarctic heath. The last mentioned habitat (5.) seems to be transitional, and its species component resembles as much subarctic birch forest as the middle oroarctic heath (habitat 7.), the yield data of which have not yet been analyzed for publication.

ACKNOWLEDGEMENTS

The research in 1976-78 was sponsored by the Academy of Finland and the Forest Research Institute. In 1983, financial support was obtained from the Emil Aaltonen Foundation. I wish to thank the sponsors and all those who have helped me in collecting and determining the research material. Special thanks are extended to Esteri and Martti Ohenoja and to Tauno Ulvinen from the Botanical Museum of Oulu University and to the staff of the Kilpisjärvi Biological Station of the University of Helsinki.

Mrs. Esteri Ohenoja, Phil. Lic., leader of the fungus yield project in Finland 1976-78, has given valuable help and advice during the whole study period.

Sincere thanks are extended to Mrs. Sirkka-Liisa Leinonen who assisted in the English translation of this article.

REFERENCES

Ainsworth, G. C., Sparrow, F. K., and Sussman, A. S., eds., 1965-73, "The fungi, an advanced treatise," I-IVB, I(1965), 748 p., II (1966), 805 p., III(1968), 758 p., IVA(1973), 640 p., IVB(1973), 504 p., New York.

Gulden, G., 1982, Soppsosiologi, en ny mykologisk forskningsretning i Norge. (Mycosociology, a new branch of mycology in Norway), *Blyttia*, 40:95-99.

______, and Lange, M., 1971, Studies in the macromycete flora of Jotunheimen, the central mountain massif of South Norway, *Norw. J. Bot.*, 18: 1-46.

Hämet-Ahti, L., 1963, Zonation of the mountain birch forests in northernmost Fennoscandia, *Ann. Bot. Sco. "Vanamo"*, 34(4): 1-127.

Heikkilä, H., 1982, Boletes from northern Finland (Lapland), *in:* "Arctic and Alpine Mycology. The First International Symposium on Arcto-Alpine Mycology," G. A. Laursen and J. F. Ammirati, eds., Barrow, Alaska, 16-23 August 1980, 316-333.

Kallio, P., 1960, Utsjoen sientistä, *Luonnon Tutkija*, 64: 38-45.

______, 1982, Aspects of northern Finnish Macromycology, *in:* "Arctic and Alpine Mycology. The First International Symposium on Arcto-Alpine Mycology," G. A. Laursen and J. F. Ammirati, eds., Barrow, Alaska, 16-23 August 1980, 410-431.

Lange, M., 1946, Mykologiske indtryk fra Lapland, *Friesia*, 3: 161-170.
______, 1957, Macromycetes III. 1. Greenland Agaricales. 2. Ecological and plant geographical studies, *Medd. Grønl.*, 1948(2): 1-125.
______, and Skifte, O., 1967,. Notes on the Macromycetes of northern Norway, *Acta Borealia*, A 23: 1-51.
Metsänheimo, K., 1981, Kilpisjärven suursienistä ja syyssienisadosta. (Summary: Larger fungi and their autumn yields at Kilpisjärvi, Finnish Lapland.), *Kilpisjärvi Notes*, 5: 1-8.
Moser, M., 1982, Mycoflora of the transitional zone from subalpine forests to alpine tundra, *in:* "Arctic and Alpine Mycology. The First International Symposium on Arcto-Alpine Mycology," G. A. Laursen and J. F. Ammirati, eds., Barrow, Alaska, 16-23 August 1980, 371-389.
______, 1983, Die Röhrlinge und Blätterpilze, *in:* "Gams, H. Kleine Kryptogamenflora IIb," 2 533 p. Fischer, Stuttgart.
Ohenoja, E., 1971, The larger fungi of Svalbard and their ecology, *Rep. Kevo Subarctic Res. Stat.*, 8: 122-147.
______, 1980, Sienisatotutkimus vv. 1976-78, *EKT-sarja*, 548: 1-42. Helsingin yliopisto.
______, and Metsänheimo, K., 1982, Phenology and fruiting body production of macrofungi in subarctic Finnish Lapland, *in:* "Arctic and Alpine Mycology. The First International Symposium on Arcto-Alpine Mycology," G. A. Laursen and J. F. Ammirati, eds., Barrow, Alaska, 16-23 August 1980, 390-409.
Petersen, P. M., 1977, Investigations on the ecology and phenology of the macromycetes in the Arctic, *Medd. Grønl.*, 199(5): 1-72.

ENDOPHYTIC FUNGI OF ALPINE ERICACEAE. THE ENDOPHYTES OF *LOISELEURIA PROCUMBENS*

Orlando Petrini

Bodenacherstrasse 86
CH - 8121 Benglen, Switzerland

Key words: *Loiseleuria procumbens*, endophytes, fungi, ecology, analysis of correspondence

ABSTRACT

Endophytic fungi have been isolated from the leaves of *Loiseleuria procumbens* (L.) Desv. from twenty sites in the Swiss Alps.

Three groups of endophytes could be distinguished. The first group comprises species which are not specific to *L. procumbens* and have been reported from other hosts as in the anamorphs of some Xylariaceae. Other taxa, so far as is known, are only from plant species belonging to closely related families and are represented by *Physalospora empetri*, and *Phyllosticta pyrolae*. A third group of fungi, which are at least host family specific, includes *Apostrasseria lunata* and the anamorph of *Godronia callunigera*.

The degree of colonisation by endophytic fungi on plants exposed to extreme ecological conditions is very likely dependent on microclimatic and ecological factors.

The distribution of endophytes can be explained partly by plant sociological and factors and host-specificity of endophytic fungi is mostly confined to the family and not to genus or species levels.

INTRODUCTION

Endophytic fungi have been found in large number of phanerogams and cryptogams (e.g., Petrini *et al.* 1979; Dreyfuss & Petrini 1984). Their occurrence in healthy-looking, living plant tissues is likely to be correlated with site specific ecological factors, some of which have been postulated in previous papers. Carroll & Carroll (1978) proposed a model for the infections of phanerogams by endophytes which takes into account humidity and elevation of the site. Luginbühl & Müller (1980) discussed the influence of plant sociological factors on the infection of plant species by endophytic fungi.

The present investigation has been conducted on *Loiseleuria procumbens* (L.) Desv., an ericaceous, prostrate mat forming evergreen undershrub which occurs in dry, stony or peaty places on acid soils of the alpine zone (1,500 to 3,000 m). The study aims to document the species composition of endophytic fungal populations within *L. procumbens* and at the same time to verify and extend the ecological models proposed in previous papers.

MATERIAL AND METHODS

Field sites and sample collection

Twenty sites were chosen to reflect ecological differences within the natural range of the species in the Swiss Alps (Table 1). Care was taken to list the presence of other ericaceous hosts and of members of closely related families such as Empetraceae and Pyrolaceae.

Table 1. Collecting sites.

Site No.	Date	Locality
1	29.6.81	Parsenn, Davos, GR
2	6.7.81	Dukantal, Davos, GR
3	7.2.82	Suretta, Splugen, GR
4	19.9.82	Splügenpass, GR
5	26.9.82	Gletsch, VS
6	29.6.83	Stillberg, Davos, GR
7	11.7.83	Zinal, VS
8	15.7.83	Albulapass, GR
9	25.7.83	Forca di Pineto, Piora, TI
10	19.7.83	Capanna Leit, Rodi, TI
11	31.7.83	Berninapass, GR
12	1.8.83	Flüelapass, GR
13	8.8.83	Furkapass, VS
14	28.8.83	Gotthardpass, TI
15	29.8.83	Radönt, Fluela, GR
16	5.9.83	Macun, Engadin, GR
17	5.9.83	Alp Lavin, Engadin, GR
18	9.9.83	Murtaröl, Ofenpass, GR
19	9.9.83	Val Sesvenna, S-charl, GR
20	9.9.83	Umbrailpass, GR

Individual branches of *Loiseleuria procumbens* were pruned by hand from 10-20 plants within each site, tagged for identification and returned to the laboratory within a few hours from the time of collection. Culturing of the material was performed within 24 hours.

Culture methods

A set of 50 healthy looking leaves was randomly sampled for cultural studies from each site. The leaves were dipped for one minute in 96% ethanol, surface sterilized for two minutes in 20 % commercial chlorox (chlorine-final concentration ca 2.5 %) and then dipped again for 30 seconds in 96% ethanol. The leaves were then cut with a sterile scalpel into two approximately equal segments, and transferred in serial order to labelled positions on 90 mm Petri plates containing 2 % malt extract agar supplemented with 50 mg/l oxytetracycline hydrochloride (Terramycin^{R}, Pfizer).

Plates were incubated at 17°C in the dark. Isolation of fungi from plates to 2 % malt extract slants was carried out by transfer of conidia or mycelial fragments. Leaf segments were scored for fungal infections for at least five weeks.

Table 2. Endophytes of *Loiseleuria procumbens*

weighted frequencies of occurrence — Anam.: anamorph of

SITE	1	2	3	4	5	6	7	8	9	10	11	12	13	14	15	16	17	18	19	20
ASCOMYCOTINA																				
Anthostomella tomicoides Sacc.	.	.	1	.	.	.	.	.	.	.	.	.	.	.	.	.	.	.	.	.
Chaetomium cochliodes Palliser	.	1	.	.	.	.	.	.	.	.	.	.	.	.	.	.	.	.	.	.
Coniochaeta discospora (Auersw.) Cain	.	.	.	1	.	.	.	.	.	.	.	.	.	.	.	.	.	.	.	.
Coniochaeta lignaria (Grév.) Massee	.	.	.	.	.	.	.	.	.	.	.	.	1	.	.	.	.	.	.	.
Coniochaeta subcorticalis (Fuck.) Munk	1	1	.	.	.	.	.	.	.	1	1	1	.	.	.	1	1	.	.	1
Coniochaeta velutina (Fuck.) Munk	.	.	.	.	.	.	.	.	.	1	.	.	.	.	.	.	.	.	.	.
Physalospora empetri Rostr.	.	.	.	.	.	.	1	.	.	.	.	.	.	.	.	.	.	.	.	.
Sporormiella australis (Speg.) Ahmed et Cain	.	.	.	1	.	.	.	.	.	.	.	.	.	.	.	.	.	.	.	.
Sporormiella intermedia (Auersw.) Ahmed et Cain	1	.	1	.	.	.	1	.	.	.	1	.	.	.	.	.	.	.	.	.
Sporormiella minima (Auersw.) Ahmed et Cain	.	.	.	.	.	1	.	.	.	.	.	.	.	.	.	.	.	.	.	.
DEUTEROMYCOTINA																				
a) *Coelomycetes*																				
Apiocarpella sp.	.	1	.	.	.	.	.	.	.	.	.	.	.	.	.	.	.	.	.	.
Apostrasseria lunata (Shear) Nag Raj	.	.	.	1	.	.	1	.	.	.	.	.	.	.	.	.	.	.	.	.
Ceuthospora latitans Grove	.	.	.	1	.	.	.	.	.	.	.	.	.	.	.	.	.	.	.	.
Coleophoma empetri (Rostr.) Petr.	.	1	.	.	.	.	.	1	.	.	1	.	1	.	.	1	.	.	.	.
Colletogloeum sp.	1	.	.	.	.	.	.	.	.	.	.	.	.	.	.	.	.	.	.	.
Cryptocline arctostaphyli Petrini	.	.	.	.	.	.	.	1	.	.	.	.	.	.	.	1	.	.	.	.
Cryptocline sp. 1	.	.	.	.	1	.	.	.	.	.	.	.	.	.	.	.	.	.	.	.
Cryptocline sp.	.	1	.	.	.	.	.	.	.	.	.	.	.	.	.	.	.	.	.	.
Cryptosporiopsis sp.	.	.	.	.	.	.	.	.	.	.	1	.	.	1	.	.	.	.	.	.
Diplodina sp.	.	.	.	.	.	.	.	.	.	.	.	.	.	.	1	.	.	.	.	.
Epicoccum purpurascens Ehrh.: Schlecht	.	.	.	1	.	.	.	.	1	.	.	.	.	.	.	.	.	.	.	.
Godronia callunigera Karst. (Anam.)	.	.	.	.	.	1	.	.	.	.	.	.	.	.	.	.	.	.	.	.
Kabatina sp.	.	.	1	.	.	.	.	.	.	.	.	.	.	.	.	.	.	.	.	.
Phoma sp.	.	.	1	.	.	.	.	.	.	.	.	.	.	.	.	.	.	.	.	.
Phomopsis sp.	.	.	1	.	1	.	.	.	.	.	.	.	.	1	.	.	.	.	.	.
Phyllosticta pyrolae Ell. et Ev.	.	1	.	.	.	.	.	.	.	.	.	.	1	.	.	.	.	.	.	.
Sirococcus sp.	.	.	.	.	.	.	.	.	.	1	.	.	.	.	.	.	.	.	.	.
Strasseria geniculata (Berk. et Br.) v Hohn.	.	.	1	1	.	.	1	1	.	.	.	.	.	.	1	1	.	.	.	.
Topospora sp.	.	.	2	1	.	.	.	.	.	.	1	.	.	.	.	.	.	.	1	.
b) *Hyphomycetes*																				
Acremonium sp.	.	.	.	.	.	.	.	.	.	.	.	.	.	.	1	.	.	.	.	.
Biscogniauxia nummularia (Bull.) O.K. (Anam.)	1	.	.	.	.	.	.	.	.	.	.	.	.	.	.	.	.	.	.	.
Botrytis cinerea Pers.: Fr.	.	1	.	.	.	.	.	.	.	.	.	.	.	.	.	.	.	.	.	.
Cladosporium cladosporioides (Fr.) De Vries	.	.	.	.	.	.	.	.	.	.	.	.	1	.	.	.	.	.	.	.
Exophiala sp.	1	.	.	.	.	.	.	.	.	.	.	.	.	.	.	.	.	.	.	.
Geniculosporium serpens Chesters et Greenhalgh	.	.	.	.	.	.	.	.	.	.	.	1	.	.	.	.	.	.	.	.
Hormonema sp.	.	.	1	.	.	.	.	.	.	.	.	.	.	.	.	.	.	.	.	.
Hypoxylon unitum (Fr.) Nitschke (Anam.)	.	.	.	.	.	.	.	.	.	.	.	.	.	1	.	.	.	.	1	.
Phialophora decumbens (Beyma) Schol-Schwarz	.	.	.	.	.	.	.	1	.	.	.	.	.	.	.	.	.	.	.	.
Phialophora spp.	.	.	1	.	.	.	.	.	.	1	1	.	1	1	.	.	.	.	.	.
Ramularia sp.	1	.	.	1	.	.	.	.	.	.	.	.	.	.	.	.	.	.	.	.
Trichoderma viride Pers.: S.F. Gray	.	.	.	1	.	.	.	.	.	.	.	.	.	.	.	.	.	.	.	.
Xylaria sp. (Anam.)	.	1	.	.	.	.	.	.	.	.	.	.	.	.	.	.	.	.	.	.
LOPR (sterile mycelium)	.	5	.	.	.	2	2	1	2	2	3	2	.	1	.	.	2	.	.	.
SSM I (sterile mycelium)	.	.	.	.	.	.	.	.	.	.	.	.	1	.	.	.	.	.	.	.
SSM 2 (sterile mycelium)	.	1	.	.	.	1	.	.	1	1	.	.	.	1	.	.	.	1	1	1
VAMY (sterile mycelium)	.	.	.	.	.	.	.	.	.	.	.	.	.	.	.	.	.	1	1	1

Fructification induction was stimulated by exposing the cultures to UV light with a 12 hour dark-light cycle.

Statistical analysis

The overall rate of infection was computed using the formula:

$$rH = \frac{\text{number of infected leaves}}{\text{number of plated leaves}}$$

and the density of infection after

$$rD = \frac{\text{number of infected segments}}{\text{total number of plated segments}}$$

for the analysis of the individual infection rates by each fungal species, a weighted score was chosen to reflect the combination of both rH and rD according to the following scheme:

		No. of strains isolated		
		0	1-20	21-40
	0	0	-	-
	1-4	-	1	-
No. of	5-8	-	2	-
leaves	9-12	-	3	4
infected	13-16	-	5	6

The relationship between overall rate of infection and the density of infection within each collecting site was analysed by means of regression analysis; whereas, the connections between sites and fungal taxa were studied using the analysis of correspondence (software package SPAD, Benzecri 1973) on a CDC-computer (computer center ETH).

RESULTS AND DISCUSSION

Only Ascomycetes and Deuteromycetes could be isolated and identified (Table 2). The absence of fungal species belonging to other classes can be explained partly by the use of no additional selective substrates.

The overall infection rates correlate well with the overall density of infection (Fig. 1, r = 0.97); no direct correlation, however, can be

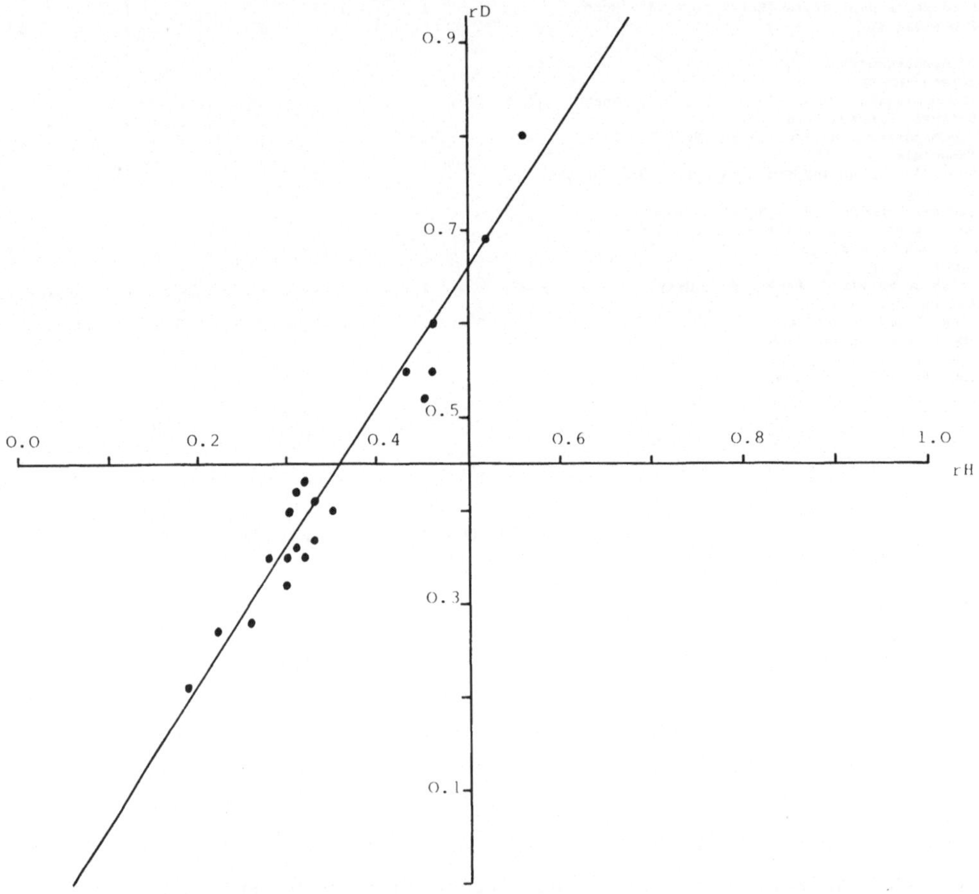

Figure 1. Regression line of the values rD (overall density of infection) vs. rH (overall infection rate).

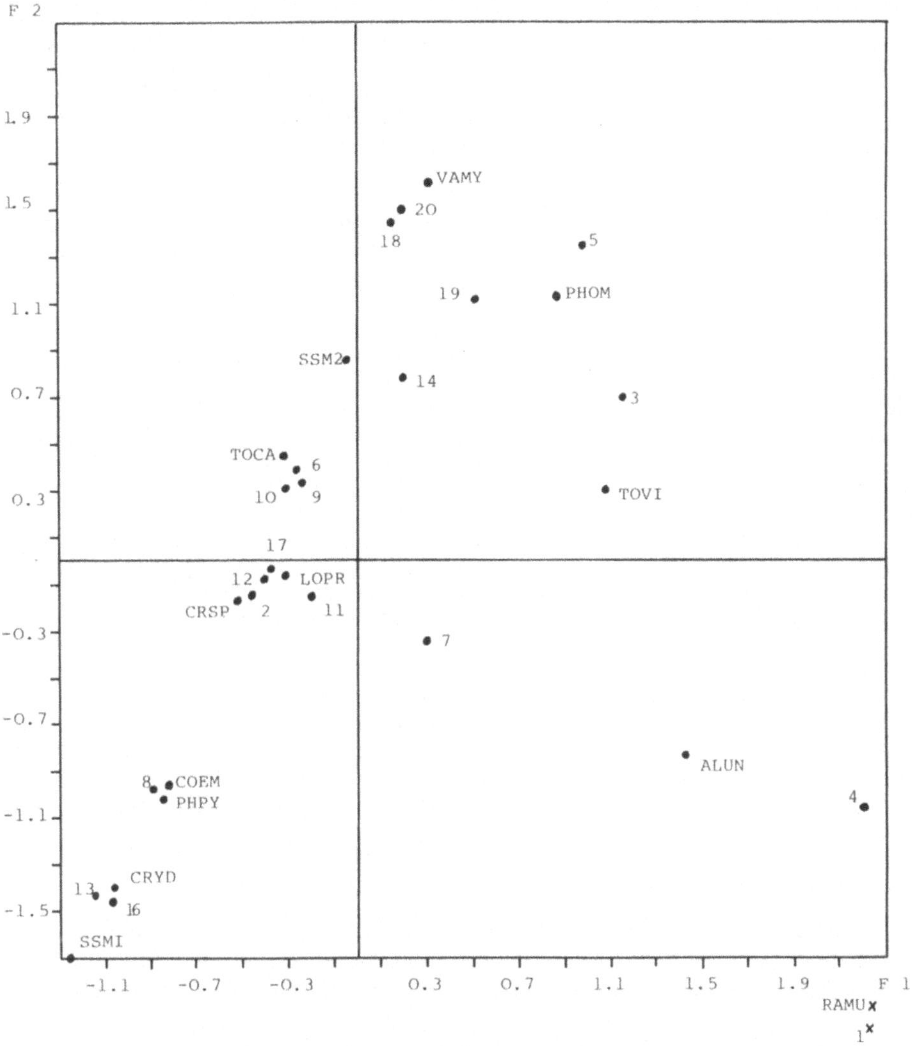

Figure 2. Analysis of correspondence performed on the 13 most frequent fungal taxa and on the 20 sites. 1-20: site number; F 1, F 2: first two factors of the analysis; ALUN: *Apostrasseria lunata*; COEM: *Coleophoma empetri*; CRYD: *Cryptocline arctostaphyli*; CRSP: *Cryptocline* sp.; TOCA: *Godronia callunigera*; PHOM: *Phomopsis* sp.; PHPY: *Phyllosticta pyrolae*; RAMU: *Ramularia* sp.; TOVI: *Topospora* sp.; LOPR, SSMI, SSM2 and VAMY: sterile mycelia. Points marked by an x have been brought nearer to the centre of gravity of the system, although their effective distance is greater than three standard deviations.

established between elevation and rate of infection, but infection rates tend to be lower in foliage collected from wind-exposed sites. Carroll & Carroll (1978) and Petrini *et al.*, (1982) suggested that differences in elevation, humidity, density of canopy cover and innate host susceptibility are likely to cause differences in the endophyte infection rates among sites. These hypotheses are only partly confirmed by the results of the present investigation.

The rather low infection rates and densities observed suggest that, although endophytes are able to colonize alpine plants growing in extreme sites, the degree of fungal colonization and their growth are dependent on ecological conditions; i.e., long-lasting snow cover, wind-exposed and cold sites as well as modest nutritional values present in the soil for the host plant. Correspondingly, the number of endophyte species would be expected to be rather low and to follow a pattern of site-specificity. However, no straightforward model be found to explain the correlation between fungal taxa and collecting sites. The analysis of correspondence performed on the 13 most frequent fungal species isolated and on the 20 sites reveals a rather homogeneous distribution with respect to most sites (Figure 2).

Only those correlations between *Ramularia* sp. and site no. 1, between the sterile mycelium VAMY and the sites no. 18 and 20, and between *Coleophoma empetri*, *Phyllosticta pyrolae* and site no. 8 deserve further attention. *Arctostaphylos uva-ursi* (L.) Sprengel and *Vaccinium vitis-idaea* (L.) Sprengel, both extremely abundant at site no. 1, are typical hosts for *Ramularia* sp. (Widler 1982; Petrini 1984); whereas, the sterile mycelium VAMY has been often isolated from *Vaccinium myrtillus* L. (Petrini, unpubl.), a species particularly frequent in sites 18 and 20. *Coleophoma empetri* and *Phyllosticta pyrolae* have been both routinely isolated from *Erica carnea* L. (Oberholzer 1982; Petrini, unpubl.), an ericaceous host which is dominant in site no. 8.

Plant sociological considerations can thus be used to explain in part the distribution of endophytic fungi, as already suggested by Luginbühl & Müller (1980). Host specificity of endophytes is mostly confined to the family and not to the genus or species level. If more hosts of the same plant family are present at the same site, the endophytes will colonize them all, very likely following a pattern of interspecific competition which will allow the ecologically better adapted fungal species to colonize most of the possible hosts and to exert an inhibitory effect on other competitors.

Host-specificity of endophytic fungi has not been adequately explained. The list of fungi isolated (Table 2) shows that three groups of endophytes can be distinguished. A first group comprises species which are not specific and have been reported from other hosts (e.g., Xylariaceae and their anamorphs and some coprophilous fungi). Other taxa appear to colonize only plant species belonging to closely related families: *Coleophoma empetri* and *Physalospora empetri* are known only from members of the Empetraceae; whereas, *Phyllosticta pyrolae* has been recorded on Pyrolaceae and Ericaceae (Oberholzer 1982). A last group includes fungi which are at least family-specific (e.g., *Apostrasseria lunata*, the anamorph of *Godronia callunigera*).

Fungi are thus very likely not to be confined to one plant species, but their specificity comprises at least members of a whole genus or of a whole family. Even plants belonging to closely related families can be infected by the same fungus species; although, a pattern of host preference can be detected.

Investigations on the nutritional relationships between endophytes and host tissues could provide further information on the host specificity and throw more light on the problem of differentiation of pathogenic fungal strains with respect to their host family.

ACKNOWLEDGEMENTS

The author wishes to thank Prof. E. Müller for his suggestions and the helpful discussions during the execution of this study. Dr. C.E. Dorworth of Sault-St. Marie, Canada, critically read the manuscript and gave helpful suggestions.

REFERENCES

Benzecri, J. P., 1973, "Analyse des données. I. La Taxonomie. II. L'analyse des correspondences," Dunod ed., Paris.

Carroll, G. C., and Carroll, F. E., 1978, Studies on the incidence of coniferous needle endophytes in the Pacific Northwest, *Can. J. Bot.*, 56: 3034-3043.

Dreyfuss, M., and Petrini, O., 1984, Further investigations on the occurrence and distribution of endophytic fungi in tropical plants, *Botanica Helvetica*, 94: 33-40.

Luginbühl, M., and Müller, E., 1980, Endophytische Pilze in den oberirdischen Organen von vier gemeinsam an gleichen Standorten wachsenden Pflanzen (*Buxus*, *Hedera*, *Ilex*, *Ruscus*), *Sydowia*, 33: 185-209.

Oberholzer, B., 1982, "Untersuchungen über endophytische Pilze von *Erica carnea* L.," Dissertation ETH Nr. 7198, 99 pp.

Petrini, O., 1984, Endophytic fungi in British Ericaceae: a preliminary study, *Trans. Br. Mycol. Soc.*

Petrini, O., Müller, E., and Luginbühl, M., 1979, Pilze als Endophyten von grunen Pflanzen, *Naturwissenschaften*, 66: 262.

Petrini, O., Stone, J., and Carroll, F. E. , 1982, Endophytic fungi in evergreen shrubs in Western Oregon: a preliminary study, *Can. J. Bot.*, 60: 789-796.

Widler, B. E., 1982, "Untersuchungen über endophytische Pilze von *Arctostaphylos uva-ursi* (L.) Sprengel (Ericaceae)," Dissertation ETH Nr. 7154, 133 pp.

THREE 4-SPORED *SACCOBOLUS* SPECIES FROM NORTH EAST GREENLAND

Henry Dissing

University of Copenhagen, Institut for Sporeplanter
ØFarimagsgade 2 D, DK-1353 Copenhagen K, Denmark

Key words: Discomycete, *Saccobolus*, Goose dung, Greenland, Arctic Canada

ABSTRACT

Three 4-spored species of the coprophilous genus *Saccobolus* (Order Pezizales) from North East Greenland are treated, viz. *Saccobolus quadrisporus*, *S. groenlandicus* sp. nov., and a third taxon for which only a preliminary description is provided. *Saccobolus groenlandicus* is also reported from Arctic Canada.

INTRODUCTION

During July and August in 1982 and 1983 the author collected Discomycetes on Ella Island (72°30'N 25°W) and around Mestersvig (72°15'N 24°W) in the southern part of the National Park in North East Greenland. About 80 species of Operculate Discomycetes (order Pezizales) were encountered. Many of these are undescribed taxa.

Three 4-spored species of the coprophilous genus *Saccobolus* are treated below. These are *Saccobolus quadrisporus* Mass. & Salm., *S. groenlandicus* sp. nov., and a third species for which a formal description will be postponed until more material is available.

Material of *Saccobolus groenlandicus* also has been obtained from Danmarks Havn (76°46'N 18°48'W), Jameson Land (71°45'N 23°W), in North East Greenland, and from Ellesmere Island (78°53'N 75°35'W), Canada.

Material and methods.

The excellent laboratory facilities in Nyhavn made it possible to prepare detailed notes on microscopic and macroscopic characters of fresh material for all species. Also, fresh goose dung was dried carefully and taken to Copenhagen, where samples were incubated on filter paper in moist chambers at room temperature in daylight.

For sectioning, *Saccobolus groenlandicus* ascocarps were isolated from dung in moist chambers and fixed according to methods described by Jensen (1982: 725). Specimens were then dehydrated and embedded in Spurr's resin and polymerized at 70°C. Sections 2 µm thick were cut on a LKB pyramitone with a glass knife. Sections were stretched in 10% acetone

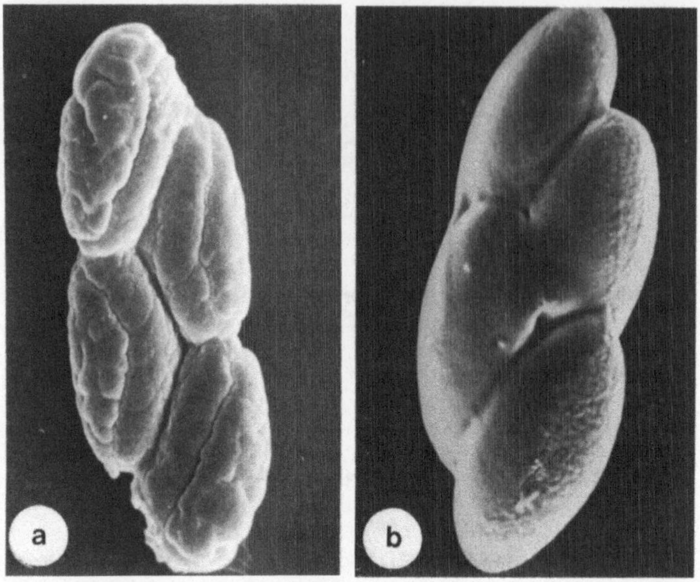

Fig. 1. SEM micrographs of spore clusters. a. *Saccobolus quadrisporus*, Gr. 83.154, b. *S. groenlandicus*, Gr. 83.149. a-b, x 1732.

on a hot plate, dried, then stained in 0.1% methylene blue, followed by 0.3% safranin in 30% ethanol. Next they were rinsed in tap water, dried and mounted in Spurr's reagent. Unless otherwise stated, the material from Greenland is deposited in the Botanical Museum, Copenhagen (C) while material from Ellesmere Island will be deposited at the University of Toronto (TRTC), with duplicates in Copenhagen (C).

Saccobolus quadrisporus Mass. & Salmon Figs. 1a, 2a, and 6a.

Ascocarps turbinate to disc-shaped, finally pulvinate, 0.3-0.6 mm broad, pale purplish to purplish brown. Asci 4-spored, 168-188 x 26-30 µm, staining pale bluish overall in Melzer's reagents. Paraphyses straight, above slightly enlarged to 6-7 µm broad, septate, content hyaline to pale brownish. Ascospores 21-23 x 10-12 µm, ellipsoid to inequilateral, at first hyaline, smooth, then pale greyish purple; when mature very dark purplish, almost opaque in the light microscop; when mature the pigment deposited in a very characteristic way, i.e., in two zones, separated by a longitudinal whitish line: half of the spore towards the center of a cluster nearly smooth, while the side towards the periphery is heavily and irregularly pigmented with nearly confluent warts. Ascospore clusters 50-55 x 17-20 µm (when the spores are displaced in two rows) or 56-58 x 14-16 µm (when displaced in one row), with two separated, clearly visible gelatinous coverings.

Material examined: East Greenland, Jameson Land, Ørsteddalen, fresh dung from Barnacle Goose, collected 17 July 1982, leg. David Boertmann. Incubated in moist chambers 11 January 1983, isolated 27 January 1983 where it was fruiting together with *Saccobolus* sp. nov. (Gr. 82.326) and an undescribed *Ascobolus* species, Gr. 82.327; - Ella Island, near Langesø, on dung from Barnacle Goose, 30 July 1983, Gr. 83.61; - ibid., fresh dung from Barnacle Goose, collected 30 July 1983. Incubated 6 January 1984, isolated 4, 6 and 13 February 1984, Gr. 83.153, 83.154 &

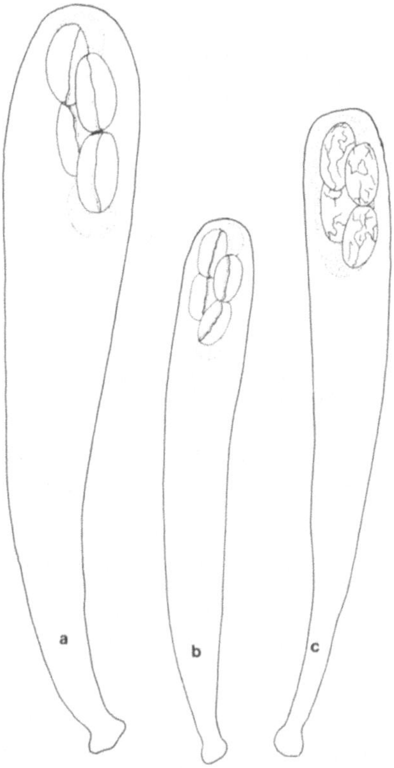

Fig. 2. Asci with spore clusters, schematic. a. *Saccobolus quadrisporus*, Gr. 82.327, b. *Saccobolus* sp. nov., Gr. 82.326, c. *S. groenlandicus*, EI. 84.116, a-c, x 495.

83.158; - Mestersvig, moist area 1 km SW of Nyhavn, on dung from Barnacle Goose, 3 August 1983, Gr. 83.92 & 83.94; ibid., 9 August 1983, Gr. 83.122.

DISCUSSION

The above description is based on notes made in Greenland on fresh material and on moist chamber material in Copenhagen. In all collections examined, the gelatinous coverings of the spore clusters were clearly visible as two separate areas. In nature a brownish amorphous substance was often deposited between cells in the excipulum and between the paraphyses in the hymenium in old apothecia.

Until now *Saccobolus quadrisporus* had been known from only two localities, viz. the type locality in Kew Garden where it was found on goose dung in November 1900 (Massee & Salmon 1901; Brummelen 1967), and from the arctic island of Spitsbergen, also on goose dung (Eckblad 1968).

By chance *S. quadrisporus* was described before *S. groenlandicus* even though the latter seems to be much more common. An example may illustrate this: one day in 1983 while working in the field in Greenland, a great number of goose dung samples (probably all from Barnacle Goose) were inspected with a hand lens. Apothecia of a *Saccobolus* could be seen on about 20 samples. A later inspection of the fresh material in the laboratory revealed that of the 20 samples only 2 were *S. quadrisporus* while 18 were identified as *S. groenlandicus*.

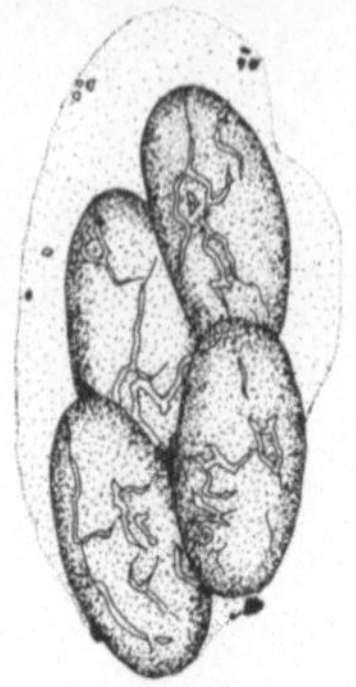

Fig. 3. *Saccobolus groenlandicus*, spore cluster with unilateral gelatinous covering, EI. 84.116, x 1072.

Saccobolus groenlandicus sp. nov. Figs. 1b, 2c, 3, 4a,b, and 5.

Carposoma primum turbinatum vel disciforme, deinde pulvinatum, 0.2-0.6 mm latum, hymenio aqueo-albido vel pallide purpureo ascis prominentibus punctato, extra glabrum infra hyphas fixatorias emittens, pallide purpureum vel - imprimis ad marginem rotundatum inconspicuum versus - vinaceum.

Excipulum ad 600 µm crassum, parietibus cellularum omnino tenuibus; pars medullaris e cellulis brevibus, hyalinis, globularibus vel hyphaceis et hyphis ramificatis, septatis ascogenibus intermixtis composita; cellulae in parte exteriore infra globulares, 6-18 µm diam., supra series plus minus manifestas, sub angulo subrecto ad superficiem directas formantes, extrorsum sensim elongatae, claviformes, 20-30 x 3-5 µm magnae, superficiales contento vinaceo insignes. Subhymenium paulum manifestum.

Hymenium 100-200 µm altum. Asci plerumque 4 spori, maturi longe prominentes, ad 140 µm longi, 20-25 µm lati, infra in bases angustas sensim attenuati, supra truncati, liquore Melzeri omnino pallide coerulescentes. Operculum 6-9 µm latum.

Paraphyses rectae, supra paulum dilatatae, 6-8 µm latae, ramificatae, septatae, brevicellules, interdum anastomosantes, hyalinae vel in summis cellulis contentum pallide vinaceum praebentes. Sporae valde cohaerentes, apicibus sese paulum tegentes, unam seriem vel duas parallelas dislocatas formantes, 16.5-18.5 x 8.5-9.5 µm magnae, ellipsoides vel inaequilaterales, primum hyalinae, deinde purpurascentes, demum colore floris Hellebori nigri, strato pigmentoso tenui, ut in Saccobolo glabro fissuras irregulares in lateribus convexis liberis praebente.

Glomus sporarum duas series formantium 32-40 x 14-16.5 µm, sporarum uniseriatarum 36-42 x 12-14 µm magnum, in altero latere strato gelatineo continuo crasso tectum.

Holotypus die 3 Augusti anni 1983 in fimo Brantae leucopsis in loco aqua transmissa uvido a Mestersvig sinus Kong Oscars Fjord 2 km in occidentem sito (lat. bor. 72°12', long. occ. 24°10') a H. Dissing et S. Sivertsen sub numero 83.95 lectus, siccus in Museo Botanico Hauniensi (C) depositus.

Ascocarps at first turbinate to disc-shaped, then pulvinate, 0.2-0.6 mm broad, hymenium watery whitish to pale purplish, dotted with protruding

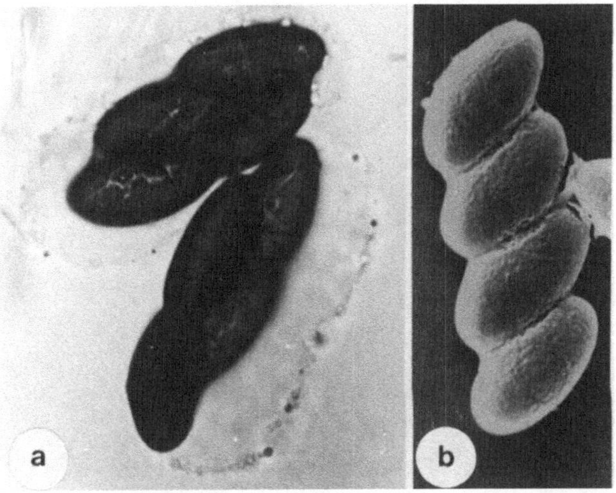

Fig. 4. Spore clusters, *Saccobolus groenlandicus*. a. LM photograph of cluster in water, EI. 84.116, b. SEM micrograph of spores displaced in one row, Gr. 83.149, a. x 1074, b. x 1247.

asci, outside glabrous, below with anchoring hyphae, pale purplish to vinaceous, especially towards the inconspicuous, rounded margin.

Excipulum up to 600 µm thick, throughout of thin walled cells; medullary excipulum of small globose to hyphae-like, short, hyaline cells, intermixed with branching, septate, ascogenous hyphae; outer excipulum below of globose cells, 6-18 µm broad, above tending to form perpendicular rows of cells which gradually elongate and become club shaped, 20-30 x 3-5 µm; outermost cells with vinaceous content. Subhymenium indistinct.

Hymenium 100-120 µm high. Asci normally 4-spored, strongly protruding when mature, then up to 140 µm long, 20-25 µm broad, below gradually tapering into a slender base, above truncate, all over staining pale bluish in Melzer's reagent. Operculum 6-9 µm broad.

Paraphyses straight, above slightly enlarged to 6-8 µm broad, branching, septate, of short cells, sometimes anastomozing, hyaline or with very pale vinaceous content in uppermost cells.

Spores strongly adhering in clusters, slightly overlapping, in one row, or in two displaced parallel rows. Individual spores 16.5-18.5 x 8.5-9.5 µm, ellipsoid to inequilateral, at first hyaline, then purplish, finally purplish brown, pigmented layer thin, with irregular crackings on the free convex side like *Saccobolus glaber*.

Spore clusters 32-40 x 14-16.5 µm (when displaced in two rows) or 36-42 x 12-14 µm (when displaced in one row), with one continuous, prominent, unilateral gelatinous covering.

Material examined: Type - Greenland, 2 km W of Mestersvig, near Nyhavn at Kong Oscars Fjord, 72°12'N 24°10'W, on dung of *Branta leucopsis* (Barnacle Goose) in moist area with seeping water, 3 August 1983, H. Dissing & S. Sivertsen, Gr. 83.95 (C, Holotype); East Greenland, Mestersvig; Gåsesøen near the airport, on dung of *Branta leucopsis*, 9

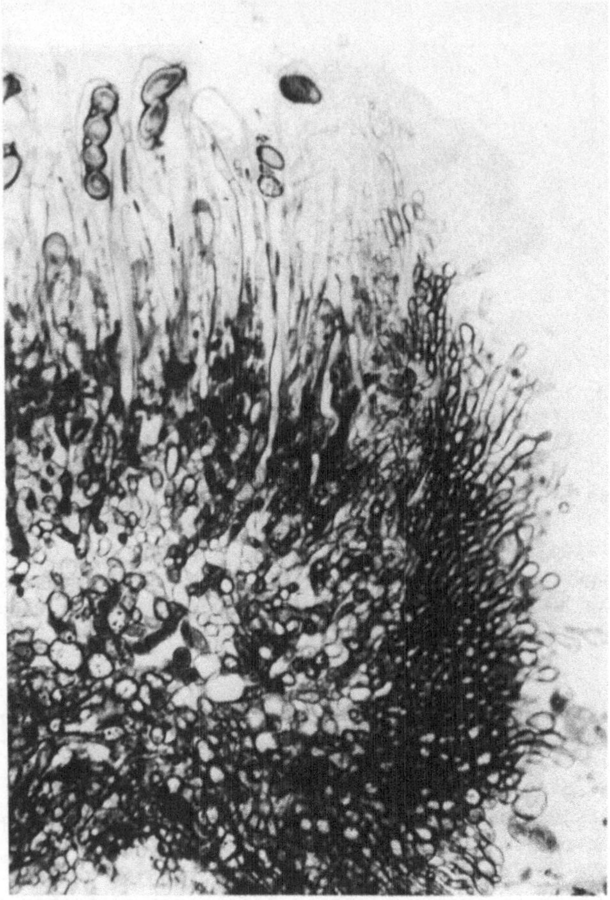

Fig. 5. *Saccobolus groenlandicus*, section of fruitbody, EI. 84.116, x 450.

August 1982, Gr. 82.137; - Store Blydal, 10 km SW of Mestersvig, on dung of goose (undetermined), 27 July 1983. Incubated 6 January 1984, isolated 24 January 1984, Gr. 83.142.

In addition more than 30 collections from Nyhavn, Gåsesøen and Store Blydal around Mestersvig have been studied. Most of these are deposited in the Botanical Museum, Copenhagen (C), but supplementary collections are deposited in Herbaria at the Universities of Trondheim (TRH) and Leiden (L).

East Greenland, Danmarks Havn; near Hvalrosodden, fresh dung from Barnacle Goose, collected 10 August 1984, leg. Børge Lauritsen. Incubated 19 January 1985, isolated 19 February 1985, Gr. 85.10. Canada, Ellesmere Island; Alexandra Fjord, 17 July 1984, EI.84.12, 84.13; - ibid., 21 July 1984, EI.84.19, 84.20; - ibid., 1 August 1984, EI.84.79, 84.85; - fresh dung from goose, collected 27 July 1984. Incubated 15 April 1985, isolated 1 May 1985, EI.84.114 and incubated 24 May 1985, isolated 20 June 1985, EI.84.116.

DISCUSSION

The above description is based on notes from fresh material of the

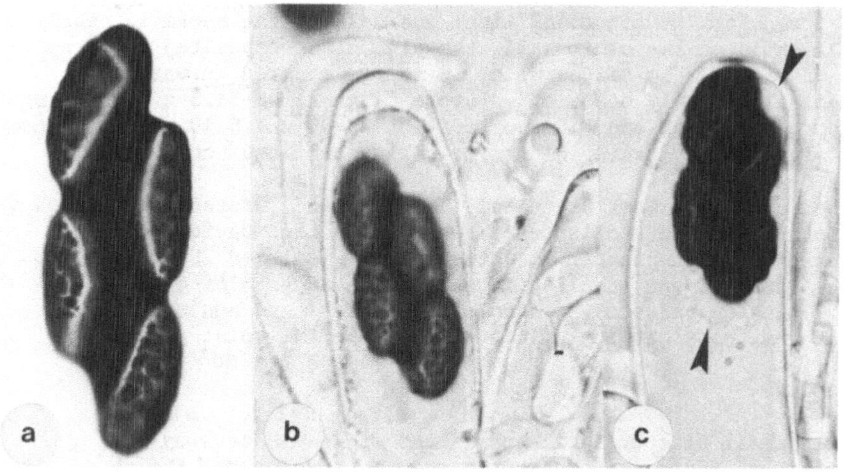

Fig. 6. LM photographs of spore clusters, a. *Saccobolus quadrisporus*, Gr. 82.327, b. *Saccobolus* sp. nov., note the curved paraphyses and bean-shaped immature spores, Gr. 82.326, c. *Saccobolus* sp. nov., ascus with spores with two separated gelatinous coverings, arrows, Gr. 82.326. a-c. x 895.

type as well as notes from material developed in moist chambers in the laboratory in Copenhagen. In nature, old apothecia will often be dark brownish, with amorphous substances deposited between cells in the excipulum and among paraphyses in the hymenium.

In *Saccobolus groenlandicus*, the asci are normally 4-spored, with the spores displaced in two rows. However, a few asci with spores in one row will be seen in almost all ascocarps. Asci with 3 or 5 spores were rarely found, and on two occasions only, asci with 8 spores in a cluster were seen (Hare dung, Gr. 83.127 and Goose dung, EI. 84.85).

Saccobolus groenlandicus can easily be distinguished from *S. quadrisporus* on characters of the size of asci, single spores and spore clusters. Besides colors of the ascocarps, characters of the spore ornamentation and gelatinous covering separate the two species.

In East Greenland, *S. groenlandicus* has been found on dung of Barnacle Goose (*Branta leucopsis*), Arctic Hare (*Lepus arcticus*) and Lemming (*Lemmus lemmus*). No collection has yet been made on dung from Pink-footed Goose (*Anser brachyrrhynchus*), which also nests there. In Ellesmere Island the goose dung most probably stems from Snow Goose (*Anser coerulescens*) or Brent Goose (*Branta bernicla*). It would certainly be interesting to study winter dung from the above mentioned geese in their respective winter quarters.

Saccobolus sp. nov. Figs. 2b and 6b.

Ascocarps 0.2-0.5 mm broad, disc-shaped to pulvinate, pale purplish to vinaceous or lilac. Asci 4-spored, 115-122 x 13-16 µm, staining pale bluish all over in Melzer's reagent. Paraphyses septate, branching, above curved, slightly enlarged to 5-6.5 µm broad. Spores 12-13.5 x 7-9.5 µm, ellipsoid to bean-shaped, at first hyaline, smooth, when mature reddish brown; in mature spores the pigment is deposited in two zones, separated by an irregular, longitudinal line: one half of the spore is nearly smooth, while on the other half, the pigment is deposited on irregular warts or ridges. In slides with old spores mounted in water the ornamentation easily peels off. Spore clusters 26-31.5 x 12.5-14 µm (when the spores are displaced in two rows) or 33-35 x 7.5-10 µm (when in one row), with two separated, clearly visible gelatinous coverings.

Material examined: East Greenland, Jameson Land, Ørsteddalen, fresh dung from Barnacle Goose, collected 17 July 1982, leg. David Boertmann. Incubated in moist chamber 11 January 1983, isolated 25 January 1983 (Gr. 82.326) where it was fruiting together with *Saccobolus quadrisporus* (82.327) and an undescribed *Ascobolus* species, and again 16 February 1983 it was where growing on filter paper in the moist chamber, Gr. 82.341.

DISCUSSION

The small size of asci, spores and spore clusters as well as characters of the spore ornamentation seem to separate the above described taxon very well from *Saccobolus quadrisporus* and *S. groenlandicus*. However, a formal description of a new species is postponed until material from nature is available.

ACKNOWLEDGEMENTS

In 1982, 1983, and 1984 travel expenses were covered largely by The Danish National Science Foundation. The Foundation also covered field work expenses in East Greenland in 1982 and 1983 while the University of Toronto covered field work expenses at Ellesmere Island in 1984. In 1982 and 1983 the Nordic Mining Cooperation Company provided excellent laboratory facilities at Nyhavn. David Boertmann (Orsteddalen) and Borge Lauritsen (Danmarks Havn) kindly collected samples of goose dung.

Joop van Brummelen made pertinent suggestions on the correct delimitation of *Saccobolus quadrisporus*. Tyge Christensen prepared the Latin diagnosis. JØrgen Fuglsang Nielsen operated the Cambridge electron microscope. Lene Christiansen and Lisbeth Haukrogh gave valuable technical assistance. Doris Thye-Petersen typed the manuscript.

Sigmund Sivertsen (1982), Henrik GØtzsche (1983), and Linda M. Kohn (1984) were brilliant co-workers during the field work.

I greatly appreciate all of their cooperation.

REFERENCES

Brummelen, J. van, 1967, A world-monograph of the genera *Ascobolus* and *Saccobolus* (Ascomycetes, Pezizales), *Persoonia, Suppl.*, 1: 1-260.

Eckblad, F. -E., 1968, The genera of the operculate Discomycetes. A re-evaluation of their taxonomy, phylogeny and nomenclature, *Nytt. Mag. Bot.*, 15: 1-192.

Jensen, J. D., 1982, The development of *Gelasinospora reticulospora*, *Mycologia*, 74: 724-737.

Massee, G., and Salmon, E. S., 1901, Researches on coprophilous fungi, *Ann. Bot.*, 15: 313-357.

ASCOMYCETES GROWING ON POLYTRICHUM SEXANGULARE

Peter Döbbeler

Institut für Systematische Botanik
Menzingerstrasse 67
D-8000 München 19, Deutschland

Key words: Bryophilous fungi, muscicolous ascomycetes, arctic-alpine, fungi, phototropism in fungi, *Polytrichum sexangulare* as host

ABSTRACT

The well-known arctic-alpine moss *Polytrichum sexangulare* Brid. (syn. *P. norvegicum* auct., Polytrichales) often forms dense mats on open siliceous soil in late snow-areas usually above the timber-line. Extensive herbarium material mainly from central and northern Europe was investigated in order to find bryophilous fungi. The following ascomycetes proved to be associated with *Polytrichum sexangulare*: *Bryochiton heliotropicus* Döbb. and *B. perpusillus* Döbb. (Dothideales) grow subcuticularly on the abaxial leaf-sides; *Lizonia sexangularis* Döbb. & Poelt (Dothideales) colonizes the antheridial cups of male plants and destroys them; *Protothelenella polytrichi* Döbb. & Mayrh. sp. nov. (affinities are uncertain, and the type species of the genus is lichenized) without any algae, emerges from the longitudinal photosynthetic lamellae of the leaves; *Gloeopeziza interlamellaris* Döbb. sp. nov. (Helotiales) with elongate and very reduced ascocarps is often completely immersed between the lamellae; *Hymenoscyphus norvegularis* Döbb. sp. nov. with apothecia inhabits all parts of the leaves. The species are keyed and described. New species are illustrated. Their European distribution is indicated. Several problems concerning their biology are discussed; e.g., phototropic reaction of asci and ascomata. The parasites apparently do not affect the growth of their hosts with the exception of the necrotrophic *Lizonia*. Infection of *Polytrichum sexangulare* by one or few of these ascomycetes is a common phenomenon. Remarkably representatives of the numerous and widespread pyrenomycetes inhabiting the lamellae interspaces of *Polytrichum* and related genera could not be detected. *Polytrichum sexangulare* has a fungus flora of its own, differing from that of other polytrichaceous hosts.

INTRODUCTION

The Polytrichaceae, including *Dawsonia*, constitute a "unique family of mosses," whose structural complexity in kind or degree is achieved by no group of bryophytes (G.L. Smith 1971: 2, 76). In some classification schemes, the well defined affinity is treated as a subclass, Polytrichidae, of its own (A.J.E. Smith 1978). The leafy gametophytic generation of these mosses is most suitable as host for numerous bryophilous non-lichenized ascomycetes of another totally different systematic position. The reason for the rich and diverse fungus flora on *Polytrichum* and allied genera seems to be that there is no quick alternation of the haploid stage. Concomitantly, the Polytrichaceae are often large and robust or even gigantic plants, which present an adequate array of microhabitats. Finally, *Polytrichum*-like mosses are a very ancient group and should be at the base in the evolution of Musci (Frey 1977). They have offered, through time, the necessary criteria for many fungi to evolve dependence on them as hosts.

Presently, there are more than 40 species of ascomycetes in six orders known to inhabit the Polytrichaceae, chiefly *Polytrichum* and *Dawsonia*, most of which have been treated by Dennis (1962), Döbbeler (1978, 1979a, 1981), and Racovitza (1940, 1946, 1959). They represent quite different biological types ranging from saprophytes to necrotrophic and as highly adapted biotrophic parasites. Some destroy the antheridial cups of male plants; others are restricted to the subcuticular region of the abaxial leaf side. Several species prefer upper, marginal or lower parts of a leaf, respectively. Memmbers of the operculate genus *Octospora* infect the subterranean rhizoids of several Polytrichaceae with complex appressoria and haustoria, sometimes inducing gemmae-like galls (Döbbeler 1979b, Döbbeler and Itzerott 1981). El Dorado are the spaces between the longitudinal photosynthetic lamellae of the adaxial leaf blade. There is no other microhabitat in Bryophytes that is more regularly occupied by fungal reproductive structures than these parts of the "pseudomesophyll," as G.L. Smith (1971: 27) aptly calls it. Approximately 20 ascomycetes with often extremely minute and partly reduced ascocarps, which in some cases do not fit the definition of "perithecium" or that of "apothecium," and have been demonstrated to grow obligately in the interspaces without inducing symptoms in any visible manner. They represent striking examples of convergent evolution in the adaptation to a specific, extraordinary and stable substrate. Scheirer and Dolan (1983), in a recently published paper on epiphytes of *Polytrichum commune* leaves, apparently are not aware of the very common phenomenon of fungi inhabiting polytrichaceous mosses.

Till now about 20 species in seven genera of Polytrichaceae have been shown to act as hosts for ascomycetes. This represents only a small percentage of the 354 species in 21 genera (Walther 1983) within the group. After studying the ascomycetes of the excellent host genus *Dawsonia* from the southern hemisphere, which has been suggested to always live symbiotically with fungi in a broad sense (Döbbeler 1981), it appeared attractive to select the holarctic truly arctic-alpine *Polytrichum sexangulare* Brid. (syn. *P. norvegicum* auct., *Polytrichastrum sexangulare* (Brid.) G.L. Smith) for a similar investigation. The only fungi demonstrated to occur on this ecologically distinguished moss are three pyrenomycetes known from a few collections (Döbbeler 1978). The main aim of the present contribution is a better idea of the occurrence, biology and distribution of the *Polytrichum* inhabiting ascomycetes and the fungi bryophili in general, which are despite their numbers and frequency, undoubtedly one of the least explored ecological niches for fungi today.

METHODS AND MATERIALS

The present work is based almost exclusively on herbarium collections of *Polytrichum sexangulare* from B (Botanisches Museum, Berlin-Dahlem), GZU (Institut für Botanik, Graz), JE (Herbarium Haussknecht, Jena), M (Botanische Staatssammlung, München), and W (Naturhistorisches Museum,

Wien). Approximately 100 specimens from different geographical regions were selected and in most cases were very carefully examined in order to find muscicolous fungi. Before scrutinizing the dried mosses with a dissecting microscope at a magnification of 40 X, a thorough rehydrating is necessary. Plants bearing ascomata had not been marked by the collectors. When found, they were separated and deposited under the fungus name in the mycological herbarium of the corresponding institution. Measurements and illustrations of microscopic details are based on mounts in lactophenol-cotton-blue. Lugol's solution was used to study the iodine reactions. Numbers following the "specimens examined" headings refer to the list at the end of the article.

KEY TO THE SPECIES

1. Ascomata perithecioid, black, forming a dense pseudostromatic layer within the antheridial cups; asci bitunicate, eightspored; spores 23-29 x 6-8,5 µm, one-septate, light brown; necrotrophic .... *Lizonia sexangularis*
1. Not with this combination of charcters 2
 2. Ascomata perithecioid with distinct ostiolum, dark 3
 2. Ascomata apothecial or apothecioid, light colored 5
3. Ascomata strongly depressed or hemispherical, less than 50 µm high; asci J-; spores one- or two-septate; subcuticular, usually on the abaxial side in upper leaf parts 4
3. Ascomata ovoid or pear-shaped, longer than 200 µm; ascus plug J + blue; spores tri-septate; superficial or emerging from the leaf lamellae .... *Protothelenella polytrichi*
 4. Spores two-septate .... *Bryochiton heliotropicus*
 4. Spores one-septate .... *Bryochiton perpusillus*
5. Ascomata in most cases completely immersed between the leaf lamellae, narrowly elliptical as seen from above, white; asci 55-85 x 11,5-18 µm; spores 13-16 x 5,5-7 µm ....*Gloeopeziza interlamellaris*
5. Ascomata superficial on both leaf sides, shortly stiped or subsessile apothecia, light colored with dark brown base; asci 39-60 x 6-8 um; spores 8,5-12 x 3-4 µm ....*Hymenoscyphus norvegularis*

DESCRIPTION OF THE SPECIES

Bryochiton heliotropicus Döbb., Mitt. Bot. Staatssamml. München 14: 210 (1978), Fig. 52.

Ascomata subcuticular at the leaves, 65-110 µm in diameter, strongly depressed, ostiolate, brown to black, glabrous, solitary. *Excipulum* seen from the outside at first with uniform, isodiametric cells, later with irregularly thickened walls and small lumina. *Paraphysoids* absent. *Asci* 25-38 x 11-15 µm, bitunicate, nearly ellipsoid or ovoid, octosporus. J - *Spores* (in lactophenol-cottonblue) 9,5-12(-13) x 3-4(-4,5) µm, (water-mounted spores of fresh material up to 15 µm long), ellipsoid, with nearly hemispherical apical cell, two-septate, at the upper cross-wall often slightly constricted, uncolored, epispore smooth. *Hyphae* subcuticular, inconspicuous.

The species is hitherto found only on *Polytrichum sexangulare*. It grows predominantly at the upper marginal or abaxial leaf-halves. As in the related *Bryochiton perpusillus* (see note under that species for differences), the ostiolum may be laterally shifted to the incident light.

Known distribution: Austria, Scotland, Norway, Sweden.

Specimens examined: 10, 12, 28, 30, 37 (for further records see Dobbeler 1978).

Bryochiton perpusillus Döbb., Mitt. Bot. Staatssamml. München 14: 226 (1978), Fig. 55.

Ascomata subcuticular at the leaves, up to 50(-70) µm in diameter, approximately hemispherical, ostiolate, brown to black, glabrous, solitary. *Excipulum* seen from the outside distinctly cellular, cells isodiametric, up to 9 µm in diametr. *Paraphysoids* missing. *Asci* 18-30 x 11-15 µm, bitunicate, ovoid or ellipsoid, octosporus, J -. *Spores* 9-11,5(-12,5) x 3-4 µm, ellipsoid, one-septate, bipolar, asymmetrical, often slightly constricted at the cross-wall. *Hyphae* (2-)3-5 µm thick, brown, forming a conspicuous subcuticular network.

This apparently cosmopolitan ascomycete has proven to occur on hepatics (*Ptilidium pulcherrimum* (G. Web.) Vainio, the type-host) and polytrichaceous mosses. It is particularly common on *Polytrichum piliferum* Hedw.

There are several characters common with *Bryochiton heliotropicus*. Both species grow subcuticularly at unchanged or slightly damaged leaves of *Polytrichum sexangulare*, avoiding the uppermost plant-region. They clearly prefer the upper parts of each leaf free of lamellae; i.e., the marginal or lower sides. Ascocarps are usually scattered and sparse. An adverse effect on the host plants cannot be observed. To distinguish the two species habitually is not always possible. Typical *B. heliotropicus* ascocarps are much larger, strongly depressed and have an indistinct cellular excipulum. Spore characters easily separate the two species.

Known distribution: on *Polytrichum sexangulare* Switzerland, Austria, Norway, Finland, (USA, Dobbeler 1978).

Specimens examined: 5, 12, 13, 14, 19, 32, 33, 40 (few further records see Dobbeler, 1978).

Lizonia sexangularis Döbb. & Poelt in Döbbeler, Mitt. Bot. Staatssamml. München 14: 313 (1978), Fig. 64,4.

Ascomata are predominantly within the male cups of *Polytrichum*, up to more than 300 µm in diameter, globose or pear-shaped, ostiolate, black, glabrous, closely aggregated and often laterally grown together. *Excipulum* seen from the outside with isodiametric cells, 10-20 µm in diameter, in longitudinal section with several layers of rectangular cells. *Paraphysoids* inconspicuous. *Asci* (75-)90-120(-130) x 15-20 µm, bitunicate, cylindrical, octosporous; J-. *Spores* (18-)23-29(-32) x (5-)6-8,5(-10,5) µm, ellipsoid, two-celled, light brown, bipolar, asymmetrical, at the cross-wall sometimes constricted, epispore smooth. *Hyphae* thick, brown, growing within the host cells, immediately below the ascocarps in a dense network replacing the internal cells of the stem. This species is restricted to *Polytrichum sexangulare*.

Presently, there are four species of *Lizonia* known to infect several species of *Polytrichum* (Döbbeler 1978), differing in the number of spores per ascus (eight or sixteen), spore size and host specificity. Common characters are necrotrophic parasitism, which results in a complete destruction of the growing points of infected plants, the organ specific restriction to the antheridial cups in male or the apical regions in female plants, and the adaptation to the periodicity of the hosts. Ascocarps develop at the same time as uninfected plants produce gametangia.

In the field, this species is visible to the naked eye because of its pseudostromatic growth within the antheridial cups of male plants. It is the only such specimens dealt with here. Female individuals having only a

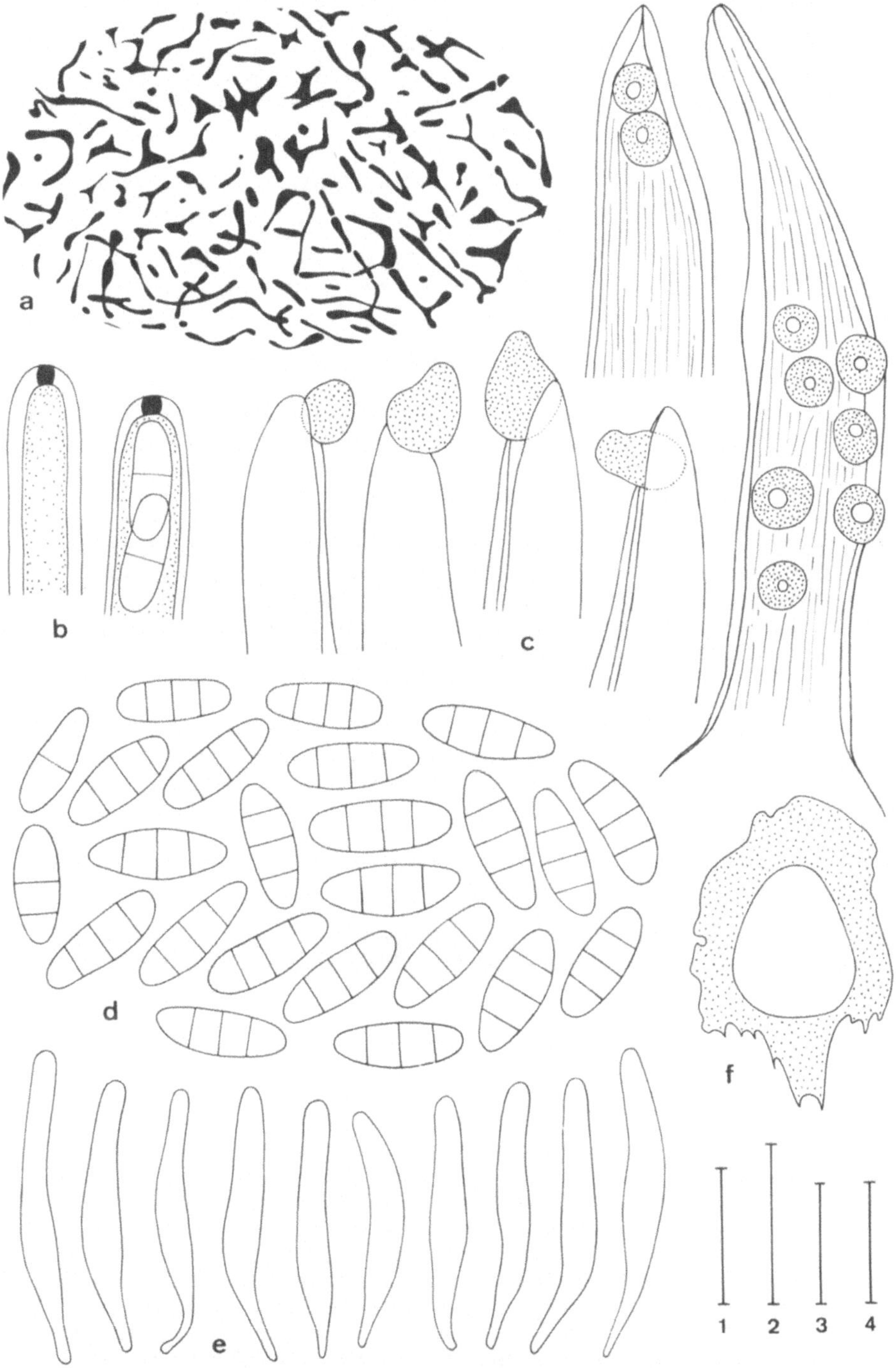

Fig. 1. *Protothelenella polytrichi* (Holotypus). a) Upper part of ascocarp wall in longitudinal section, hyphal lumina black. b) Tips of immature asci with J + blue apical structures. c) Habit sketches of ascomata on host leaves. d) spores, the two left ones atypical. e) Asci from one ascocarp. f) Non median, longitudinal section of an ascocarp which has developed on the leaf-lamellae. - Scale 1 = 30 µm (e), 2 = 0,5 mm (c), 3 = 10 µm (a,b,d), 4 = 100 µm (f).

few ascocarps in the apical region are much more difficult to detect and probably poorly represented in the collections. The question, as to whether or not infected plants lacking antheridial cups are female or sterile males could definitely be decided. Several times sporophyte bearing, and therefore female, plants were inhabited by the fungus at the top of a lateral branching,

Known distribution: Switzerland, Austria, Italy, Czechoslovakia, Scotland, Norway, Sweden, USSR.

Specimens examined: 1, 2, 3, 4, 12, 13, 14, 15, 17, 18, 20, 25, 27, 29, 31, 32, 34, 37, 42, 44 (in addition to several collections recorded by Dobbeler 1978).

Protothelenella polytrichi Döbb. & Mayrhofer sp. nov. (Fig. 1).

Ascomata 210-285 µm longa, 150-200 µm lata, plerumque ovoidea ad pyriformia, griseonigra vel subnigra, basin versus minus colorata, externe strato hyalino tecta, glabra, superficie verrucosa; statu sicco semper nigra, splendida. Ostiolum 40-70 µm diametiens. Paries ascomatum lateraliter 20-40 µm, apicaliter usque ad 70 µm crassus; e hyphis crassitunicatis reticulatim ramosis et anastomosantibusque luminibus persubtilibus compositus. Paraphysoidea filiformia, 1,5 µm crassa, copiosa, parietibus ascomatum ubique connecta. Asci 60-85 x 8-11 µm, subcylindrici, saepe parte media vel basali nonnihil dilatati, in pedem plus minusve longiorem gradatim attenuati, octospori. Sporae 10-13,5 x 4-5 µm, ellipsoidales, triseptatae, incoloratae, dimidiis plerumque subaequalibus, ad septa non constrictae, episporio laevi. Hyphae tenues, non coloratae, super intraque cellulas foliorum hospitis repentes. Algae deficientes. Jodum parietes ascorum et gelatinam hymenii dilute caeruleos vel subrubros, structuram apicalem in apicibus ascorum distincte caeruleam tingens.

Habitat parasitice sparsim et singulariter in foliis vivis, apices eorum praeferens, *Polytrichi sexangularis*.

Typus: Austria, Tyrol, Samnaun-Group, areas of late snow-lie eastwards close to the Furglersee above Serfaus, ca. 2450 m, July 1976 J. Poelt and R. Moberg (Holotypus GZU).

Etymology: refers to the host genus *Polytrichum*.

Ascomata 210-285 µm long (without basal ledges), 150-200(-250) µm thick, ovoid or pear-shaped, rarely almost globose, greyish black to blackish, in the lower part less colored, with a more or less distinct hyaline layer, without hairs or bristles, surface coarsely warted, in a dried state always black and glossy. *Ostiolum* 40-70 µm in diameter, sometimes visible as a light spot under a binocular. *Excipulum* laterally 20-40 µm, above 50-70 µm thick, cartilaginous, consisting of strongly branched and anastomosing hyphae forming an irregular network, hyphae thick-walled with delicate lumina scarcely reaching 1 um in diameter. *Paraphysoids* filiform, ca. 1,5 µm thick, copious, mainly in the lower part with ramifications and anastomoses, everywhere connected with the inner wall of the excipulum. *Asci* (55-)60-85(-90) x 8-11 µm, not typical bitunicate, subcylindrical, usually in the middle or below slightly expanded, often curved, gradually contracted in a shorter or longer foot, octosporous. *Spores* 10-13,5(-15) x (3,5-)4-5(-5,5) µm, ellipsoid, tri-septate, uncolored, bipolar, symmetrical or more often slightly asymmetrical, not constricted at the cross-walls, epispore smooth. *Hyphae* hyaline, without any contact to algae, up to 3 um in diameter, near the ascomata sometimes forming a dense reticulum, running superficially, often preferring the anticlinal walls, above the cells of the host leaves, as well as within single cells and then only about 1,5 um thick, hyphae not completely occupying the host cells. *Iodine reaction*: Ascus walls and hymenial jelly J + faintly blue varying to dirty reddish, mature asci with a strong seemingly layered apical structure; KOH pretreatment gives the same reactions.

Infected leaves of the host *Polytrichum sexangulare* are unchanged and apparently healthy or now and then conspicuously damaged. The nearly always scanty ascocarps usually develop singly or by twos on each leaf. Only a few times have the maximum number of seven per leaf and 20 per plant been observed. Ascomata occur in the upper region of a plant extending one centimeter below the growing point, here again especially at the uppermost leaf parts. The adaxial side of the leaf blade is clearly chosen more often than the abaxial one.

Known distribution: Switzerland, Austria, Sweden, USSR.

Specimens examined: 3, 10, 11, (Holotypus), 12, 16, 37, 38, 43.

The small elliptical ascocarp initials of *Protothelenella polytrichi*, emerging from the longitudinal leaf-lamellae, look like laterally compressed pyrenomycetes which regularly occuy this habitat in other Polytrichaceae. Fully grown they bear at the base several ledges or better wedges (teeth in longitudinal sections) due to the lamellae.

It is noteworthy that the ascomata are often laterally attached to the substrate. Moreover the ostiola are frequently shifted phototropically towards the incident light. Both facts can be interpreted as adaptations to enable spore discharge in the open air.

In seven of the eight available collections of *Protothelenella* the *Polytrichum* is also infected by *Lizonia sexangularis*, which seems more than mere accident. Otherwise, most specimens of *Lizonia* are not associated with *Protothelenella*. Sometimes both species grow in the immediate vicinity of each other - *Lizonia* dense layer within the antheridial cups and *Protothelenella* here and there at its borders.

The genus *Protothelenella* Räs. is dealt with by Mayrhofer and Poelt in a paper on the lichen genus *Microglaena* sensu Zahlbruckner in Europe (in preparation), an affinity which seems to belong to the Pertusariaceae (Eriksson 1981: 96). The biologically heterogenous taxon *Protothelenella* is remarkably comprised of members with distinct curstose thalli, or reduced thalli as found in the type species *P. sphinotrinoidella* (Nyl.) Mayrh. & Poelt, unpubl. They may also be completely missing a thallus as in the moss-inhabiting *P. polytrichi*, having no contact with algae. Hafellner (1983) points out that it is a rare event for a lichen mycobiont to abandon its connection to its phycobiont and grow saprophytically because most species apparently cannot "forget" their phylogenetic origin. The new and parasitic *Protothelenella*, however, succeeded in entering a fungus-moss-symbiosis free of algae.

Gloeopeziza interlamellaris Döbb. sp. nov. (Figs. 2-4).

Ascomata inter lamellas foliorum *Polytrichi* omnino immersa vel nonnihil protrudentia, 95-140 µm alta, 90-180 µm longa, 18-50 µm lata, incolorata, glabra, superne visa ambitu elliptico. Excipulum reductum, e hyphis paraphysoideis et in gelatina iacentibus, luminibus 1-2 µm crassis, vix ramosis formatum; apicaliter hyphae excipuli ut multae paraphyses

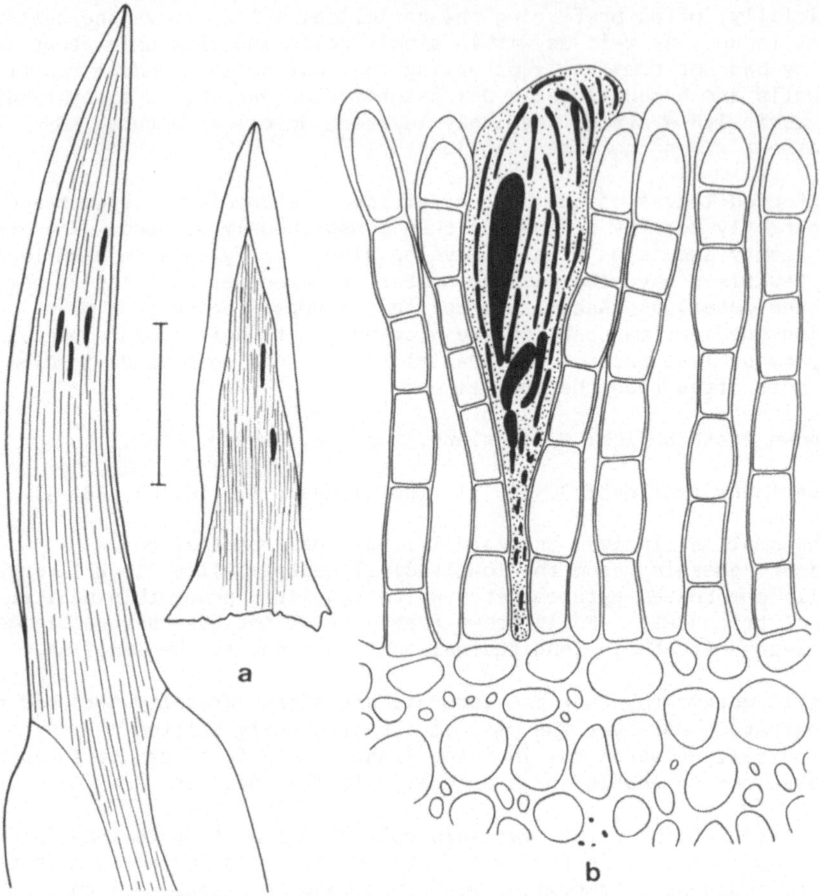

Fig. 2. *Gloeopeziza interlamellaris* (Holotypus). a) Habit sketches of ascocarps (black) between the leaf-lamellae; scale = 0,5 mm. b) Vertical section of an immature ascocarp developing between the leaf-lamellae; scale = 40 µm.

flexuosae et ascos laxe tegentes. Paraphyses filiformes, 1-2 µm crassae, apicaliter non dilatatae. Asci 55-85 x 11,5-18 µm, unitunicati, cylindrici usque ad plerumque claviformes, octospori; maturitate asci extensi ex epihymenio distincte prominentes et saepe ad lumen versus distincte flexuosi. Jodum anulum in apice ascorum caeruleum tingens. Sporae 13-16 x 5,5-7 µm, ellipsoidales, unicellulatae, incoloratae, episporio laevi. Hyphae 1-2 µm crassae, non vel rarior dilute coloratae, supra cellulas lamellarum repentes et parietes cellularum superiorum trichomatum ex lamellis basaliter sparsae orientium, infestantes.

Habitat sparsim inter lamellas foliorum validorum superiorum plantarum *Polytrichi sexangularis*.

Typus: Austria, Tyrol, Samnaun-Group, areas of late snow-lie eastwards close to the Furglersee above Serfaus, ca. 2450 m, July 1976 J. Poelt and R. Moberg (Holotypus GZU, Isotypus M).

Etymology: from Latin "inter" and "lamella", referring to the ascomata between the lamellae of the host leaves.

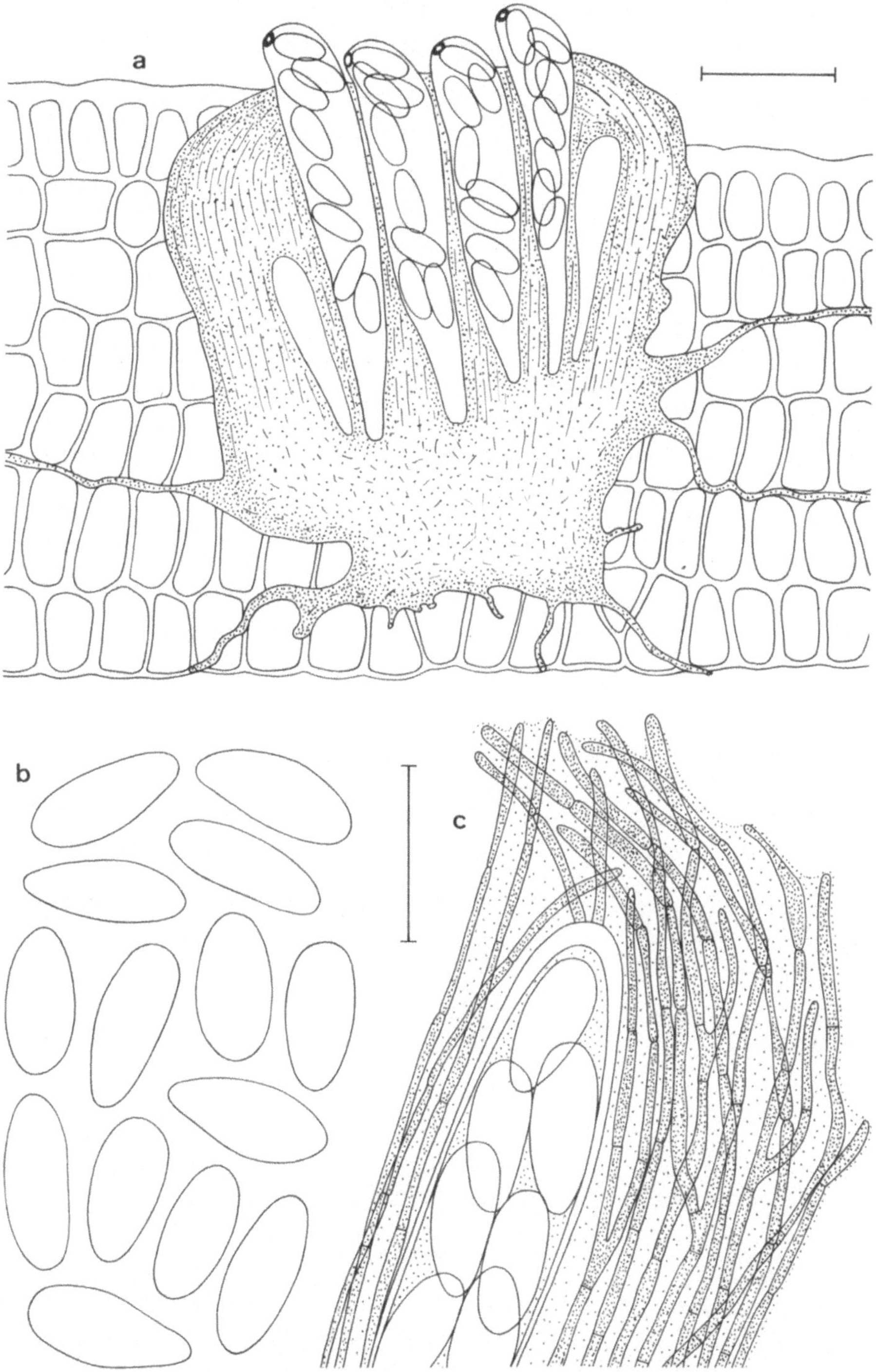

Fig. 3. *Gloeopeziza interlamellaris*. a) Optical longitudinal section of an ascocarp, fully mature asci protruding above the hymenium, their tips with iodine stained rings (black) showing to the incident light from the left; scale = 30 µm (Germany, Diesbachscharte). b) Spores. c) Upper right part of an ascocarp with an enclosed ascus and excipular texture; scale = 15 µm (b, c Holotypus).

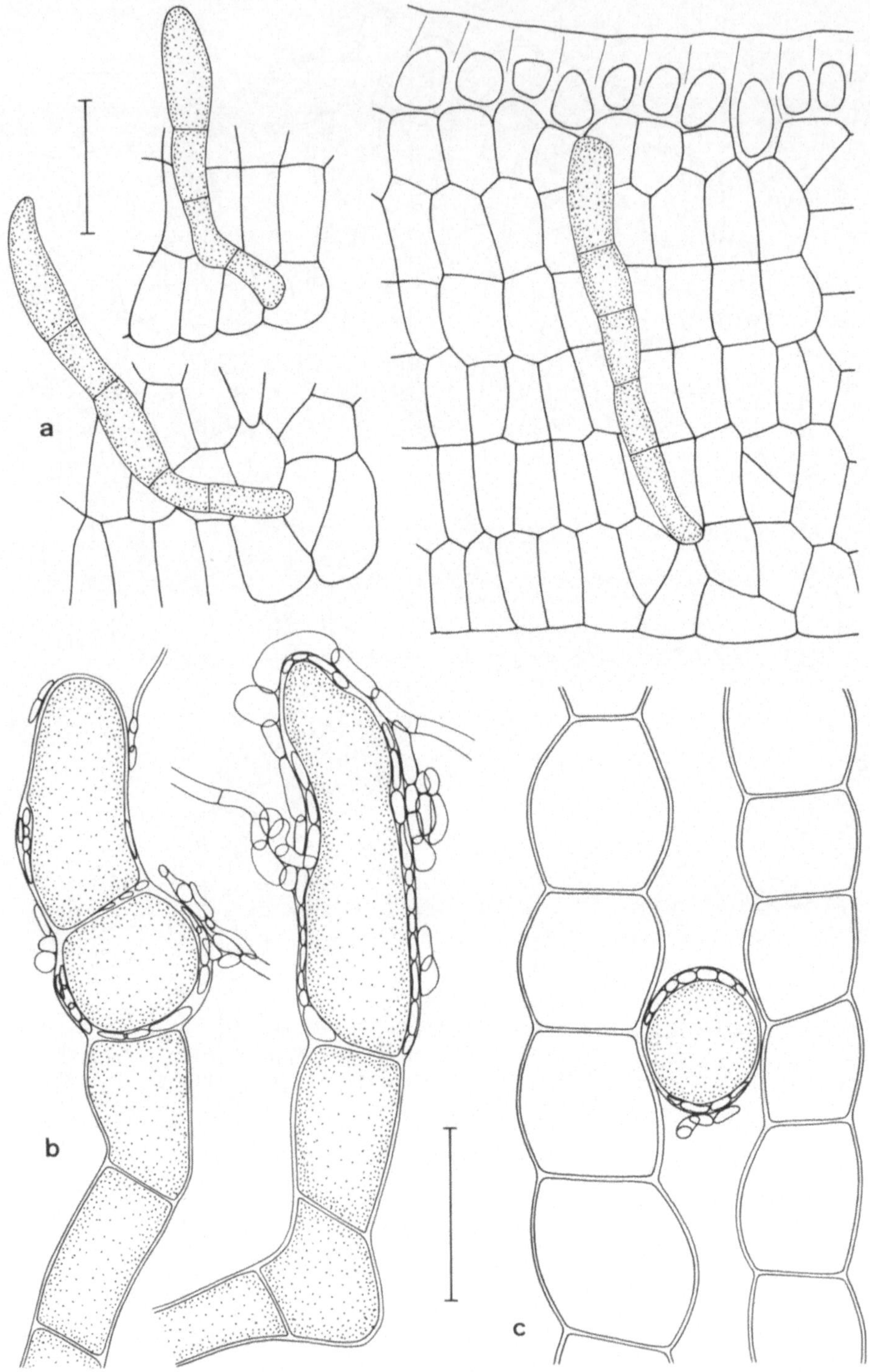

Fig. 4. *Gloeopeziza interlamellaris* (Holotypus). a) Habit sketches of uninfected trichomes born at the basal cells of the leaf-lamellae; scale = 30 µm. b) Hyphae growing within the walls of the apical trichome cells. c) Cross section of a hyphal infected trichome and adjacent leaf-lamellae. - Scale = 15 µm.

Ascomata completely immersed between the leaf lamellae of *Polytrichum* or projecting up to 35(-50) µm, 95-140(-170) µm high, (55-)90-180(-230) µm long, 18-50(-65) µm broad, small elliptical as seen from above, approximately wedge-shaped in cross section, the adjacent lamellae often with apical overgrowths, uncolored, glabrous. *Excipulum* reduced especially alongside the lamellae, composed of extended, paraphyses-like hyphae, with 1-2(-3) µm wide lumina, scarcely branched and lying in a gel, hyphae and many paraphyses in the uppermost part of the ascocarp curved, forming a kind of epihymenium covering the asci, subhymenial tissue of irregular and shorter celled hyphae. *Paraphyses* filiform, 1-2 µm in diameter, not expanded above, not distinguishable from the hyphae of the excipulum. *Asci* 55-85 x 11,5-18 µm, unitunicate, cylindrical or usually club-shaped, octosporous, only few per ascocarp in a mature state. *Iodine* stains the apical ring blue up to 5 µm in diameter. *Spores* (12-)13-16(-17) x (5-)5,5-7 µm, ellipsoid, one-celled, uncolored, often slightly bipolar, asymmetrical, and with a smooth episporium. *Hyphae* ca. 1-2 µm thick, not or rarely a little colored, thin walled, running superficially on the cells of the leaf-lamellae sometimes preferring the anticlinal walls; furthermore, delicate undulate hyphae in outline are infecting the cell walls of the trichomes between the lamellae.

Ascomata develop sparsely and individually in the higher plant region of *Polytrichum sexangulare*, up to 1,5 cm below the growing point. The host is not affected in any visible way.

Known distribution: Switzerland, Germany, Austria, Italy, Czechoslovakia, Scotland, Norway, Sweden, USSR, USA.

Specimens examined: 4, 5, 6, 7, 8, 9, 10, 11 (Holotypus), 12, 14, 15, 17, 19, 20, 21, 22, 23, 24, 25, 26, 28, 30, 34, 35, 36, 37, 39, 41, 42, 43, 44, 45.

Gloeopeziza interlamellaris, the most common species encountered on *Polytrichum sexangulare*, seems to be more frequent than the 32 records suggest. Ascocarps are difficult to detect, partly owing to the smallness and obscure occurrence of the unpigmented fruit bodies, and partly due to the nearly always scattered infection. Occasionally, only one ascocarp is detectable per plant. A few times the maximal number of seven fruit bodies per leaf could be observed. In many cases, microscopic screening of the leaves is necessary. To treat them with iodine is favorable because the blue ascus plugs disclose the fungus.

Large ascocarps reaching the costa have flat bases. Hyphae run away only from the lateral lower parts. If ascocarps are less deeply immersed, the hyphae originate from the whole basal region of the irregularly limited fruit bodies.

Growth of the otherwise superficial mycelium within the cell walls of the uniseriate, cylindrical to almost club-shaped trichomes, here and there arising from the basal cells of the lamellae, is remarkable. Colonized are only the apical thin walled and unpigmented parts of these hair-like structures. Here again the exposed sides not adjacent to the lamellae are predominantly infected. The apical cells are distinguished by a dense plasmatic content, which is not invaded by hyphae. Apparently, occurrence and function of the trichomes is not clear. Goebel (1930: 944) observed "Keulenhaare" at the lamellae in water cultures of *Polytrichum* and suggested that they might play a role in mucous secretion.

The positively phototropic reaction of the asci, seen in most cases, is characteristic. At maturity they elongate and protrude strongly above the epihymenial level. Often their tips with the iodine-blue colored apical ring bend to the incident light. This phenomenon, so far reported only in Pezizales (Ingold 1966, 1971: 14), is even demonstrable in immature asci, that are enclosed by epihymenial hyphae.

The new species surely belongs to the Helotiales, though not fitting Kimbrough's (1970: 123) characterization of the order "by ascocarps in which the hymenium ... is borne in a superficial apothecium." Strictly speaking, the ascocarps are neither superficial nor are apothecia as cup- or saucer-like ascomata (Hawksworth et al. 1983). Their development in the small spaces between the longitudinal leaf-lamellae forces the peculiar habit of the ascocarps and allows a serious reduction of the sterile tissue surrounding the hymenium.

Using the keys of the gelatinous Leotiaceae following Korf (1973) and Carpenter (1981) leads to the bryophilous genera *Gloeopeziza* Zuk. and *Mniaecia* Boud. *M. jungermanniae* (Nees: Fr.) Boud., type of the last mentioned taxon, is quite different. There are, however, important characters in common with the hepaticolous *G. rehmii* Zuk., type species of the genus *Gloeopeziza*. According to the carefully illustrated original description by Zukal (1891), *G. rehmii* growing on *Blepharostoma trichophyllum* (L.) Dum. has small sessile apothecia up to 200 um in diameter. The simple constructed excipulum is composed of paraphyse-like hyphae embedded in a gel found in all parts of the fruit body. The inoperculate asci produce eight hyaline one-celled spores. Iodine reaction is not indicated. Apart from the immersed, elongated ascocarps of *G. interlamellaris*. *G. rehmii* differs only in few points: The paraphyses do not form a kind of epihymenium and the hyphae are loosely associated with algae resulting in a microscopic thallus. This feature should be reinvestigated. It seems to be more justified to refer the new species to *Gloeopeziza* than to any other known genus.

Affinities to the helotialean *Dawsicola neglecta* Döbb., an obligate parasite of the lamellar interstices of *Dawsonia*, do not exist. This monotypic genus differs in having an almost completely reduced excipulum, branched paraphyses, ascus pores without iodine reaction and subhymenial tissues that stain reddish brown in iodine (Döbbeler 1981).

Hymenoscyphus norvegularis Döbb. sp. nov. (Figs. 5, 6).

Apothecia 70-160 µm lata et fere similiter alta, turbinata et sessilia vel breviter stipitata, glabra, subalba usque ad dilute cana, basin versus fusca. Excipulum apicem versus e hyphis incoloratis usque ad 2,5 µm latis longitudinaliter extensis (textura porrecta) formatum; in media parte cellulae excipuli paene isodiametricae et saepe difficile visibiles; basis vel stipes hyphis fuscis irregularibus formatus. Paraphyses filiformes, 1-2 µm crassae, vix ramosae. Asci 39-60 x 6-8 µm, unitunicati, claviformes, in pedem longiorem gradatim attenuati, octospori. Anulus jodo in apice ascorum caerulescens. Sporae 8,5-12 x 3-4 µm, anguste ellipsoidales, unicellulares aut rarior septa una praeditae, incoloratae, dimidiis aequalibus vel subaequalibus, episporio laevi. Hyphae inconspicuae, 1-2 µm crassae, fuscae vel subcoloratae, praecipue intra cellulas foliorum hospitis crescentes.

Habitat sparsim et plerumque singulariter in lateribus ambobus foliorum superiorum vivorum vel inferiorum veterumque *Polytrichi sexangularis*.

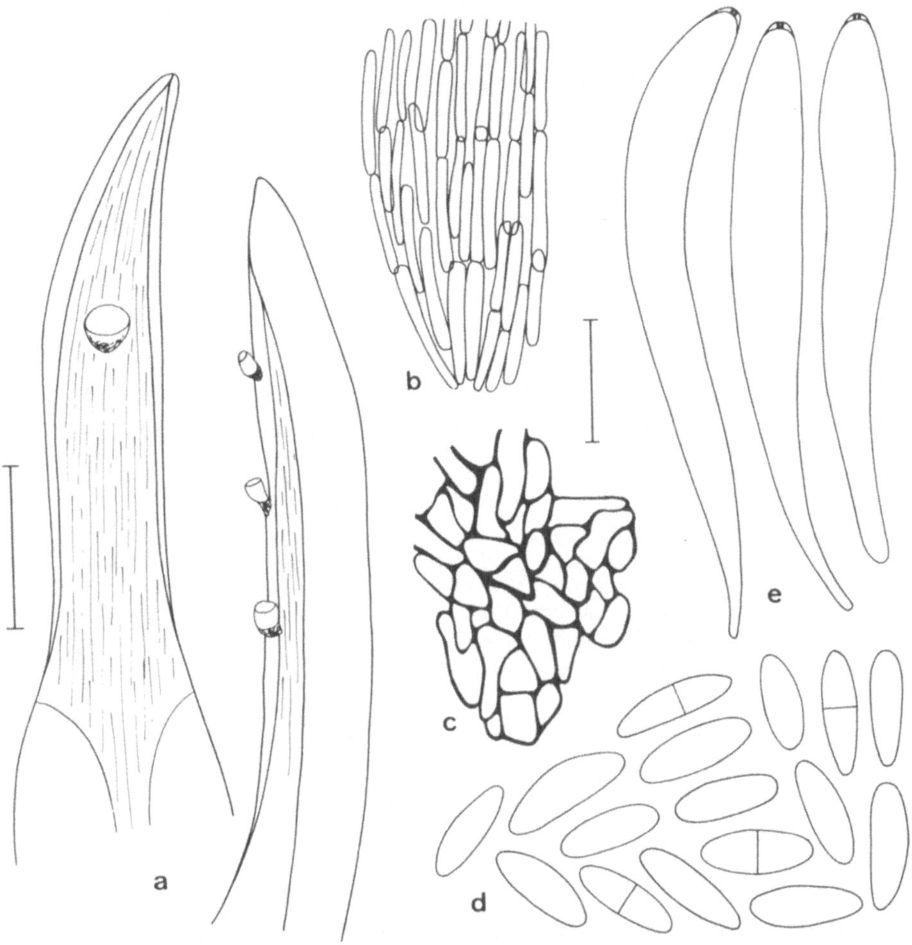

Fig. 5. *Hymenoscyphus norvegularis* (Holotypus). a) Habit sketches of apothecia on host leaves; scale = 0,5 mm. b) Upper outer excipulum texture. c) Outer texture of the stipe or basal ascocarp region. d) Spores from one ascocarp. e) Mature asci with J + blue apical rings. - Scale = 10 µm.

Typus: Austria, Tyrol, Stubaier, Alps, slopes between Dortmunder Hütte and Birchkogel, 2450-2600 m, 17 September 1977 E. Albertshofer (Holotypus Do 2964 in M).

Etymology: an artificial composition derived from the specific synonymous host names, *Polytrichum norvegicum* and *P. sexangulare*.

Apothecia 70-160(-250) µm broad and nearly as high, top-shaped and sessile or shortly stalked, hymenium plane, without prominent margin, glabrous, almost white to light grey, with dark brown basal part; easily breaking off. *Excipulum* in the upper part composed of uncolored, extended, parallel hyphae up to 2,5 µm wide (textura porrecta); base or stipe with irregular, short celled and darkly pigmented hyphae, region between stipe and excipulum consisting of more or less isodiametric, slightly colored, cells and often difficult to observe. *Paraphyses* filiform, 1-2 µm in diameter, apically neither curved nor expanded, scarcely branched. *Asci* 39-60 x 6-8(-9) µm, unitunicate, claviform, straight or somewhat curved, gradually contracted in a rather long foot, octosporous. *Iodine* distinctly stains an apical ring blue. *Spores* (8-)8,5-12(-13) x 3-4 µm, narrowly ellipsoid, one-celled, rarely two-(three-)celled, uncolored, bipolar, symmetrical or nearly so, epispore smooth. *Hyphae* inconspicuous, 1-2 µm in diameter, brown or almost uncolored, near the base of the ascocarp often forming a limited reticulum and superficially running above the host cells over short distances as well as growing within single cells, without occupying them completely.

The apothecia develop sporadically at both sides, tips, margins, and/or even the sheaths of leaves of *Polytrichum sexangulare*.

Known distribution: Switzerland, Germany, Austria, Italy, Czechoslovakia.

Specimens examined: 5, 8, 10, 12 (Holotypus), 13, 14, 17, 20, 22.

Hymenoscyphus norvegularis seems to be a saprophytic or weakly parasitic species. Obviously, the leaves are often already damaged before infection, especially the lower ones, up to 3 cm below the growing point. Discolored regions at the points of attachment were not observed, but numerous minute, light, circular perforations could be observed after the removal of an ascocarp from the leaf. They represent hyphae penetrating the periclinal walls of the host cells. After passage, they enlarge to the usual diameter within the cell lumina.

The apothecia are not restricted to a special part of the leaf. If originating at the adaxial leaf side, they usually emerge from the lamellar interstices. The brown initial stages look like perithecia. Though fruit bodies are nearly always scattered and solitary, a few times more than 50 initials per leaf have been counted. Often the ascocarps do not develop perpendicularly to the leaf surface, but are bent parallel to the leaves, allowing for a more efficient spore discharge.

The new species is habitually distinguished by its small, light colored apothecia with dark bases and the occurrence on *Polytrichum sexangulare*, the first record of this moss as a host for discomycetes except for *Gloeopeziza interlamellaris*. The few other Helotiales known to inhabit polytrichaceous mosses; e.g., *Durella polytrichina* (Karst. and Starb.) Racov. (Racovitza 1940, 1946) and *Pezizella polytrichi* Dennis (1962), are clearly different.

DISCUSSION

Polytrichum sexangulare grows "in often dense mats on open, moist or wet siliceous soil in late snow-areas, ... in the farthest north or in the alpine belt of the mountains, rarely below the tree-limit" (Nyholm 1969). The typical and easily known moss may occupy abundant areas (Gjaerevoll 1950: 410, Mårtensson 1956) and develop characteristic "Mooswiesen" (Herzog 1926: 179). In paleoecological studies *P. sexangulare* is used as an obligate indicator of snowbeds (Miller 1984: 1221). Geissler (1982: 174) points out that in such places snow cover is prolonged and the growing season varies from one to three months, differing from year to year. It is interesting to note that all collections cited below were gathered within ten weeks between 9 July and 17 September.

Very careful screening of most of about 100 herbarium specimens of *Polytrichum sexangulare* from different geographical regions revealed

Fig. 6. *Hymenoscyphus norvegularis* (Holotypus). a) Cells of abaxial leaf side just below a detached apothecium, note the numerous hyphal pegs penetrating the cell walls. b) Intracellular mycelium. - Scale = 20 µm.

that several fungi are often associated with that moss. Forty-five collections of the host show one or up to all of the six described ascomycetes (together 90 specimens, including several collections recorded by Döbbeler 1978). The assumption is justified that about half of all collections of *P. sexangulare* deposited in herbaria are inhabited by at least one species of ascomycete. It is demonstrated once more that bryological herbaria harbour a vast amount of bryophilous fungi, accidentally gathered with their hosts, unknown to the collectors and often unknown to science. Fungal infections of *Polytrichum* gametophytes or populations in the field, however, must be significantly higher than proved. Gatherings in herbaria are sometimes scanty and fungi are often difficult to demonstrate because of the sporadic occurrence of ascomata. In such cases, they are only detectable by chance. I do not hesitate to state that several of the indicated ascomycetes are always present in larger European populations of *Polytrichum sexangulare*. Only one species has a detrimental effect on its substratum, viz *Lizonia sexangularis*, which destroys the plant's growing points. All other parasites apparently do not effect growth and sexual reproduction of their hosts.

The biotrophic *Gloeopeziza interlamellaris* which is hidden between the leaf lamellae, is the most often encountered species (32 collections), followed by *Lizonia sexangularis* (20 collections), which needs male plants with antheridial cups for colonization. Mutual stimulation or exclusion of species could not be observed apart from *Protothelenella polytrichi*, which is nearly always associated with the necrotrophic *Lizonia*. It should be indicated that the fungus flora of *P. sexangulare*, even in its European range, is not completely considered. Some ascomycetes found once or very few times have been omitted. It is not clear whether or not these fungi only attack *P. sexangulare* as *Bryodiscus arctoalpinus* Döbbeler and Poelt (1974) occasionally does or whether they are extremely rare specialists.

Interpretation of the geographical distribution of the fungal parasites within the host range has to note that mainly central and northern European collections of the moss were available. There was not a single potential host collection from Siberia, Greenland or North America, except for one infected specimen from the Cascade Mountains. There is reason to believe that most of the fungi inhabiting *P. sexangulare* are widely distributed as are many other bryophilous fungi, and that they occupy the entire holarctic area of the host.

The bryophilous habit implies an array of physiological and morphological adaptations to the unique substratum. Of course, that is true for the ascomycetes on *Polytrichum sexangulare*. Special characters in connections with the arctic-alpine occurrence of the host cannot be detected as demostrated for arctic fungi in general by Savile (1972) and for arctic-alpine muscicolous basidiomycetes by Redhead (1984). The striking phototropic reaction of different parts of ascomata, missing only in the exposed *Lizonia sexangularis*, and fruiting predominantly in the upper leaf halves allows spore discharge in the open air from infected leaves within dense moss carpets. In *Protothelenella polytrichi* and *Hymenoscyphus norvegularis*, ascocarps are often laterally attached or curved in the latter species; in both *Bryochiton* species and in *Protothelenella* the ostiola can be laterally shifted, and in *Gloeopeziza interlamellaris* the ascus tips often show a + phototropism towards the direction of incident light. Adaptations in connection with efficient spore dispersal are common phenomena in fungi found on mosses and liverworts, thus offering excellent examples of convergent evolution.

Comparison of the fungi growing on *Polytrichum sexangulare* with those of other polytrichaceous hosts reveals that only *Bryochiton perpusillus* is known from other hosts, and it even attacks hepatics. Surprisingly, there are no pyrenomycetes in the lamellar interstices that so often infect members of *Polytrichum* and *Dawsonia*. As far as is known, *P. sexangulare* has a fungus flora of its own.

EXAMINED SPECIMENS OF *POLYTRICHUM SEXANGULARE* AND THEIR FUNGI

Switzerland:
1 Helvet, (hb. F. Brandis in B). *Lizonia sexangularis*.

2 Gotthard, August 1839, Schimper (W). *Lizonia sexangularis.*
3 Fibbia, St. Gotthard, 2400 m, 7 August 1881, J. Weber (B). *Lizonia sexangularis, Protothelenella polytrichi.*
4 Rhaetia, Albula, 2439 m, 8 August 1886, H. Graef (B). *Lizonia sexangularis, Gloeopeziza interlamellaris.*
5 Engadin, Bernina Pass, Weiger Sea, 2287 m, 15 August 1880, (GZU). *Bryochiton perpussilus, Gloeopeziza interlamellaris, Hymenoscyphus norvegularis.*

Germany:
6 Allgäu, above the Wildsee by Hinterstein, 1810 m, 13 July 1887, Holler (W). *Gloeopeziza interlamellaris.*
7 Allgau, foot of the Kastenkopf above the Wildsee, 1800 m, September 1904, J. Ziegler (Fl. exsicc. Bav.: Bryoph., 473. *Polytrichum sexangulare*), (M). *Gloeopeziza interlamellaris.*
8 Upper Bavaria, Geigelstein-Region, Rossalpe eastern of Rossalpenkopf, ca. 1720 m, 11 September 1920, H. Paul (M). *Gloeopeziza interlamellaris, Hymenoscyphus norvegularis.*
9 Upper Bavaria, Steinernes Meer, Diesbachscharte (borderland to Salzburg), ca. 2200 m, 12 August 1920, H. Paul (M). *Gloeopeziza interlamellaris.*

Austria:
10 Tyrol, Samnaun-Group, northwest slope below the Medrigjoch, south of the Ascher Hütte, ca. 2400 m, 8 September 1972, J. Poelt (GZU). *Bryochiton heliotropicus*, (*Lizonia sexangularis*, Holotypus GZU, Dobbeler 1978: 314), *Protothelenella polytrichi, Gloeopeziza interlamellaris, Hymenoscyphus norvegularis.*
11 Tyrol, Samnaun-Group, eastwards close to the Furglersee above Serfaus, ca. 2450 m, July 1976, J. Poelt and R. Moberg (GZU). (*Bryochiton heliotropicus*, Holotypus GZU, Döbbeler 1978: 210, *Lizonia sexangularis* GZU, Döbbeler 1978: 314), *Protothelenella polytrichi* (Holotypus GZU), *Gloeopeziza interlamellaris* (Holotypus GZU).
12 Tyrol, Stubaier Alps, slopes between Dortmunder Hütte and Brichkogel, 2450-2600 m, 17 September 1977, E. Albertshofer. *Bryochiton heliotropicus* (Dö 5103 in M), *B. perpusillus* (Dö 5108 in M), *Lizonia sexangulare* (Dö 2960 in M), *Protothelenella polytrichi* (Do 5106 in M), *Gloeopeziza interlamellaris* (Dö 5104 in M), *Hymenoscyphus norvegularis* (Holotypus, Dö 2964 in M).
13 Tyrol, Hohe Mut southern Gurgl, August 1878, F. Arnold (M). *Bryochiton perpusillus, Lizonia sexangularis, Hymenoscyphus norvegularis.*
14 Salzburg, Pinzgau, Felber Tauern, 2000-2800 m, 13 September 1932, J. Baumgartner (hb. F.J. Widder in GZU). *Bryochiton perpusillus, Lizonia sexangularis, Gloeopeziza interlamellaris, Hymenoscyphus norvegularis.*
15 Kärnten, Hohe Tauern, Hoher Sadnig, southeastern Döllach in the Moll-Tal, 28 July 1935, F.J. Widder (GZU). *Lizonia sexangularis, Gloeopeziza interlamellaris.*
16 Steiermark, in the Eiskaar by Schladming, 1829-2134 m, 29 July 1869, J. Breidler (GZU). (*Bryochiton heliotropicus*, Döbbeler 1978; 212, *Lizonia sexangularis*, Döbbeler 1978: 314), *Protothelenella polytrichi.*
17 Steiermark, Schladminger Tauern, Klafferkogel, 2300 m, 6 September 1932, K. Redinger (B). *Lizonia sexangularis, Gloeopeziza interlamellaris, Hymenoscyphus norvegularis.*
18 Steiermark, Dürrmoos by St. Nicolai in the Solk, 2000 m, 16 august 1886 J. Breidler (B). *Lizonia sexangularis.*

19 Steiermark, Wölzer Tauern, slopes southwest of the Planneralm between Goldbachscharte and Karlspitze, 2000-2080 m, 18 July 1972, J. Poelt. *Bryochiton perpusillus* (Dö 5038 in GZU), *Gloeopeziza interlamellaris* (Dö 5037 in GZU).

Italy:
20 South Tyrol, Presanella massif, at Lago nero behind Pinzolo, 2230 m, 17 September 1903, J. Baumgartner (W). *Lizonia sexangularis*, *Gloeopeziza interlamellaris*, *Hymenoscyphus norvegularis*.

Czechoslovakia:
21 Bohemia, Montes Krkonoše, in valle rivuli Bílé Labe apud hospitium Luční bouda, ca. 1300 m, September 1948, Zd. Pilous (Musci čechoslov. exsicc., ed. Zd. Pilous, 637. *Polytrichum sexangulare*), (W). *Gloeopeziza interlamellaris*.
22 Bohemia, Riessen Gebirge, on the banks of the Weisswasser, 1440 m, 22 August 1930, R. Vaněk (GZU). *Gloeopeziza interlamellaris*, *Hymenoscyphus norvegularis*.
23 Slovakia, Montes Liptovské hole, in cacumine montis Plačlivé, ca. 2126 m, August 1947, Zd. Pilous (Musci cechoslov. exsicc., ed. Zd. Pilous, 329. *Polytrichum sexangulare*), (W). *Gloeopeziza interlamellaris*.
24 Slovakia, Tatra Magna, prope lacum Spišské pleso, ca. 2000 m, 9 July 1951, J. Šmarda (Cryptog. čechoslov. exsicc., 150. *Polytrichum sexangulare*), (W). *Gloeopeziza interlamellaris*.
25 Slovakia, Tatra Magna, at the Langensee, 1880 m, v. Bosniacki (M). *Lizonia sexangularis*, *Gloeopeziza interlamellaris*.

Scotland:
26 Inverness, Aonach Mor, 13 July 1898, H.N. Dixon (JE). *Gloeopeziza interlamellaris*.
27 Ben Lawers, 23 August 1836, (W). *Lizonia sexangularis*.
28 Ben Mac Dui, 16 August 1836, (Wilson, Musci Brit., 211. *Polytrichum sexangulare*), (W). *Bryochiton heliotropicus*, *Gloeopeziza interlamellaris*.
29 Ben Avon, Braemar, July 1844, (M). *Lizonia sexangularis*.

Norway:
30 Troms. Nordreisa. Gabrus, 20 August 1891, H.W. Arnell (W). *Bryochiton heliotropicus*, *Gloeopeziza interlamellaris*.
31 Sør-Trøndelag. Oppdal. Knutshø, July/August 1865, S. Berggren (W). *Lizonia sexangularis*.
32 Sør-Trøndelag, Oppdal. Nystuhø, 1300 m, August 1907, Winter (JE). *Bryochiton perpusillus*, *Lizonia sexangularis*.
33 Oppland. Dovre: in monte Snoehetta, 9 August 1884, P. Olsson (W). *Bryochiton perpusillus*.
34 Hordaland. Ulvik. Finse, snowbed at Finsevann, 21 July 1916, G. Samuelsson (B). *Lizonia sexangularis*, *Gloeopeziza interlamellaris*.
35 Telemark. Odda, between Røldal and Seljestad, July 1908, Winter (JE). *Gloeopeziza interlamellaris*.

Sweden:
36 Torne Lappmark: Jukkasjärvi par., Torneträsk-area, N.W. of Njutum, ca 625 m, 4 August 1916, G. Samuelsson (W). *Gloeopeziza interlamellaris*.
37 Torne Lappmark: Jukkasjärvi par., southern slopes of Nuolja above Abisko, ca. 1000 m, 22 August 1980, H. Hertel and P. Döbbeler. *Bryochiton heliotropicus* (Dö 4858 in M), *Lizonia sexangularis* (Dö 4715 in M), *Protothelenella polytrichi* (Dö 4863 in M), *Gloeopeziza interlamellaris* (Dö 5031 in M).
38 Jämtland: Handöl par., Snasahögarna, 13 July 1923, S. Arnell (W). *Protothelenella polytrichi*.

39 Härjedalen: Storsjö par., Mt. Helag, August 1913, H. Smith (W). *Gloeopeziza interlamellaris.*

Finland:

40 Enl, Kilpisjärvi, Malla, 4 August 1951, W. v. Neuenstein (M). *Bryochiton perpusillus.*

USSR:

41 Lapponia murmanica, peninsula piscatorum, Bumans-fjord, 12 August 1885, V.F. Brotherus (357. *Polytrichum sexangulare*), (W). *Gloeopeziza interlamellaris.*

42 Peninsula Kolaensis, montes Chibiny, mons Vudjavrczorr, 15 August 1956, R.N. Schljakov (Hep. Musci URSS exsicc., 163. *Polytrichum sexangulare*), (B). *Lizonia sexangularis, Gloeopeziza interlamellaris.*

43 N.W. Altai, Lumultinsky Belok, 14 August 1923, W. Saposhnikow and E. Nikitina (M). (*Bryochiton heliotropicus*, Dobbeler 1978: 212, *Lizonia sexangularis* Döbbeler 1978: 314), *Protothelenella polytrichi, Gloeopeziza interlamellaris.*

44 N.W. Altai, Sea Karagol, 18 August 1923, W. Saposhnikow and E. Nikitina (M). *Lizonia sexangularis, Gloeopeziza interlamellaris.*

USA:

45 Washington, Mt. Rainier, ca. 1768 m, 8 September 1898, J.A. Allen 97, (Moss. Casc. Mount., 78. *Polytrichum sexangulare*), (M). (*Bryochiton perpusillus*, Döbbeler 1978: 233), *Gloeopeziza interlamellaris.*

ACKNOWLEDGEMENTS

I wish to thank the directors and curators of B, GZU, JE, M, and W for the loan of material and for allowing me to separate fungus-infected mosses from their valuable bryological collections. Without their generosity, this study would not have been possible. Thanks also go to Dr. H. Roessler (München) for checking the Latin descriptions, to Dr. H. Mayrhofer (Graz) and Prof. Poelt (Graz) for information on *Protothelenella*. I feel particularly grateful to the editor Dr. G.A. Laursen (Fairbanks) for linguistic correction of this paper.

REFERENCES

Carpenter, S. E., 1981, Monograph of *Crocicreas* (Ascomycetes, Helotiales, Leotiaceae), *Mem. New York Bot. Gard.*, 33: 1-290.

Dennis, R. W. G., 1962, New or interesting British Helotiales, *Kew Bull.*, 16: 317-327.

Döbbeler, P., 1978, Moosbewohnende Ascomyceten I. Die pyrenocarpen, den Gametophyten besiedelnden Arten, *Mitt. Bot. Staatssamml. München*, 14: 1-360.

______, 1979a, Moosbewohnende Ascomyceten III. Einige neue Arten der Gattungen *Nectria*, *Epibryon* und *Punctillum*, *Mitt. Bot. Staatssamml. München*, 15: 193-221.

______, 1979b, Untersuchungen an moosparasitischen Pezizales aus der Verwandtschaft von *Octospora*, *Nova Hedw.*, 31: 817-864.

______, 1981, Moosbewohnende Ascomyceten V. Die auf *Dawsonia* vorkommenden Arten der Botanischen Staatssammlung München, *Mitt. Bot. Staatssamml. München*, 17: 393-473.

Döbbeler, P., and Itzerott, H., 1981, Zur Biologie von *Octospora libussae* und *O. humosa*, zwei im Moosprotonema wachsenden Pezizales, *Nova Hedw.*, 34: 127-136.

Döbbeler, P., and Poelt., J., 1974, Beiträge zur Kenntnis moosbewohnender Discomyceten I. Die Gattung *Bryodiscus*, *Svensk Bot. Tidsk.*, 68: 369-376.

Eriksson, O., 1981, The families of bitunicate ascomycetes, *Opera Botanica*, 60: 1-220.

Frey, W., 1977, Neue Vorstellungen über die Verwandt-schaftsgruppen und die Stammesgeschichte der Laubmoose, *in:* "Beiträge zur Biologie der niederen Pflanzen," pp. 117-139, W. Frey, H. Hurka and F. Oberwinkler (Herausg.), Stuttgart, New York, G. Fischer Verlag.

Geissler, P., 1982, Alpine communities, *in:* "Bryophyte ecology," pp. 167-189, A. J. E. Smith ed.

Gjaerevoll, O., 1950, The snow-bed vegetation in the surroundings of Lake Torneträsk, Swedish Lappland, *Svensk Bot. Tidsk.*, 44: 387-440.

Goebel, K., 1930, "Organographie der Pflanzen, insbesondere der Archegoniaten und Samenpflanzen," 2. Teil, 3. Aufl. Jena, Verlag von G. Fischer.

Hafellner, J., 1983, Studien über lichenicole Pilze und Flechten II. *Lichenostigma maureri* gen. et spec. nov., ein in den Ostalpen haufiger lichenicoler Pilz (Ascomycetes, Arthoniales), *Herozgia*, 6: 299-308.

Hawksworth, D. L., Sutton, B. C., and Ainsworth, G. C., 1983, "Ainsworth and Bisby's Dictionary of the fungi," Commonwealth Mycol. Institute, Kew, Surrey.

Herzog, T., 1926, "Geographie der Moose," Jena, Verlag G. Fischer.

Ingold, C. T., 1966, Aspects of spore liberation: violent discharge, *in:* "The fungus spore," pp. 113-132, M. F. Madelin ed., Butterworths, London.

______, 1971, "Fungal spores. Their liberation and dispersal," Clarendon Press, Oxford.

Kimbrough, J. W., 1970, Current trends in the classification of Discomycetes, *Bot. Rev. (Lancaster)*, 36: 91-161.

Korf, R. P., 1973, Discomycetes and Tuberales, *in:* "The Fungi. An Advanced Treatise," pp. 249-319, G. C. Ainsworth, F. K. Sparrow and A. S. Sussman eds., Acad. Press, New York and London.

Mårtensson, O., 1956, Bryophytes of the Tornetrask area, Northern Swedish Lappland. II, *Musci. K. Svenska Venenskapsakad. Avh.*, 14: 1-321.

Miller, N. G., 1984, Tertiary and Quaternary fossils, *in:* "New Manual of Bryology," Vol. 2: 1194-1232, R.M. Schuster ed.

Nyholm, E., 1969, "Illustrated moss flora of Fennoscandia. II," Musci, Fasc. 6, Stockholm.

Racovitza, A., 1940, *Niptera polytrichina* n. sp. (champignon ascomycète) saprophyte sur *Polytrichum formosum* Hedw., *Bull. Sect. Sci. Acad. Roumaine*, 23(1): 24-27.

______, 1946, Notes mycologiques, *Bull. Sect. Sci. Acad. Roumaine*, 29(1): 50-77.

______, 1959, Étude systématique et biologique des champignons bryophiles, *Mém. Mus. Natl. Hist. Nat., sér. B, Bot.*, 10 (fasc. 1): 1-288; pl. 1-84.

Redhead, S. A., 1984, *Arrhenia* and *Rimbachia*, expanded generic concepts, and a reevaluation of *Leptoglossum* with emphasis on muscicolous North American taxa, *Can. J. Bot.*, 62: 865-892.

Savile, D. B. O., 1972, Arctic adaptations in plants. *Agric. Can. Monograph*, 6: 1-81.

Scheirer, D. C., and Dolan, H. A., 1983, Bryophyte leaf epiflora: an SEM and TEM study of *Polytrichum commune* Hedw., *Amer. J. Bot.*, 70: 712-718.

Smith, A. J. E., 1978, "The Moss Flora of Britain and Ireland," Cambridge etc.: Cambridge Univ. Press.

Smith, G. L., 1971, Conspectus of the genera of Polytrichaceae, *Mem. New York Bot. Gard.*, 21(3): 1-83.

Walther, K., 1983, Bryophtina, Laubmoose, *in:* "A. Engler's Syllabus der Pflanzenfamilien," Kapitel V, 2, 13. Aufl., Herausg. J. Gerloff and J. Poelt, Berlin, Stuttgart: Gebr. Borntraeger.
Zukal, H., 1891, Halbflechten, *Flora,* 74: 92-107.

NORDIC JUNCICOLOUS MYCOSPHAERELLAE

Lennart Holm and Kerstin Holm

Institute of Systematic Botany
University of Uppsala, P.O. Box 541
S-751 21 Uppsala, Sweden

Key words: *Mycosphaerella*, *Juncus*, Nordic, *Monascostroma*

ABSTRACT

Mycosphaerellae are very important constituents of the rich mycoflora on *Juncus*. The most common species is *M. perexigua* (Karst.) Johans. Originally described from Spitzbergen, on *J. biglumis*, it has turned out to be widely distributed also in lowland areas and on many species of *Juncus*. It is generally recognized by its narrow spores with two distinct oil globules in each cell. Next in importance is *M. juncellina* Munk, so far only known from the type collection (Denmark, *J. squarrosus*). In macroscopic appearance, the latter is very much like *M. perexigua* but it is easily recognized by the large spores. It is found to be fairly common all over Scandinavia on several *Juncus* species. However, it is not well positioned in *Mycosphaerella*, as it possesses interascal threads. It seems to be more closely related to *Monascostroma bacillifera* (Karst.) O. Erikss. The name *Monascostroma* is untenable, as it is based on an anamorph. Other *Mycosphaerellae* are sometimes encountered, like *M.* cfr. *recutita* (Fr.) Johans. and the plurivorous *M. tassiana* (De Not.) Johans.

INTRODUCTION

The rushes, *Juncus* spp., harbour a very rich micromycete flora, which is still far from being fully explored. This is evident by the many interesting finds of new or little known species recently published by Nannfeldt (i.e., 1976, 1984). Frequently dominating in this microflora is *Mycosphaerella perexigua*. Dead and dying culms of *Juncus* are often densely dotted by its minute ascocarps. Ironically, this condition has been hardly noticed, and, on the whole, the juncicolous *Mycosphaerellae* have been seldom reported. It is significant that in Oudeman's Enumeratio, no *Mycosphaerella* is recorded on *J. conglomeratus*. Only one, the polyphagous *M. tassiana*, is reported from *J. effusus*. Nevertheless, those two rushes are regularly infested by *M. perexigua*. It is to be noted that both rushes are very common throughout large areas of Europe, and that their tall, persistent culms are easily collected in quantity. *Mirabile dictu* the widespread *M. perexigua* was first described on material from Spitzbergen. It is noteworthy that our knowledge of the microfungi on *Juncus*, to a large extent, is based on the studies of Arctic and alpine collections.

The juncicolous *Mycosphaerellae* are thus a largely neglected group, a condition which we find quite comprehensible in light of our experience. They are truly exasperating as they are seldom found well developed and as a rule, the ascocarps are either immature or empty. When dealing with unripe material, as is the usual case, one must remember that the spores will change considerably when maturing, not only in size but also in form, turning more obtuse at the ends and becoming relatively thicker. Moreover, there is undoubtedly an appreciable variation not due to age. All this is distressing, as *Mycosphaerellae* are mainly diagnosed by spore characters.

Owing to these difficulties, we have not been able to definitely classify all forms encountered. This can hardly be done without extensive cultural work, which is outside the scope of this work. However, we have arrived at certain ideas about variation patterns and as they are based comparatively on a very large amount of material, we think it appropriate to present our results here, emphasizing their partly preliminary nature.

The material at hand (about 100 collections on 15 species of *Juncus*) represents at least four different *Mycosphaerellae*. One is the plurivorous *M. tassiana*. This species plays a subordinate role, has been met with rather infrequently in a few collections only, and is mainly from the Arctic. *M. tassiana* deviates macroscopically from the other *Mycosphaerellae* by its larger and somewhat conical ascocarps. It is not dealt with further in this reporting.

Another characteristic species is *M. juncellina*, easily recognized microscopically by its large, ± banana-shaped spores. Until now, it was only known from the type collection, but is now turning out to be fairly common.

By far the most common is a form group which we treat as *M. perexigua*. It is characterized by the comparatively narrow spores that usually have two distinct guttules per cell. Finally, we have come across a very small-spored species in a few collections, which we tentatively identify as *M. recutita*.

For the present study we have scrutinized more than 300 collections, mainly Scandinavian, representing 15 species of *Juncus*. Less than a third of them yielded an identifiable *Mycosphaerella*. If not stated otherwise, the material cited is preserved in UPS.

Mycosphaerella perexigua (Karst.) Johans.

Öfvers. K. Vet.-Akad. Förhandl. 1884(9): 166. (1884) - *Sphaerella perexigua* Karst., Öfvers. K. Vet.-Akad. Förhandl. 1872(2): 107 (1872). Types: Spitzbergen, "In foliis Junci biglumis siccis ad Adventbay et Nordfjorden m. Aug. 1868." (UPS).
= *Mycosphaerella vogesiaca* (Syd.) Kirschst., Ann. Myc. 34: 202 (1936) *Sphaerella vogesiaca* Syd., Ann. Myc. 9: 556 (1911). - Type: France, Alsace, Hohneck, *Juncus conglomeratus*, 13.VII.1910, H. Sydow (= Syd., Myc. germ. 979) (S! UPS!).

Exs.: Petr., Fl. Bohem. Mor.. II(1): 1251 ("*M. vogesiaca*") - Syd., Myc. germ. 979 ("*Sph. vogesiaca*").

Figures 1-26.

Ascocarps subepidermal, densely scattered, often ± covering wide areas of the host, subglobose, minute, usually 30-60 µm diam., mostly seated on a profuse mycelium of contorted brownish hyphae, 3-4 µm thick. *Asci* 2-10(-15), saccate, 25-45 x 12-24 µm, 8-spored. *Spores* fusiform,

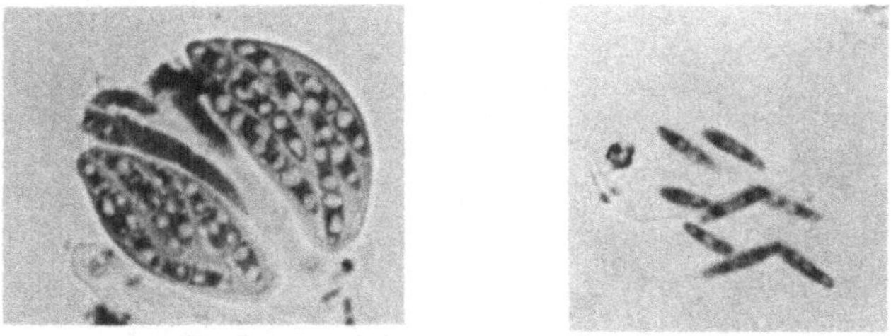

Figs. 1-2. *Mycosphaerella perexigua*, x 500. Fig. 1. on *Juncus arcticus*, Holm 2658 - asci with almost mature spores. Fig. 2. same coll. - immature spores.

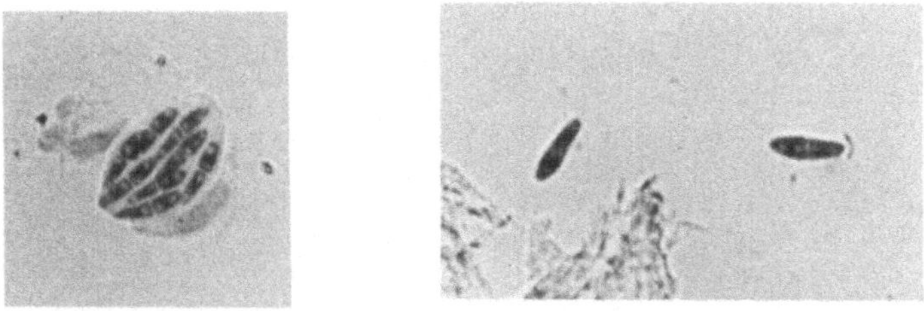

Figs. 3-4. *Mycosphaerella perexigua*, x 500. Fig. 3. on *Juncus arcticus*, Nannfeldt 1450 - ascus and semimature spores. Fig. 4. same coll. - more mature spores, unusually short and plump.

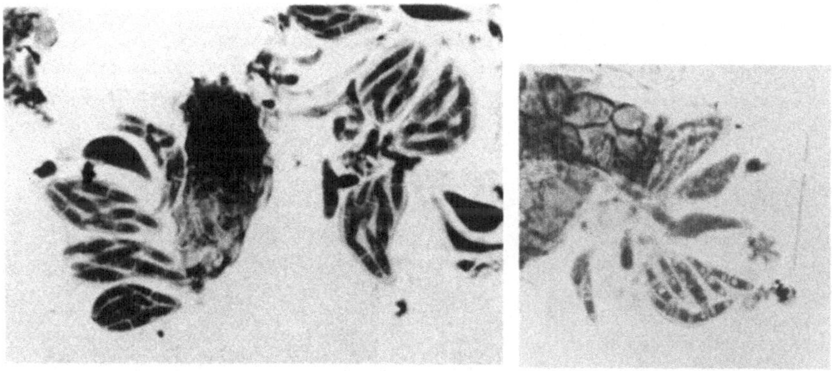

Figs. 5-6. *Mycosphaerella perexigua*, x 500. Fig. 5. on *Juncus balticus*, Nannfeldt 1508 - asci and immature spores with inconspicuous oildrops. Fig. 6. on *J. balticus*, Holm 2716 - ascus and semimature spores with distinct guttules.

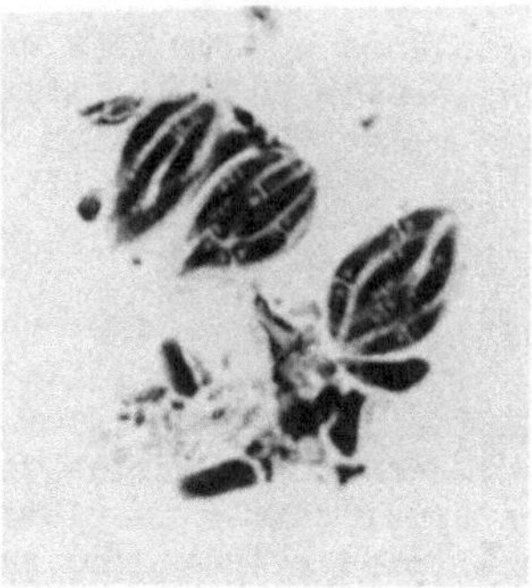 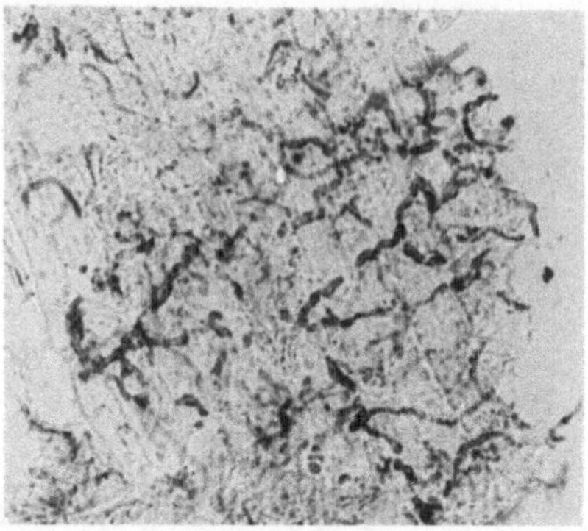

Figs. 7-8. *Mycosphaerella perexigua*, x 500. Fig. 7. on *Juncus biglumis*, syntype - immature, pointed spores. Fig. 8. on *J. biglumis*, Norway, Th. Fries - mycelium.

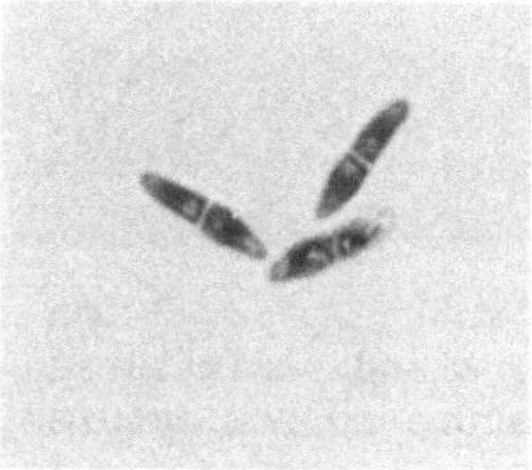 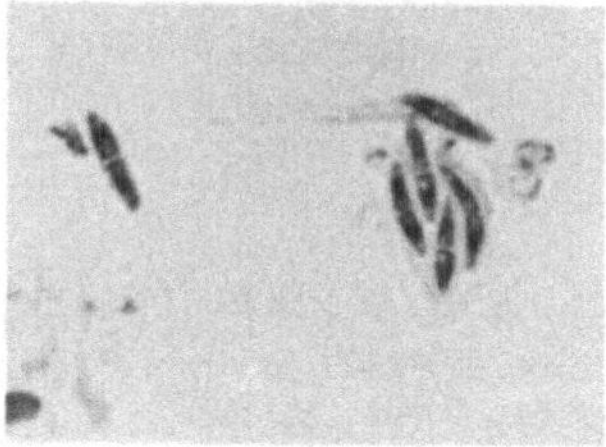

Figs. 9-10. *Mycosphaerella perexigua*, x 500. Fig. 9. same collection as in fig. 8 - mature spores. Fig. 10. same coll. - immature spores.

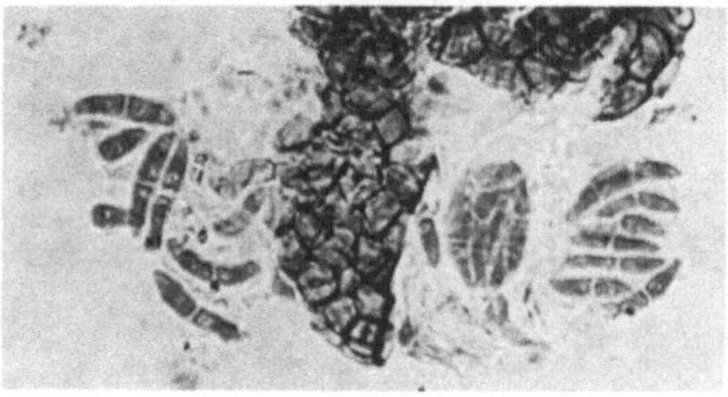 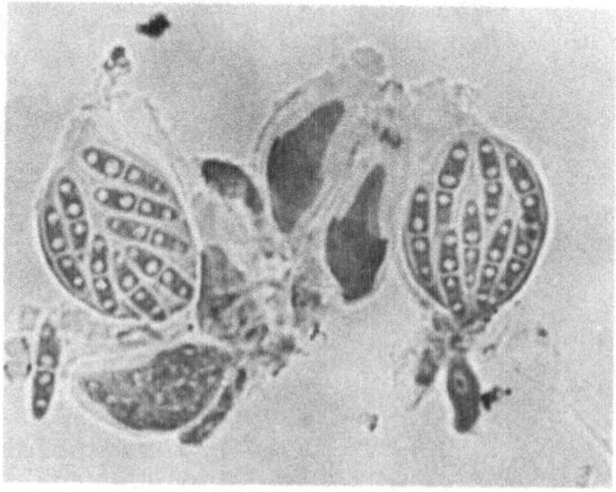

Figs. 11-12. *Mycosphaerella perexigua*, x 500. Fig. 11. on *Juncus biglumis*, Nov. Zemlya, Th. Holm - rather mature spores. Fig. 12. on *J. bufonius*, Fagerstrom - asci with mature spores.

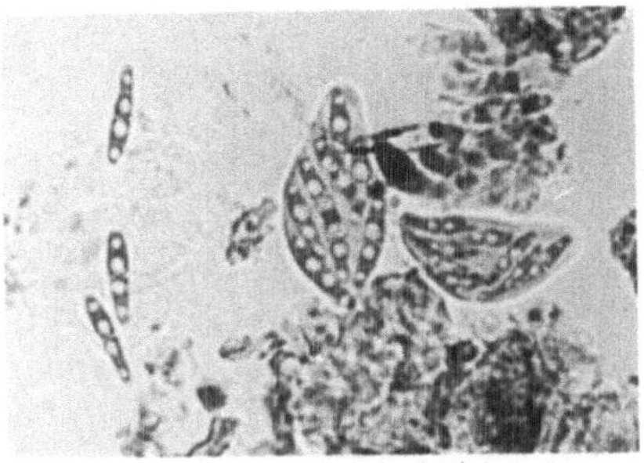

Figs. 13-14. *Mycosphaerella perexigua*, x 500. Fig. 13. on *Juncus conglomeratus*, Holm 723a - almost mature spores. Fig. 14. on *J. conglomeratus*, Holm 934 - immature spores.

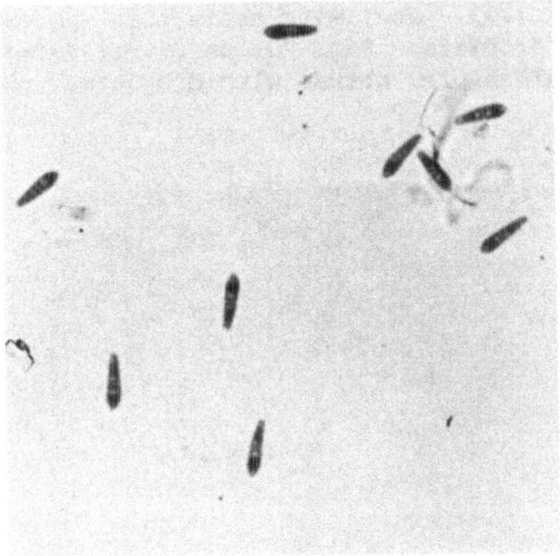

Figs. 15-16. *Mycosphaerella perexigua*, x 500. Fig. 15. on *Juncus conglomeratus*, Syd., Myc. germ. 979 - clavate spores, immature?. Fig. 16. same coll. - spore of "ordinary type."

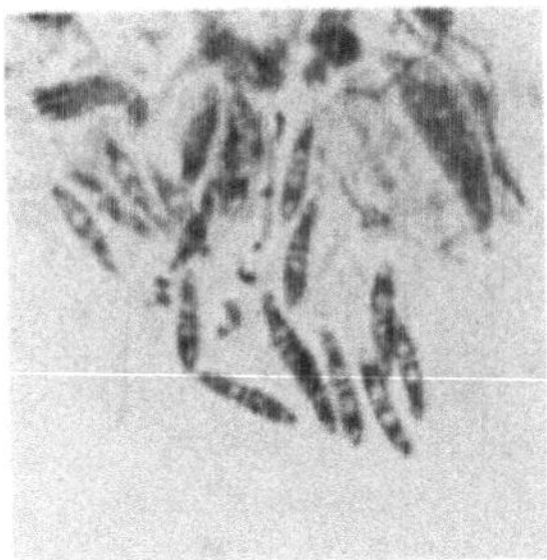

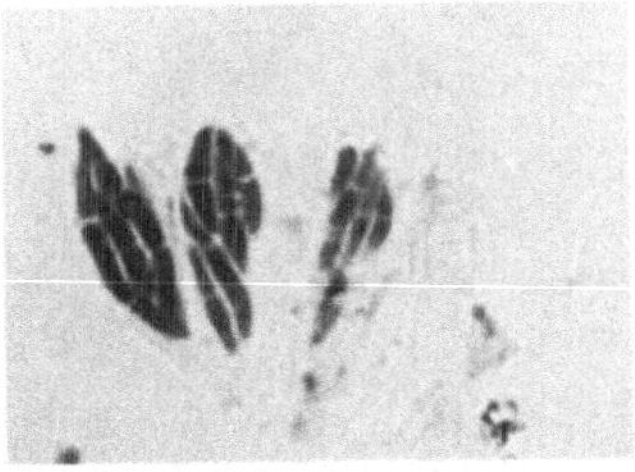

Figs. 17-18. *Mycosphaerella perexigua*, x 500. Fig. 17. on *Juncus effusus*, Holm 1499e - probably rather mature spores. Fig. 18. on *J. filiformis*, Alanko - immature spores.

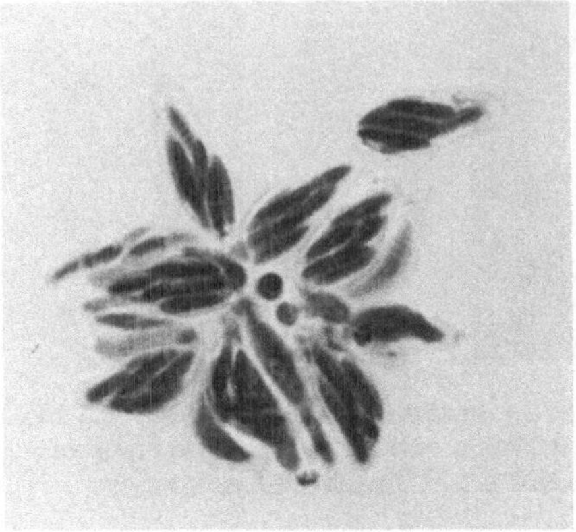

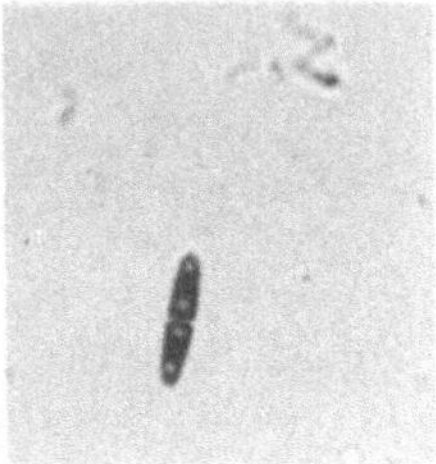

Figs. 19-20. *Mycosphaerella perexigua*, x 500. Fig. 19. on *Juncus filiformis*, Holm 1733 - asci with semimature spores, rather obtuse, without droplets. Fig. 20. on *J. filiformis*, Holm 2693 - rather mature spore, obtuse with droplets.

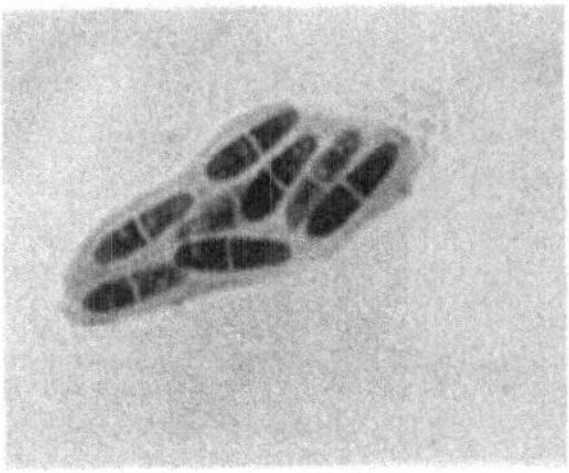

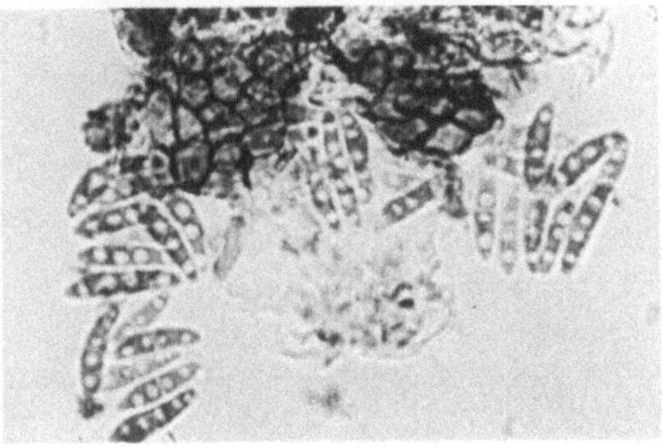

Figs. 21-22. *Mycosphaerella perexigua*, x 500. Fig. 21. on *Juncus filiformis*, Heikinheimo - ascus with unusually short, obtuse spores. Fig. 22. on *J. trifidus*, Holm 2688 - mature spores.

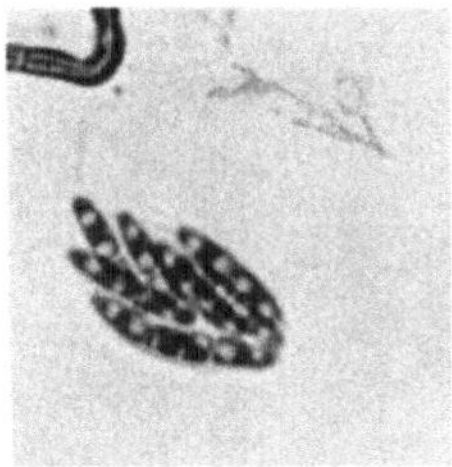

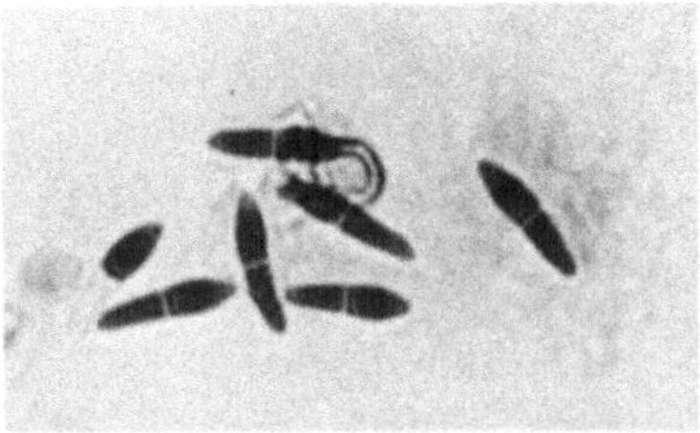

Figs. 23-24. *Mycosphaerella perexigua*, x 500. Fig. 23. same collection as in fig. 22 - smaller, less mature? spores. Fig. 24. on *Juncus trifidus*, Sweden, Mt. Nuolja, Alm - mature? spores, with rather pointed ends, without droplets.

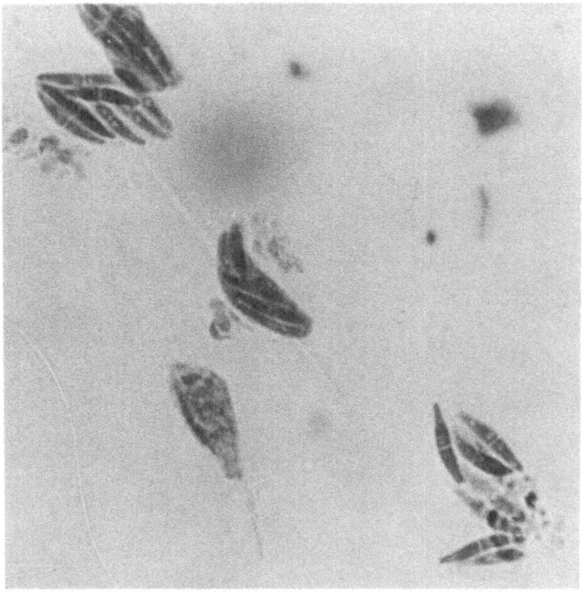

Fig. 25. *Mycosphaerella perexigua*, x 500. Fig. 25. on *Juncus triglumis*, Norway, Dovre, Th. Fries - large, but rather immature spores with pointed ends.

subcylindric or sometimes slightly clavate, (10-)14-20(-22) x 3.5-4.5(-5) µm, with a median or slightly supramedian septum, usually with two oil droplets in each cell, hyaline (or at last with faintly brownish, verruculous walls).

On various Cyperaceae and Juncaceae.

Anamorph: unknown if any. According to Lind (1928) *M. perexigua* "is often found associated with *Septoria punctoidea*, which it greatly resembles."

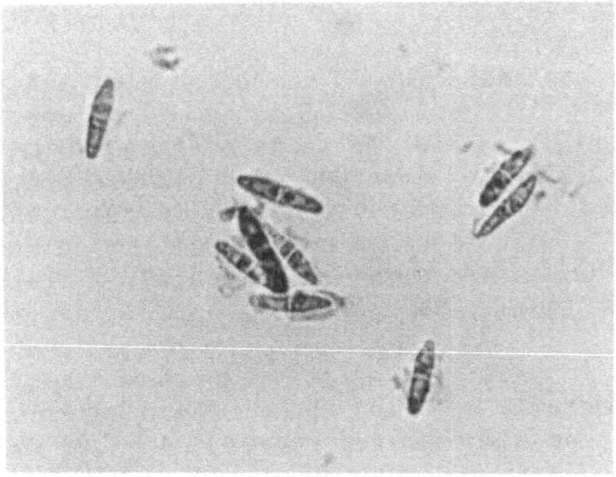

Fig. 26. *Mycosphaerella perixigua* on *J. triglumis*, Alm & Smith - rather short and obtuse spores, probably mature, x 500.

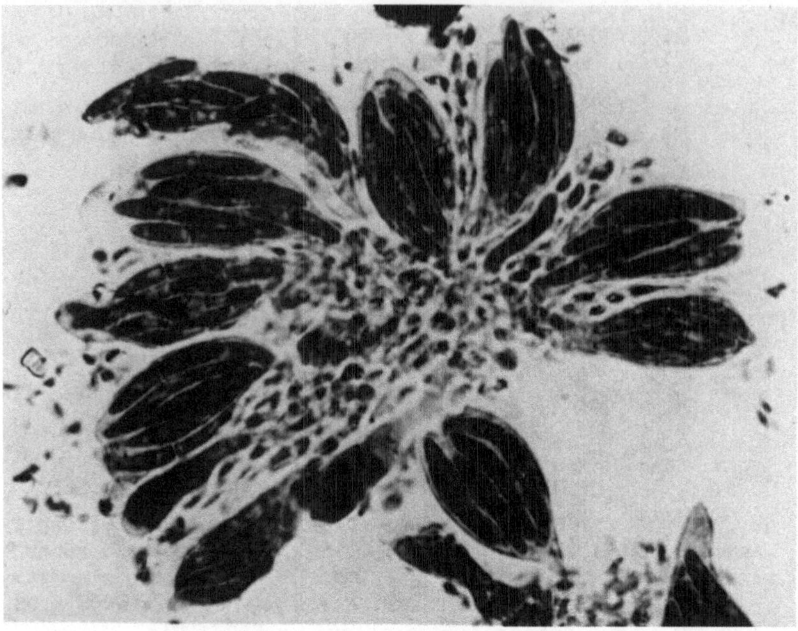

Fig. 27. *Mycosphaerella juncellina* on *Juncus arcticus*, Holm 694a - rather mature asci and spores, x 500. Note the interascal threads.

M. perexigua, as here conceived, is a variable and perhaps heterogenous taxon comprising rather unlike forms that are connected "par enchaînement." It is difficult to find a single diagnostic criterion common to all of them. Barr (1959, 1972) has rightly drawn attention to some significant traits: the conspicuous mycelium and the terminally pointed spores with distinct oil globules.

In regard to spore characters, one must be aware of the considerable changes occurring during spore development. At a very early stage the spores always have pointed ends. Semi-mature spores, which are mostly observed, frequently are still ± pointed (cfr. e.g. figs. 3, 16, 24). However, immature but nevertheless rather blunt spores are sometimes encountered (figs. 4, 15, 20).

The spores are generally fusoid to oblong, but slightly clavate spores do occur; e.g., in the type collection of *M. vogesiaca* having fusiform spores (figs. 15, 16). It is interesting to note that Kirschstein (1938 p. 355) has observed such clavate spores in "*Sphaerella vogesiaca*." Moreover, in Munk's description of *M. perexigua* (Munk 1957) that is based on one collection from *J. conglomeratus*, the spores are given as "clavate", a statement hardly supported by Munk's figure 120b.

The spore size and also the relative spore width is variable (cfr. figs. 21 & 25). Generally speaking the southern material has, on average more obtuse and perhaps also shorter spores. Another variable, but an important character, is the occurrence of oil droplets. As a rule they are very striking, particularly in mature spores (cfr. figs. 1, 12, 22). They are generally distinct even in younger spores, but remarkable exceptions have been found. Exceptions were found in several collections on *J. trifidus* (cfr. fig. 24) and the only one on *J. castaneus*.

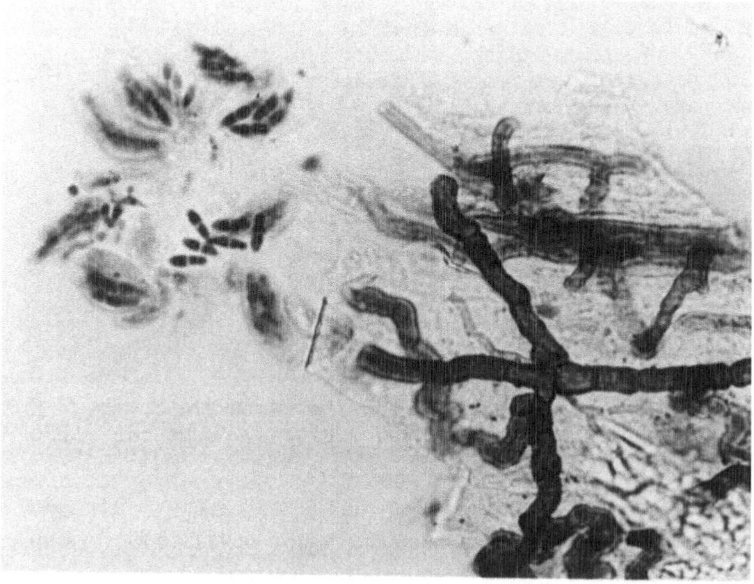

Fig. 28. *Mycosphaerella ?recutita*, x 500. Fig. 28. on *Juncus conglomeratus*, Holm 2745c - mycelium, asci and spores.

M. perexigua was described from Spitzenbergen and the remaining few finds so far reported are almost all from northern regions (Barr, op. cit.; Lind, 1934). Apart from the collections of "M. vogesiaca", the only exception known to us, without question, is the record of a Danish find on *J. conglomeratus* by Munk (1957 p. 314). In our opinion, *M. perexigua* is not a particularly northern species, nor is it restricted to *Juncus*, even if rushes probably are principal hosts. The species also occur on *Luzula* and various *Cyperaceae* (cfr. Barr 1972). It seems to be generally saprophytic, but we have seen two collections on *Luzula pilosa* where the fungus apparently is a parasite, occurring as grayish spots on otherwise green leaves.

Material examined: on:

Juncus acutus: ENGLAND, Scilly Islands, Bryher, 12.IX.1978, Holm 1503.

J. alpinus: FINLAND: Kar. austr., Ky, 13.VII.1934, Ulvinen (H).

J. arcticus: ICELAND: Árnessýsla, Thingvellir, 20.VII.1971, Holm 20e-71. NORWAY, Finnmark, Porsanger, Oldereidaneset, 17.VIII.1982, Holm 2658a. SWEDEN, Torne Lappmark, Abisko, 25.VII.1928, Nannfeldt 1450c.

J. articulatus: SWEDEN, Jämtland, Handöl, 2.VIII.1951, Nannfeldt 11656i.

J. balticus: NORWAY: Nordland, Bjerkvik, Saegnes, 28.VII.1928, Nannfeldt 1508 Troms, Kvalöya, 22.VIII.1982, Holm 2716. SWEDEN: Norrbotten: Nederkalix, Bodon, 14.VI.1981 & 20.VI.1981, G. Sandberg. Pitsund, 6.VIII.1909, E. Marklund. Skåne, Vittskövle, VII.1867, Th. Fredriksson.

J. biglumis: SPITZBERGEN: Nordfjorden, 11.VIII.1868, Th. Fries (type) Kap Thordsen, 18.VII.1890, Björling. NOVAYA ZEMLYA: Lille Olonje, 23.VIII.1882, Th. Holm. NORWAY: Dovre, Knutsho, 22.VIII.1863, Th. Fries. SWEDEN: Torne Lappmark, Mt. Nissontjåkko, 30.VII.1920, G. Björkman.

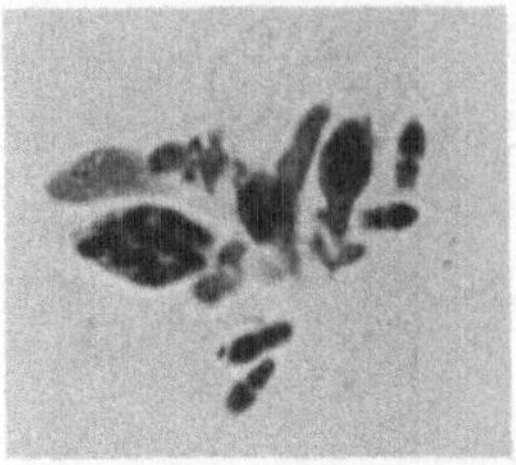

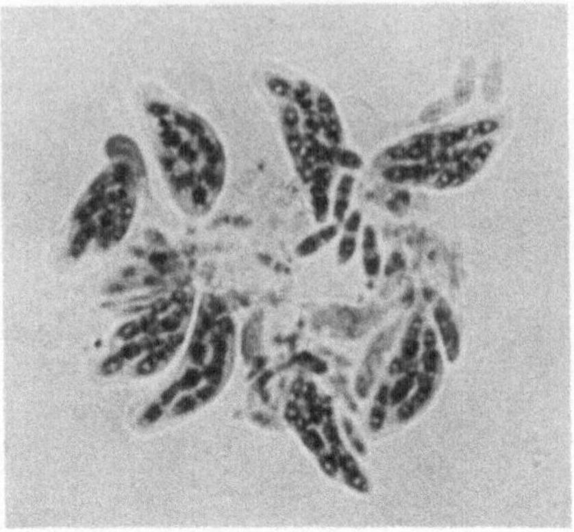

Figs. 29-30. *Mycosphaerella ?recutita*, Holm 2745c. Fig. 29. Asci and eguttlate spores, x 500. Fig. 30. same collection - asci with guttulate spores.

J. bufonius: FINLAND: Kar. austr., Aspo, 5.VIII.1950, Fagerström (H).
J. castaneus: NORWAY, Sör-Tröndelag, Oppdal, 22.VIII.1973, Holm 84b.
J. conglomeratus: SWEDEN (13 collections), FRANCE (1).
J. effusus: SWEDEN (5), FINLAND (2), ENGLAND (2), GERMANY (1), CZECHOSLOVAKIA (1)
J. filiformis: NORWAY (1), SWEDEN (6), FINLAND (1).
J. trifidus: NORWAY (4), SWEDEN (5), FINLAND (1).
J. triglumis: GREENLAND: Nigerdleq, 18.VIII.1966, B. Jørgensen & S. Larsen NORWAY: Dovre, Vaarstien, 5.VIII.1863, Th. Fries. SWEDEN: Torne Lappmark, Bjorkliden, 20.VII.1939, Alm & Smith. Jämtland, Ås, 29.VI.1925, Asplund.

Mycosphaerella juncellina Munk

Dansk Bot. Arkiv 17(1): 315 (1957). - Type: Denmark, Laesö, *Juncus squarrosus*, 24.VII.1902, J. Lind (C).

Fig. 27.

Ascocarps often densely scattered, ± covering large parts of the host infested culms or leaves, subepidermal, subglobose 50-80 µm diam., *Interascal* short-celled threads present. *Asci* saccate, with an apically strongly thickened wall, 40-70 x 15-25 µm, 8-spored. *Spores* ± parallel, subcylindric to bananashaped, (20-)24-30 x (4-)5-6 µm, with a median or slightly submedian septum, hyaline with several large oil drops without distinct form; upper spore cell somewhat thicker and slightly pointed.

Saprophytic in *Juncus* spp.

The aspect of this species is very similar to *M. perexigua*. They can hardly be distinguished macroscopically, but the spores will permit safe determination. *M. juncellina* is obviously not at all rare. It is amazing that this characteristic species has been almost completely overlooked. For that reason, we list all collections as to where it has been found.

The true systematic position of the species is doubtful. Certainly it is not well accommodated in *Mycosphaerella*, as is indicated by the interascal threads. The spores are atypical for the genus too. As pointed out to us by Dr. Ove Eriksson, the species is reminiscent of *M. bacillifera* (Karst.) Lind, found only on *Scheuchzeria palustris*. The kinship seems beyond doubt. The latter species was recently transferred to *Monascostroma* by Eriksson (1982), mainly on account of the interascal threads. This assignment has much to recommend it, but must be disputed on formal grounds anyway, as the genus *Monascostroma* seems untenable:

Monascostroma was erected by von Höhnel (1918) for the sole species *Hendersonia innumerosa* Desm., which von Höhnel identified with an ascomycete known as *Mycosphaerella pheidasca* Schröt. Desmazières' original description does not support such an interpretation, but refers clearly to an anamorph. Certainly, *M. pheidasca* is present (intermixed) in the type material (PC), but there is also evidence of a pycnidial fungus matching the description. This anamorph is well known on *Juncus* and is treated by Groves (1935) as *Stagonospora innumerosa*. The name *Monascostroma*, consequently, formally applies to a coelomycete and is preferably dropped.

A more material problem is what taxonomic importance should be assigned to the interascal threads. If these are considered remnants of stroma "tissue" (and not pseudo-paraphyses) it will make little difference as to whether or not such threads are present. It would rather be a question of the degree of resorption of this "tissue". If so, it seems reasonable to allot a provisional place for *M. juncellina* in the *Didymellina* group of *Mycosphaerella*, as was done for *M. bacillifera* by Muller & Poelt (1963).

Material examined: on:

Juncus arcticus: NORWAY, Hedmark, Folldal, 1.VII.1975, Holm 694a.
J. articulatus: SWEDEN, Uppland, Börstil, 29.IV.1980, Holm 2036d.
J. balticus: NORWAY, Troms, Kvalöya, 22.VIII.1982, Hom 2716d. SWEDEN, Norrbotten, Nederkalix, Bodön, 14.VI.1981 & 20.VI.1981. G. Sandberg. ESTONIA, Osel, 22.VI.1899, Skottsberg & Vestergren.
J. compressus: SWEDEN, Uppland, Börstil, 29.IV.1980, Holm 2057a.
J. conglomeratus: FINLAND, Nyland: Tuusula, 26.V.1968, Alanko 6623 (H). Espoo, 5.V.1935, Heikinheimo (H). Kerava, 24.VI.1929, Heikinheimo (H). Helsinki, 16.IX.1945, Kujala (H). SWEDEN, Västergötland, Ledsjö, 19.VIII.1981, Holm 2453. Uppland, Uppsala-Näs, 10.VII.1976, Holm 893a. Vastmanland, Tärna, 23.VII.1978, Holm 1462a. Dalarna, Grytnäs, 23.VII.1978, Holm 1461b.
J. effusus: FINLAND, Kar. austr., Vehkalahti, 27.V.1978, Fagerström (H). Nyland, Lojo, 23.IV.1983, Holm, 2853a. SWEDEN, Västergötland, Ledsjö, 19.VIII.1981, Holm 2441. Södermanland, Södertälje, 3.V.1949, Ph. Johansson. Uppland, Alsike; 21.I..1975, Holm 430c. 11.VI.1976, Holm 858c. 7.VII.1980, Holm 2209a. Funbo, 17.IV.1980, Holm 2024a, Hagby, 12.V.1983, Holm 2875.
J. filiformis: SWEDEN, Dalarna, St. Kopparberg, 3.VI.1979, Holm 1733a.
J. gerardii: NORWAY, Finnmark, Kongsfjorden, 10.IX.1864, Th. Fries. Troms, Tönsvik, 21.VIII.1982, Holm 2692e. SWEDEN. Södermanland, Dalarö, 27.VII.1975, Holm 712a.
J. squarrosus: DENMARK, Bornholm, 22.VI.1971, B. Jonsell 2521. Laesö, 24.VII.1902, Lind (C, type).
J. trifidus: SWEDEN, Torne Lmk, Abisko, 9.VIII.1917, Asplund.

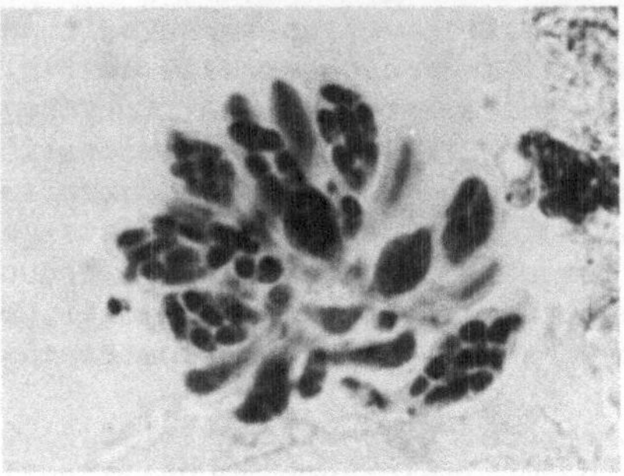

Fig. 31. *Mycosphaerella ?recutita* on *Juncus effusus*, Fagerstrom - asci and eguttulate spores, x 500.

Mycosphaerella cfr. *recutita* (Fr.) Johans.

Figs. 28-31.

A few times we have come across a *Mycosphaerella* which is tentatively identified as *M. recutita*. In gross appearance it is similar to *M. perexigua*, but deviates by having smaller and plumper spores, 8-10 x 3-3.5 µm with or without guttules; asci 24-30 x 8-12 µm.

According to M. Barr (1972), *M. recutita* is common on various monocotyledons and can thus be expected to occur on *Juncus*. We would not be surprised if *M. minor*, on dicots, should turn out to be identical.

Material examined: on:

Juncus alpinus: FINLAND: Kar, austr.: Sippola, 11.VIII.1969, Fagerström (H).

J. conglomeratus: SWEDEN, Uppland, Uppsala-Näs, 2.X.1982, Holm 2745c.

J. effusus: FINLAND, Kar. austr., Miehikkälä, 21.V.1977, Fagerström (H).

REFERENCES

Barr, M., 1959, Northern Pyrenomycetes I. Canadian Eastern Arctic, *Contrib. Inst. Bot. Univ. Montréal*, 73.

_____, 1972, Preliminary studies on the Dothideales in temperate North America, *Contrib. Univ. Mich. Herb.*, 9(8).

Eriksson, O., 1982, Notes on ascomycetes and coelomycetes from NW Europe, *Mycotaxon*, 15: 189-202.

Grove, W. B., 1935, "British stem and leaf-fungi, I," Cambridge Univ. Press, 489 pp.

Höhnel, F., von, 1918, Uber *Hendersonia (Piestospora) innumerosa* Desmazières, *(Mycol. Fragm. CCLXXVI). Ann. Myc.*, 16: 159-160.

Kirschstein, W., 1938, Sphaerellaceae, *Krypt.-Fl. Mark Brandenburg*, 7: 305-448.

Lind, J., 1928, The Micromycetes of Svalbard, *Skrifter om Svalbard og Ishavet* 13.

_____, 1934, Studies on the geographical distribution of arctic circumpolar Micromycetes, *Kgl. Danske Vidensk. Selskab Biol. Medd.*, 11(2).

Munk, A, 1957, Danish Pyrenomycetes, *Dansk Bot. Arkiv.*, 17.

Müller, E,. and Poelt, J., 1963, *Sphaerella bacillifera* Karst. Ein nordischer Ascomycet in Mitteleuropa, *Mitt. Bot. Staatssamml. München*, 5: 135-138.

Nannfeldt, J. A., 1976, *Micropeziza* Fck. and *Scutomollisia* Nannf. nov. gen. (Discomycetes Inoperculati), *Bot. Not.*, 129: 323-340.

_____, 1984, Notes on *Diplonaevia* (Discomycetes Inoperculati) with special regard to the species on *Juncus*, *Nord. J. Bot.*, 4 (in press).

NEW SVALBARD FUNGI

Seppo Huhtinen

Department of Biology, University of Turku
SF-20500 Turku, Finland

Key words: Svalbard, Ascomycetes, Basidiomycetes

ABSTRACT

A list of 25 higher fungi is presented from Svalbard (Norway), mostly with detailed descriptions and illustrations. A new genus, *Polaroscyphus*, is described for a hyaloscyphaceous taxon with rounded ascus apices and disarticulating spores. An undescribed, large spored taxon of *Conocybe*, belonging to the *C. pubescens* group, is briefly treated. Two collections of *Marasmius kallioneus* are reported. Brown spores are shown to be a character present in Hyaloscyphaceae. This specimen is conspecific with *Psilocistella obsoleta* (Vel.) Svrček, the type of which is shown to have juvenile apothecia only. Most of the material is reported for the first time from Svalbard and many are new to the Arctic. Other species in the list are: *Helvella pocillum* Harmaja, *Leucoscypha hetieri* (Boud.) Rifai, *Octospora humosa* (Fr.: Fr.) Dennis, *O. melina* (Vel.) Dennis & Itzerott, *O. moravecii* Khare, *Scutellinia minor* (Vel.) Svrček, *Ciboria polygoni-vivipari* Eckbl., *Crocicreas culmicola* (Desm.) Carpenter, *C. cyathoideum* (Bull.) Carpenter var. *cacaliae* (Pers.) Carpenter, *Cudoniella clavus* (Alb. & Schw.: Fr.) Dennis, *Psilachnum acutum* (Vel.) Svrček, *P. inquilinum* (Karst.) Dennis, *Trichopezizella nidulus* (Schmidt & Kuntze: Fr.) Raitv. var. *hystricula* Karst., *Bolbitius* cf. *variicolor* Atk., *Conocybe magnicapitata* Orton, *Flagelloscypha kavinae* (Pilát) W.B. Cooke, *Inocybe praetervisa* Quél., *I. leucoblema* Kühner, *Melanoleuca cognata* (Fr.) Konr. & Maubl., *Rickenella fibula* (Bull.: Fr.) Raith., and *Typhula culmigena* (Mont. & Fr.) Berthier.

INTRODUCTION

For a long time, Svalbard remained mycolgoically terra incognita. Papers by Kobayasi et al. (1968) and Ohenoja (1971) marked a change. However, this "new era" resulted in no larger a treatment then from studies based on the author's own personal collections. Newer contributions are based on fragmentary material only. As Svalbard is one of the northernmost land masses in the Arctic, mycologically quite virgin and easily accessible by modern jet services, these islands should offer an irresistable temptation. In 1983 I visited the area together with Prof. Denise Lamoure and Dr. G.A. Laursen. During the ten days between late July and early August we collected material in the vicinity of Longyearbyen (78°14'N, 15°38'E). A large amount of material has been collected recently by Drs. M. Lange and G. Gulden.

In Table 1, some additions extracted from the recent literature on higher fungi dealing partly, or more rarely, exclusively with material from the islands are listed. Out of these studies only two concentrate on Svalbard mycoflora, namely that by Reid (1979) and partly that by Watling (1983). Newer contributions are to be found elsewhere in the present volume and are here omitted.

TABLE 1. Some additions to higher mycoflora of Svalbard, extracted from recent literature.

ASCOMYCOTINA	
Bryochiton microscopicus Döbb. & Poelt	Döbbeler (1978)
B. monascus Döbb. & Poelt	Döbbeler (1978)
B. perpusillus Döbb.	Döbbeler (1978)
Cainiella borealis Barr	Holm (1975)
Crocicreas gramineum (Fr.) Fr. var. *gramineum*	Carpenter (1981)
Epibryon polysporum Döbb.	Döbbeler (1978)
E. diaphanum Döbb.	Döbbeler (1979)
Gibberia barriae Holm & Holm	Holm & Holm (1980)
Gnomoniella hyparctica (Lindl) Barr	Holm (1975)
Helvella arctoalpina Harmaja	Harmaja (1977)
H. dryadophila Harmaja	Harmaja (1977)
H. lacunosa Afz.: Fr.	Dissing (1966)
Hysteronaevia advena (Karst.) Nannf.	Nannfeldt (1984)
H. clavulifera Nannf.	Nannfeldt (1984)
H. kobayasii Nannf.	Nannfeldt (1984)
H. luzulicola Nannf.	Nannfeldt (1984)
H. lyngei (Lind) Nannf.	Nannfeldt (1984)
Hysteropezizella diminuens (Karst.) Nannf.	Nannfeldt (1984)
Mycosphaerella cassiopes Barr	Holm (1975)
Physalospora hyperborea Bäumler	Holm (1975)
Saccobolus quadrisporus Mass. & Salm.	Eckblad (1968)
Scleropleella hyperborea (Fuck.) Holm	Holm (1975)
Sporormia americana Griff.	Reid (1979)
Venturia subcutanea Dearn.	Watling (1983)
BASIDIOMYCOTINA	
Agaricus macrosporus (Møll. & Schaeff.) Pilát	Watling (1983)
Anellaria semiovata (Sow.: Fr.) Pears. & Dennis	Skifte (1979)
Collybia cf. *obscura* Favre	Reid (1979)
Cystoderma arcticum Harmaja	Harmaja (1984)
C. adnatifolium (Peck) Harmaja	Harmaja (1984)
Exobasidium cassiopes Peck	Nannfeldt (1981)
E. hypogenum Nannf.	Nannfeldt (1981)
Galerina pseudocerina Smith & Singer	Reid (1979)
G. pseudombrophila Kühn.	Watling (1983)
Hebeloma marginatulum (Favre) Bruchet	Watling (1983)
H. versipelle (Fr.) Gill.	Watling (1983)
Inocybe cf. *fuscomarginata* Kühn.	Watling (1983)
I. cf. *maleconii* Heim	Reid (1979)
Laccaria striatula (Peck) Peck	Reid (1979)
Lactarius cf. *theiogalus* Bull.: Fr.	Watling (1983)
Leptoglossum retirugum (Bull.: Fr.) Rick.	Høiland (1976)
Omphalina cf. *cupulatoides* Orton	Reid (1979)
O. pseudomuralis Lamoure	Reid (1979)
Peniophora pithya (Pers.) J. Erikss.	Arvidsson (1979)
Russula nitida (Pers.: Fr.) Fr.	Reid (1979)

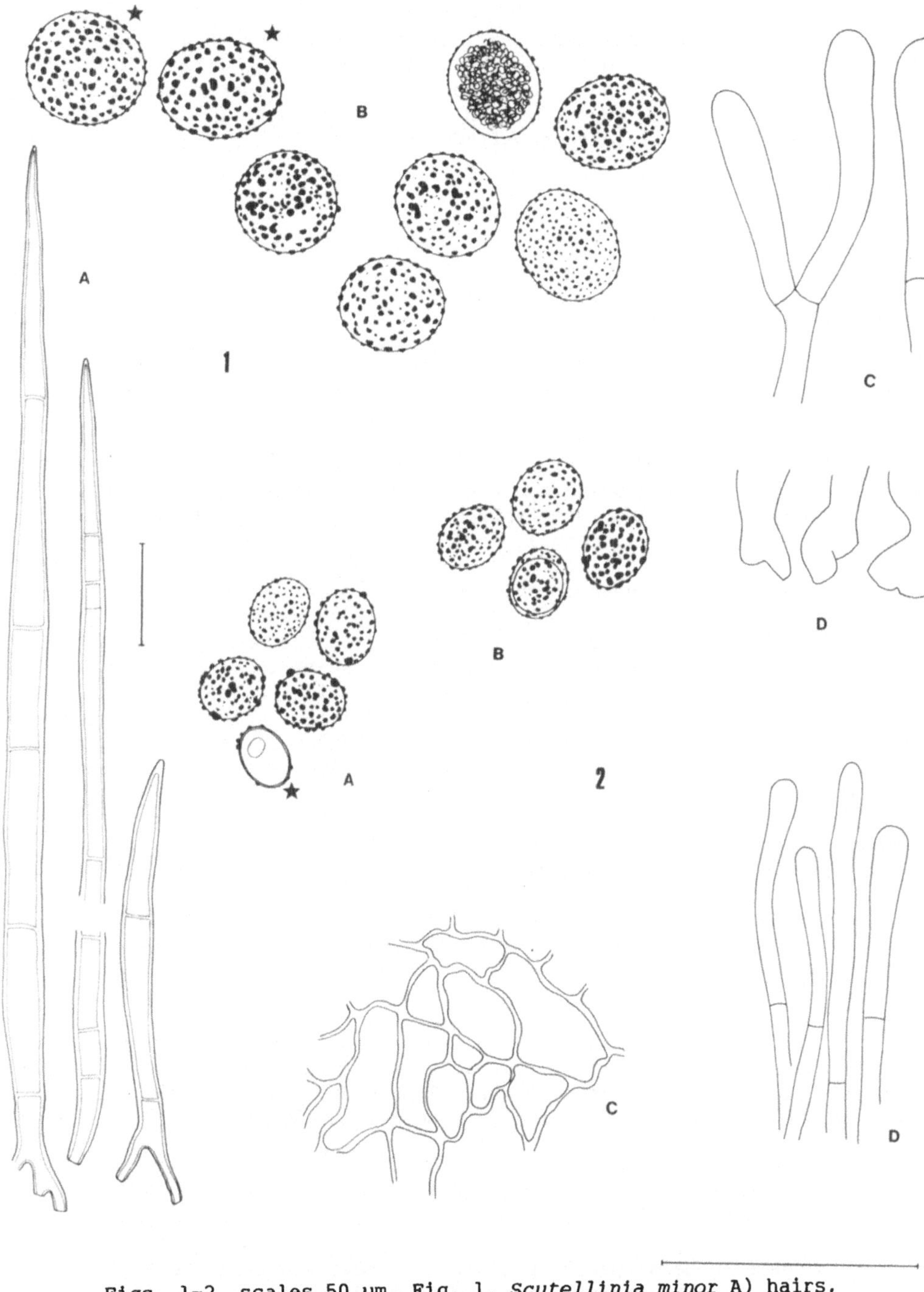

Figs. 1-2, scales 50 µm. Fig. 1. *Scutellinia minor* A) hairs, B) spores, marked in Congo red, others in cotton blue, C) paraphyses, D) base of asci. Fig. 2. *Octospora moravecii* A) spores of 83/220 in cotton blue, marked juvenile, B) spores of 83/179, C) ectal excipulum in lactic acid, D) paraphyses.

Present material was studied using X1500 magnification. All characters and illustrations, unless otherwise stated, are based on dried material. The camera lucida drawings were made from mounts in cotton blue or Melzer's reagent. A different mountant are specifically mentioned. Spore measurements exclude ornamentation. For basidiomycetous fungi spores were measured from a piece taken from middle of the lamella, unless otherwise stated. Color codes refer to the following books: Kornerup & Wancher (1961: eg. 6A5), Küppers (1978: eg. S20-Y50-M20) and Cailleux (1981: eg. P50). Material is deposited in TUR.

LIST OF SPECIES

Ascomycotina (Pezizales)
Helvellaceae

Helvella pocillum Harmaja

Kongsfjorden, Blomstrandhalvöya, Ny London, meadow, 23. August 1966 Heikkilä (sub. *H. dryadophila* Harmaja, det. S. Huhtinen).

This collection with somewhat immature apothecia was kindly redetermined by Dr. Harmaja. It shows the typical characters of *H. pocillum*; e.g., almost sessile apothecia, dark coloring, pleurorhyncous and thick walled asci, and large spores often reaching 25–28(-30) µm in length (Dr. Harmaja, pers. comm.).

The proportion of these conspicuous and probably submature spores may easily be misinterpreted. I noticed the spores to swell and burst immediately when Melzer's reagent is added. Soon the mounts show typical, large, oily globules present in abundance. It seems to me that only a small proportion of these large spores survive this treatment, and that in cotton blue the swelling is diminished. In addition to the large spores, there are seemingly mature spores present in abundance that measure 20–25 X 12-14.5 µm (L/W 1.5/1.7). The present collection is the third of *H. pocillum* (cf. Harmaja 1977).

Pyronemataceae
Leucoscypha hetieri (Boud.) Rifai — Fig. 3

Apothecia up to 1 mm in diameter when fresh, disc of different shades of orange, margin white. *Hairs* hyaline, varying in size and shape; long hairs narrow, undulating, somewhat tapering, up to 200 µm long, wall 1-2 µm thick, structurally uniform, multiseptate. *Asci* 160-230 X 14-15 µm. *Spores* ellipsoid-ovate, 15-18 X 9-11 µm, L/W 1.5-1.7, smooth, uniguttulate, wall acyanophilous, 0.5 µm thick, the guttule may show orange color in lactic acid. *Paraphyses* clavate, gradually widened towards apex, where 5-7 µm wide, basally 2.0-2.5 µm wide, wall 0.4 µm thick. Nuclei indistinctly stained in acetocarmine even after KOH pretreatment.

Campsite at the airfield, abundant in an old fireplace, amongst *Funaria* sp., pH 5.8, 2 August 1983, Huhtinen 83/180.

This seems to be a species previously unreported from the Arctic. It is noteworthy that the material has nuclei which stain only weakly in acetocarmine being in no way different from those of a majority of *Octospora* species. The difference of nuclear staining between these two genera is not as distinguishable as previously emphasized. Furthermore, the genus *Octospora* now includes at least one species with carminophilic nuclei (Dissing & Sivertsen 1983).

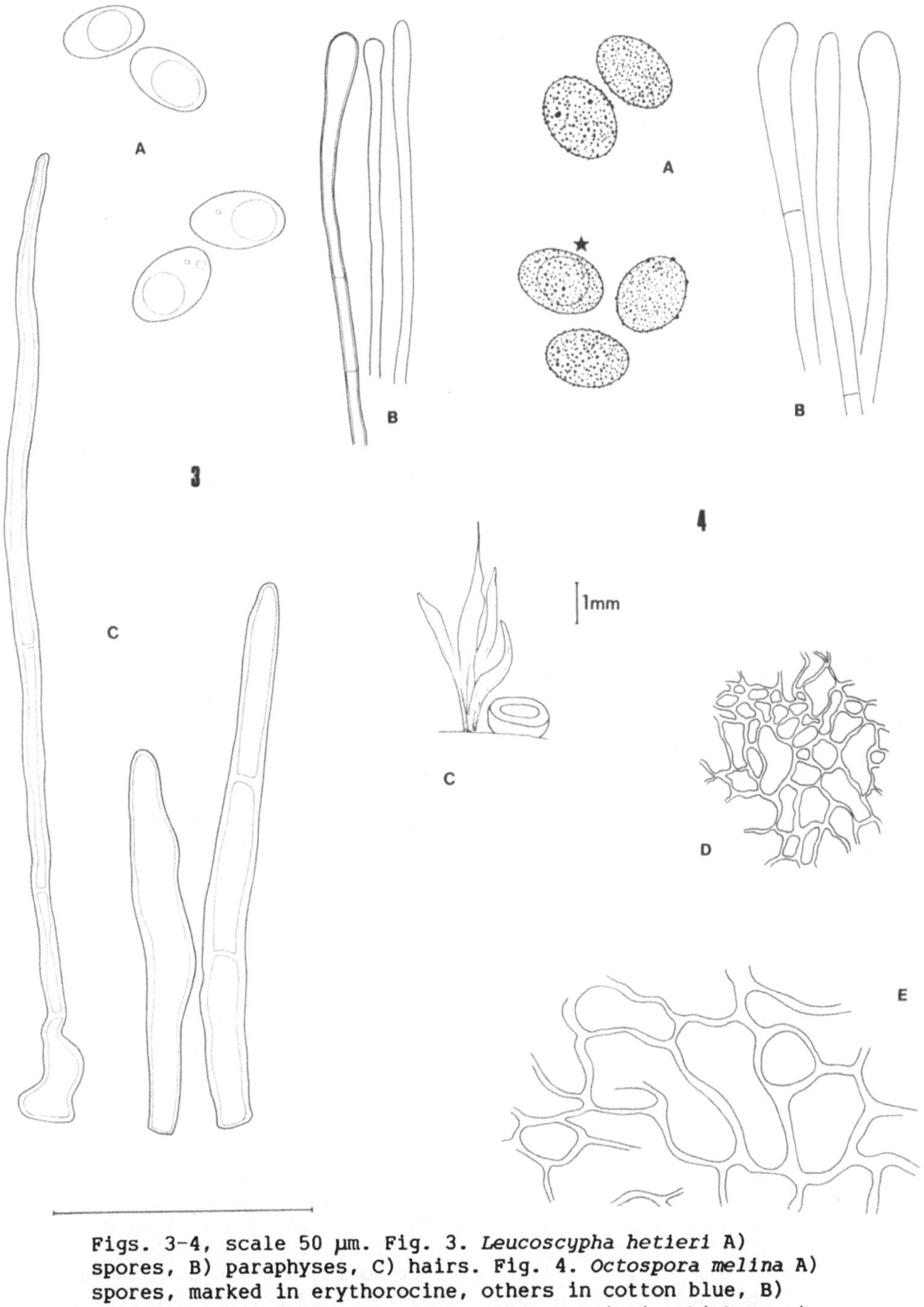

Figs. 3-4, scale 50 μm. Fig. 3. *Leucoscypha hetieri* A) spores, B) paraphyses, C) hairs. Fig. 4. *Octospora melina* A) spores, marked in erythorocine, others in cotton blue, B) paraphyses, C) apothecium with a characteristic thick margin, D) medullary excipulum in lactic acid, E) ectal excipulum in lactic acid.

Octospora humosa (Fr.: Fr.) Dennis

Campsite at the airfield, peaty soil mixed with charcoal dust, 25 July 1980, Huhtinen.

The collection is similar with material from subarctic Canada (Huhtinen 1985a). Spores appear to be always perfectly smooth (X1500) but closer scrutiny reveals minute roughness at each end. Of course this might be nothing more than an intermediary stage when the cyanophilous outer layer starts to disappear towards spore maturity. However, a delicate reticulum has been observed in one northern collection using SEM (Dissing 1982).

Judging from the description and illustrations, the specimen reported by Kobayasi et al. (1968) as *O. leucoloma* Hedw.: Fr. falls under the modern circumscription of *O. humosa*. My view is mainly based on the rounded spore ends and large size of the apothecia (cf. Dennis & Itzerott 1973, Itzerott 1981).

Octospora melina (Vel.) Dennis & Itzerott Fig. 4

Apothecia up to 1.5 (2) mm in diameter when fresh, first cupulate with a thick margin, later flattened (margin, in proportion, less prominent), sessile, first dull orange, later bright orange. Ectal excipulum of angular, irregular cells measuring 10-35 X 6-15 µm. Medullary excipulum of similar, but with smaller cells; both layers with clearly thickened walls. Margin composed of clavate, somewhat thick walled hyphae. *Asci* up to 280 X 17 µm, with indistinct crozier, base long, narrow. *Spores* ellipsoid, 16-17 X 10.8-12.0 µm, L/W 1.4-1.5, uniguttulate, de Bary bubbles lacking; ornamentation consisting of small warts, typically 0.5 µm wide and high, rarely exceeding 1 µm in either dimension, nonconfluent, regular, clearly visible in optical section (X1500), in general appearance of even size. *Paraphyses* clavate, 3 µm wide below, up to 8 µm at their apices, terminal cell 40-70 µm long, nuclei carminophobic.

Airfield, brook margin, somewhat clayey ground, 29 July 1983, Huhtinen 83/122. Village area, ditch bank, pH 4.0, 5 August 1983, Huhtinen 83/251.

This species has recently been treated by Dennis & Itzerott (1973), Svrček (1979a), Itzerott (1981), and by Dissing & Sivertsen (1983), and is now shown to occupy the more northern zones. When young, apothecia have a characteristically thick margin. In the only totally mature apothecium seen, the margin was less prominent and the overall coloring brighter than that of juvenile apothecia. According to Itzerott (1983) this rare species is parasitic on rhizoids of Bryaceae.

Octospora moravecii Khare Fig. 2

Apothecia minute, up to 1 mm in diameter when fresh, orange throughout, margin thin, fimbriate, flanks somewhat whitish pruinose. Ectal excipulum of prismatic cells, 10-30 X 5-15 µm, walls clearly thickened. Medullary excipulum of similar structure, but cells smaller. Margin composed of clavate cells. *Asci* up to 250 X 14-18 µm. *Spores* broadly ellipsoid to nearly globose, 12-14 X 10-12 µm, L/W 1.1-1.3, uniguttulate, de Bary bubbles rapidily lost in Melzer's reagent; ornamentation consisting of distinct warts of uneven size, typically 0.5-1.0 µm wide, up to 0.8 µm high, roundish, round-topped, mixed with scattered, similar but larger, confluent warts, up to 2 µm wide and high, ornamentation lost in KOH-pretreated acetocarmine mounts, in which one moderately colored area by the guttule can be seen. *Paraphyses* clavate, 3 µm wide below, 4-6 µm at their apices, terminal cell 35-60 µm long.

Campsite at the airfield, growing on *Pohlia* sp., 4 August 1983, Huhtinen 83/220. Campsite, old fireplace, together with *Leucoscypha hetieri*, pH 5.8, 2 August 1983, Huhtinen 83/179.

Two of the recent treatments of *O. moravecii* mention the connection to *Pohlia* (Dissing 1982, Dissing & Sivertsen 1983). *Pohlia* is also listed in the collection data of Schumacher's material, but the collection was made from bare sand without an apparent connection (Schumacher 1979). Judging from the still few collections the species seems to be montane to alpine. The larger warts on the spores that characterize the present collection are also seen in the illustrations by Moravec (1969, as *O. wrightii*) and Dissing & Sivertsen (1983).

Scutellinia minor (Vel.) Svrček — Fig. 1

Apothecia shallowly cupulate, up to 8 mm in diameter when fresh, disc bright orange-red, flanks orange, covered with either totally dark brown hairs or with hairs whose upper half are totally hyaline. Longest *hairs* restricted to the margin, 300-600 X 15-20 µm, base simple or bifurcate, only rarely more complex, walls 2.5-3.0 µm, many hairs show tendency to be closely septate, septal intervals being in places only 10-20 µm; hairs on lower flanks shorter, usually 200-250 µm long, sparsely septate, accompanied by abundant, blunt surface hairs. Cells of ectal excipulum globose, 30-60 µm in diameter. *Asci* 200-300 X 21-25 µm, base with indistinct croziers, wall 0.8-0.9 µm thick in lactic acid. *Spores* globose to subglobose, the former 18-19 µm in diameter (excluding ornamentation), the latter measuring 18.3-21.3 X (15.8-)17.0-19.8 µm, L/W 1.1-1.2, in ammoniacal Congo red reaching 21-23 X 18.5-19.5 µm, L/W similar, contents spumose when mature, one de Bary bubble frequent in lactic acid, juvenile spores may show a loosening perisporium in heated lactic acid; ornamentation consisting of mainly separate warts, up to 1 um wide and 0.4-0.8(-1.0) µm high, roundish, round-topped, ridges lacking. *Paraphyses* clavate, 9-13 µm at their apices, terminal cell 30-60 µm long.

Campsite at the airfield, very abundant, on bare ground with abundant charcoal dust, or buried amongst mosses, or even amongst *Splachnum ampullaceum*, pH 5.6-6.6, all sites with seeping water, 4 August 1983, Huhtinen 83/230, 83/234, 83/236, 83/237.

Coming from a 20 X 20 m area, the four collections still have different ecology. Especially noteworthy is the occurrence on old dung amongst *Splachnum*.

The material is distinguished from *S. trechispora* (Berk. & Br.) Lamb. on the basis of spore characters. Although many of the mature spores are globose or are seen as globose from an apical view, an equal proportion of spores are subglobose. Ornamentation consists of smaller warts than in *S. trechispora*. I have also compared my collections with a fragment of Petersen's material (Dissing 1982) from Greenland (Gr. 73.106) and conclude they are similar. Some discs were found to be problematical. In these, a majority of spores were globose. And though the ornamentation was again consisting of smaller warts than is generally seen in those of *S. trechispora*, it apparently falls into Kullman's (1982: Tab. 7, Figs. 1,2) concept of the species. This spore variation is represented by a small minority of the abundant collections comprising ca. 200 apothecia.

Svrček (1971) synonymized *S. subglobispora* Svrček & Moravec with *S. minor* and the key provided states size range of spores as similar to present collections. However, the type collection seems to have smaller apothecia and spores, and shorter hairs, adding some further variation to *S. minor* (Svrček & Moravec 1969). Hair color varied even in closely seated apothecia of the Svalbard collections. Partial hyaline hair color can be seen in other brown-haired taxons from the Arctic. This trend has been noted also by Dr. Dissing (pers. comm.).

In his key, Moravec (1974) groups *S. minor* among the species with a spiny or truncate spore ornmentation. His material of *S. arenosa* (Vel.) Le Gal (syn. *S. hrabanovi* (Vel.) Svrček, cf. Svrček 1979a) had spores quite similar to the present taxon, but differs in much shorter hairs measuring 110-135 µm in length only. At the same time, Moravec suspected the conspecificity between *Lachnea arenosa* Vel. and *S. arenosa* (Vel.) Le Gal., the latter of which has hairs reaching 800-1000 um in length (Le Gal 1966). Taking into consideration Svrček's (1971) characterization of warts in *S. minor* and the differences towards *S. arenosa*, the former taxon best embraces the present collections.

Ascomycotina (Helotiales)
Sclerotiniaceae

Ciboria polygoni-vivipari Eckbl. Fig. 12

Apothecia up to 2 mm in diameter when fresh, dark brown with a lighter brown disc. Ectal excipulum of globular to angular to more elongated cells perpendicular to the surface, end cells clavate to pyriform, cells with scattered brown encrustation. Stipe cortex of textura porrecta, hyphae 4-6 µm wide, medulla of narrower hyphae; clavate, hair-like hyphal ends frequent. Medullary excipulum of textura porrecta in the flanks, in sections easily distorted to remind one of textura intricata, hyphae 3-5 µm wide, brown encrusted; in central part of cup medullary excipulum of textura intricata. Subhymenium amyloid. *Asci* 160-180 X 12 µm. J+ without KOH pretreatment, base with indistinct croziers. *Spores* uniseriate, ellipsoid-inequilateral, 13-17 X 6-7 µm, L/W 2.2-2.7, biguttulate, guttules 4-5 µm in diameter in Melzer's reagent. *Paraphyses* clavate, 2 µm wide below, typically 5 µm wide above, wall slightly thickened, brown encrusted.

Airfield, on buried bulbils of *Polygonum viviparum*, 29 July-5 August 1983, Huhtinen 83/113, 83/120, 83/166, 83/254. The bulbils were usually soft and partly sclerotized, but on one occasion a true sclerotium enclosing dead plant tissue was found. Its connection with the present species is somewhat questionable. This minute species was found to be common in the area, but it escapes the eye being brown and minute.

In addition to ecology, the material is characterized by bicolored cups when dried. The upper flanks are dark brown in contrast to lower parts which are lighter brown. In micro section this can be seen as a very abundant, dark brown pigment between the hymenium and ectal parts. From the margin it goes ca. 160 um downwards. Although first appearing as smooth, the apothecia are typically minutely scurfy due to the characteristic, pyriform to clavate "hairs" occurring on the flanks and stipe.

Eckblad's (1969) original description was found to differ in some aspects from present collections. The type shows more elongate and irregular spores, narrower paraphyses and hair-like hyphae on the stipe a feature which I could not see in my collections. After studying the type

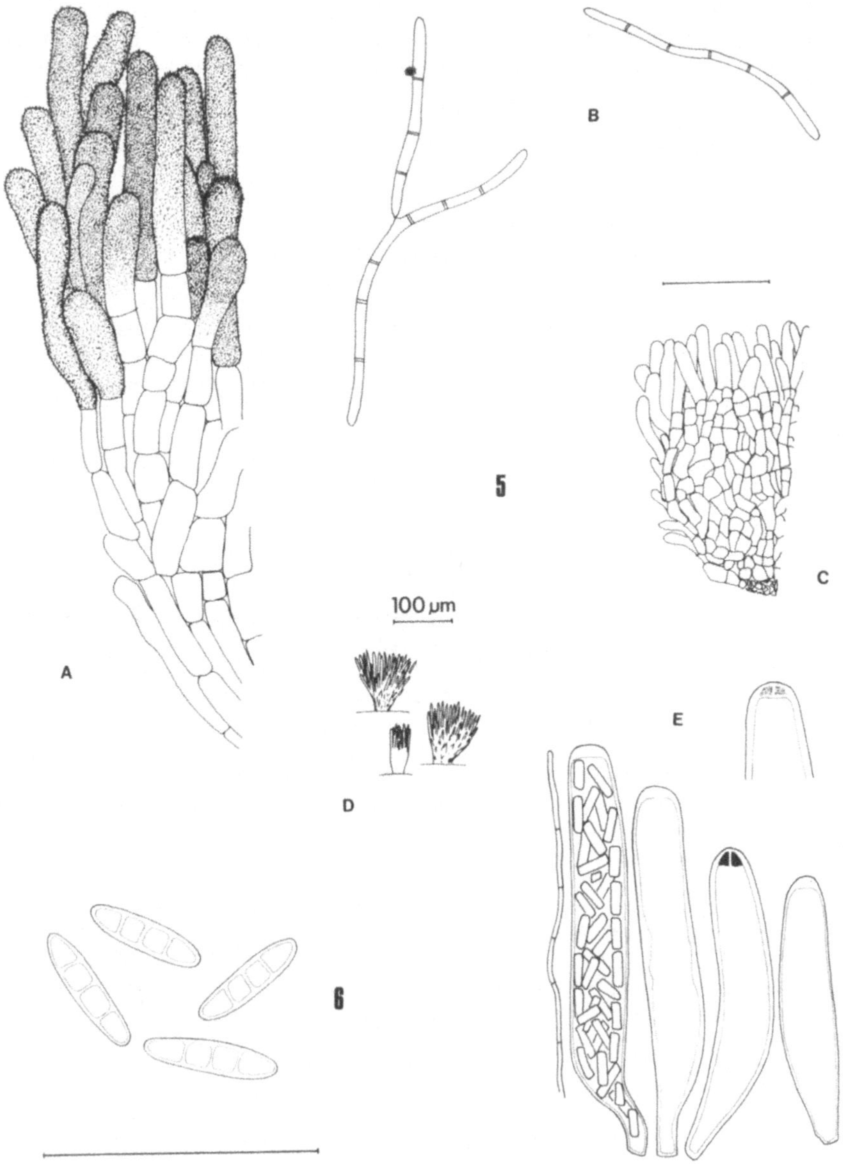

Figs. 5-6, scales 50 µm. Fig. 5. *Polaroscyphus spetsbergianus* A) hairs at the margin, B) spores, found free in a squash mount, C) schematic surface view, D) apothecia, E) asci and paraphysis, showing younger asci to the right. Fig. 6. *Crocicreas culmicola*, spores in cotton blue.

of *C. polygoni-vivipari* (in O) I found some other minor differences. The typically bicolored cups were not seen in the type, but characterized the Greenland collection, I also studied (Korf 1982). Since the type comes from a more southern location than these later collections, the increased pigmentation might mark a climatic transition. In addition, spore guttulation is prominent in Svalbard material when compared to that of the type. All collections have minutely scurfy apothecia.

Kohn (1979) lists another species growing on *Polygonum*; i.e., *Sclerotinia polygoni* Rehm. Dr. Schumacher has studied the types of both species and concludes them to represent different species (Schumacher, pers. comm.).

Hyaloscyphaceae

Polaroscyphus gen. nov.

Genus ordinis Helotialium positione incerta, probabiliter familiae Hyaloscyphacearum. Habitu externu speciebus reductis generis Dasyscyphi similis est, sed apice asci rotundato et in iodo non colorato et sporis rumpentes differt. Parte exteriori albopilosa, pili cylindrati, hyalini, ubique dense spinulati vel in basi glabri. Excipulum externum hyalinum, cellulis prismaticis consistens. Sporae filiformes, septatae, hyalinae, in asco rumpentes. Paraphyses filiformes, simplices.

Species typica generis: Polaroscyphus spetsbergianus Huhtinen. Etymology of the generic name. - Referring both to the northern area where the collection was made and to the substrate.

Polaroscyphus spetsbergianus gen. et sp. nov. Fig. 5

Apothecia minuta, cylindrata vel cyathiformia, sicca usque ad 90 um lata, 130 µm alta, in margine hyalina, pilosa, in basi pallide brunnea, anguste sessilia. Pilia usque ad 45 X 7 µm, hyalini, cylindrati vel clavati, ubique dense minuter spinulosi, raro basi laeva, non septati, non amyloidei. Excipulum externum textura prismatica, cellulis 10-15 X 4-5 µm vel 15 X 12 µm, tenuiter tunicatis. Asci late cylindrati, tetraspori, 50-73 X 7-12 µm, post solutionem kalii amyloidei, in apice crassotunicati, rotundati, porus angustus, in basi non uncinatus. Sporae filiformes, 50-62 X 1.8-2.0 µm, multiseptate, hyalinae, crebro in asco rumpentes. Paraphyses filiformes, 1 µm latae, ascos non superantes.

Holotypus: Norway, Svalbard, Longyearbyen, Björndalen, on leaves of *Salix polaris*, 3 August 1983, Huhtinen 83/201 (TUR).

Apothecia discoid, cyathiform, and up to 90 µm wide and 130 µm high when dried, white and hairy at the margin, faintly brown below when dried, subsessile with a short, narrowing base. *Hairs* up to 45 X 7 µm at the margin, shorter below, hyaline, cylindrical to clavate, 4-6 µm wide below, apex up to 8 µm wide, densely and regularly spinulose, either totally or more rarely base smooth, the spinulose part aseptate, not amyloid or dextrinoid, not exuding any amorphous matter, appears cyanophilous due to the dense, cyanophilous covering of spinules, but wall uncolored in cotton blue; spines typically 0.3-0.8 µm high, also coloring in Congo red, not dissolving in strong KOH, spinules often minute or lacking at the base, best developed at the extreme apex. Ectal excipulum colorless under the microscope, of clear textura prismatica, cells ca. 10-15 X 4-7 µm below, often more isodiametric just below the hairs, measuring ca. 7 X 7 µm or 9 X 9 µm, wall thin or up to 0.6 µm thick in lactic acid. *Asci* 50-73 X 7-12 µm, broadly cylindrical with parallel flanks, four-spored, always with a rounded apex, J- without KOH pretreatment, J+ after it, apex up to 2-3 um thick in lactic acid when young, pierced by a narrow pore, asci arising from a small and inconspicuous crozier, but when mature base typically simple, wall 0.8 µm thick in lactic acid, not coloring in Congo red, maturation successive. *Spores* filiform, 50-62 X 1.8-2.0 µm, multiseptate, hyaline, either situated in more or less regular rows in asci or entangled, disarticulating at an early stage while still in asci, seldomly seen free and unfragmented, obviously capable of budding; spore

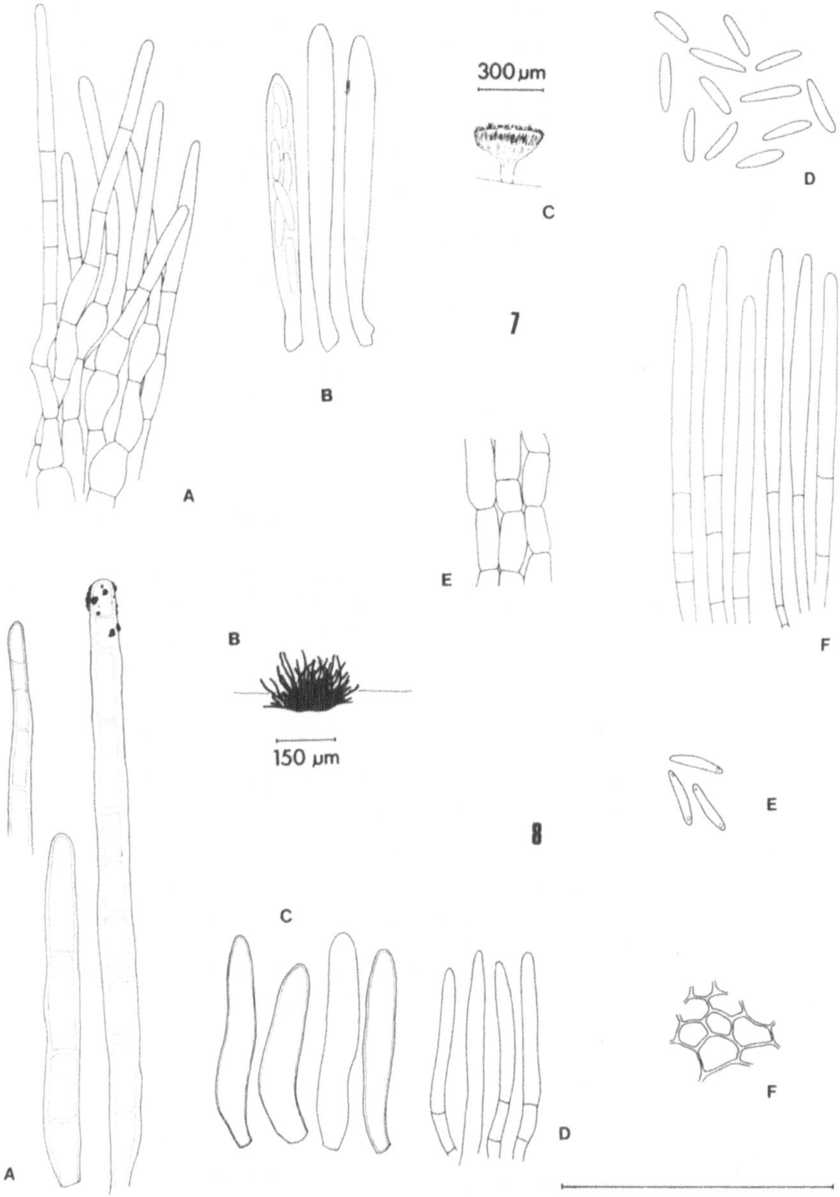

Figs. 7-8, scale 50 µm. Fig. 7. *Psilachnum acutum* A) hairs at the margin, B) asci, C) apothecium, D) spores, E) ectal excipulum, F) paraphyses. Fig. 8. *Trichopezizella nidulus* var. *hystricula* A) hairs, B) apothecium, C) asci, D) paraphyses, E) spores, F) ectal excipulum; A-E in 5% KOH, F in lactic acid.

fragments variable in size, 4-9 X 2-3 µm in Congo red, with a central globule.
Paraphyses filiform, 1 µm wide, simple, not exceeding the asci.

The character combination presented above makes it impossible to place this taxon in to any existing, modern genus. The asci are unitunicate with poridieal dehiscence, and the potential relatives of this new genus are to be searched among the Helotiales. The taxonomic position is somewhat uncertain, but Hyaloscyphaceae is here considered most probable.

The asci have firm walls. The always rounded apex is maximally 3 um thick, but in age it becomes much thinner and sometimes indistinguishable in thickness from the flanks. There is certain similarity with asci of *Perrotia*, but the J+ reaction, spores and different hairs exclude the possible relationship. The asci of *Polaroscyphus spetsbergianus* can be considered as an expression of the trend that filiform spores are associated with cylindrical asci with a thickened apex (Sherwood 1977).

The sequence of spore septation was not observed by me. Very young spores are laborious to locate and septation takes place early. The spores disarticulate frequently while still in the asci, and free, whole spores are rarely seen in mounts. Disarticulating, filiform spores are commonly met in Ostropalean fungi, but there is no other obvious relationship with *Polaroscyphus*, which in any case has an isolated taxonomic position in modern classification (cf. Sherwood 1977).

Another collection of this new species was found ca. 4 km away on leaves of *Salix polaris* (Blomsterdalen, 5, August, 1983, Huhtinen 83/252a).

Psilachnum acutum (Vel.) Svrček Fig. 7

Apothecia shortly stipitate, up to 300 µm in diameter when dried, white when fresh, stipe 100 X 50 µm when dried. *Hairs* 40-60 µm long at the margin, 10-20 µm on lower flanks, 3 µm wide, smooth in lactic acid, somewhat tapering or cylindrical, always blunt, 3-4 septate. Ectal excipulum of textura prismatica, cells 8-14 X 5-8 µm close to the margin, hyaline, thin walled in lactic acid. *Asci* 46-60 X 5.0-5.5 µm, base often with indistinct croziers. *Spores* elliptic subfusoid to cuneiform, 7-10(-12.8) X 1.8-2.0 µm, aseptate, indistinctly multiguttulate in cotton blue. *Paraphyses* almost similar to hairs, cylindrical to somewhat lanceolate, 3.0-3.5 µm wide, apically 1 um narrower but still blunt, basally 2-3 septate, exceeding the asci by 5-20 µm.

Airfield, Blomsterdalen, on culms of *Calamagrostis neglecta*, 5 August 1983, Huhtinen 83/270.

Except for the long asci, this collection corresponds with Svrček's (1979b) description of the species. Svrček studied many other collections agreeing with the type, and also material reported by Dennis (1949) which had short asci. Thus, this character needs further attention when northern material is collected again. *P. acutum* is previously unreported from the Arctic.

Psilachnum inquilinum (Karst.) Dennis

Bjorndalen, on *Equisetum arvense*, 3 August 1983, Huhtinen 83/199.

This collection corresponds in all critical respects to material from subarctic Canada illustrated and treated by the author (Huhtinen 1985a).

Psilocistella obsoleta (Vel.) Svrček Figs. 9-11

Apothecia shortly but clearly stipitate, separate to confluent, regular to lobed, 200-300(-500) µm in diameter when dried, disc flat, more

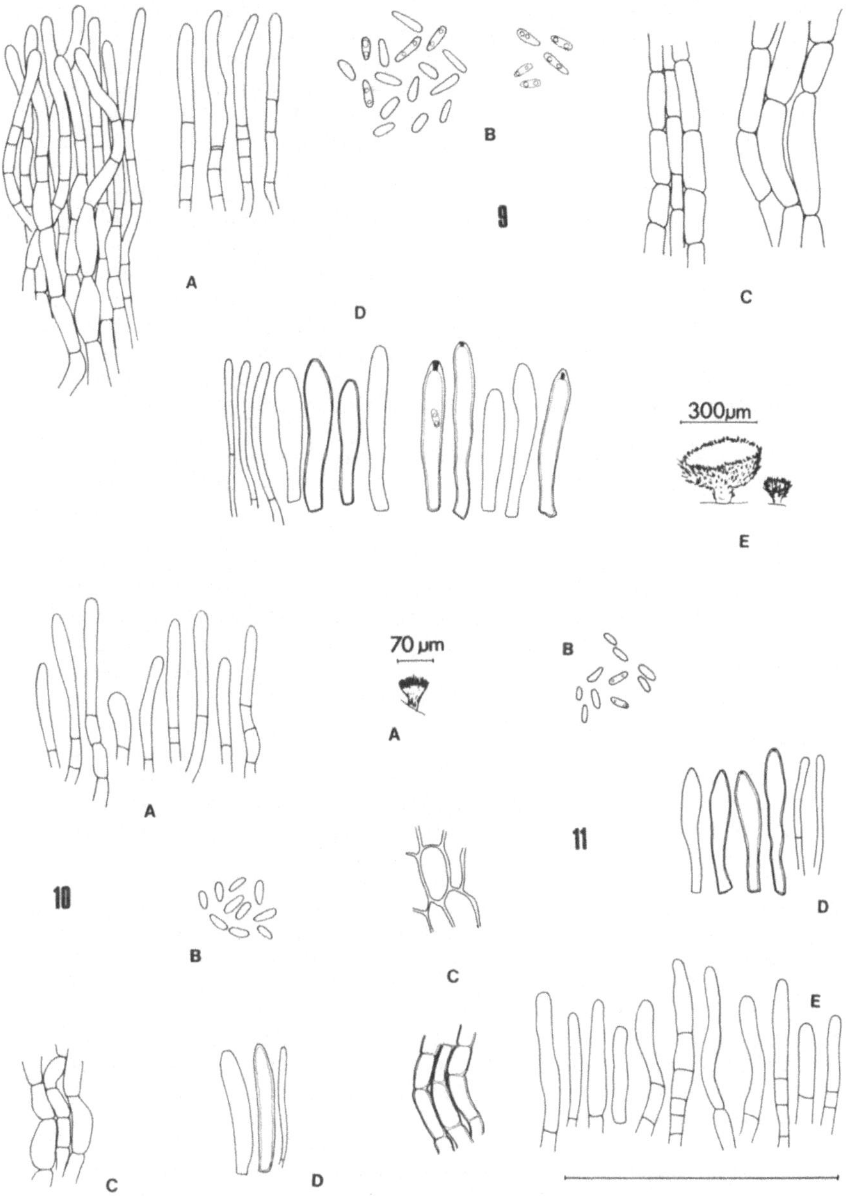

Figs. 9-11, scale 50 µm. Fig. 9. *Psilocistella obsoleta*, mature apothecia of 83/134 A) hairs at the margin, B) spores (brown), C) ectal excipulum, D) asci and paraphyses, E) apothecia. Fig. 10. *Psilocistella obsoleta*, juvenile apothecia of 83/134 A) hairs at the margin, B) spores (hyaline or nearly hyaline), C) ectal excipulum, D) asci and paraphysis. Fig. 11. Type of *Psilocistella obsoleta* (PR 150906) A) apothecium, B) spores (hyaline or nearly hyaline), C) ectal excipulum, D) asci and paraphyses, E) hairs at the margin.

rarely convex, first white in fresh condition, later green (S70-C50-Y99), disc yellowish to greenish grey (appro. S70-C20-Y40) in dried condition, uneven; apothecia very soft and easily detached from the substrate when dried. *Hairs* smooth also in water mounts, cylindrical to slightly widening apically, originating from excipular cells of similar width so that the delimitation of hairs is questionable, 30-60 X 2.0-3.0 µm, 0-4 septate, terminal cell 15-28 µm long, walls thin, hyaline, inamyloid, not dextrinoid. Ectal excipulum of textura prismatica, cells in smallest apothecia measuring 6-10 X 3-4 µm, in mature apothecia 10-20 X 3-5 µm, hyaline, typically more roundish and yellow brown under the microscope at stipe base, walls thin. *Asci* 21-29(-39) X 3.5-5.0 µm, clearly J+ without KOH pretreatment, base with indistinct croziers, walls typically somewhat thickened (0.5 µm). *Spores* cylindrical, subfusoid to cuneiform, 3.8-5.8 X 1.0-1.9 µm, first hyaline and free, later becoming strongly pigmented, brown already in asci, often glued together, aseptate, with 0 to 2 guttules, smooth but sometimes roughened by the abundant pigment which is insoluble in KOH. *Paraphyses* equalling the asci, cylindrical, somewhat irregular, 1.0-2.5 µm wide, septate, often branched 20-25 µm below the apex, thin walled, smooth, not pigmented.

Village area, old dump above Nybyen, old board lying on the ground, 30 July 1983, Huhtinen 83/134. The site offered a sheltered and warm habitat.

Based on hair characters the collection keys into a rather recent genus, *Psilocistella*, established by Svrček (1977) for hyaloscyphaceous species with cylindrical hairs and paraphyses. The genus embraces six species (Svrček 1977, 1978, 1979b, 1983). None share the surprising combination of characters present in my collection. Nevertheless, after studying the type of *P. obsoleta* (PR 150906), I consider it conspecific with my specimens. The type consists of very minute, juvenile apothecia only. These have still nearly hyaline spores. In all critical aspects they are identical with the juvenile apothecia of this Svalbard collection. Furthermore, the production of brown pigment on spores can be observed in the type too. The more mature spores are already glued together and have a light brown tinge, which is rather easily seen using high quality optics. The differences found were meager. The type has a yellow brown excipulum with slightly thickened walls and the hairs are slightly wider. I consider these differences nonsignificant at the species level.

My view of the identity of these two collections alters the concept of the genus *Psilocistella* drastically. To my knowledge this is the first hyaloscyphaceous taxon that clearly has brown spores. I also have another species of the genus with dark brown spores. This collection was made in SW Finland and will be treated in a separate paper. Hence, I do not consider brown spores and not even a green disc anomalous for the genus. However, additional and more mature Czech material of *P. obsoleta* would conclusively show how consistent these pigments are. Were one to search for a common reason for the dark spores, it might be that both of my collections were made late in the season.

This is the first report of the genus *Psilocistella* from the Arctic and the second reported collection of its type species.

Trichopezizella nidulus (Schmidt & Kuntze: Fr.) Raitviir var. *hystricula* (Karst.) Haines — Fig. 8

Apothecia up to 200 µm in diameter when dried, often 100-150 µm only, sessile, hairy, dark brown. Hairs up to 130 µm long, frequently reaching 100 µm, up to 8 µm wide, frequently 6 µm wide below, only slightly tapering, straight, dark brown except for the apex, which may be slightly widened, often bearing crystals and resinous matter, 6-9 septate when over 100 µm, walls 1 µm thick, septae thick (0.8 µm) seldom thin except in the apex. Excipulum of textura angularis, walls brown, smooth or encrusted, thickened, cells 5-8 µm across. *Asci* 30-40 X 5-6 µm, base typically ill defined, J+ after KOH pretreatment, base simple. *Spores* cylindric to subfusoid, 9.0-10.5 X 1.0-1.8 µm. *Paraphyses* narrowly lanceolate only, 2.5-3.0 µm wide, blunt to somewhat tapered at their apices, basally multiseptate, walls slightly thickened, equalling the asci or exceeding by 5 µm only.

Carolinedalen N of Longyearbyen, on *Carex lachenalii*, 1 August 1983, Huhtinen 83/148.

This taxon was previously known only from the type locality (Haines 1974). Its typical features are the long spores, narrow paraphyses, and substrate. My collection supports Haines' view. Only in the paraphyses is more variation added to var. *hystricula*. Long spored specimens of var. *nidulus* are also sporadically seen (Haines 1974; also Jaap Fungi Sel. Exs. 409 in TUR), but the characters of var. *hystricula* seem to be stable enough to merit separation.

Leotiaceae

Crocicreas culmicola (Desm.) Carpenter Fig. 6

Old dump above Nybyen, on grass culms, 30 July 1983, Huhtinen 83/130a.

The collection is well characterized by the 3-septate, hyaline spores surrounded by a gelatinous sheath and measuring 21-26 X 5.5-6.0 µm. The margin is fimbriate and covered by abundant crystals. Asci measure 130-170 X 16-21 µm and are J+ without KOH pretreatment.

The hyphae which cover the ectal excipulum are darker than stated by Carpenter (1981) and Stadelmann (1978), resulting in much more brown apothecia than usual. Another difference can be seen in spores which are somewhat wider than in earlier reported material. Spore walls are acyanophilous and spore septae seem to have a clear pore. The paraphyses are hyaline to very pale brown at their somewhat irregular apices; in some discs, however, they are dark brown.

This change in apothecial coloring might be caused by a change in the amount of solar radiation received. Material studied by Carpenter (1981) was primarily from lowland sites.

Crocicreas cyathoideum (Bull.) Carpenter
var. *cacaliae* (Pers.) Carpenter

Old dump above Nybyen, on culms of *Poa alpigena* f. *vivipara*, 30 July 1983, Huhtinen 83/130.

At the species level my determination was confirmed by Dr. Carpenter. According to him, this specimen is one of those with intermediate pigmentation between var. *cyathoideum* and var. *cacaliae*. The ability to produce deeply pigmented apothecia results in placing the collection to the latter variation.

Two slightly deviating characters were found. The spores measured 7.0-8.8 X 1.9-2.0 µm. Nearly all apothecia are totally covered by hyaline crystals.

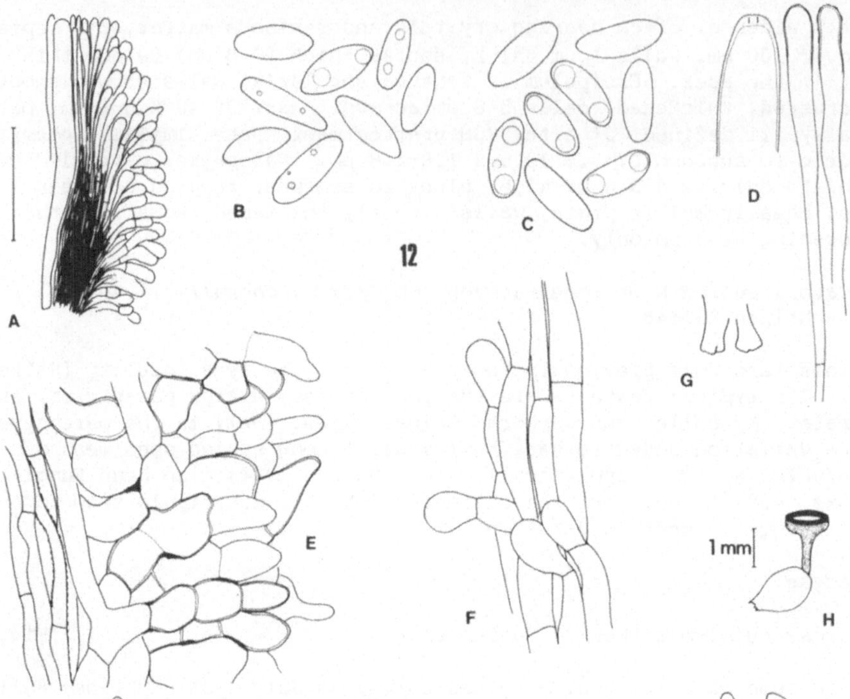

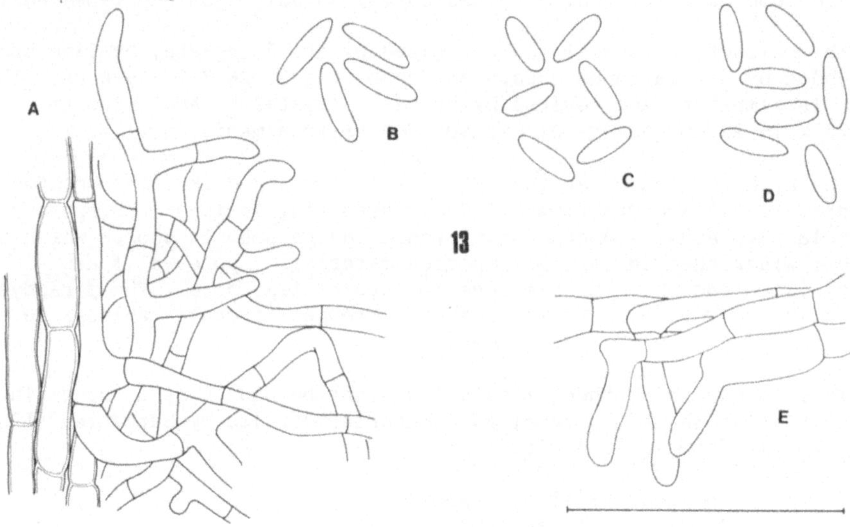

Figures 12-13, scales 50 µm. Fig. 12. *Ciboria polygoni-vivipari* A) section of margin showing abundant pigmentation, B) spores of the type, C) spores of Svalbard material, D) ascus and paraphyses, E) ectal excipulum, F) stipe surface, G) bases of young asci, H) apothecium in dried condition; all except B from Svalbard collections. Fig. 13. *Cudoniella clavus* A) stipe surface of 83/216, B) spores of 83/216, C) spores of 83/151, D) spores of 83/186a, E) hair-like hyphal ends on cup flanks, collection 83/151.

Cudoniella clavus (Alb. & Schw.: Fr.) Dennis Fig. 13

Carolinedalen N of Longyearbyen, on wet *Carex* litter, 1 August

1983, Huhtinen 83/151. Bjordalen, 3 August 1983, Huhtinen 83/186a. Airfield, 4 August 1983, Huhtinen 83/216.

This species occupies wet depressions and small ravines, sheltered from the desiccating wind. These collections have larger spores than material from subarctic Canada (Huhtinen 1985a). Spores are aguttulate or show faint guttules only. A new character can be seen in collection 83/216, in which the apothecia have somewhat hairy stipes. The collections show a clear wine red color on the ectal excipulum in Melzer's reagent.

Basidiomycotina

Clavariaceae

Typhula culmigena (Mont. & Fr.) Berthier

Village area, old dump above Nybyen, on culms of *Poa alpigena* f. *vivipara*, 30 July 1983, Huhtinen 83/126.

The specimen is well characterized by the 3-lobate spores measuring c. 5 X 4 µm (Berthier 1976). The clavula is pure white and ca. 2 mm high, and surrounded by a gelatinous sheath observable under the microscope. The stipe is devoid of a basal sclerotium and its cortex is composed of gelatinized, clamped, frequently branched hyphae. Rosette-like crystals are abundant on the stipe. They measure 10-12 µm in diameter. More scattered, smaller, roundish lumps od crystals can be found on the hymenium. Other characters match Berthier's (1976) description of the species.

This minute species is known from western Europe, Morocco and Australia, and is here reported for the first time from the Arctic and from any northern locality. Most of the earlier records are from gramines, but *T. culmigena* has even been found on *Fraxinus* leaves (Dr. Berthier), pers. comm.).

Tricholomataceae

Marasmius kallioneus Huht.

This taxon was described as new in a separate paper (Huhtinen 1985b). It was first treated in the literature by Ohenoja (1971) as *Marasmius* sp. As I managed to find two additional collections from Svalbard, which indicates the species is probably not rare. Lamoure et al. (1982) reported the same taxon from Greenland.

Melanoleuca cognata (Fr.) Konr. & Maubl. Fig. 14

Pileus 7-10 cm in diameter when fresh, with (faded?) yellow-brown colors, centre more brown; when dried 34-48 mm in diameter rather unicolorous, mat clay brown (Y50-M40-C40, close to 5D5). *Stipe* only slightly shorter or longer than cap diameter, 2.5-3.0 mm wide above when dry, widening to 6-10 mm at base, yellow brown when fresh, lacking the grey component of pileus color, when dried apically of gill color, basally darker, brownish (Y60-M40-C30), furrowed. *Gills* yellowish when fresh, with a slight red tinge, dried gills dark cream (Y30-M10-C00, 4A4). *Spores* hyaline, ellipsoid, 8.5-10.8(-11.5) X 4.8-5.6(-6.0) µm when measured from a gill mount, typical L/W 1.7-1.9, average of 40 spores 9.3 X 5.2 µm, average L/W 1.8; from cuticle, spores are more regular in size and shape, 8.9-9.8 X 5.3-6.0 µm, typical L/W 1.5-1.7, average 9.2 X 5.8 µm, average L/W 1.6; spore wall ca. 0.7 µm thick, cyanophilous, warts

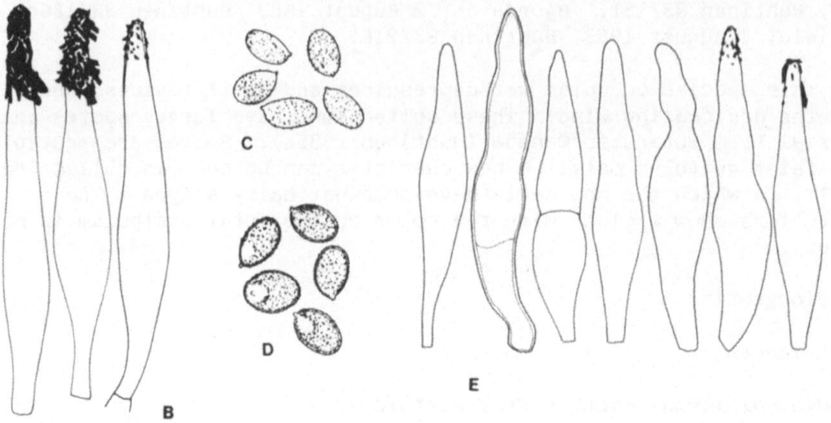

14

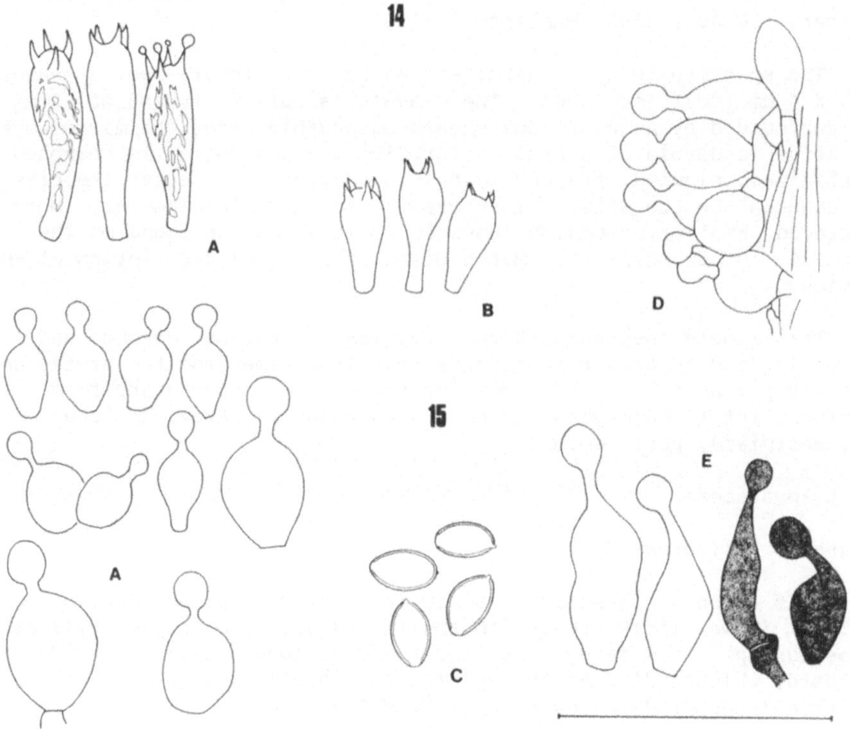

Figs. 14-15, scale 50 µm. Fig. 14. *Melanoleuca cognata* A) basidia, B) pleurocystidia, C) spores from gill mount, D) spores from pileus cuticle, E) cheilocystidia. Fig. 15. *Conocybe magnicapitata* A) cheilocystidia, B) basidia, C) spores, D) caulocystidia, E) pileocystidia.

strongly amyloid, c.a. 0.3 µm high, scattered, grouped only at the margin of this clearly delimitated plaque, ornamentation lost in 10 percent KOH. *Basidia* 30-38 X 8.5-11.0 µm, four-spored, frequently filled with resinous matter becoming reddish orange in Melzer's reagent. *Pleurocystidia* scattered, fusoid, 59-75 X 9-10 µm, typically strongly encrusted. *Cheilocystidia* fusoid, but somewhat more irregular in shape,

50-60(-80) X 8-11 µm, often smooth or only minutely encrusted. *Caulocystidia* and *pileocystidia* not seen. *Pileus cuticle* of interwoven, cylindrical to inflated, smooth hyphae, mostly 5-15 µm wide. *Pileus trama* of up to 20 µm wide, inflated, thin-walled hyphae, separating and reviving slowly. All hyphae from basal mycelium to hymenium clampless.

Airfield, moist heath with *Salix* and *Dryas*, 2 August 1977 Kallio & Heikkilä.

Since no specific odor and taste was annotated by the collector, material of *Melanoleuca adstringens* (Pers.: Fr.) Konr. was also studied for comparison. Two collections, both annotated to have either a disagreeable taste or odour, were studied (Norway, Finnmark, 7 Sept. 1970 Ulvinen, TUR; Federal Republic of Germany, Frauenalp, 9 Sept. 1968 Bresinsky, M). The latter collection was earlier studied and cited by Bresinsky & Stangl (1977). Of the characters stressed to be typical to *M. adstringens* by these authors, the present collection only shows two; i.e., the short stipe and the tendency for the cheilocystidia to be apically somewhat undulating.

Studying collections of *M. cognata* (Fr.) Konr. & Maubl. from Finland and those cited by Bresinsky & Stangl (1977), I found a rather wide variation in both these features. *M. cognata* has cheilocystidia showing such a wide variation that no delimitation to those of *M. adstringens* can be made. The Norwegian collection of that species has regular and rather wide cheilocystidia. There is no difference in spore size between the two taxa. And stipe length is likely to be regulated by the arctic environment.

On the basis of the light, yellow brown cap color, yellowish gills, presence of cystidia, and the apparent lack of a distinct odour, I conclude the present specimens to belong to *M. cognata* in section Cognatae Kühner (Bresinsky & Stangl 1977, Kühner 1978). The orange-red coloring of basidia in Melzer's reagent is a character irregularly present, lacking in Bresinsky's collection of *M. adstringens* but present in other collections studied here.

Flagelloscypha kavinae (Pilát) W.B. Cooke

Blomsterdalen, on leaves of *Salix polaris*, 5 August 1983 Huhtinen 83/252.

The collection was kindly determined by Prof. R. Agerer, Münich.

Rickenella fibula (Bull.: Fr.) Raith.

Airfield, somewhat clayey brook bank with small mosses, 2 August 1983 Huhtinen 83/157.

With the following character combination, this specimen falls clearly within the modern concept of the species (cf. Clémençon 1982): cap bright orange, stipe long and narrow, more yellowish; the subcapitate, 50-70 µm long cystidia are abundant on stipe, cap and lamellae; basidia four-spored measuring 13-20 X 4.0-4.5 µm; spores nearly cylindrical measuring 5-6 X 2.0-2.6 µm; clamps present.

Bolbitiaceae

Bolbitius cf. *variicolor* Atk. Fig. 17

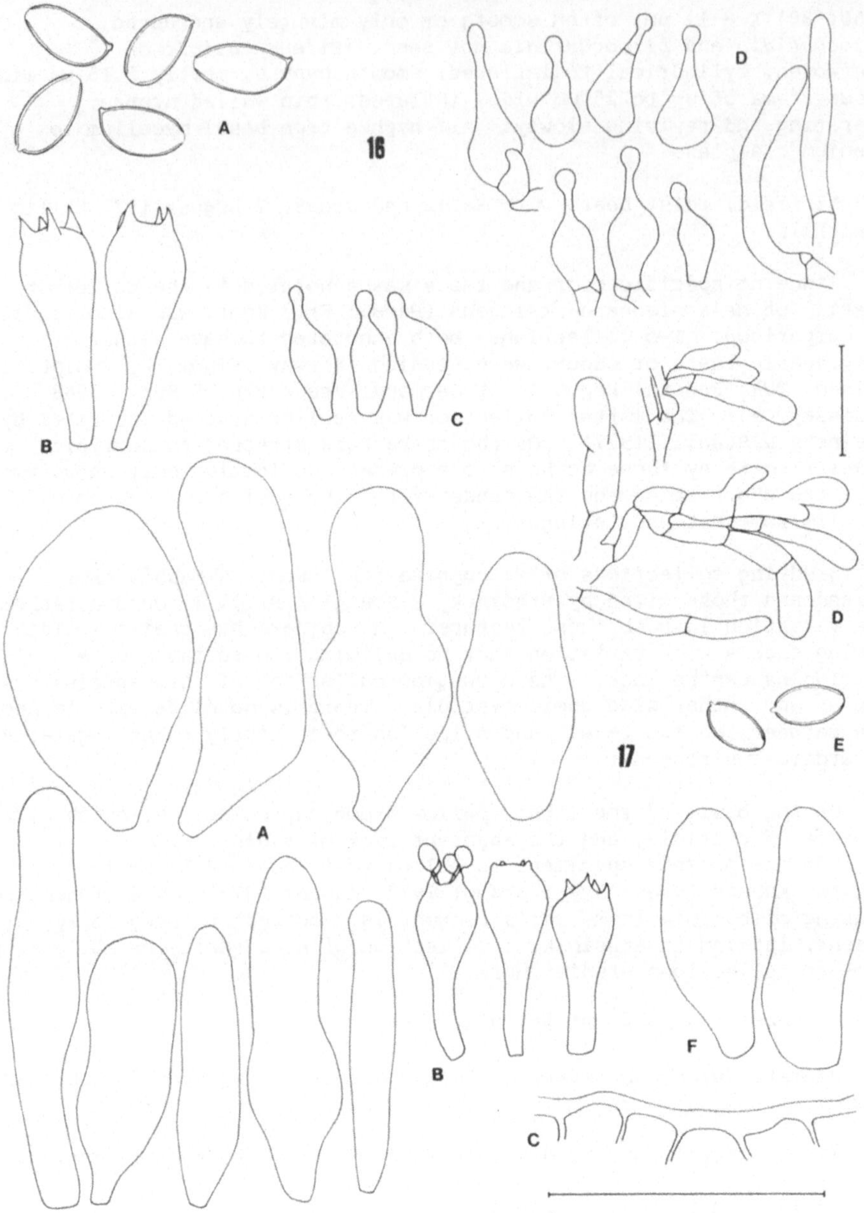

Figs. 16-17, scales 50 µm. Fig. 16. *Conocybe* sp. A) spores, B) basidia, C) cheilocystidia, D) caulocystidia. Fig. 17. *Bolbitius* cf. *variicolor* A) cheilocystidia, B) basidia, C) gelatinized layer of cuticle, in cotton blue, D) caulocystidia, E) spores, F) pleurocystidia.

Pileus with a dark brown disc, viscid, margin lighter brown, translucent striate, diameter up to 2 cm when dried, margin grooved, fragile, disc black, otherwise pileus brown with an olivaceous tinge (Y60-M50-C40). When fresh, *stipe* white, apex pure yellow, base also with some yellow color, longitudinally striate, pruinose; when dried unicolorous, yellowish (Y40-M20-C00), slightly longer than cap diameter.

Mature *gills* light brown when fresh, somewhat darker than Y90-M60-C40 when dried, young gills yellowish when dried (Y50-M20-C00). *Spores* inequilaterally elliptic in sideview, smooth, 11-13 X 6-7 µm, L/W 1.7-1.8(-1.9), average of 25 spores 11.8 X 6.7 µm, average L/W 1.8. *Basidia* 4-spored, abruptly clavate, 30-38 X 10-12 µm covered with a strongly cyanophilous coating around the sterigmata. *Cheilocystidia* abundant, large, hyaline, broadly cylindrical, clavate-ventricose to more rarely rounded, up to 70 X 19 µm or 65 X 13 µm. *Pleurocystidia* scattered, clavate-cylindrical, ca. 43 X 16 µm. *Caulocystidia* abundant on the upper part of stipe, very variable in shape and size, clavate-cylindrical, thin walled, entangled, free to appressed, 60-100 X 8-11 µm. *Pileus cuticle* with a 3 µm thick, gelatinized layer (in cotton blue). All hyphae clampless.

Longyearbyen, at the village, 4 August 1977, Kallio & Heikkilä.

This collection comes close to *B. variicolor* Atk. according to Dr. Watling. Although the field notes state cap color as brown, the olivaceous tinge might have been present, at least in youth. The bright yellow color or stipe is especially noteworthy. Deviations in microscopical features from Watling's (1982) description are restricted to cheilocystidia, which in the present collection are much more prominent.

I also studied a collection of *B. variicolor* (BRD, Karlsruhe-Durlach, September 1974 Schwobel) cited by Kriegelsteiner (1983). Macroscopically it differs from Svalbard material in darkness of cap color. Colors are alike at the margin, but the blackish disc and otherwise darker brown caps in arctic collections mark a difference. Spore size and shape are alike. According to the description by the collector, cheilocystidia are twofold: roundish, measuring 20-60(-75) um in diameter and smaller, lageniform. They revive hardly at all and remain collapsed, so the comparison is difficult. Limits of variation are broad in both collections and I consider the differences meager. But a marked difference is found in basidia. In Schwöbel's collection they are remarkably small even for a member of Bolbitiaceae, measuring only 16-20 X 10-13 µm.

Finnish material of the species was available for study (Finland, Turku, 31 July 1984, Huhtinen 84/94). This rather abundant collection is in perfect agreement with the description of *B. variicolor* by Watling (1982). Olivaceous tinges were especially prominent on caps, the discs were always dark brown. In contrast to Schwobel's collection, these shades are well preserved when dried. In aged caps the green colors are lost and the overall coloring turned to dark brown and corresponds well with Svalbard material, which has very little green when dried.

Roundish cheilocystidia measuring 30-50 µm across were seen in abundance in mounts of fresh material, mixed with clavate and ventricose cells. Basidia were found intermediate between the two earlier collections. In my collection the total range in length is 18-32 µm and a majority of basidia measured 20-25 µm.

The restricted coverage of *B. variicolor* in the literature and the differences seen between these three collections are the reason for the preliminary determination of Svalbard specimens. However, these collections are so closely linked that only further material will most likely verify the conspecificity.

Conocybe magnicapitata Orton Fig. 15

Pileus up to 1 cm in diameter when fresh, conico-convex, when dried, the colour is Y60-M50-C50 (somewhat lighter than 5E6). *Stipe* 40/1 mm when fresh, yellowish brown above, darker brown below, the uppermost 15 mm minutely pruinose. *Gill* colour Y90-M60-C50 (6E7) when dried. *Spores* elliptic, 9-11 X 5.0-6.5 µm, L/W 1.6-1.7, average 10.0 X 6.0 µm (50 spores), average L/W 1.7; smooth, germ pore prominent, notably darkening in alkali. *Basidia* 4-spored, 17-30 X 9-10 µm, base occasionally ill-defined, lacking a basal clamp. *Cheilocystidia* lecythiform, 22-26(-30) X 10-18 µm, head 5.0-8.0 µm. *Caulocystidia* of the same type, somewhat more robust, 25-30 X 14.0-15.5(-18.0) µm, head 6-10 µm. *Pileocystidia* rather abundant, lecythiform, 25-30(-42) X 7-13 um, head 5-7 µm, often totally brown. *Pileus cuticle* hymeniform, the clavate elements are thin to notably thick walled. Hyphae clamped in gill trama, cuticle and stipe, but clamps seem to be infrequent and are often inconspicuous. No needle-shaped crystals were formed in ammoniacal mounts.

Gramineous site at the village, clayey ground, 28 July 1980, Kallio & Huhtinen.

Dr. Watling kindly verified my determination. Although I found some characters deviating from Watling's (1982) treatment, this specimen falls within the limits of variation for *C. magnicapitata*. A reference collection from Perthshire, Scotland (23 September leg. & det. Watling in TUR) also showed clamped hyphae and the lecythiform pileocystidia. The pileus color was different between these two collections, but not significantly. Dr. Watling, after checking additional collections of *C. magnicapitata*, verified the occurrence of pileocystidia in this species. (Watling, pers. comm.).

Conocybe sp. Fig. 16

Same site as the previous species, no dung observed around, 30 July 1983, Huhtinen 83/132.

This collection consists of only one fruit-body. It represents an undescribed taxon. My view was kindly verified by Dr. Watling. The species is a member of the *C. pubescens* group being characterized by lecythiform caulocystidia mixed with hair-like cells, a smooth pileus and large spores. Though sparse or lacking in the upper portion of the stipe, these hair-like cells become abundant toward the base. Spores measure (15-)16-18(-18.8) X 8.5-9.5(-9.8) µm, with a L/W -ratio of 1.7-1.9. In face view they are elliptic, in side view slightly elongate to amygdaliform. A germ pore is always prominent. The lecythiform cystidia on stipe and hymenium are alike with heads only 3 µm wide. In contrast to the other species in stirps *Pubescens*, the cap is totally smooth (cf. Watling 1982).

The different shades of brown in dried condition are as follows: cap Y60-M40-C20-30, stipe Y90-M60-C50 (6E7), gills Y60-M40-C30 (5D6). No crystals are formed in ammoniacal mounts. Hyphae are clamped. The collection shows no rhizoids, no distinct basal bulb, nor a pseudorhiza.

Due to the scarce material I refrain from describing this as new, though its character combination would justify doing so.

Cortinariaceae

Inocybe leucoblema Kühner Figs. 18,21

Pileus up to 2 cm in diameter when fresh, with a broad umbo, dull yellowish brown, mat, structure strongly appressed-fibrillose throughout,

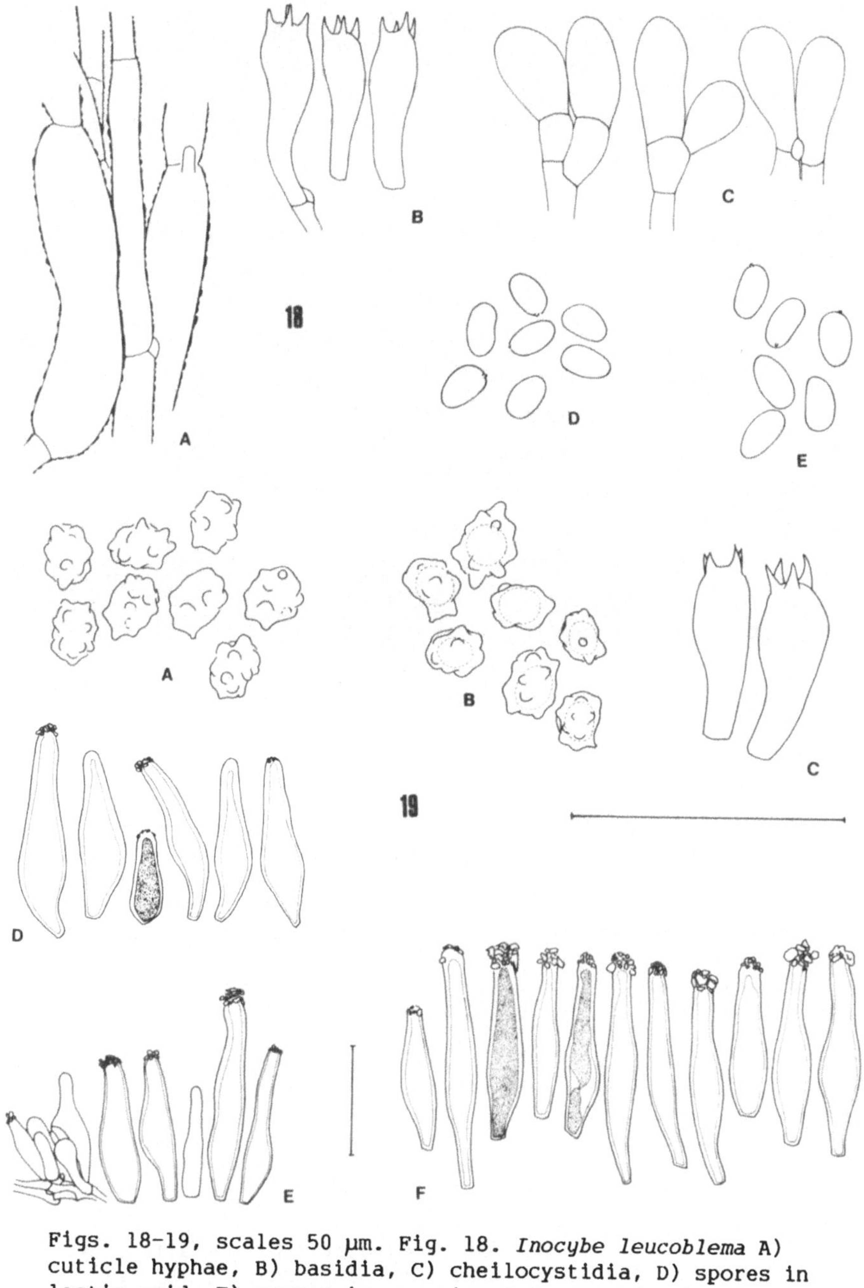

Figs. 18-19, scales 50 µm. Fig. 18. *Inocybe leucoblema* A) cuticle hyphae, B) basidia, C) cheilocystidia, D) spores in lactic acid, E) spores in ammoniacal Congo red. Fig. 19. *Inocybe praetervisa* A) spores in lactic acid, B) spores in KOH, C) basidia, D) pleurocystidia, E) caulocystidia, F) cheilocystidia.

margin entire, not rimose, somewhat inrolled; when ca. 8 mm wide pileus nearly totally covered with prominent white velum, remaining at the margin of expanded caps as prominent white bands, ca. 1 mm apart from the extreme margin, also fragmentarily present more centrally; dried caps unicolorous, pale brown (N75, 5D6), mat. Trama moderately thick also at

Fig. 20. *Inocybe praetervisa*, Huhtinen 83/249. Photo by collector. Scale 1 cm.

the margin. *Odor* slightly spermatic. *Lamellae* adnate to subdecurrent, horizontal, L = 30–40, light brownish when fresh, edge minutely uneven sub lenta, whitish; when dried, gills brown (P69). *Stipe* 20 mm long, 4 mm wide, slightly widening toward base, somewhat lighter than cap, basally white due to the abundant tomentum, white fibrils characterize upper stipe, becoming darker brown when handled. *Spores* oblong to subreniform with a slightly tapered apex and a frequently blunt base, 9.0–10.3 X 5.0–5.9 µm, average of 30 spores 9.6 X 5.2 µm, average L/W 1.8, smooth, apiculus minute, spores with a very minute germ pore. *Basidia* 4-spored. *Cheilocystidia* very rare, broadly clavate, 15–30 X 8–11 µm. *Pileus cuticle* composed of parallel, narrow to inflated, brown encrusted hyphae, measuring ca. 45–60 X 5–12 µm, constricted at septae, encrustation 0.2–0.3 µm high. Velar hyphae 4–6 µm wide, hyaline, smooth. Stipe cortex of similar but generally shorter hyphae, only very minutely encrusted to smooth. Clamp connections present in all parts of the fungus.

Airfield, on fine fluvial soil accumulated by a small brook, already vegetated by e.g. *Salix polaris*, 29 July 1983, Huhtinen 83/115.

There is only one critical character in which the present specimens deviate from *Inocybe leucoblema* Kühner. That is the nearly total lack of morphologically different sterile cells on gill edges. In all collections studied for comparison (France, Pralognan, 18 September 1971, leg. Bresinsky, det. Kühner; Federal Republic of Germany, Bubingen, 23 June 1974, leg. & det. Stangl, nr. 1066, both in M; and numerous Finnish collections), such cells are very abundant. But all the other characters, and above all the abundant white velum remaining on expanded caps too, clearly make the collection *I. leucoblema* (cf. Kühner 1955). Coloring is somewhat darker than in the reference collection determined by Prof.

Fig. 21. *Inocybe leucoblema*, Huhtinen 83/115. Photo by collector. Scale 1 cm.

Kühner, but not exceeding that observed in Finnish collections of the species.

Lack of cheilocystidia is known also for another species of group *Dulcamarae*; i.e., in *I. dulcamara* var. *homomorpha* Kühn. This taxon is not, however, characterized by a truly prominent velum as is the present taxon. Occasionally scattered, morphologically distinguishable cells can be found in the collection, but it is not possible to decide whether the sterile cells are basidioles or not. The germ pore is truly minute even with X1500 magnification and comprises a slight decrease in wall thickness only.

In Finland, *I. leucoblema* is known from the calcareous coniferous sites of S. Finland to the subalpine birch forests of northernmost Finland (J. Vauras, pers. commun.). According to Prof. Kühner (pers. comm.), this species is not uncommon above the timberline and is known from such sites in Switzerland, Norway and Sweden.

Inocybe praetervisa Quél. Figs. 19,20

Pileus up to 22 mm in diameter when fresh, broadly conical, umbonate, ochraceous brown when fresh, umbo relatively broad, with nonseparated fibrils, showing indistinct, appressed squamules, fibrils well separated at the margin (as in *I. fastigiata*) and becoming strongly entangle, pileus brown (Y70-M50-C40, N67) when dried. *Stipe* 30 mm high, 3 mm wide when fresh, first pure white, later becoming brown between the white striate, devoid of red tinges, most stipes turn brownish when dried, base bulbous, but bulb not abruptly marginate; the amount of caulocystidia rapidly decreasingfrom the middle part downwards, basal parts nearly smooth. *Gills* yellowish in age, quite concolorous to somewhat more yellow than pileus in dried material, edge easily attaining dark brown color. Odor not distinctive. *Spores* oblong, tuberculate, 9.5-12.8(-14.0) X 7.0-9.0(-10.0) µm, L/W 1.2-1.5, average of 30 spores 11.0 X 8.1 µm, average L/W 1.4; tubercules numerous, 1-2 µm wide, 1-2(-3) µm high, spores uniguttulate in KOH. *Basidia* 4-spored, 33-50 X 10-14 µm. *Cheilocystidia* narrowly ventricose to fusoid, typically with a narrow base, 60-100(-110) X (10-)12-18 µm, wall reaching 3 µm above, 2 µm in the middle part, often with yellow to deep orange-brown contents in KOH, typically strongyl encrusted. *Pleurocystidia* shorter and wider, ventricose, 50-70(-100) X 13-22 µm, encrusted or smooth, they may also be colored. *Caulocystidia* extremely variable in shape and size, clavate, cylindrical to ventricose, +/- encrusted, walls thick or thin. *Pileus cuticle* of encrusted hyphae, at the margin strongly encrusted and measuring 70-120 X 7-15 µm. All hyphae are clamped.

Airfield, arctic heath with *Salix polaris* and a well developed moss carpet, 5 August 1983, Huhtinen 83/249.

According to Stangl (1983), the main characters of *Inocybe praetervisa*, separating it from *I. mixtilis* (Britz.) Sacc., are 1) a more fibrous cuticle; 2) stipes turn brownish in age and when dried; 3) not so abruptly delineated basal bulb; 4) larger spores and basidia; and 5) more elongated cystidia. Some of these characters were already stressed by Kühner & Romagnesi (1953). My collection has all of the characters of *I. praetervisa* with some minor deviations. Cheilocystidia are even more fusoid than seen in the illustrations by Stangl (1977, 1983) and Alessio & Rebaudengo (1980). However, Favre (1955) had material with elongated cystidia, being close to present material.

Another noteworthy character is seen on gill edges, which in the oldest caps turns brown. Under the microscope this brown zone is observed in the deep orange-brown contents of cystidia.

Inocybe praetervisa var. *rufofusca* Favre was reported from Svalbard by Ohenoja (1971). This taxon, as *I. pseudohiulca* Kühner, has a cap darker brown than in the present material.

ACKNOWLEDGEMENTS

This study was supported by the Academy of Finland. I am also indebted to all the colleagues who have helped me in so many ways.

REFERENCES

Alessio, C. L., and Rebaudengo, E., 1980. Inocybe, *Iconographia Mycologica*, 29, Suppl. 3. 367 pp., 100 pl. Trento.

Arvidsson, L., 1979, *Peniophora pithya* funnen på Spetsbergen, *Svensk. Bot. Tidskr.*, 72: 293-294.

Berthier, J., 1976, Monographie des *Typhula* Fr., *Pistillaria* Fr. et genres voisins, *Bull. Soc. Linn. Lyon, Numero Spec.*, vol. 45: 1-213.

Bresinsky, A., and Stangl, J., 1977, Beiträge zur Revision M. Britzelmayrs "Hymenomyceten aus Sudbayern" 13. Die Gattung *Melanoleuca* unter besonderer Berücksichtigung ihrer Arten in der Umgebung von Augsburg, *Zeitschr. f. Pilzk.*, 43: 145-173.

Cailleux, A., 1981, "Code des couleurs des sols."

Carpenter, S. E., 1981, Monograph of *Crocicreas* (Ascomycetes, Helotiales, Leotiaceae). *Mem. New York Bot. Gard.*, 33: 1-290.

Clémençon, H., 1982, Kompendium der Blätterpilze. Europäische omphalinoide Tricholomataceae, *Z. Mykol.*, 48: 195-237.

Dennis, R. W. G., 1949, A revision of the British Hyaloscyphaceae with notes on related European species, *Mycol. Pap.*, 32: 1-97.

Dennis, R. W. G., and Itzerott, H., 1973,*Octospora* and *Inermisia* in Western Europe, *Kew Bull.*, 28: 5-23.
Dissing, H., 1966, The Genus *Helvella* in Europe, with Special Emphasis on the species Found in Norden, *Bot. Arkiv.*, 25: 1-172.
______, 1982, Operculate discomycetes (Pezizales) from Greenland, p. 53-81, *in*: "Arctic and Alpine Mycology. The First International Symposium on Arcto-Alpine Mycology," 559 pp., G. A. Laursen and J. F. Ammirati eds., University of Washington Press, Seattle and London.
Dissing, H., and Sivertsen, S., 1983, Operculate Discomycetes from Rana (Norway) 4. *Octospora hygrohypnophila*, *Peziza prosthetica* and *Scutellinia mirabilis* spp. nov., *Nord. J. Bot.*, 3: 415-421.
Döbbeler, P., 1978, Moosbewohnende Ascomyceten 1. Die pyrenocarpen, den Gametophyten besiedelnden Arten, *Mitt. Bot. Staatssamml. München*, 14: 1-360.
______, 1979, Moosbewohnende Ascomyceten 3. Einige neue Arten der Gattungen *Nectria*, *Epibryon* und *Punctillum*, *Mitt. Bot. Staatssamml. München*, 15: 193-221.
Eckblad, F. -E., 1968, The genera of the Operculate Discomycetes. A Re-evaluation of their Taxonomy, Phylogeny and Nomenclature, *Nytt Mag. Bot.*, 15: 1-191.
______, 1969, Contributions to the Sclerotiniaceae of Norway, *Friesia*, 9: 4-9.
Favre, J., 1955, Les champignons supérieurs de la zone alpine du Parc National Suisse, *Ergeb. Wiss. Untersuch. Schweiz. Nationalparks*, 5: 1-212.
Haines, J. H., 1974, Notes on the genus *Trichopezizella* with descriptions of new taxa, *Mycologia*, 66: 213-241.
Harmaja, H., 1977, A revision of the *Helvella acetabulum* group (Pezizales) in Fennoscandia, *Karstenia*, 17: 45-58.
______, 1984, *Cystoderma adnatifolium* and *C. arcticum* n.sp. in Spitzbergen, *Karstenia*, 24: 31-32.
Høiland, K., 1976, The genera *Leptoglossum*, *Arrhenia*, *Phaeotellus*, and *Cyphellostereum* in Norway and Svalbard, *Norw. J. Bot.*, 23: 201-212.
Holm, L, 1975, Taxonomic Notes on Ascomycetes 8. Microfungi on *Cassiope tetragona.*, *Svensk Bot. Tidskr.*, 69: 143-160.
Holm, L., and Holm, K., 1980, Microfungi on *Cassiope* (*Harrimanella*) *hypnoides*, *Norw. J. Bot.*, 27: 179-184.
Huhtinen, S., 1985a, Mycoflora of Poste-de-la-Baleine, northern Québec. Ascomycetes. *Naturaliste can.* 112: 473-524.
______, 1985b, *Marasmius kallioneus*, a new Arctic species, *Mycol. Helvetica*, 1(5): 341-351.
Itzerott, H., 1981, Die Gattung *Octospora* mit besonderer Berücksichtigung der Pfalzer Arten, *Nova Hedwigia*, 34: 265-280.
______, 1983, *Octospora melina*, ein seltener Gallbildner, *Agarica*, 8:108-114.
Kobayasi, Y., Tubaki, K., and Soneda, M., 1968, Enumeration of the Higher Fungi, Moulds and Yeasts of Spitsbergen, *Bull. Nat. Sci. Mus. Tokyo*, 11: 33-75.
Kohn, L. M., 1979, A monographic revision of the genus *Sclerotinia*, *Mycotaxon*, 9: 365-444.
Korf, R. P., 1982, Inoperculate discomycetes of the arctic and alpine zones of Finnmark, Lapland, and Greenland, p. 27-34, *in*: "Arctic and Alpine Mycology. The First International Symposium on Arcto-Alpine Mycology," 559 pp., G. A. Laursen, and J. F. Ammirati eds., University of Washington Press, Seattle and London.
Kornerup, A., and Wanscher, J. H., 1961, Värien kirja. 260 pp. WSOY. Porvoo. Krieglsteiner, G.J. 1983. Über neue, seltene, kritische Makromyzeten in der Bundesrepublik Deutschland. 4, *Z. Mykol.*, 49: 73-106.

Kullman, B., 1982, A revision of the genus *Scutellinia* (Pezizales) in the Soviet Union, *Scripta Mycol.,*. 10: 1-158.
Kühner, R., 1955, Compléments à la "Flore Analytique." 6. *Inocybe* goniosporés et *Inocybe* acystidiés. Espèces nouvelles ou critiques, *Bull. Soc. Mycol. France,* 71: 169-201.
______, 1978, Agaricales de la Zone Alpine. Genre *Melanoleuca* Pat., *Bull. Soc. Linn. Lyon,* 47: 12-52.
______, and Romagnesi, H., 1953, "Flore Analytique des Champignons supérieurs," 556. pp.
Küppers, H., 1978, "DuMont's Farben-Atlas," 163 pp., DuMont Buchverlag, Koln.
Lamoure, D., Lange M., and Petersen, P. M., 1982, Agaricales found in the Godhavn area, West Greenland, *Nord. J. Bot.,* 2: 85-90.
Le Gal, M., 1966, Un *Scutellinia* peu commun: *Scutellinia arenosa* (Vel.) Le Gal nov. comb., *Bull. Soc. Mycol. France,* 82: 623-626.
Moravec, J., 1969, Některé operkulátní diskomycety nalezené v okresech Mladá Boleslav a Jičín, Some operculate discomycetes from the districts of Mladá Boleslav and Jičín (Bohemia), *Česká Myko.,* 23: 222-235.
______, 1974, Several operculate Discomycetes from Greece and remarks on the genus *Scutellinia* (Cooke) Lamb. emend. Le Gal., Nekolik operkulátních diskomycetu z Řecka a poznamky k rodu *Scutellinia* (Cooke) Lamb. emend. Le Gal., *Česká Mykol.,* 28: 19-25.
Nannfeldt, J. A., 1981, *Exobasidium,* a taxonomic reassessment applied to the European species, *Symb. Bot. Upsal.,* 23: 1-72.
______, 1984, *Hysteronaevia,* a new genus of mollisioid Discomycetes, *Nord. J. Bot.,* 4: 225-247.
Ohenoja, E., 1971, The larger fungi of Svalbard and their ecology, *Rep. Kevo Subarctic Res. Stat.,* 8: 122-147.
Reid, D. A., 1979, Some fungi from Spitsbergen, *Rep. Kevo Subarctic Res. Stat.,* 15: 41-47.
Schumacher, T., 1979, Notes on taxonomy, ecology, and distribution of operculate discomycetes (Pezizales) from river banks in Norway, *Norw. J. Bot.,* 26: 53-83.
Sherwood, M. A., 1977, The ostropalean fungi, *Mycotaxon,* 5: 1-277.
Skifte, O., 1979, Storsopp på Svalbard, *Ottar,* 110-112: 29-39.
Stadelmann, R. J., 1978, Beitrag zur Kenntnis der Discomyceten-Gattung *Belonioscypha* Rehm, *Nova Hedwigia,* 30: 815-833.
Stangl, J., 1977, Die eckigsporigen Risspilze (3), *Zeitschr. f. Pilzk.,* 43: 131-144.
______, 1983, *Inocybe praetervisa* Quél. und *Inocybe mixtilis* (Britz.) Sacc. - eine Gegenüberstellung wichtiger. Trennungsmerkmale, *Agarica,* 8: 18-22.
Svrček, M., 1971, Tschechoslowakische Arten der Diskomyzetengattung *Scutellinia* (Cooke) Lamb. emend. Le Gal (Pezizales) 1, Československé druhy rodu *Scutellinia* (Cooke) Lamb. emend. Le Gal (Pezizales). 1, *Česká Mykol.,* 25: 77-87.
______, 1977, New or less known Discomycetes. 6, Nové nebo méně známé diskomycety. 6, *Česká Mykol.,* 31: 193-200.
______, 1978, New or less known Discomycetes. 9, Nové nebo méně známé diskomycety. 9, *Česká Mykol.,* 32: 202-204.
______, 1979a., A taxonomic revision of Velenovský's types of operculate discomycetes (Pezizales) preserved in National Museum, Prague, *Arta Mus. Nat. Pragae,* 32B: 115-194.
______, 1979b, New or less known Discomycetes. 10, Nové nebo méně známé diskomycety. 10, *Česká Mykol.,* 33: 193-206.
______, 1983, New or less known Discomycetes. 12, Nové nebo méně známé diskomycety. 12, *Česká Mykol.,* 37: 65-71.
Svrček, M., and Moravec, J., 1969, Species novae Discomycetum (Pezizales) e Bohemia. Nové druhy operkulátních diskomycetu z Čech., *Česká Mykol.,* 23: 156-159.

Watling, R., 1982, Bolbitiaceae: *Agrocybe*, *Bolbitius* & *Conocybe*, *British Fungus Flora* 3, 138 pp. Edinburgh.

______, 1983, Larger cold-climate fungi, *Sydowia*, 36: 308–325.

PHAEOSPHAERIA IN THE ARCTIC AND ALPINE ZONES

Adrian Leuchtmann

Department of Microbiology
Swiss Federal Institute of Technology
CH-8092 Zürich, Switzerland

Key words: Ascomycetes, Arctic and Alpine, Loculoascomycetes

ABSTRACT

Sixteen arctic-alpine species of *Phaeosphaeria* are presented in a key with short descriptions and notes on host plants, distribution, and conidial states. Illustrations of all the species are also included.

The geographical distribution of the arctic-alpine species forms three groups: species restricted to the Alps, species restricted to the Arctic, and species found both in the Alps and in the Arctic. This may be explained by the relation of the fungi to the host plants.

The genus *Phaeosphaeria* shows many ecological adaptations to Arctic-alpine conditions: the simplification of life cycle and breeding system, the formation of thick, deeply pigmented walls of ascomata and ascospores, and ascospores coated with gelatinous sheaths.

INTRODUCTION

Phaeosphaeria Miyake is an ascomycetous genus, abundant in species, belonging in the Loculascomycetes. Characteristics of *Phaeosphaeria* are: small, perithecioid, thin-walled ascomata: bitunicate, cylindrico-clavate asci surrounded by pseudoparaphyses; ellipsoidal to fusiform, phragmosporous (rarely dictyosporous) and usually pigmented ascospores. These parasitic microfungi form ascomata mainly on dead stems and leaves of Poaceae, Cyperaceae, and other monocotyledons, and rarely also on dicotyledons (Caryphyllaceae) and pteridophytes. Many representatives of this genus are conspicuously host-specific, found only in the distribution area of the host plant. Some species are specialized also on alpine or Arctic plants.

A revision of some northern species (especially on Poaceae) was made by Holm (1957) and Eriksson (1967). The entire genus was taxonomically detailed in a recent work (Leuchtmann 1984) based on morphological and cultural studies. In the present article, the *Phaeosphaeria* species of the arctic and alpine zones are presented in a key, with short descriptions and notes on host plants, distribution, and conidial states. Distribution and some characteristics of adaptation to arctic-alpine conditions are discussed in the second part of the article.

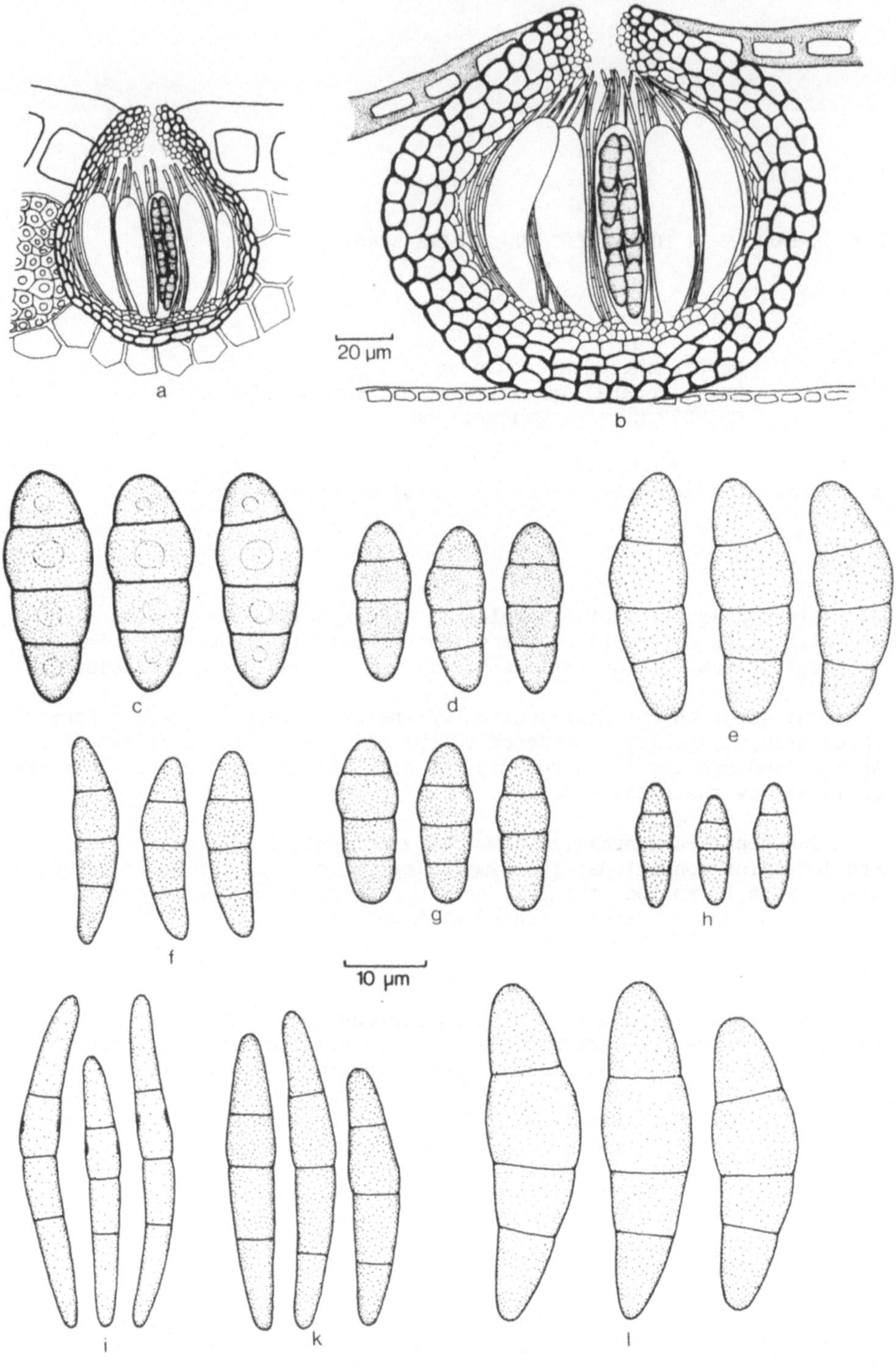

Fig. 1. a-b. ascomata: a. *Ph. lutea*; b. *Ph. alpina*, c-l. ascospores: c. *Ph. oreochloae*; d. *Ph. alpina*; e. *Ph. caricinella*; f. *Ph. tofieldiae*; g. *Ph. microscopica*; h. *Ph. lutea*; i. *Ph. juncicola*; k. *Ph. silenes-acaulis*; l. *Ph. dennisiana*.

1. Ascospores with 3 transverse septa. 2
1. Ascospores with more than 3 transverse septa 10
2. On monocotyledons. 3
2. On Caryophyllaceae. 9
3. Ascospores more than 6 times as long as wide. Ascomata globose or depressed, 90-200 µm diam., mostly very thick-walled (up to 40 µm). Asci broadly cylindrical, 50-60 X 12-15 µm. Ascospores 4-seriate, pale yellowish; end segments longer than the middle segments; 30-46 x 4-4.5 µm. On *Juncus* (mainly *J. trifidus*), rarely on *Luzula* or *Carex*, arctic-alpine to subalpine. Anamorph absent. (Fig. 1,i)
 1. *Ph. juncicola* (Rehm) Holm
3. Ascospores less than 6 times as long as wide 4
4. Ascospores with granular ornamentations. 5
4. Ascospores smooth. 6
5. Ascospores at least 18 µm long. Ascomata subglobose, 100-150 µm diam., thin-walled. Asci cylindrical, 60-85 x 13-18 µm. Ascospores 2-seriate, brown to yellow-brown, with granular ornamentations 18-23 x 6-7.5 µm; gelatinous sheath thick, entire. On Poaceae or *Carex*, arctic-alpine to subalpine. Anamorph absent. (Fig. 1,g)
 2. *Ph. microscopica* (Karst.) O. Eriksson
5. Ascospores not more than 16.5 µm long. Ascomatat globose to ovate, 60-90 µm diam., very thin-walled. Asci cylindrico-clavate, 48-60 x 10 µm. Ascospores 2-seriate, yellowish, with granular ornamentations, 14-16.5 x 4-5 µm; gelatinous sheath thick, entire. On *Luzula lutea*, alpine in the Alps. Anamorph absent.(Fig. 1,a,h)
 3. *Ph. lutea* Leuchtmann
6. Wall of the ascomata up to 40 µm thick, composed of large isodiametric cells. 7
6. Wall of the ascomata not more than 20 µm thick, composed mostly of strongly flattened cells (the arctic *Ph. caricinella* occasionally shows thick walls too). 8
7. Ascospores ellipsoidal, at most 23 µm long. Ascomata subglobose, 100-160 µm diam., thick-walled. Asci broadly cylindrical, 50-75 x 15-17 µm. Ascospores irregularly 2-seriate, yellow-brown, thick-walled, smooth, 18-23 x 6.5-8.5 µm; gelatinous sheath thick, umbilicate on both ends of the spore. On Poaceae and other monocotyledons, alpine to subalpine in the Alps. Anamorph: *Stagonospora*. (Fig. 1,b,d; 3,d)
 4. *Ph. alpina* Leuchtmann
7. Ascospores usually clavate, at least 23 µm long. Ascomata subglobose, 100-180 µm diam., thick-walled. Asci ellipsoidal to broadly cylindrical, 70-80 x 25 µm. Ascospores irregularly 2 to 3-seriate, yellow-brown to pale grey-brown, thick-walled, smooth, 23-31 x 9-11 µm; gelatinous sheath thick, didymous, umbilicate on both ends of the spore. On *Oreochloa disticha*, alpine in the Alps. Anamorph absent. (Fig. 1,c) 5. *Ph. oreochloae* Leuchtmann
8. Ascospores not more than 3.5 times as long as wide. Ascomata subglobose, 150-250 µm diam., mostly thin-walled (Arctic forms also thick-walled). Asci cylindrico-clavate, 90-115 x 15-25 µm. Ascospores irregularly 2-seriate, pale yellow-brown, smooth, 27-35(38) x 9-11 µm; gelatinous sheath thick, entire. Mainly on *Carex*, rarely on Poaceae and Juncaceae, arctic to subarctic. Anamorph: *Stagonospora*. (Fig. 1,e; 3,a)
 6. *Ph. caricinella* (Karst.) O. Eriksson

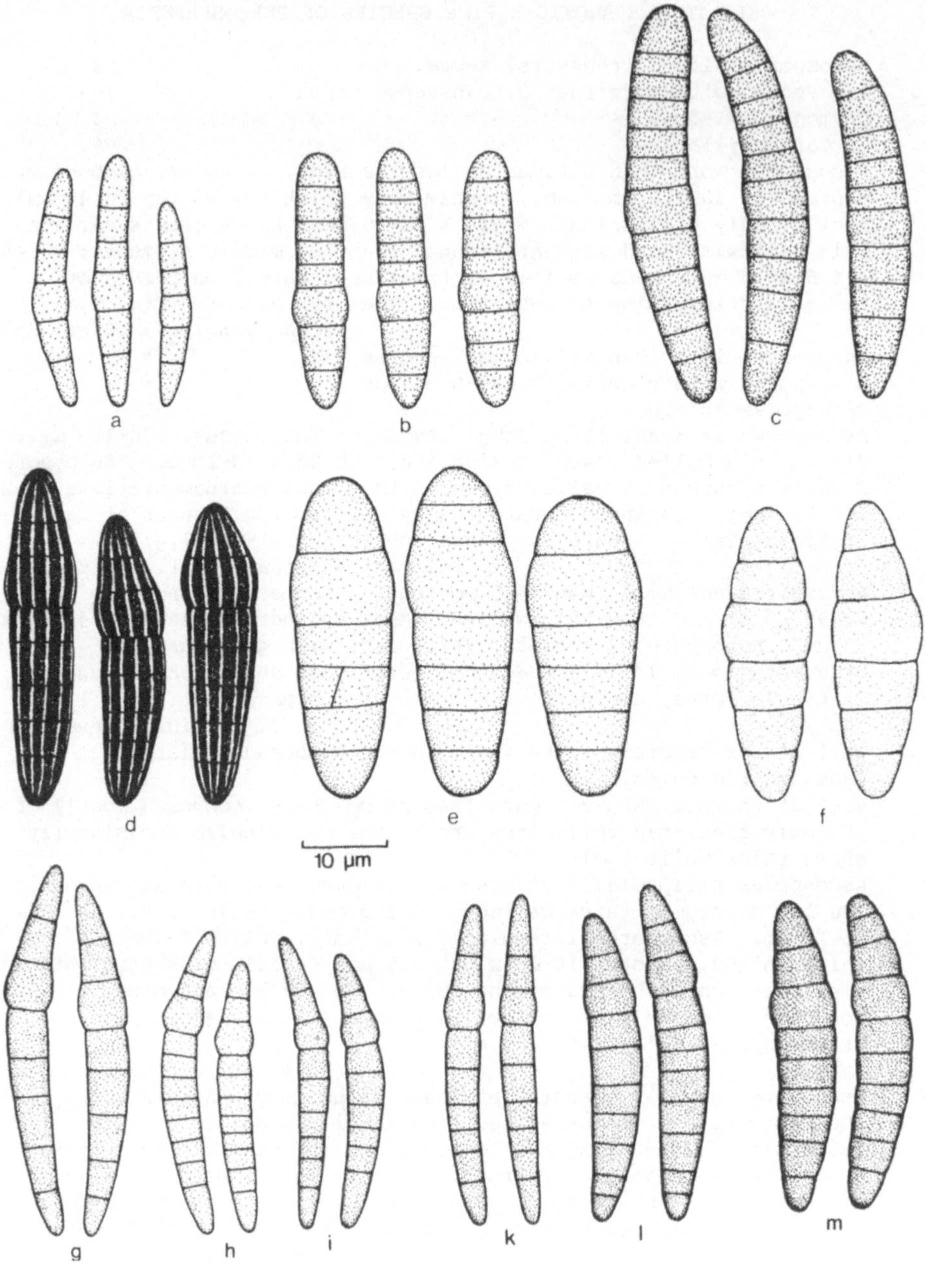

Fig. 2. ascospores: a. *Ph. nardi*; b. *Ph. lindii*; c. *Ph. equiseti*; d. *Ph. pleurospora*; e. *Ph. hierochloes*; f. *Ph. insignis*; g-l. *Ph. herpotrichoides* (different forms); m. *Ph. volkartiana*.

8. Ascospores at least 4 times as long as wide. Ascomata globose, 60-125 µm diam., thin-walled. Asci broadly cylindrical, 45-60 x 10-13 µm. Ascospores 2 to 3-seriate, pale yellow-brown, smooth, 20-25 x 4.5-5.5 µm; gelatinous sheath thick, entire. On dead leaf tips of *Tofieldia calyculata*, alpine to montane in the Alps. Anamorph absent. (Fig. 1,f)

7. *Ph. tofieldiae* (Müller) Leuchtmann

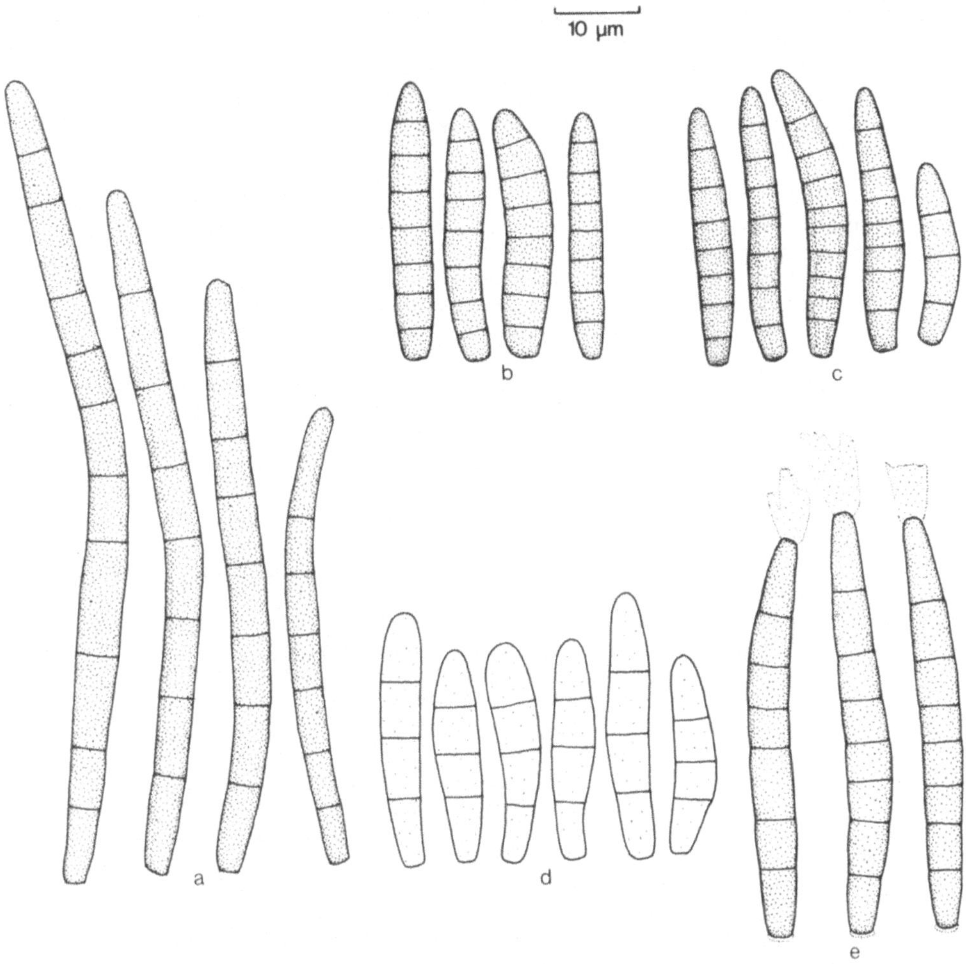

Fig. 3. conidial states (*Stagonospora*) in pure culture; a. *Ph. caricinella*; b. *Ph. herpotrichoides*; c. *Ph. volkartiana*; d. *Ph. alpina*; e. *Ph. insignis*.

9. Ascospores at least 5 times as long as wide, without an inflated segment. Ascomata globose to ovate, 50-80(100) µm diam.; Asci narrowly ellipsoidal, 45-65 x 13-17 µm. Ascospores irregularly packed, yellow-brown, smooth or slightly granular, 27-40 x 5-6.5 µm; gelatinous sheath thin, entire. On leaves of *Silene acaulis* and alpine *Minuartia* species, arctic-alpine, frequent. Anamorph absent. (Fig. 1,k)

8. *Ph. silenes-acaulis* (De Not.) Holm

9. Ascospores less than 5 times as long as wide, second segments slightly inflated. Ascomata globose, 110-160 µm diam.; Asci ellipsoidal to cylindrical, 70-83 x 22-27 µm. Ascospores irregularly 3-seriate, pale yellow-brown, smooth, 33-43 x 9.5-12 µm; gelatinous sheath thick, entire. On *Minuartia sedoides*, alpine, in the Alps and Scotland, rare. Anamorph absent. (Fig. 1,l)

9. *Ph. dennisiana* Leuchtmann

10. Primary septum of the ascospores median; both hemispores with the same number of transverse septa. Ascomata globose, 120-200 µm diam.; Asci clavate, 100-130 x 20-27 µm. Ascospores irregularly 2 to 3-seriate, pale yellow-brown, smooth, 5-septate (rarely, an additional septum in one of the end segments), 32-45 x 9.5-14 µm; gelatinous sheath thick, entire. On Poaceae and *Carex*, Arctic-alpine. Anamorph: *Stagonospora*. (Fig. 2,f; 3,e)

10. *Ph. insignis* (Karst.) Holm

10. Primary septum of the ascospores not median; the hemispores with different number of transverse septa. 11

11. Proximal hemispore with the greater number of transverse septa; inflated segment, if present, supramedian. 12

11. Distal hemispore with the greater number of transverse septa; inflated segment submedian 14

12. Ascospores with distinct longitudinal stripes. Ascomata globose to ovate, 80-160 µm diam.; Asci broadly cylindrical to ellipsoidal, 65-100 x 18-22 µm. Ascospores mostly irregularly packed, 5 to 8-septate, 30-42 x 7.5-9 µm; walls composed of brown, later nearly black, fillets; gelatinous sheath thick, entire. On Poaceae, Cyperaceae and other monocotyledons, alpine to subalpine in the Alps. Anamorph absent. (Fig. 2,d)

11. *Ph. pleurospora* (Niessl) Leuchtmann

12. Ascospores without stripes. 13

13. Ascospores broadly clavate, distal hemispore mostly 1-septate. Ascomata globose, 200-260 µm diam.; Asci broadly clavate, 100-125 x 30-35 µm. Ascospores irregularly 2 to 3-seriate, pale yellow-brown, smooth, 5 to 6-septate, 35-42 x 12-15 µm; gelatinous sheath thin, entire. On *Hierochloe alpina*, in the Arctic. Anamorph not known. (Fig. 2,e)

12. *Ph. hierochloes* (Oudem.) O. Eriksson

13. Ascospores fusiform, distal hemispore 2 to 3-septate. Ascomata pyriform, 200-350 µm diam., with a long, neck-like ostiolum. Asci cylindrico-clavate, 80-120 x 10-13 µm. Ascospores 2 to 3-seriate, yellow-brown, with distinct granular ornamentations, 7 to 8-seriate, usually slightly constricted at the septa, 32-45 x 7-8.5 µm; gelatinous sheath thick, entire. On alpine *Trisetum* species, rarely on other Poaceae, alpine in the Alps. Anamorph: *Stagonospora*. (Fig. 2,m; 3,c)

13. *Ph. volkartiana* (Müller) Hedjaroude

This taxon belongs to the *Ph. herpotrichoides*-complex (cf. Leuchtmann 1984), of which other forms also occur occasionally in the alpine zone.

14. Ascospores with 5 transverse septa.Ascomata globose, 130-240 µm diam., in mountainous regions usually very thick-walled. Asci cylindrico-clavate, 65-80 x 8-10 µm. Ascospores 2-seriate, pale yellow-brown, smooth, 5-septate; forth segment only indistinctly inflated, usually much longer than wide; 23-29 x 3.5-4.5 µm; gelatinous sheath in fresh conditions, composed of 3 parts. On *Nardus stricta*, Arctic-alpine to montane. Anamorph absent. (Fig. 2,a)

14. *Ph. nardi* (Fr.) Holm

14. Ascospores with more than 5 transverse septa. 15

15. Ascospores at least 36 µm long, without a distinctly inflated segment. Ascomata globose, 200-300 µm diam.; Asci cylindrico-clavate, 110-140 x 13-15 µm. Ascospores brownish, slightly granular, 7 to 11-septate, (28)36-48 x 6-7 µm; gelatinous sheath not observed. Mainly on *Equisetum variegatum*, Arctic to subarctic. Anamorph not known. (Fig. 2,c)

15. *Ph. equiseti* (Karst.) L. & K. Holm

15. Ascospores at most 32 µm long, with a distinctly inflated segment. Ascomata and Asci similar to *Ph. equiseti*. Ascospores pale brownish, smooth, 7-septate, 25-32 x 6-7 µm; gelatinous sheath not observed. On *Equisetum variegatum* and *E. scirpoides*, arctic to subarctic. Anamorph not known. (Fig. 2,b)

16. *Ph. lindii* (L. & K. Holm) Leuchtmann

DISTRIBUTION

Among the Arctic-alpine species of the genus *Phaeosphaeria*, three groups may be distinguished relative to the geographical distribution: species restricted to the Alps, species restricted to the Arctic, and species found both in the Alps and in the Arctic (Table 1). This distribution pattern may be partially explained by the relation of the fungi to the host plants.

Phaeosphaeria species are generally considered to be saprophytes, because they form fructifications on the dead tissues of the host plant. Nevertheless, a distinctive host specificity has been observed for many species. These species are found only in the distribution area of the host plant concerned: in the Alps, *Ph. oreochloae* on the alpine *Oreochloa disticha*, *Ph. lutea* on *Luzula lutea*, *Ph. tofieldiae* on *Tofieldia calyculata*, *Ph. volkartiana* on *Trisetum distichophyllum*, *Ph. dennisiana* on *Minuartia sedoides*; in the Arctic, *Ph. hierochloes* on the arctic *Hierochloe alpina*; and in both distribution areas, *Ph. nardi* on the Arctic-alpine *Nardus stricta*, *Ph. juncicola* on *Juncus trifidus*, *Ph. silenes-acaulis* on *Silene acaulis*.

The attack of these host specific fungi is believed to begin on living leaves and stems, since only living plants are able to react selectively to infection of fungi. The parasites may grow at first inside the living tissues as endophytes without symptoms, as has been demonstrated for many similar fungi (e.g., Petrini & Müller 1979). Thus, the plants colonizing new territories may be accompanied by endophytic fungi.

Species which are not found in every distribution area of their host plants are usually polyphagous on a wide host range; in the Alps, *Ph. alpina* on Poaceae and other monocotyledons, and in the Arctic, *Ph. caricinella* on different *Carex* species and other Cyperaceae. These endemic species probably have evolved only in recent times and have not yet overcome the existing distribution barriers. On the other hand, there are also polyphagous species which are found in the Alps as well in the Arctic: *Ph. microscopica*, *Ph. insignis*, and *Ph. herpotrichoides* agg.

ARCTIC-ALPINE ADAPTATION

In the Alps, as well as in the Arctic, the climate is mainly characterized by short summers and low mean temperatures. To survive under such conditions, fungi must adapt in many ways. Savile (1972) discussed in detail such adaptations with fungi from the Arctic. The genus *Phaeosphaeria* demonstrates, in an ideal way, that many of his observations are current also with fungi found in the Alps. *Simplification of the life cycle* may guarantee a regular sporulation which is needed for new infections (and thus for survival of the fungi), even during the short vegetation period. One of these simplifications is the suppression of the conidial state (anamorph).

Table 1. Distribution and formation of anamorph in culture of the arctic-alpine *Phaeosphaeria* species.

	Alps		Arctic		Anamorph
	alpine/subalpine		arctic/subarctic		
Ph. alpina	+	+			+
Ph. caricinella			+	+	+
Ph. dennisiana	+				–
Ph. equiseti			+		not cultured
Ph. herpotrichoides	+	+*)	+	+	+
Ph. hierochloes			+		not cultured
Ph. insignis	+		+		+
Ph. juncicola	+	+	+	+	–
Ph. lindii			+		not cultured
Ph. lutea	+				–
Ph. microscopica	+	+	+		–
Ph. nardi	+	+*)	+	+	–
Ph. oreochloae	+				–
Ph. pleurospora	+	+			–
Ph. silenes-acaulis	+		+		–
Ph. tofieldiae	+	+*)			–
Ph. volkartiana	+				+

*) also at lower altitudes

Among 14 (arctic-) alpine species which were cultivated by the author, only five species (36 per cent) formed the anamorph (belonging to the genus *Stagonospora*) in pure culture (Table 1). In non-alpine species, 17 of a total of 28 cultivated species (61 per cent) produced an anamorph (Leuchtmann 1984). In alpine fungi the percentage of anamorph producing species is obviously much lower than that found for Arctic species.

According to Savile (1972), another adaptation is *simplification of the breeding system*. Self-compatibility forces the development of the sexual state (ascospores) and favors new establishment after dispersal over a great distance. However, self-incompatibility seems to be rare both in alpine and non-alpine species. In cultural studies with single-ascosporic cultures of 45 *Phaeosphaeria* species, only two were self-incompatible (Leuchtmann 1984, Rapilly et al. 1973). Yet both were non-alpine species.

As a consequence of this breeding system the occurrence of a great number of morphologically different forms is observed within some species, as found in the *Ph. herpotrichoides* complex (Fig. 2,g-l). The lack of recombination and the unchanged proliferation of any mutation may have favored this phenomenon.

Many alpine *Phaeosphaeria* species have *thick, deeply pigmented walls* of ascomata and ascospores. Besides protection against UV-radiation, the main functions of pigmentation are (according to Savile 1972) to increase temperature inside the fungal cell by better absorption of sunlight and the reduction of desiccation. This ecological adaptation is conspicuously demonstrated within groups of closely related species. The Arctic-alpine *Ph. microscopica*, for instance, forms deeply pigmented ascospores with granular ornamentations, while the closely related lowland species *Ph. culmorum* has pale and smooth ascospores. Similar differences exist between some alpine forms of the *Ph. herpotrichoides* complex and the lowland forms (Fig. 2,g-l).

When compared to other species of the genus, the alpine *Ph. alpina* (Fig. 1,b) and *Ph. oreochloa* and the arctic *Ph. caricinella* form distinctly thicker ascomata walls. *Ph. nardi* has thick-walled ascomata only at higher altitudes. However, thin-walled ascomata are found in alpine species too (e.g., *Ph. lutea*, Fig. 1,a).

The ascospores of most *Phaeosphaeria* species are coated with a conspicuous *gelatinous sheath*. Shape and division of the sheath are very characteristic in fresh condition for some species and may even have taxonomic importance. Gelatinous sheaths may favor the adherence on suitable substrates during the dispersal of the ascospores. This is especially important in cases where conidial states are lacking.

ACKNOWLEDGEMENTS

I wish to thank Prof. Dr. E. Müller, Zürich, for his stimulating ideas and for reading the manuscript. Thanks are due also to Ms. J.A. Manguson, Zürich, for correcting the English.

REFERENCES

Eriksson, O., 1967, On graminicolous pyrenomycetes from Fennoscandia. II. Phragmosporous and scolecosporous species, *Ark. Bot.*, 6: 381-440.

Holm, L., 1957, Étude taxonomique sur les Pléosporacées, *Symb. Bot. Upsal.*, 14(3): 1-188.

Leuchtmann, A., 1984, Über *Phaeosphaeria* Miyake und andere bitunicate Ascomyceten mit mehrfach querseptierten Ascosporen, *Sydowia*, 37: 75-194.

Petrini, O., and Müller, E., 1979, Pilzliche Endophyten am Beispiel von *Juniperus communis* L., *Sydowia*, 32: 224-251.

Rapilly, F., Foucault, B., and Lacazedieu, J., 1973, Étude sur l'inoculum de *Septoria nodorum* Berk. (*Leptosphaeria nodorum* Mueller(agent de la septoriose du blé, *Ann. Phytopathol.*, 5: 131-141.

Savile, D. B. O., 1972, "Arctic adaptations in plants," Canad. Dept. of Agriculture Monograph No. 6.

SARCOLEOTIA GLOBOSA (SOMMERF.: FR.) KORF, TAXONOMY, ECOLOGY AND DISTRIBUTION

Trond Schumacher

Department of Biology, Division of Botany
University of Oslo, P.O. Box 1045 Blindern, 0316 Oslo 3
Norway

and

Sigmund Sivertsen

The Museum, Botanical Department
University of Trondheim, N-7000 Trondheim, Norway

Key words: *Sarcoleotia globosa*, taxonomy, distribution

ABSTRACT

Sarcoleotia globosa is circumscribed and retained as the only species in the genus *Sarcoleotia* Imai. Its ecology and distribution are discussed. *S. globosa* is arctic and boreo-oroarctic in distribution; a single record from Tierra del Fuego in the Southern hemisphere is within the Cool Temperate zone.

INTRODUCTION

Sarcoleotia globosa is a soil inhabiting, inoperculate discomycete, which was described from Saltdal, Northern Norway, as *Mitrula globosa* by Sommerfelt (1826). For more than 130 years the species appeared to be an endemic to the Scandinavian region, because the only material available were Sommerfelt's original specimens in the University herbaria of Oslo and Uppsala. Mycological exploration of arcto-alpine habitats in Argentina, Eurasia and North America during the past 25 years has changed this picture. *Sarcoleotia globosa* has now been found in many new localities, thus considerably extending the distributional area of the species. Based on the new collections, a more correct distributional pattern can be drawn. It is now possible to better understand the infraspecific variation in the species and to delimit it from closely related taxa.

The purpose of the present study was to examine the taxonomical status, ecology and distribution of *S. globosa*. The taxonomical status of the genus and the other taxa which have been referred to *Sarcoleotia* are also briefly discussed.

MATERIALS AND METHODS

Microscopical investigations in the present study were made both on fresh and dried specimens. The latter were rehydrated in water. Dried specimens from the following museums and herbaria have been examined (herbarium abbreviations are those of Holmgren, Keuken & Schofield, 1981: AMNH, C, FH, H, L, O, OULU, TRH, TROM, TUR and UPS). Studies of apothecia were made using squash mounts, microtome sections and hand sections. Fresh apothecia were fixed in FAA 70 percent for at least 24 hours then dehydrated in a gradual butyl-alcohol series, embedded in paraffin wax and sectioned at 8 um. Sections were stained and photographed in Safranin - Fast Green using the staining technique outlined by Johansen (1940: 81). Observations were also made in H_2O, Melzer's Reagent and in methyl blue in lactic acid (Cotton blue). Photomicrographs were taken with a camera mounted on a Zeiss WL microscope. Drawings were made with the aid of a Wild drawing tube. A biometrical study of the spore morphology was carried out. A scatter diagram was plotted using average lengths and widths of 10 spores from each collection. This means that each dot represents the average spore size of one collection. Twenty-five Nordic collections and four additional taxonomically important ones have been plotted.

SARCOLEOTIA GLOBOSA (Sommerf.: Fr.) Korf, Phytologia 21: 206. 1971.

- = *Mitrula globosa* Sommerf., Suppl. Fl. Lapp., p. 287, pl. 3, fig. 3. 1826.
- = *Geoglossum globosum* Sommerf.: Fr., Elench. Fung. 1: 234. 1828.
- = *Leotia globosa* Sommerf., *In* sched. herb. osloensis.
- = *Corynetes globosus* (Sommerf.: Fr.) Durand, Ann. Myc. 6: 417. 1908.
- = *Microglossum globosum* (Sommerf.: Fr.) Imai, J. Fac. Agr. Hokkaido Imp. Univ., Sapporo 45: 192. 1941.

= *Cudonia osterwaldi* P. Henn., Verh. Bot. ver. Prov. Brandenburg 46: 118. 1905.

= *Sarcoleotia nigra* S. Ito & Imai, *In* Imai, Trans. Sapporo Nat. Hist. Soc. 13: 182. 1934.

- = *Leotia nigra* S. Ito & Imai, Proceed. Jap. Assoc. Adv. Sci. 7: 148. 1932 (*nomen nudum*).

= *Cudonia clandestina* Rahm, Schw. Z. Pilzk. 44: 172. 1966.

- = *Sarcoleotia clandestina* (Rahm) Rahm, Schw. Z. Pilzk. 53: 42. 1975 (*nomen nudum*).

= *Sarcoleotia platypoda* (DC.: Fr.) Maas G. (ut *S. platypus* (DC.: Pers.) Maas G. *fide* Maas Geesteranus, 1966), Koninkl. Nederl. Akademie van Wetensch. Amsterdam, Proc., Ser. C, 69: 191. 1966.

- = *Helvella platypoda* DC., Fl. franc. 5: 29. 1815.
- = *Leotia platypoda* DC.: Fr., Syst. mycol. 2(1): 28. 1822.

Fruitbodies stipitate, at first capitate, then pileate, up to 50 mm high, fleshy, consisting of a fertile head and a stipe. Fertile head obovate to subglobose, remaining so or more commonly becoming flattened and depressed in center; margin at first strongly inrolled and adherent to stipe; hymenium then apparently continuous with the stipe, in fully mature specimens the margin is receded from the stipe so as to reveal an annular cavity between pileus and stipe ('with a sterile roof of the cap') (Fig. 1.A.).

Pileus 2-12 mm broad, 2-8 mm high, dark sepia to chestnut brown (drying black), underneath on receptacle greyish brown in expanded specimens. Stipe 5-35 mm long, 0.5-2 mm broad, terete or slightly flattened, sometimes with a shallow longitudinal groove on one or two sides, occasionally stipe expanded above to 4 mm wide, glabrous or minutely floccose, distally permanently floccose, greyish brown to dark sepia brown, becoming more whitish towards the base. Medullary excipulum of the fertile portion principally of a loose textura intricata with hyphae 2.5-5 um broad.

Subhymenium of compact agglutinated, slightly interwoven hyphae. Margin of expanded specimens (ectal excipulum) many cells wide of textura porrecta to textura prismatica, innermost cells 2-4 um broad, outermost cells 4-10 x 10-24 um, brown-walled. Ectal excipulum of stipe of textura porrecta, hyphae 3.0-5.8 um wide, coherent, in places the hyphae of the outer layer of ruptured into short segments and with the loose ends curled outwards (Fig. 1B). Medullary excipulum of less coherent textura porrecta, the hyphae running in bundles and interspersed by cavities. Asci clavate, inoperculate, 8-spored, 70-155 x 7.2-11 um, ascospore distinctly J+. Ascospores obliquely biseriate, clavate to subfusiform, straight or slightly curved, smooth, hyaline, at first unicellular and multiguttulate, then principally 1-2 septate or occasionally with 3-5 septae, 22.0-45.5 x 3.0-5.8 um (Fig. 1D). Paraphyses 1.7-2.2 um broad, slightly enlarged to 3.5 um above, septate, apices curved or hooked, occasionally almost straight, cells brown-walled, filled with oleaginous matter (Fig. 1C).

SPECIMENS EXAMINED

TYPE: Norway

Nordland. Saltdal. In arena. 9/1819 and 9/1823 Chr. Sommerfelt, in sched. *Leotia globosa* (O-*holotype* of *Mitrula globosa*; UPS ex herb. E. Fries - *isotype*, slide of *holotype* ex. herb. O).

OTHER SPECIMENS:

Norway:

Hedmark. Stor-Elvdal. Myrstad, Glåma. 7.10.1976 A. Pedersen & T. Schumacher (O).

Oppland. Dovre. Grimsdalen. Kvannbekken. 12.9.1982 T. Schumacher & K. Østmoe (O). Dovre. Grimsdalen. Buåi. 17.9.1983 T. Schumacher & K. Østmoe (O). Dovre. Grimsdalen. Tverråi. 6.8.1984 T. Schumacher, S. Sivertsen & K. Østmoe (O-2 coll.). Dovre. Grimsdalen. Verkenseter 9.8.1984 T. Schumacher, S. Sivertsen & K. Østmoe (O). Vågå. Krokåi in Slådalen 8.8.1984 T. Schumacher, S. Sivertsen & K. Østmoe (O).

Hordaland. Ulvik. Finse. N of Hardangerjøkelen. 10.8.1960 F.-E. Eckblad (O).

Sør-Trøndelag. Midtre Gauldal. SW Amdalvolltjern. 16.9.1972 Å. Erlandsen (TRH). Oppdal. Vinstradalen. Ryphuskollen. 11.1.1984 T. Schumacher, S. Sivertsen & K. Østmoe(O).

Nord-Trøndelag. Røyrvik. Lake Namsvatnet at Vierma. 6.9.1969 S. Sivertsen (TRH).

Nordland. Hattfjelldal. Børgefjell National Park. Storskavlbekken. 30.8.1969 K.I. Flatberg & S. Sivertsen (TRH). Rana. Granlund. Glomåga. 6.9.1975 H. Dissing (TRH). Rana. 1 km S Rausandaksla. 21.9.1974 S. Sivertsen (TRH). Rana. 1 km NW of Reinforshei 3.9.1975 S. Sivertsen (TRH). Rana. Hammerneset. 8.9.1976 A. Pedersen (O). Rana. Dunderlandsdalen. Ørtfjellmoen. 19.9.1974 S. Sivertsen (TRH), 11.9.1976 G. Gulden (O). Rana. Virvassdalen. Virvasselven River. 29.8.1981 S. Sivertsen (TRH). Rana. Virvassdalen. Beveråa. 7.9.1975 T. Schumacher (O). Rana. Virvassdalen. Blerekelva. 7.9.1975 H. Dissing (O, TRH). Fauske. Blåmannsisen W, at Leirelva. 26.8.1967 S. Sivertsen (TRH).

Troms. Målselv. Holt. 30.8.1964 S. Sivertsen (TROM).
Finnmark. Alta. Tomasbakken at Alta River. 13.8.1961 F.-E. Eckblad (O, UPS, TUR). Kautokeino. Øvre Anarjåkka National Park. Elvkrokfjellet. 9.8.1966 S. Sivertsen (TROM). Kautokeino. Mazejåkka. 15.8.1978 H. Dissing (TRH). Porsanger. Lakselv. Lakselv hotel. 16.8.1961 F.-E. Eckblad (O, UPS). Tana. Polmak River. 12.8.1963 E. Kankainen (O, TUR).

Sweden:

Jämtland. Åre. Storlien. 20.8.1983 J. Nitare (Herb. J. Nitare, TRH) Åre. Handöl, at Handolsforsen. 15.8.1984 J. Nitare (Herb. J. Nitare, UPS).

Finland:

Kuusamo. Posio. Pernu. Korouoma at Kurttajoki River. 19.8.1977. T. Ulvinen & M. Ohenoja (OULU). Juuma, NE-side of the village. 20.9.1975 T. Ulvinen (OULU).
Enontekiön Lappi. Enontekiö. Porojärvi. Porovuoma at Waltijoki River. 20.8.1961 S. Sivertsen (O). Enontekiö. Kuttanen. 13 km SEE of Kaaresuvanto. Palovuoma. 23.9.1968 T. Ulvinen (OULU, UPS, H).

Iceland:

S. Thing. 6346 Grafarlönd. 22.8.1974 Hördur Kristinsson (AMNH). S. Thing. 5747 Sandmuladalur. 25.8.1976 H. Hallgrimsson (AMNH).

Greenland:

Sydprøven. 30.9.1971 P.M. Petersen -71.162 (C). Frederikshåb. Equluit. 19.8.1973 P.M. Petersen -73.314 (C, as *Sarcoleotia platypus*).

Canada:

Northwest Territories. Keewatin. Rankin Inlet. Kudlulik Peninsula. Melvin Bay. 16.8.1971 M. Ohenoja (OULU). Québec. Poste-de-la- Balaine, grid 2126. 8.8.1982 S. Huhtinen (TUR).

Japan:

Hokkaido. Kushiro. Mt. Meakan. 25.9.1935 E. Homma (S - ex herb. S. Imai, as *Sarcoleotia nigra*).

Germany:

Brandenburg. Marchia. Röntgenthal. Buch. 15.10, 6.11.1904 K. Osterwald (Jaap- Fungi selecti exs. no. 128, Rabenhorst-Pazschke - Fungi europaei et extraeuropaei no. 4466, Rehm-Ascomyceten no. 1576 (S, as *Cudonia osterwaldi*).

Belgium:

Neighbourhood of Brussels, no date, Bommer & Rousseau (FH- ex herb. Patouillard, as *Leotia platypoda*).

Netherlands:

Asten, Ospeler Peel. 9.10.1965 C. Bas et al. 4600 (L).

LITERATURE RECORDS:

NORWAY (Sommerfelt 1826 - as *Mitrula globosa*; Imai 1940, 1955, Nannfeldt 1942, Eckblad 1963, Kallio & Kankainen 1966 - as *Corynetes globosus*). FINLAND (Ulvinen 1976 - as *Sarcoleotia globosa*). GREENLAND (Korf & Gruff 1981, Korf 1982 - as *Sarcoleotia globòsa*; Petersen & Korf 1982 - as *Sarcoleotia globosa* and *S. platypus*). JAPAN (Imai 1934, 1941 - as *Sarcoleotia nigra*). GERMANY (Hennings 1905 - as *Cudonia osterwaldi*). BELGIUM (Patouillard 1886 - as *Leotia platypoda*). NETHERLANDS (Maas Geesteranus 1966 - as *Sarcoleotia platypus*).

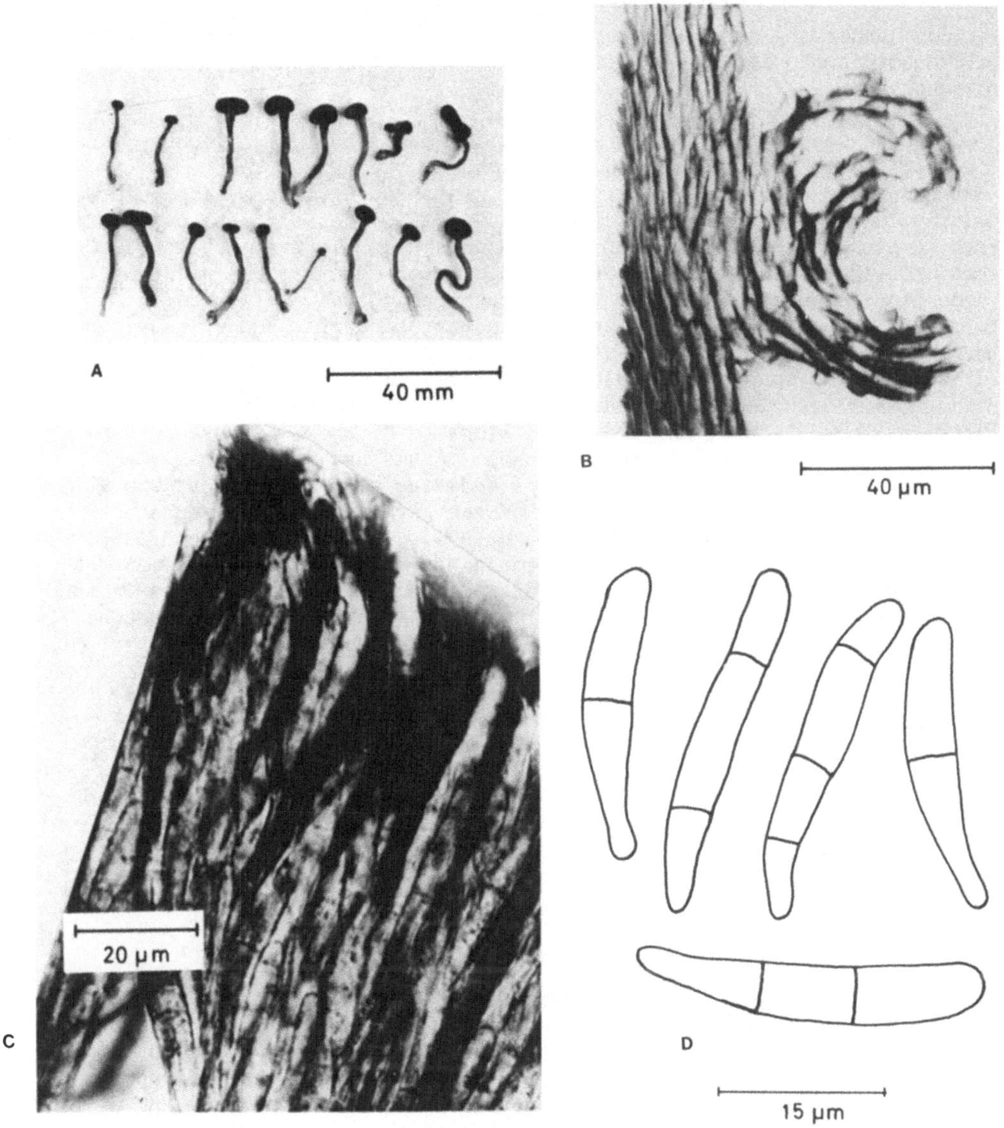

Fig. 1. *Sarcoleotia globosa*. A. Ascocarps of variable shape and size, from coll. 8.8.1984 Schum., Siv. & Østmoe (O); B. detail of stipe with outer portion ruptured and curled outwards; C. detail of hymenium, asci with ascospores and fascicles of paraphyses; D. ascospores, coll. 7.9.1975 Schum. (O).

SWITZERLAND (Rahm 1966 - as *Cudonia osterwaldi* and *C. clandestina*, Rahm 1975 - as *Corynetes globosus* and *Sarcoleotia clandestina*, Müller 1977, Irlet 1984 - as *Sarcoleotia globosa*). ARGENTINA (Gamundi 1979 - as *Sarcoleotia nigra*).

SPECIES DELIMITATION

Sarcoleotia globosa is characterized by a pileate, nongelatinous, castanean brown fruitbody with a greyish black, more or less terete stipe. Under the lense the curved, brownish paraphyses and the hyaline, clavate to subfusiform, non- to few - (3-4) septate ascospores are diagnostic.

Spore septation is a character which has been widely discussed in connection with *S. globosa* and its taxonomic synonyms. The multi-septate ascospores of *S. globosa*, which among others were pointed out my Imai (1940), made Maas Geesteranus (1966) hesitant as to include the species in the synonomy of *S. platypoda* (see discussion below). On the other hand, the multiseptate ascospores of *Cudonia osterwaldi* were according to Maas Geesteranus (1966) rather suggestive of those in *S. globosa*. Rahm (1975) emphasized the larger ascocarps, longer asci and granularly to guttulate, non-septate ascospores of *S. globosus* as distinguishing characters against *C. osterwaldi*. In an earlier paper Rahm (1966) described a new taxon, *Cudonia clandestina*, from the Swiss Alps and found it distinct from *S. globosa* (= *C. osterwaldi* sensu Rahm 1966) on the basis of a larger ascocarp, longer asci and unicellular ascospores of the former. *Cudonia clandestina* also was compared to *S. platypoda* and kept distinct because of the permanently, non-septate ascospores in the former species (Rahm 1975). Maas Geesteranus (1966) studied the type specimen of *Sarcoleotia nigra* and found the ascospores occasionally 3 to 4 - septate, which exceeded the number of septa in the ascospores of *S. platypoda* observed by him. Nevertheless, he included *S. nigra* in the synonymy of *S. platypoda* (Maas Geesteranus 1966).

Table 1. Number of septa in the ascospores of *S. globosa*. (based on counts of 10 ascospores from 26 collections), n = 260.

Number of septa	*% of ascospores*
0	18.8
1	43.1
2	28.8
3	8.5
4	0.8

As evident from our studies, the septation of the ascospores in *S. globosa* varies considerably (cp. Table 1.). Generally, young specimens tend to have non- to one-septate ascospores, while fully expanded specimens usually have one or more septa per ascospore. Mature specimens with unicellular ascospores also have, however, been observed, and we find the spore septation to be a variable character of little diagnostic value. It should be emphasized that it is difficult to observe the septa of hyaline ascospores in unstained preparations, which may explain some of the discrepancies in the spore statements in the literature. Examination of authentic specimens of *C. osterwaldi* and *S. nigra* also showed variation in the spore septation from specimen to specimen.

Another variable character is the spore size. A scatter diagram showing the distribution of 30 OTUs (operational taxonomic units) with spore length plotted against spore width is given in Fig. 2. The plots include authentic specimens of *C. osterwaldi* and *S. nigra*, and European material interpreted as *S. platypoda* by Maas Geesteranus (1966). A fairly continuous variation is found. It is noted, though, that the collections of *S. nigra* from Japan and *S. platypoda* from The Netherlands are extremes that might call for infraspecific recognitions. However, the number of collections from these areas are limited, which make it impossible to conclude whether or not these slight variations are taxonomically significant. Based on our microscopic examinations and the biometrical study of spore morphology, we recognize only one species.

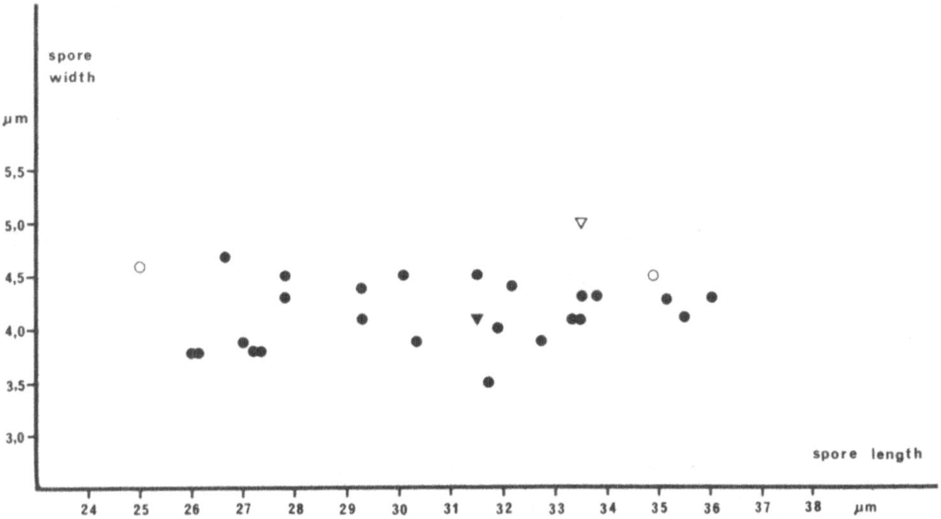

Fig. 2. *Sarcoleotia globosa*, scatter diagram of length/width of the ascospores. Each dot represents an average of 10 measurements per specimen; (solid circle) nordic collections of *S. globosa*, (solid triangle) *C. osterwaldi*, holotype, (empty triangle) *S. nigra*, paratype, (empty circle) *S. platypus*, material in L and FH.

NOMENCLATURE

Maas Geesteranus (1966) adopted the name *Helvella platypoda* DC. when he recorded the above cited Belgian and Dutch specimens as *Sarcoleotia platypus*. *Helvella platypoda* DC. is an older name than *Mitrula globosa* Sommerf., and it could possibly be used for the present species, as was suggested by Maas Geesteranus (1966). However, we have found several reasons to depart from this solution.

According to De Candolle (1815), *H. platypoda* is gelatinous, somewhat leathery in consistency, with an irregular plicate-undulate brownish head and a compressed, white stipe ('platypoda'). This is indicative of a small *Helvella*, or perhaps a *Leotia* or *Cudonia* species. Fries (1822) adopted De Candolle's species as *Leotia platypoda*. Sommerfelt's specimens, which were seen by Fries, were on the other hand referred to *Geoglossum* (Fries 1828). This indicates that Fries himself never suspected a relationship between the two species. Furthermore, *H. platypoda* was described from the province of Grasse in Southern France, which is far outside the known distributional area of *S. globosa*. Apparently, no authentic specimen of *H. platypoda* exists, while the type specimen of *S. globosa* has been preserved, thus linking the original and Friesian concept of the species. Under these circumstances we have found it unwise to give nomenclatorial preference to *H. platypoda*, even if a neotype could have been selected. Consequently, we have adopted the name *S. globosa* for our species, and *S. platypoda* is here considered a *nomen dubium*.

Cudonia clandestina is, on the basis of the description and drawings, put in synonymy with *S. globosa* (Rahm 1966, 1975).

GENERIC ASSIGNMENT AND CIRCUMSCRIPTION OF THE GENUS SARCOLEOTIA

The generic assignment of *S. globosa* has been disputed since the

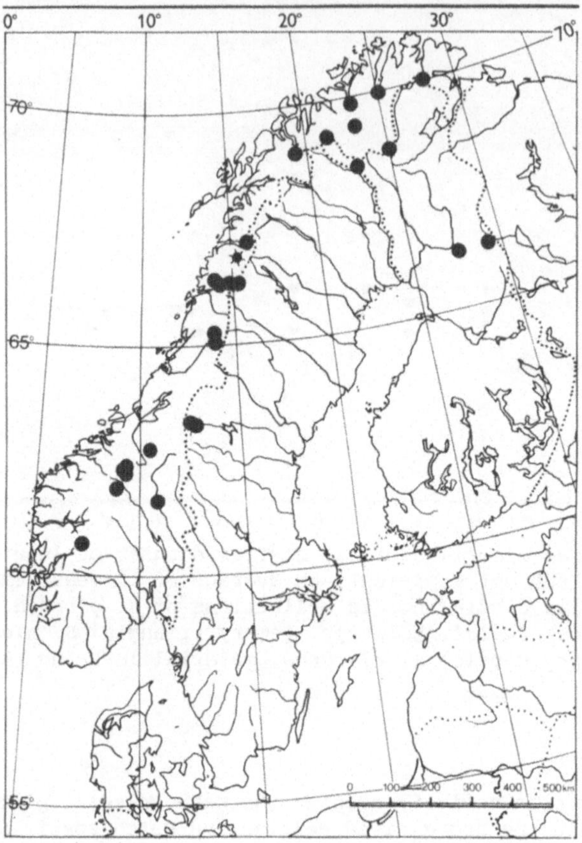

Fig. 3. Distribution of *S. globosa* in Fennoscandia (spoked circle = type locality).

days of its discovery. Sommerfelt was apparently uncertain about its taxonomical position himself; the species was described as *Mitrula globosa*. However, on the herbarium sheet it was referred to as *Leotia* in Sommerfelt's own handwriting (O - ex herb. Sommerfelt). E.M. Fries received some specimens from Sommerfelt and referred the species to *Geoglossum* (Fries 1828). Durand (1908) studied the isotype in Uppsala and transferred it to *Corynetes*, a disposition concurred with by most subsequent authors (Imai 1940, 1955, Nannfeldt 1942, Eckblad 1963). Imai also tentatively referred *M. globosa* to the genus *Microglossum* (Imai 1941), but later on he abandoned this solution and again referred it to *Corynetes* (Imai 1955). Eckblad (1963) pointed out, however, that with regard to shape the species was rather out of place in *Corynetes*. He also drew attention to *Sarcoleotia nigra*, which in shape and colour appeared rather similar to *C. globosus*. Joining the two species in one genus, however, was out of the question due to the dissimilar microscopical features in *S. nigra* as reported by Imai (1934, 1941).

Maas Geesteranus (1964), after having studied Norwegian material of *C. globosus*, concluded that it was 'not a *Corynetes* at all, differing from the species of that genus both in gross morphology and structurally.' He also studied authentic material of *Sarcoleotia nigra* and found it identical to European material interpreted to be *Leotia platypoda* DC. by Patouillard (1886). Based on these findings,

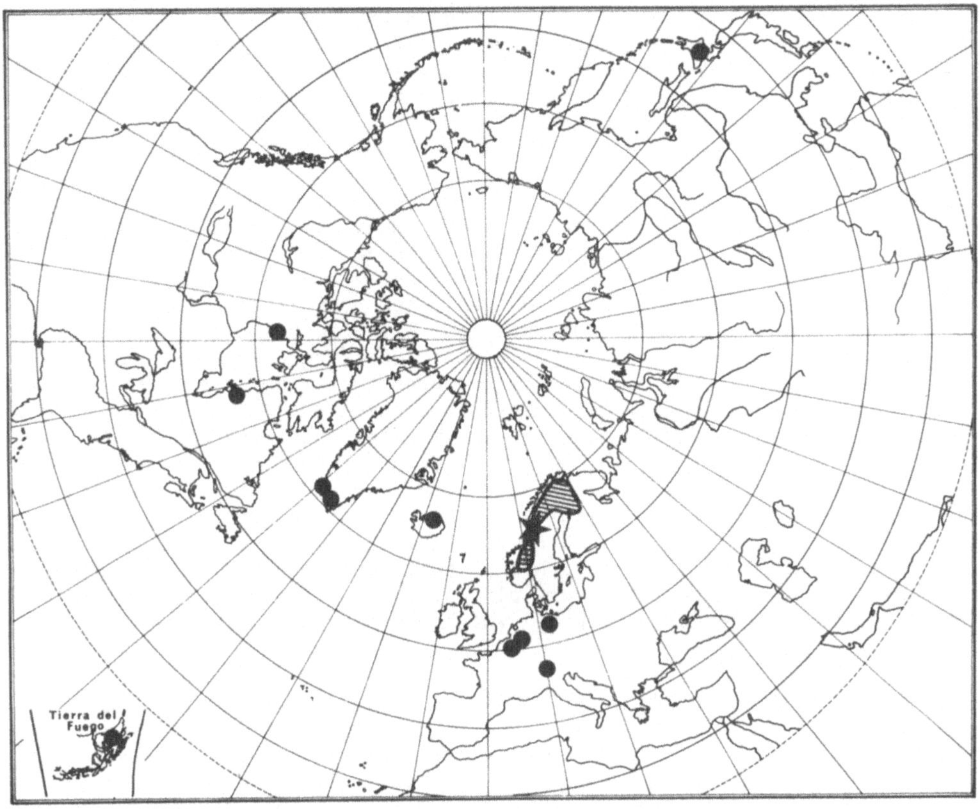

Fig. 4. World distribution of *S. globosa* (star = type locality).

Patouillard's specimens were accommodated in *Sarcoleotia* as *S. platypus* (Maas Geesteranus 1966). The emended description of *S. nigra* provided by Maas Geesteranus (1966), eliminated the main obstacles of bringing *C. globosus* and *S. nigra* (= *S. platypus* sensu Maas G.) together. *Sarcoleotia* was maintained as a "leotiaceous" genus with a pileate fruitbody, having a margin; the stipe being fleshy, non-gelatinous, of textura porrecta throughout; the asci with pore bluing in Melzer's reagent; ascospores hyaline, 1-4-celled and clavate to subfusiform in shape; and the paraphyses being coloured, brownish, straight or somewhat curved above, almost uniform in width throughout (Maas Geesteranus 1966). Maas Geesteranus (1966) concluded: "it is beyond doubt that *C. globosus* belongs to the genus *Sarcoleotia*." Because of the long and slender stipe of *C. globosus*, the dark brown colour of the head and stipe, and the fact that Imai (1940) never suspected the relationship between *S. nigra* and *C. globosa*, Maas Geesteranus (1966) hesitated to include *C. globosus* in his concept of *S. platypus*. A formal transfer of *C. globosus* to *Sarcoleotia* was made by Korf (1971), who apparently accepted the additional species *S. platypoda* in the genus as well (Korf 1973, Petersen & Korf 1982). These taxa are here considered to be identical. *Cudonia clandestina* was also tentatively referred to *Sarcoleotia* (Rahm 1975), however, it was in disaccordance with the Code, and it is therefore not validly published. This taxon is regarded as synonymous to *S. globosa* by us. Another taxon which has been referred to *Sarcoleotia* is *S. turficola* (Boud.) Dennis

(Dennis 1971). It was previously treated among the ombrophiloid, gelatinous members of the genus *Coryne* (= *Ascocoryne* Korf). However, as demonstrated by Dennis (1971), this species lacks the gelatinous layer typical of *Ascocoryne*. For this reason, and because he did not want to erect another new, monotypic genus in the Leotiales, the species was tentatively referred to *Sarcoleotia* Imai (Dennis 1971). We are at present unable to decide where this species belongs. However, the possibility of a common generic assignment with *S. globosa* is excluded. The possibility that *Nothomitra cinnamomea* Maas G. (Maas Geesteranus 1964) represented young specimens of *S. globosa* made us restudy authentic specimens of *N. cinnamomea* (L - 962.271-144, *holotype*). The dried specimens are light-coloured (yellowish brown), concolorous on head and stipe, with hymenium being continuous with the stipe (no sterile roof of the head). Consequently they are not closely related to *S. globosa*, even though there are microscopical features that might indicate some kind of relationship. We agree with Maas Geesteranus (1964) in the erecting of a separate genus to accommodate the above cited specimens.

Accordingly, *Sarcoleotia* Imai is here maintained with *S. globosa* as the only species. The pileate, fleshy, non-gelatinous ascocarp, having a distinct margin, the none- to few-septate hyaline ascospores, and the uniformly, distally curved paraphyses indicate an affinity to the genus *Cudonia* Fr., the latter, however, also is characterized by inamyloid asci and long, acicular ascospores which bud off conidia while still within the ascus.

Originally referred to the family Geoglossaceae (Imai 1934, 1941), *Sarcoleotia* has now been transferred to the Leotiaceae due to the presence of a margin on the ascocarp (Maas Geesteranus 1966). The presence or absence of a margin between the fertile head and the stipe of the ascocarp has been proposed as a main criterion in the demarcation of Leotiaceae and Geoglossaceae, the former family being characterized by having a margin, the latter by the lack of a margin (Maas Geesteranus 1966). Following Maas Geesteranus (1966) and Korf (1973) *Sarcoleotia* and its close relatives in *Cudonia* and *Leotia* are now accommodated in the family Leotiaceae.

ECOLOGY

Based on our own field observations supplemented by an examination of the herbarium collections and the given notes on the labels of the envelopes of the dried specimens, the site types of the 38 Nordic (Greenland included) collections have been summarized in Table 2.

The sites have been assigned to three main habitat groups, i.e., vegetation influenced by some kind of natural physical or chemical perturbations, vegetation disrupted or created by man, or vegetation on apparently undisturbed ground. Most collections are from disturbed habitats, having been subjected to erosion, inundation by water, burning or cryopedological processes in the many arctic and oroarctic localities. River beds, banks and slopes of intermittent brooks and rivulets turn out as especially favourable sites; all having been influenced by the inundation and erosion of flood water. The river bank localities in the southern parts of Norway are all in boreal and oroarctic river sections, from 240 to 1320 m. Here, as well as along the many recorded arctic river bed localities, *S. globosa* is growing in all subzones of the banks (cp. Schumacher 1978). Common bryophyte and liverwort associates on the banks of rivers, brooks and rivulets are *Blasia pusilla*, *Jungermannia* species, *Philonotis fontana* and *Pohlia gracilis*; on the upper parts of the banks *Polytrichum* species also are characteristic associates. In

Table 2. The occurrence of *S. globosa* in different habitat types

Site	Number of collections
A. *Vegetation influenced by physical/chemical perturbations*	
River bed (of these from glacial rivers 6)	13
Rivulet/brook	6
Burnt site	3
B. *Vegetation having been disrupted by man*	
Path/road	4
Gravel pit	2
C. *Vegetation on undisturbed ground*	
Minerotrophic fens	2
Dwarf shrub heath/forest floor	4
Gravel and sandy soil (unspecified)	2
Unknown	2

one collection, from the upper inundation zone of the bank of the Virvasselven River, the moss remnants accompanying the specimens of *S. globosa* were especially rich, including *Blasia pusilla*, *Scapania subalpina*, *Aongstroemia longipes*, *Bryum* sp., *Dichodontium pellucidum*, *Distichium* sp., *Drepanocladus uncinatus*, *Campylium stellatum*, *Hypnum lindbergii*, *Leptobryum pyriforme*, *Meesia uliginosa*, *Blindia acuta*, *Onchophorus virens*, *O. wahlenbergii*, *Pohlia filum*, *P. wahlenbergii*, and *Philonotis tomentella*. On the river bank of Glama in the middle boreal zone of eastern Norway (240 m), *S. globosa* was growing on coarse sand and gravel in a carpet of putrefying sheets of *Nostoc* among shoots of *Pohlia gracilis* and *Jungermannia* sp. These 'algal' environments seem rather appropriate to the type locality of *Cudonia osterwaldi* in Germany, such as described by Hennings (1905). Although not manifest in Table 1, the cryopedological phenomena of frost upheaval and solifluctions are environmental factors, which create and modulate many of the arctic and oroarctic microhabitats of *S. globosa*. The records of *S. globosa* from the Canadian Arctic also seem to fit the characteristic habitat types of the species in Fennoscandia; the specimens were collected on the edge of a ditch-like frost crack amongst *Aulacomnium palustre*, *Distichium capillaceum*, *Riccardia pinguis* and *Eurhynchium pulchellum*, and on sandy soil in between *Equisetum arvense*, *Bryum* sp. and *Polytrichum* sp. on a river bed.

Three of the examined collections are from old, burnt sites amongst pioneer mosses such as *Bryum capillare*, *Pohlia nutans* and *Ceratodon purpureus*. An old fireplace also was recorded as the typical habitat of *Cudonia clandestina* from Austria (Rahm 1966). Six of the Nordic collections are from localities where the soil and vegetation have been disrupted by man, i.e. roadsides, paths and gravel pits with an open, early succesional pioneer vegetation. Only six out of the 38 collections noted in Table 1 are from apparently undisturbed ground. Two collections are from moss carpets in rich, minerotrophic fens with recorded bryophyte associates such as *Sphagnum warnstorfii*, *S. fuscum*, *Calliergon stramineum* and *Paludella squarrosa*, while the remaining are from dwarf shrub heaths of *Juniperus* and *Empetrum* (3 coll.) and from the ground in a *Pinus* forest (1 coll.). Whether the actual collecting sites have been in closed vegetation, such as interpreted here, or rather within 'scars' in a disrupted vegetation cannot be assumed based on the incomplete information on the herbarium labels of many of the specimens. The many new collections of *S. globosa* in recent years, however, makes it reasonable to conclude that *S. globosa* prefers habitats which have been subjected to some kind of natural or man-made perturbations. Such habitats are known to be favourable sites to the soil-inhabiting operculate discomycetes as well (Petersen 1967, 1982, Schumacher 1978).

The majority of the collections of *S. globosa* are from calcareous areas and from rich soil types. This, together with the affinity of *S. globosa* to burnt sites, known to have an alkaline soil reaction (Petersen 1971), indicates a preference or ability to colonize soils of neutral to alkaline reactions. The Dutch record, however , is maintained as on "poor, peaty soil among mosses" (Maas Geesteranus 1966). It remains to be seen if *S. globosa* is a species which may colonize soils of different nutritional qualities in different climatic subzones of its distributional area.

DISTRIBUTION

The distribution of *S. globosa* is shown in Figs. 3 and 4. A continuous distribution for *S. globosa* in Central and Northern Fennoscandia is drawn in Fig. 3. Based on the many records in the area, it is concluded that *S. globosa* has an oroarctic and northern boreal distribution in Fennoscandia (cp. Ahti, Hämet-Ahti & Jalas 1968). The majority of finds are from the oroarctic subzones. The other than Fennoscandian records also seem to confirm this phytogeographical picture; the specimens from Iceland, Greenland, the Canadian tundra, the Swiss Alps and the mountains of Hokkaido, Japan, are either from oroarctic areas, low Arctic or from the northern boreal vegetational subzone (cp. Hämet-Ahti 1981). A single record within the Cool Temperate Zone of the Southern Hemisphere from Tierra del Fuego, Argentina, is reported by Gamundi (1979). The Fennoscandian distribution together with the occurrence of *S. globosa* in Eurasia and North America north of 60° latitude may indicate that *S. globosa* has a more or less transcontinental, northern circumpolar boreo-oroarctic and arctic distribution.

At present the temperate lowland localities in Belgium, in the Netherlands and in Germany turn out as somewhat aberrant outposts in the distributional area of *S. globosa* Still more searching for the species in typical lowland-habitats of Middle and Northern Europe is necessary before definite conclusions about its distribution can be drawn.

ACKNOWLEDGEMENTS

We would like to thank Mr. J. Nitare and Mr. S. Huhtinen for having placed their Swedish and Canadian finds at our disposal, and for various information on the matter. The directors and curators of the herbaria AMNH, C, FH, H, L, O, OULU, S, TROM, TUR, and UPS are thanked for loan of material. Mr. Arne A. Frisvoll, Trondheim, has determined Bryophyta from a number of the collections.

REFERENCES

Ahti, T., Hämet-Ahti, L., and Jalas, J., 1968, Vegetation zones and their sections in northwestern Europe, *Ann. Bot. Fenn.*, 5: 169-211.

De Candolle, M., 1815, "Flore Francaise, ou descriptions succinctes de toutes les plantes qui croissent naturellement en France. Vol. 5," Desray, Paris.
Dennis, R. W. G., 1971, New or interesting British Microfungi, *Kew Bull.*, 25: 335-374.
Durand, E. J., 1908, The Geoglossaceae of North America, *Ann. Myc.*, 6: 387-477.
Eckblad, F. -E., 1963, Contributions to the Geoglossaceae of Norway, *Nytt Mag. Bot.*, 10: 137-158.
Fries, E., 1822, "Systema mycologicum," 2(1): 1-274.
Fries, E., 1828, "Elenchus Fungorum," 1: 1-238.
Gamundi, I. J., 1979, Subantarctic Geoglossaceae. II, *Sydowia*, 32: 86-98.
Hämet-Ahti, L., 1981, The boreal zone and its biotic subdivisions, *Fennia*, 159(1): 69-75.
Hennings, P., 1905, Zwei neue Cudonieen aus der Umgebung Berlins, *Verh. Bot. verein Prov. Brandenb.*, 46: 115-119.
Holmgren, P. K., Keuken, W., and Schofield, E. K., 1981, Index Herbariorum. Part 1, *Regnum Vegetabile*, 106: 1-452.
Imai, S., 1934, Studies on the Geoglossaceae of Japan, *Trans. Sapporo Nat. Hist. Soc.*, 13: 179-184.
Imai, S., 1940, The Geoglossaceae of Norway, *Ann. Myc.*, 38: 268-278.
Imai, S., 1941, Geoglossaceae Japoniae, *J. Facul. Agr. Hokkaido Imp. Univ.*, Sapporo, 45: 155-264, pl. 6-10.
Imai, S., 1955, Contributiones ad studia monographica Geoglossacearum. II, *Yokohama Nat. Univ. Sci. Reports, Sec. II*, no. 4: 1-11.
Irlet, B., 1984, Ein Beitrag zur Discomycetenflora der alpinen Stufe der schweizer Alpen, *Mycologia Helvetica*, 1: 129-143.
Johansen, D. A., 1940, "Plant Microtechnique," McGraw-Hill Book Co., New York, 523 pp.
Kallio, P., and Kankainen, E., 1966, Additions to the mycoflora of northernmost Finnish Lapland, *Ann. Univ. Turku Ser. A, II*, 36: 177-210.
Korf, R. P., 1971, Some new discomycete names, *Phytologia*, 21: 201-207.
Korf, R. P., 1973, Discomycetes and Tuberales, *in*: G. C. Ainsworth, F. K. Sparrow, and A. S. Sussman, eds., "The fungi - an advanced treatise, vol. IV," New York, pp. 249-319.
Korf, R. P., 1982, Inoperculate discomycetes of the arctic and alpine zones of Finnmark, Lapland, and Greenland, *in*: G. A. Laursen, and J. F. Ammirati, eds., "Arctic and alpine mycology: The First International Symposium on Arcto-Alpine Mycology,"Univ. of Washington Press, Seattle & London, pp. 27-37.
Korf, R. P., and Gruff, S. C., 1981, Discomycetes exsiccati, fasc. IV, *Mycotaxon*, 13: 5-15.
Maas Geesteranus, R. A., 1964, On some white-spored Geoglossaceae, *Persoonia*, 2: 81-96.
Maas Geesteranus, R. A., 1966, On *Helvella platypus* DC., *Koninkl. Nederl. Akademie van Wetenschappen, Amsterdam, Proc., Ser. C*, 69: 191-203.
Müller, E., 1977, Zur Pilzflora des Aletschwaldreservats (Kt. Wallis, Schweiz), *Beitr. Krypt. Fl. Schweiz.*, 15(1): 1-126.
Nannfeldt, J. A., 1942, The Geoglossaceae of Sweden (with regard also to the surrounding countries), *Ark. Bot.*, 30A(4): 1-67, 5 pl.
Patouillard, N., 1886, "Tabulae analyticae fungorum. Descriptions et analyses microscopiques des champignons nouveaux, rares ou critiques," Fasc. 5 (no. 401-500): 181-232.
Petersen, P. M., 1967, Studies on the ecology of some species of Pezizales, *Bot. Tidsskr.*, 62: 312-322.
Petersen, P. M., 1971, The macromycetes in a burnt forest area in Denmark, *Bot. Tiddsskr.*, 66: 238-248.

Petersen, P. M., 1982, The ecology and distribution of soil inhabiting Pezizales in Western Greenland, *in*: G. A. Laursen, and J. F. Ammirati, eds., "Arctic and alpine mycology: The First International Symposium on Arcto-Alpine Mycology," Univ. of Washington Press, Seattle & London, pp. 334-348.

Petersen, P. M., and Korf, R. P., 1982, Some inoperculate Discomycetes and Plectomycetes from West Greenland, *Nord. J. Bot.*, 2: 151-154.

Rahm, E., 1966, Geoglossaceae im Hochtal von Arosa, *Schw. Z. Pilzk.*, 44: 165-179, 2 pl.

Rahm, E., 1975, Geoglossaceae im Hochtal von Arosa (II), *Schw. Z. Pilzk.*, 53: 40-43.

Schumacher, T., 1978, Operculate discomycetes (Pezizales) on river banks in Norway, *Norw. J. Bot.*, 25: 207-220.

Sommerfelt, S. C., 1826, "Supplementum Florae Lapponicae," Christianiae, 332 pp. 3 Tabs.

Ulvinen, T., ed., 1976, "Suursieniopas," Suomen sieniseura, Helsinki, 359 pp.

THE GENUS GALERINA ON SVALBARD

Gro Gulden

Botanical Museum, University of Oslo
Trondheimsvn. 23 B, N-0562 Oslo 5, Norway

Key words: *Galerina*, Arctic-alpine agarics, Svalbard mushrooms

ABSTRACT

Twelve species of *Galerina* are recorded from Svalbard. Eight of them are new to the archipelago. The examined collections (322 nos.) are almost exclusively from the west coast of Spitsbergen between the Bellsund district and Norskeöya, and from the fjord districts. Eleven collections are from Kong Karls Land.

Short descriptions and drawings of diagnostic microscopic features are presented. There are also scanning micrographs of spores, information on habitats, known distributions of species, and a key for determination.

The *Galerina* flora of Svalbard comprises an element of species considered to have their ecological optimum under Arctic-alpine conditions, an element with their optimum in boreal regions but with outliers in Arctic and in alpine areas, and an element of species more or less equally distributed from temperate to Arctic-alpine regions. Lignicolous, sphagnicolous and litter decomposing species are rare or absent, probably due to scarcity of suitable substrates.

INTRODUCTION

During the last few decades, several mycologists and other persons visiting Svalbard have brought home agaric material, but few papers describing that material have appeared. Notable exceptions are Hagen (1950), Kobayasi et al. (1968), Ohenoja (1971), Reid (1979), and Skifte (1979).

The aim of this study has been to describe the genus *Galerina* as it occurs on Svalbard. *Galerina* is mentioned by Singer (1954) as one of the most important genera in Arctic areas. Species like *G. pseudomycenopsis* (syn. *G. moelleri*) and *G. arctica* are among the commonest of all agarics in the Arctic. In July and August 1981 I collected on Svalbard for a fortnight. Unfortunately this was just at the start of the agaric season that year. However, other mycologists, and especially Mr. O. Skifte and Mr. J. Stordal, have very generously supplied material. Their collections constitute the bulk of the material examined. I believe the material is fairly representative for Spitsbergen, the main island of Svalbard.

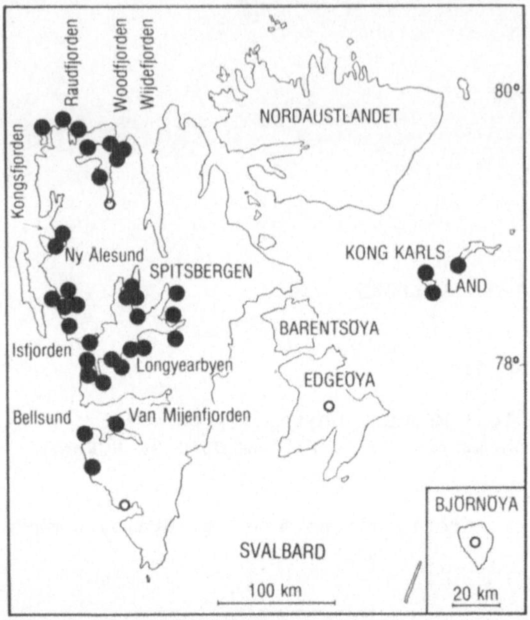

Fig. 1. Map of Svalbard showing localities from which material has been examined (dots). Circles denote literature records not examined. Note a single dot often represents more than one locality.

Four species of *Galerina* are previously known from Svalbard, viz., *G. pseudomycenopsis* Pilát (= *G. moelleri* Bas), *G. pseudocerina* Smith & Singer, *G. pumila* (Pers.: Fr.) Singer, and *G. mniophila* (Lasch) Kühner, according to Skirgiello (1961, 1968), Ohenoja (1971) and Reid (1979). Other records of *Galerina* species from Svalbard; i.e., *G. hypnorum* (by Lindblom 1841, Karsten 1872 and Dobbs 1942), *G. mniophila* and *G. spartea* (by Michelmore 1934) and *G. embolus* (by Karsten 1872, from Björnöya) lack sufficient information to establish the identity of the species.

Materials and Methods

The material consists of 322 collections, made by various mycologists and other persons during eight different years. The places visited are indicated in fig. 1. The main collectors are indicated in the text by their initials thus:

GG = Gro Gulden
HH = Heli Heikkilä
PK = Paavo Kallio
DL = Denise Lamoure
GAL = Gary A. Laursen
JM = Jon Markussen
EK = Esteri (Kankainen) Ohenoja
OS = Ola Skifte
JS = Jens Stordal

Vertical sections through the pileipellis and sometimes the whole pileus including the lamellae, were made by hand. Hymenial structures were studied in squash preparations. Spores have always been measured in preparations of the pileipellis. This is to ensure that only mature spores are considered. Spores found on the stipe or in spore prints will probably always include some immature spores shed as a result of picking. Numbers in parentheses before spores sizes are the number of spores, the number of fruitbodies, and the number of collections examined, in this order. The numbers in italics are spore size means. The spore colors are those in preparations of 2 percent KOH; i.e., considerably darker than when observed in H_2O. The term *pore* is used for an abrupt, circular thinning of the spore wall, which in profile appears as a small and pale column. A *callus* is present when a more gradual attenuation of the spore wall takes place and the apical area is more diffuse. The spore wall layers are described in accordance with Pegler & Young (1971, fig. 32). The *plage* is a spot in the suprahilar region where the perisporium is ruptured and exosporial ornamentation is exposed.

The "Diagnostic microscopic features" specified for each species do not include character stages that are normal in the genus, such as basidia 4-spored or clamps present, but gives the deviating stages: basidia 2-spored, clamps absent, etc.

For the scanning microscopy (SEM), small segments of the pileus with intact lamellae were soaked overnight in distilled water after a short soak in ethanol. Pieces were then transferred, first to 45 percent acetic acid, then to a 1:1 glacial acetic acid - 100 percent ethanol solution, and finally to three successive baths of pure 100 percent ethanol. A critical point (drying) treatment and coating (sputter) by gold in an argon atmosphere (2+2 min.) followed. A JEOL Scanning Microscope, Model JSM-35C, was used for observations and photographs. For each species, spores were studied on the pellicle and the hymenium.

TAXONOMIC SECTION

Synopsis of the recognized taxa

Genus: *Galerina* Earle

Subgenus: *Tubariopsis* (Kühner ex Bas) Smith & Singer emend. Gulden
Section: *Tubariopsis*

1. *G. clavata* (Velen.) Kühner
2. *G. arctica* (Singer) Nezd.

Section: *Tibiicystidinae* (Smith & Singer) Gulden

3. *G. pseudocerina* Smith & Singer

Subgenus: *Phaeogalera* (Kühner) Gulden
Section: *Phaeogalera*

4. *G. stagnina* (Fr.) Kühner

Subgenus: *Naucoriopsis* Kühner ex Gulden
Section: Naucoriopsis

5. *G. pseudomycenopsis* Pilát apud Pilát & Nannf.

Subgenus: *Galerina*
Section: *Mycenopsis* Smith & Singer

6. *G. hypnorum* (Schrank.: Fr.) Kühner
7. *G. calyptrata* Orton
8. *G. antheliae* Gulden
9. *G. mniophila* (Lasch) Kühner
10. *G. pseudomniophila* Kühner
11. *G. pumila* (Pers.: Fr.) Singer
12. *G. embolus* (Fr.) Orton

Key to the known species of *Galerina* on Svalbard

1. Cystidia tibiiform; spores never with plage; clamps present or absent 2
1. Cystidia not tibiiform; plage present in all ornamented spores 4
2. Clamps present; spores coarsely ornamented; cystidia relatively slender with narrow necks (1.5–2.5 µm) and small heads (3–5 µm) .... *3. G. pseudocerina*
2. Clamps absent; spores verrucose to practically smooth 3
3. Spores verrucose, yellow brown to rusty brown, not collapsed on hymenium; cystidia relatively stout, necks 2.5–6 µm, heads 5–12 µm .... *1. G. clavata*
3. Spores practically smooth, golden yellow brown, mostly collapsed on hymenium; cystidia relatively slender, neck 2–3 µm, head (2–)4–6 µm .... *2. G. arctica*
4. Spores smooth, without plage; cystidia mostly slightly ventrally inflated, neck often flexuous or repeatedly inflated/constricted, apex inflated or not, often with an attenuated tip above the head (apex beaked or lanceolate) 5
4. Spores minutely to distinctly ornamented, plage present; cystidia ventricose with cylindric to fusoid necks, apex obtusely rounded to capitate 7
5. Spores ellipsoid with apical pore, often truncate, biconvex in profile view; cystidia rounded or subcapitate at apex .... *4. G. stagnina*
5. Spores amygdaliform in profile view and with suprahilar applanation, pore absent; cystidia often beaked at apex 6
6. Spores 10.3-*11.7*-13 x 6.8-*7.4*-8.3 µm, yellow brown; pileus ochre or yellow brown to fulvous; lamellae ascendant-adnate, normal to subdistant .... *11. G. pumila*
6. Spores 12-*12.7*-14 x 6.8-*7.7*-8.5 µm, dark yellow brown to rusty brown with a tendency of the perisporium to loosen; pileus dark brown; lamellae horizontal, adnate, often with tooth, distant .... *12. G. cf. embolus*
7. Cystidia ventricose and acute, obtuse or a little capitate at apex (up to 7.5 µm) 8
7. Cystidia ventricose-capitate, heads up to 15 µm, sometimes of the size of the ventral portion 10
8. Pleurocystidia present; spores marbled to rugulose, rusty brown in KOH, with plage and pore; stipe with a membraneous ring (some remnants usually present); pileus yellow brown to red brown ....*5. G. pseudomycenopsis*
8. Pleurocystidia absent; spores almost smooth and rather pale brown in KOH; stipe ± fibrillose from veil; pileus ochre to greyish olive brown, often with more hyaline greyish apex than margin 9
9. Cystidia mostly somewhat capitate; spores yellow brown in KOH; lamellae ochre .... *10. G. pseudomniophila*

9. Cystidia with cylindric neck without apical inflation (or slight in some); spores dull brown in KOH, rather pale; lamellae dull brownish .... *9. G. mniophila*

10. Spores 7.7 µm or broader, perisporium not loosening .... *8. G. antheliae*

10. Spores up to 7.7 µm broad, calyptrate or not 11

11. Cystidia moderately inflated at apex (4-7 µm); spores pale, yellow brown, with indistinct plage, almost smooth, perisporium not loosening .... *10. G. pseudomniophila*

11. Cystidia mostly distinctly capitate (apex 5-15 µm); spores rusty brown with distinct plage, marbled, perisporium loosening or not .. 12

12. Spores calyptrate, distinct ears in many spores; habitat restricted (?) to *Sphagnum* spp. .... *7. G. calyptrata*

12. Spores without perisporial loosening or perisporium only forming small blisters; habitat on *Dicranum* and other mosses ....*6. G. hypnorum*

1. *Galerina clavata* (Velen.) Kühner, Encycl. Myc. 7: 171, 1935.
Fig. 2 and Pl. I, figs. 1-2.

Synonym: *Galera fragilis* Velen. var. *clavata* Velen. Česke Houby p. 548, 1921.

Misapplied name: *Galerina heterocystis* (Atk.) Smith & Singer 1957 and Smith & Singer 1964.

Diagnostic microscopic features: *Spores* (151/11/10): (9.6-)10.2-*13.0*-15.5(-17.4) x 6-7.5-8.7(-9.7) µm, amygdaliform to subphaseoliform in profile, ± ellipsoid in face view, verrucose, yellow brown to rusty brown, plage and pore/callus absent, content without large oil drop(s). *Cheilocystidia* tibiiform, 28.8-43.2 x 7.3-15.6 x 2.5-6 x 5-12 µm; *pileocystidia* few or absent, *caulocystidia* numerous, extending down to lower portions of the stipe. *Clamps* absent.

This is one of the most common *Galerina* species on Svalbard, but it has not been previously recorded. It is a typical species of the high Arctic wet tundra where it grows on various mosses.

The spore size varies considerably, but continually, and attempts to separate two taxa on this basis were unsuccessful. The spore wall is relatively thick and the ornamentation coarse, mainly in the form of ± discrete warts. I have not seen spores as large, dark and coarsely ornamented in Norwegian material as seen in the extremes from Svalbard. The SEM micrographs of spores from this and the two other species of subgenus *Tubariopsis*; i.e., *G. arctica* and *G. pseudocerina* do not reveal any differences between spores in the hymenium and spores deposited on the pileus surface, and no plage can be seen.

The closest relative of *G. clavata* on Svalbard is *G. arctica* which differs by having smooth, golden yellow, thin walled spores that are mostly seen collapsed in preparations of the hymenium.

Smith & Singer (1957, 1964) introduced the name *G. heterocystis* for *G. clavata*. According to their examination of the type, the spores in *G. heterocystis* are smooth to weakly ornamented, usually with a yellowish content in KOH. Cheilocystidia, as described and depicted by them, are also rather different from those found in European material, which should accordingly retain the name *G. clavata*. Kühner's

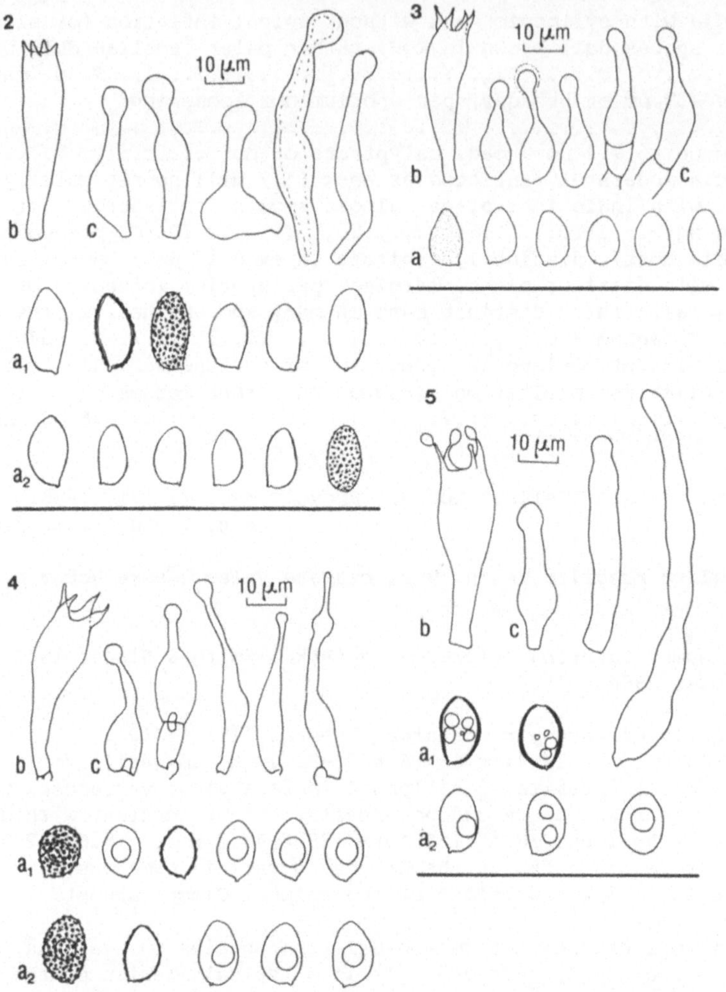

Fig. 2. *Galerina clavata*. Spores (a_1 and a_2), basidium (b), and cheilocystidia (c); the second from the right has become thick-walled. (a_1, b and c from OS 216, a_2 from JS 11633). Fig. 3. *Galerina arctica*. (GG 99/81). Spores (a), basidium (b), and cheilocystidia (c). Fig. 4. *Galerina pseudocerina*. Spores (a_1 and a_2), basidium (b), and cheilocystidia (c). (a_1, b and c from GG 160/81; a_2 from GAL 2743). Fig. 5. *Galerina stagnina*. (ML 14.08.82). Spores (a), basidium (b), and cheilocystidia (c).

description (1935) of *G. clavata* (Velen.) forme *tetrasporic* and Velenovský's original description of *G. fragilis* var. *clavata*, as translated into Latin by Pilát (1948) cover the species well. No authentic material of *G. fragilis* var. *clavata* Velen. exists (Svrček, 1984 in litt.).

Habit and Habitats: Most finds of *G. clavata* on Svalbard are from moist, wet or very wet habitats in moss dominated vegetation or in more grassy communities. It occurs gregariously or even subcaespitosely,

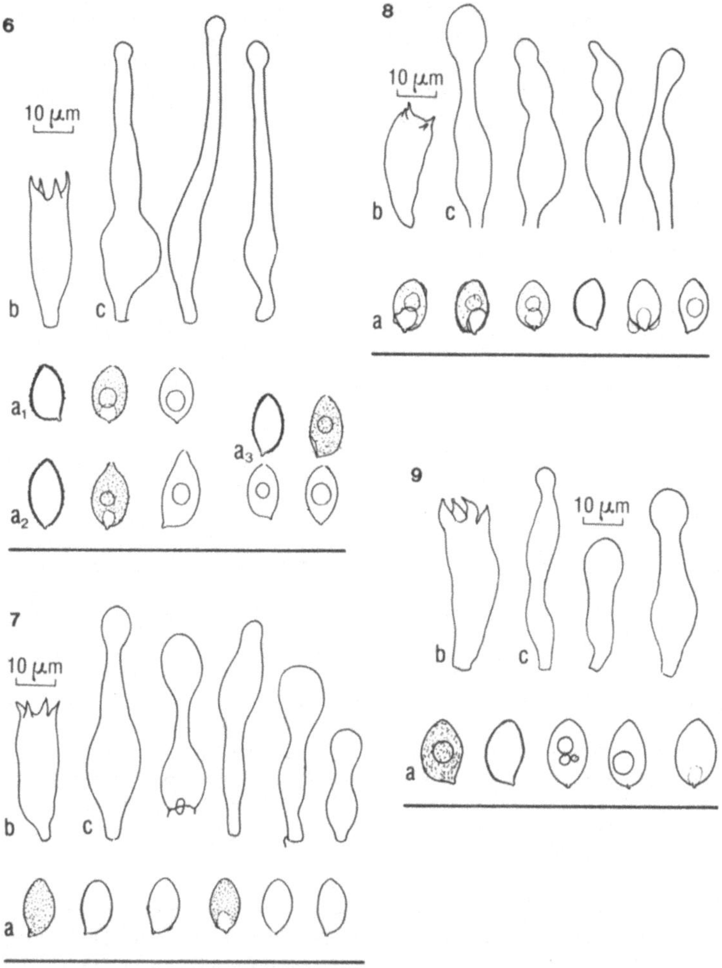

Fig. 6. *Galerina pseudomycenopsis.* Spores (a_1-a_3), basidium (b), and cheilocystidia (c). (a_1, b and c from GG 254/81, a_2 from OS 498, and a_3 from SH 20). Fig. 7. *Galerina hypnorum.* (OS 346). Spores (a), basidium (b), and cheilocystidia (c). Fig. 8. *Galerina calyptrata.* (OS 184). Spores (a), basidium (b), and cheilocystidia (c). Fig. 9. *Galerina antheliae.* (GAL 2693). Spores (a), basidium (b), and cheilocystidia (c).

rarely solitarily. Bryophytes noted are: *Calliergon stramineum, C. sarmentosum, Drepanocladus revolvens, D. uncinatus, Tomentohypnum nitens* and *Mnium* sp. The most commonly noted phanerogams are: *Ranunculus hyperboreus, Cochlearia officinalis* and *Saxifraga* spp. It was often found together with *Leptoglossum lobatum, Galerina arctica* and *G. pseudomycenopsis*. *Season*: Appears early in the agaric season, collected from 8 July to 25 August. Collections are from eight different years.

Distribution: Several finds have been made on the west coast and in the fjord districts of Spitsbergen, between Dunderbukta in the Bellsund district (77°28'N) and Reinsdyrflya (79°45'N).

Galerina clavata has an almost world wide distribution and has been recorded several times from alpine, subarctic and Arctic areas. It is not among the five species mentioned by Horak (1982) from the Antarctic Botanical Region. Apparently members of subgenus *Tubariopsis* do not occur in the Antarctic or Subantarctic Region (Horak l.c., Pegler et al. 1980).

Material examined: OS 112, 135, 136, 137, 151, 192, 192A, 193, 193A, 194, 195, 213, 214, 215, 216, 243, 307, 308, 309, 310, 311, 445, 743, 755, 854; leg. N. Foged 25 August 1958 (TROM). - Leg. EK and/or PK 12 August 1966 (2 coll.), 13 August 1966, 15 August 1966 (2 coll.), 18 August 1966, 23 August 1966; leg. HH and /or PK 31 July 1977, 1 August 1977 (TUR). - JS 11580a, 11580b, 11604, 11633, 11781; leg. Forfang 28 July 1960; GG 100/81, 147/81, 155/81, 232/81, 243/81, 256/81, 261/81, 262/81, 314/81, 315/81; JM 14/84 (O): - DL 83-96 (LY).

2. *Galerina arctica* (Singer) Nezd., Mik. Fitopat. 16: 209, 1982.
Fig. 3 and Pl. I, figs. 3-4.

Synonyms: *Cortinarius arcticus* Singer, Bot. Mater. Otdela Sporovych Rast. Bot. Inst. Komarov Akad. Nauk SSSR 4: 15, 1938. *Galerina griseipes* Kühner, Bull. Soc. Myc. Fr. 88: 153, 1972.

Diagnostic microscopic features: *Spores* (71/7/6): (9.7-)10.2-*11.4*-12.6(-13.2) x 5.3-*6.8*-7.3(-9) µm, ± ovoid in face view, in profile amygdaliform to wedge-formed, biconvex, thin walled and mostly seen collapsed in preparations of the hymenium, contour smooth to almost smooth, surface or content minutely granulose, golden yellow to yellow amber, plage and pore/callus absent, large oil drop(s) absent. *Cheilocystidia* tibiiform, 25-35 x 7-11 x 2-3 x (2-)4-6 µm; *pileocystidia* scattered, *caulocystidia* numerous, extending down to lower portion of stipe. *Clamps* absent.

Next to *G. pseudomycenopsis*, this is the *Galerina* species most frequently collected on Svalbard. It is a typical species of the high Arctic wet tundra. In the field it has mostly been confused with *G. clavata*, which also has a fully pruinose stipe and is common in the same types of habitats. *Galerina arctica* is possibly an even more hygrophilous species than *G. clavata*. A full description of the species is found in Kühner (1972b) under the name *G. griseipes*.

Galerina arctica was originally described as *Cortinarius arcticus* by Singer (1938) who examined Arctic material from the Soviet Union preserved in ethanol. His diagnosis is incomplete, lacking information on cystidia and clamps. The identity of this species as a *Galerina* was later established by Nezdoyminogo (1982), who found tibiiform cheilocystidia and quite smooth spores in the type material in Leningrad. Clamps were absent. According to Nezdoyminogo and Melnik (the curator of fungi in Leningrad), the type material of *Cortinarius arcticus* Singer is "quite destroyed" (pers. communications, in litt.). Singer (in litt. 11.12.1984) has kindly informed me that he is "convinced now that Nezdoyminogo is right that this is a *Galerina*." He further shares my opinion that it is the same species as described later by Kühner (1972b) as *G. griseipes*. Since some important structures may still be seen in the holotype, I refrain from assigning a neotype.

From my descriptions of fresh material and from notes with some of the collections, a few amendments to Kühner's (1972b) description seem appropriate: The pileus is usually between 10 and 15 mm wide, but can

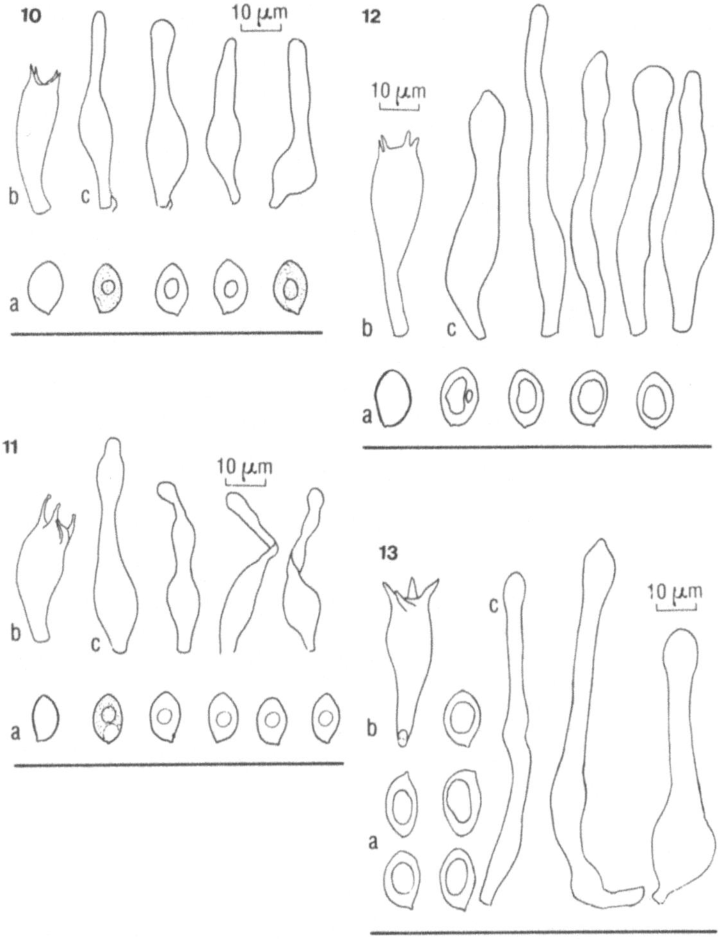

Fig. 10. *Galerina mniophila*. (GG 260/81). Spores (a), basidium (b), and cheilocystidia (c). Fig. 11. *Galerina pseudomniophila*. (Sundling & Falkinger 16.08.60). Spores (a), basidium (b), and cheilocystidia (c). Fig. 12. *Galerina pumila*. (DL 31.07.83). Spores (a), basidium (b), and cheilocystidia (c). Fig. 13. *Galerina* cf. *embolus*. (GG 346/81). Spores (a), basidium (b), and cheilocystidia (c).

reach at least 26 mm. It is generally brightly coloured and often with a more brownish centre and a paler margin. Rather often it is acutely umbonate and papillate and only faintly translucently striate. Sometimes a fatty, sticky or subviscid surface has been noted. The lamellae in dried specimens are characteristically warm ochre to cinnamon brown. The stipe is usually distinctly paler than the pileus and flushed with more ochre at the apex or in the upper half than over the lower portion, which often is almost hyaline or more occasionally greyish. A very scanty veil has been noted between pileus and stipe in the smallest buttons of one collection (OS 323). The flesh is brittle and without farinaceous smell or taste. The SEM micrographs of the spores (Pl. I, figs. 3-4) show a very faint ornamentation and no plage.

Galerina arctica is best recognized on the basis of the pale, thin-walled, smooth, and frequently collapsed spores. Kühner (1972b) emphasized the canescence of the stipe base and the presence of small inter- or epicellular, brown to bistre granules in the basal part of the stipe, but these features also occasionally occur in *G. clavata* on Svalbard and in *G. laevis* (Pers.) Singer (= *G. graminea* (Velen.) Kühner) as observed by Kühner (op. cit.). The greying of the stipe is not a constant feature of *G. arctica*, not even as a microscopic character, in the Svalbard material. *Galerina arctica* has larger spores than the other species in sect. *Tubariopsis* with smooth spores. The spore form is characteristic. Several spore characters, size of cheilocystidia and the presence of pileocystidia distinguish *G. arctica* from *G. clavata*.

Habits and Habitats: Generally gregarious, occasionally subcaespitose or solitary, on and between mosses in moist to very wet habitats, even in running water. Many collections are from sites along water margins, specifically near brooklets, pools and springs. Some collections are from well manured sites under bird roosting cliffs. Mosses noted are: *Calliergon sarmentosum*, *Philonotis fontana*, *Tomentohypnum nitens*, *Aulacomnium palustre*, *Drepanocladus* sp. and *Mnium* sp. The most frequently noted phanerogams are: *Saxifraga cernua*, *Cardamine nymani*, *Cochlearia officinalis*, *Phippsia algida*, *Cerastium regelii* and *Saxifraga rivularis*. *Galerina arctica* frequently has been found together with *G. clavata*, *G. pseudomycenopsis* and *Leptoglossum lobatum*. *Season*: Appearing early in the agaric season, collected from 10 July to 26 August. Collections are from eight different years.

Distribution: Finds have been made in several places on the West coast of Spitsbergen, from the Bellsund area (77°30'N) to Ytre Norskeøya at Raudfjorden (79°50'N), and in the fjord districts. There are also collections from Kong Karls Land.

Galerina arctica seems to be primarily an Arctic-alpine mushroom; for instance, it was the most commonly collected *Galerina* in the Arctic Soviet Union after *G. pseudomycenopsis* and *G. pseudocerina*, according to Nezdoyminogo (1982) who reports it from Franz Josef Land, Novaya Zémlya, Severnaya Zémlya, Dickson Island, eastern coast of Taimur Peninsula and from the New Sibirian Islands. Many of these places and also Kong Karls Land are within the area of the polar deserts (Aleksandrova 1980) with the most severe of climatic conditions. In France, it is known from alpine sites and from the upper part of the forest belt (Kühner 1972b). After my study on the alpine *Galerina* species of Norway appeared (Gulden 1980), I found the species twice in the Finse area, in a late-melting eutrophic snow-bed meadow in the mid-alpine belt (GG 701/83 and 703/83).

Material examined: OS 131, 134, 149, 150, 154, 160, 161, 285, 286, 287, 288, 289, 290, 291, 323, 324, 325, 326, 327, 328, 329, 330, 341, 350, 357, 367, 389, 408, 450, 451, 456, 458, 473, 474, 505, 618, 619, 733, 735, 742; leg. N. Foged 26 August 1958 (TROM). - EK 18 August 1966, 24 August 1966; leg. HH & PK 30 July 1977, 1 August 1977; leg. SH & PK 118, 136 (TUR). - JS 11561, 11593, 11594, 11595, 11600, 11601, 11602, 11610, 11617; GG 99/81, 104/81, 107/81, 108/81, 109/81, 111/81, 112/81, 127/81, 146/81, 164/81, 166/81, 167/81, 192/81, 196/81, 218/81, 219/81, 253/81, 257/81; JM 20/84, 33/84, 46/84 (O). - GAL 2688, 2732 (AL).

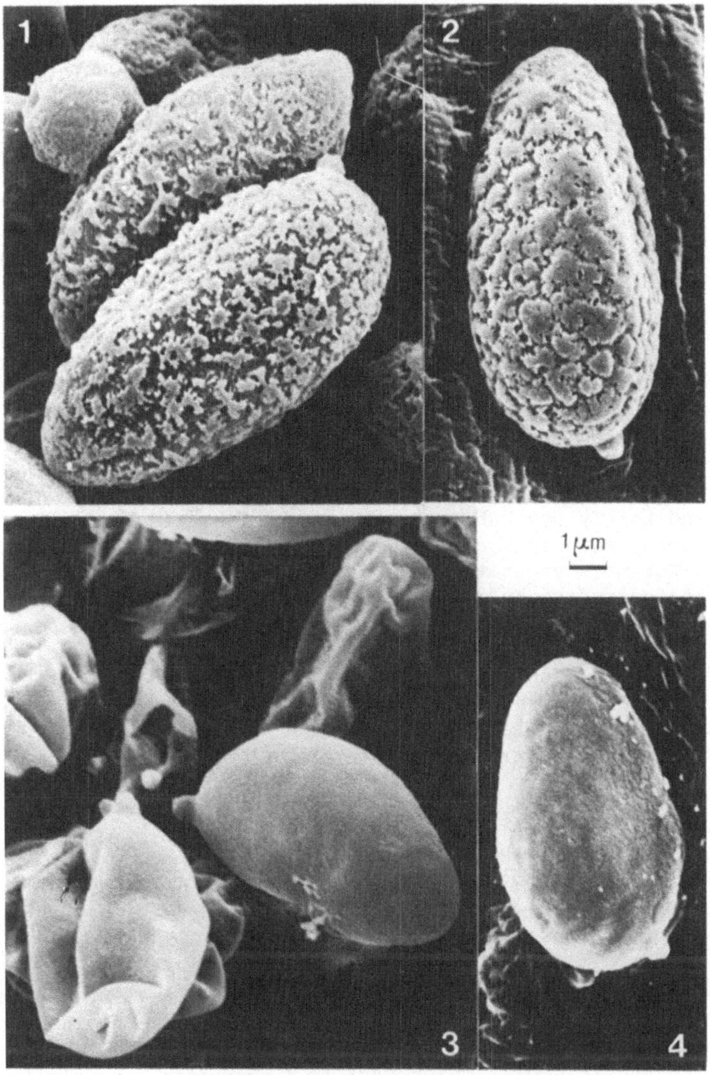

Plate I: Figs. 1-2. *Galerina clavata* (GG 155/81). 1. Spores in hymenium. 2. Spore deposited on pileipellis. Figs. 3-4. *Galerina arctica* (GG 99/81). 3. Spores in hymenium. 4. Spore deposited on pileipellis.

3. *Galerina pseudocerina* Smith & Singer, Mycologia 50: 483, 1958.
Fig. 4, Pl. II, figs. 5-6 and Pl. III, fig. 7.

Diagnostic microscopic features: *Spores* (61/6/6): 10-*12.2*-13.5(-14.5) x 7-*8.5*-9.5 µm, nearly subglobose to ellipsoid, blunt or sometimes nearly truncate at apex, biconvex in profile, coarsely verrucose to patchy ornamented, dark tawny to rusty brown, plage absent, apex sometimes as if a callus or pore were present, large oil drop(s) present. *Basidia* 4-spored, 25-42 x 9.7-12 µm. *Cheilocystidia* tibiiform, 24-45 x 6-9.6 x 1.5-2.5 x 3-5 µm, with relatively long and thin necks; *dermatocystidia* restricted to upper half of stipe. *Clamps* present.

This is the most common *Galerina* species in drier habitats on Svalbard. It can be recognized in the field by its dark, only faintly striate pileus, the pale stipe and the horizontal and subdistant lamellae. On some occasions it has been confused with *G. pseudomycenopsis* or mixed with this. The coarsely ornamented and dark spores are distinctive for the species.

The size and shape of the spore vary considerably in the material. Attempts, however, to distinguish two taxa were not successful. The basidia in the Svalbard material are always 4-spored and longer than indicated by Smith & Singer (1964), viz., 25-42 x 9.7-12 µm. Plage as reported by Smith & Singer (1958) and pleurocystidia as reported by Reid (1979) are not present (cf. Pl. II, fig. 6). The farinaceous smell and taste, which can be very distinct, was sometimes absent or escaped notice. The absence of plage and the presence of tibiiform cystidia were the main reasons for transferring this species into the subgenus *Tubariopsis* (Gulden 1980).

Habit and Habitats: Generally found in small groups on moss in eutrophic and fairly dry habitats. Several times it was found together with *Dryas octopetala* and *Salix polaris*. Three times it was collected in moist or wet moss vegetation. Twice it was growing together with a *Calvatia* species and seven times together with *Cortinarius* (*Dermocybe*) *polaris* Hoiland. The latter species is classified among the Nordic species of *Cortinarius* subgen. *Dermocybe* as occurring on the top of an Arctic-lowland gradient, a rich-poor gradient, and in the middle of a xeric-wet gradient (Høiland 1984). The type of habitats, even moist ones, agree well with those described for the species by Kuhner (1966 and 1972a). *Season*: Appears early in the agaric season, collected between 18 July and 17 August. Collections are from five different years.

Distribution: Reid (1979) records one find of this species from Svalbard. It now has been found scattered on the west coast of Spitsbergen between Isfjorden and Ytre Norskeøya (79°50'N) and eastwards to inner Isfjorden (Sassenfjorden). At Kapp Wijk (78°05'N) it was collected at 600 m, which is in the high Arctic vegetation region as defined by Eurola (1968) or the polar desert as defined by Aleksandrova (1980). Most previous records of *G. pseudocerina* are from Arctic or alpine habitats. Nezdoyminogo (1982) reports the species as the most common in the Arctic Soviet Union next to *G. pseudomycenopsis* (as *G. moelleri*). However, Kühner (1972a) also records a find from "étage montagnard" at 800 m in France. Urbonas & Matelis (1983) record it from pine forests in Lithauen (Pinetum myrtilletosum). I have collected the species once in a rich boreo-nemoral pine forest in South Norway (GG 654/67).

Material examined: OS 403, 430, 489, 499 (TROM). - JS 11603, 11605, 11635, 11698, 11710, 11719; GG 114/81, 134/81, 160/81, 162/81, 165/81, 202/81, 235/81, 255/81, 339/81 (O). - DL 83-29 (LY).

4. *Galerina stagnina* (Fr.) Kühner, Encycl. Myc. 7: 187, 1935.
Fig. 5 and Pl. III, fig. 8.

Synonym: *Phaeogalera stagnina* (Fr.) Kühner, Bull. Soc. Myc. Fr. 88: 141, 1972.

Diagnostic microscopic features: *Spores* (16/2/2): 12.1-*13.8*-14.5 x 7.3-*8.3*-9.7 µm, bluntly ellipsoid to ovoid, biconvex

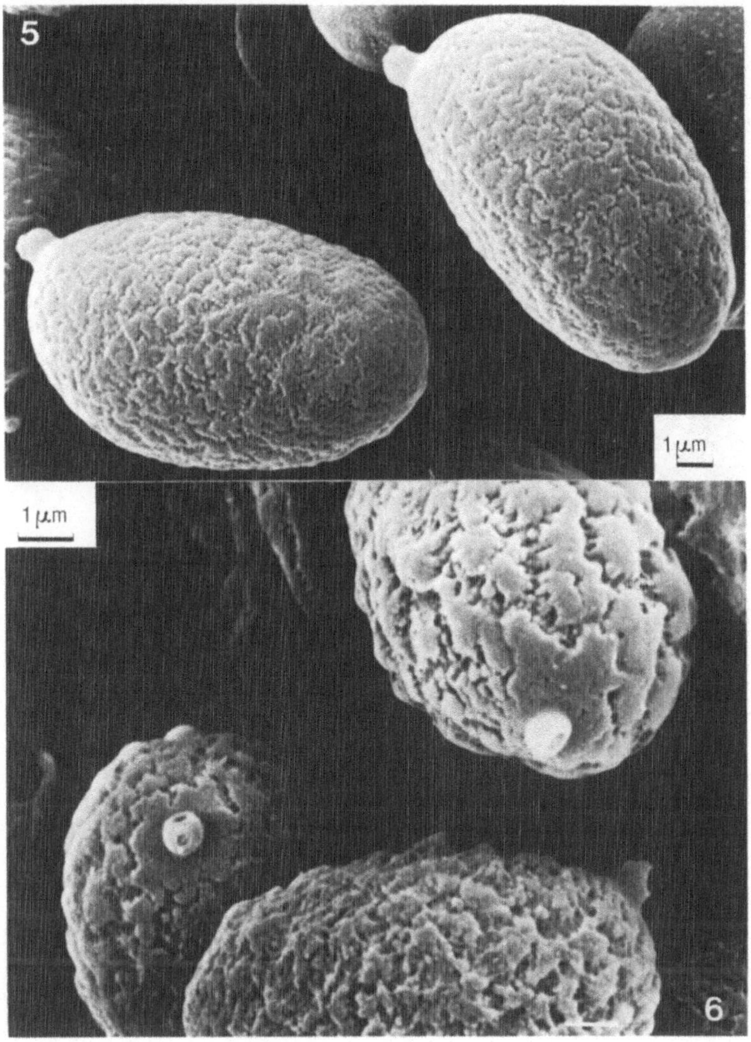

Plate II: Figs. 5-6. *Galerina pseudocerina* (GAL 2743). Spores in hymenium.

in profile view, apex ± truncate with distinct apical pore, smooth, dark yellow brown, plage absent, inamyloid. *Cheilocystidia* 45-82 x 6-8(-15) x 3-7 x 6.5-12 µm, mostly narrowly lageniform with (sub)capitate tips, often with more constrictions/inflations, and sometimes without ventral inflation; *dermatocystidia* restricted to stipe apex.

The Svalbard material corresponds in all essential features, pileipellis included, to *Phaeogalera stagninoides* (Orton) Pegler & Young as recently described by Reid (1984) who restudied type material. This, however, I believe is hardly different from *G. stagnina*. Reid distinguishes the two on gross morphology and slight differences in spore size with the smaller spores from *G. stagnina* (11.0-14.0(-15.0) x 7.0-8.0 µm) and the larger from *G. stagninoides* (12.0-18.2 x 7.0-9.5(-10.0) µm). As previously stated (Gulden 1980), *G. stagnina* is morphologically very variable and I find problems in distinguishing taxa near *G. stagnina*, at least in the way proposed.

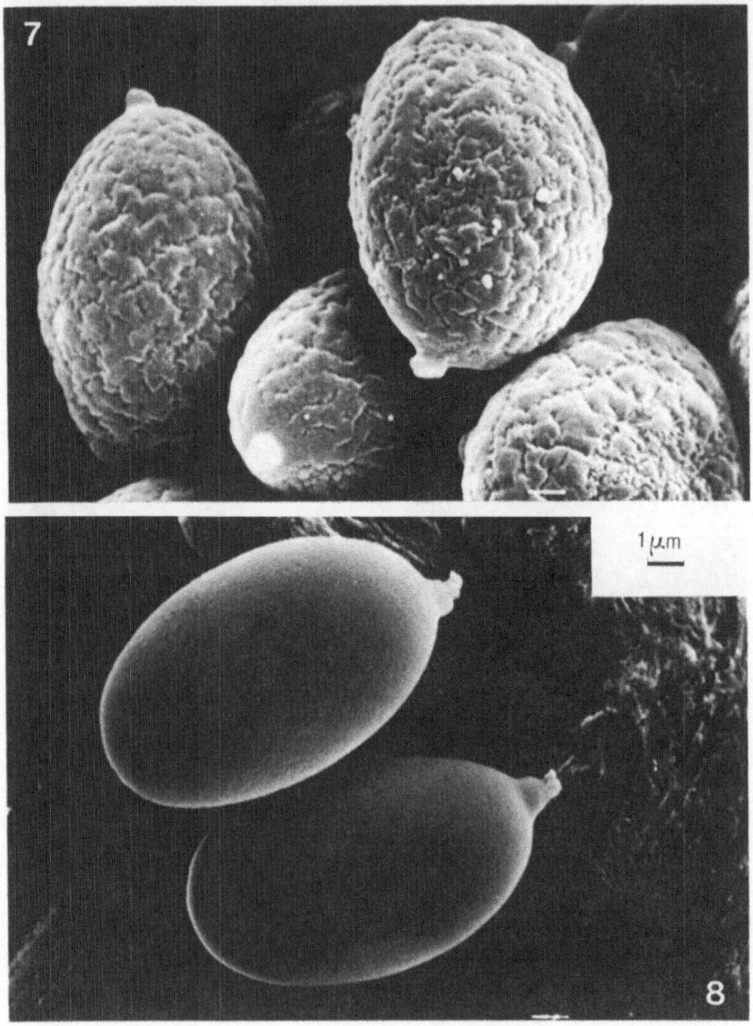

Plate III: Fig. 7. *Galerina pseudocerina* (GAL 2743). Spores deposited on pileipellis. Fig. 8. *Galerina stagnina* (ML 14.08.82). Spores deposited on pileipellis.

Habit and Habitats: One to few specimens together. Collections are from a marsh or swamp, a snow-bed at the margin of small, low moss cushions and in moist moss in a snow-bed in scree field. The vegetation on Svalbard, with very little acid mires, probably renders few favorable habitats for the species.

Distribution: *Galerina stagnina* is mainly a species of boreal, low alpine and subarctic areas. It is rare in lowland habitats in Central Europe as in Arctic regions, which seem to be beyond its optimal area. *Galerina stagnina* is known also from the Subantarctic Botanical Region (Pegler et al. 1980).

Material examined: Spitsbergen: Colesbukta, 14 August 1982, M. Lange (C). - St. Jonsfjorden, NE of Lövlibreen, 1 and 2 August 1960, JS 11667 11670 (O).

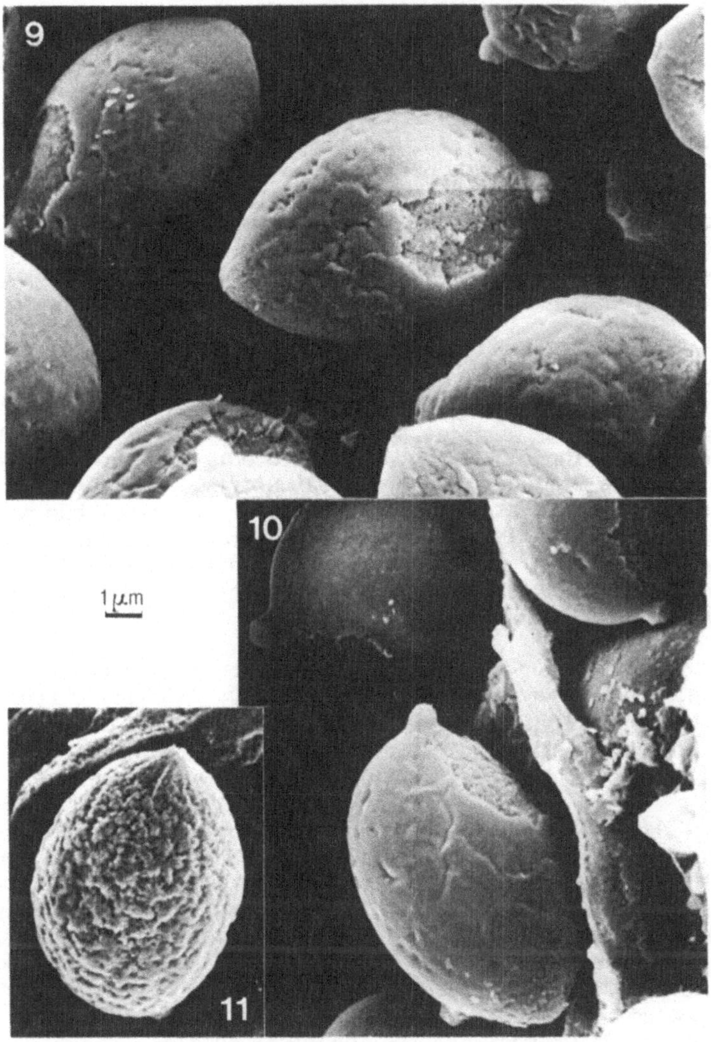

Plate IV: Figs. 9-11. *Galerina pseudomycenopsis*. 9. Spores in hymenium (GG 254/81). 10. Spores deposited on pileipellis (GG 91/81). 11. Spore deposited on pileipellis (GG 254/81).

5. *Galerina pseudomycenopsis* Pilát apud Pilát & Nannf., Friesia 5: 19, 1954. Fig. 6 and Pl. IV, figs. 9-11.

Synonyms: *Galera pumila* (Fr.) Favre f. *oreina* Favre, Ergeb. Wiss. Unters. Schweiz. Nat. Parks 5 (NF) 33: 204, 1955. *G. moelleri* Bas, Persoonia 1: 310, 1960.

Misapplied names: *Pholiota pumila* Fr. by Møller 1945, *Galerina pumila* (Fr.) by M. Lange 1957, *Galerina pseudopumila* Orton by Pegler & Young 1972, Reid 1979.

Diagnostic microscopic features: *Spores* (52/5/4): 9.7-*11.0*-13.5 x 6.8-*7.1*-8 µm, broadly amygdaliform to ovoid or citriniform, with apex rounded or snout-like outdrawn, with small pore, rusty brown, marbled to rugulose, plage distinct, without perisporial loosening. *Cheilocystidia* ventricose-fusoid with blunt or (sub)capitate tips, 37-90 x 6-14.5 x 3.6-5 x 3.5-7 µm; *pleurocystidia* few to numerous; *dermatocystidia* restricted to stipe apex.

The numerous collections of this species (142 or almost 45 percent of the collections) indicate that this is a species well adapted to the harsh conditions on Svalbard. It is one of the most common agarics on Svalbard and a typical species of the high Arctic wet tundra. All the investigators have collected this species and it has been recorded from Svalbard by Skirgiello (1961, 1968, as *Pholiota pumila*), by Ohenoja (1971 as *G. moelleri*) and by Reid (1979, as *G. pseudopumila*).

Galerina pseudomycenopsis is extremely variable in gross morphology. The pileus normally measures 1-2 cm, but pilei up to 2.8 cm are found. The stipe is up to 2.4 mm thick. On the other hand, tiny specimens with 3-5 mm broad pilei and almost filiform stipes exist in the material examined. The pileus can be translucently striate or not. Normally there is a well developed, membranous ring, but sometimes only minute veil remnants are present on the stipe or on the pileus margin. The moist pileus is usually some shade of red brown, but can be yellow brown to dark grey brown. Spore size varies considerably and sometimes seems to be different from one specimen to another. I will not exclude that more taxa could be involved.

In SEM the spore surface appears almost smooth with coarse cracks (Pl. IV, fig. 9-10) or wrinkled-rugulose (Pl. IV, fig. 11). Probably the latter appearance results from a faint loosening of the perisporium. The plage is ornamented just as it is in all the other examined species where a plage is present.

Gulden (1980 p. 233) presented a description of isotype material of *G. pseudomycenopsis* and removed it from sect. *Porospora* of subgenus *Galerina*, where it was placed by Smith & Singer (1964), into the subgenus *Naucoriopsis*. Since the veil in this material was poorly developed as also was stated in the original description by Pilát, it was assumed that the species belonged in stirps *Cedretorum*. Having now seen the great and apparently continuous variation in material of what I previously called *G. moelleri* on Svalbard, I no longer doubt that *G. pseudomycenopsis* and *G. moelleri* are identical species. A re-examination of the isotype, where spores deposited on the pileus cuticle were measured, gave the size 9-*9.9*-11(-12.5) x 6.3-*6.6*-7.0 µm. A thin upper layer of the pileipellis consisted of very few pale and narrow hyphae mixed with pale or little incrusted normally wide hyphae, apparently resting in a thin gelatinous matrix.

The name *G. pseudomycenopsis* antedates *G. moelleri* Bas and *G. pseudopumila* Orton, both published in 1960. The latter name is not connected to a description (Orton 1960 p. 176); only a reference to *Agaricus pumilus* in Fries 1828 p. 29 is given. The identity of this species is not clear, but it is very probable that it is not that of an alpine-Arctic species.

Galerina pseudomycenopsis is the only *Galerina* known from Svalbard with a ring. (*Galerina stagnina* can sometimes have veil remnants forming a belt on the stipe.) It is also the only known member of subgenus *Naucoriopsis* on Svalbard. In alpine areas of Norway and Central Europe further terricolous or muscicolous *Naucoriopsis* species such as *G. unicolor* (Vahl: Fr.) Singer and *G.* (*Pholiotina*) *alpina* (Horak) Horak & Watling occur. *Galerina pseudomycenopsis* differs from

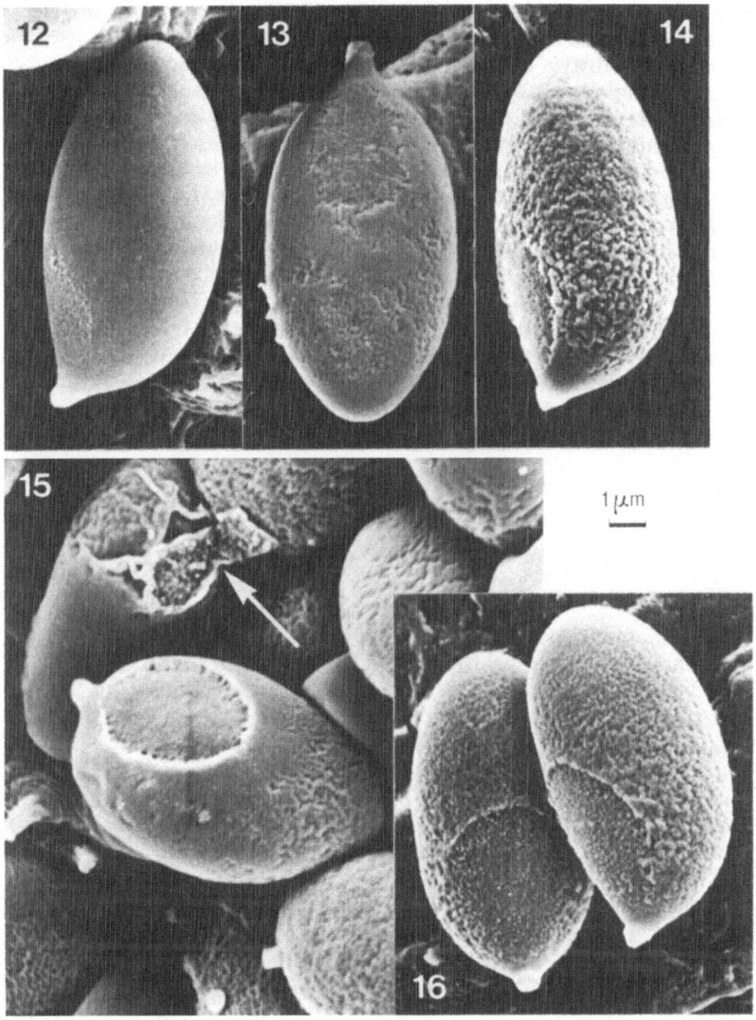

Plate V: Figs. 12-14. *Galerina hypnorum* (OS 346). 12. Spore in hymenium. 13-14. Spores deposited on pileipellis. Figs. 15-16. *Galerina calyptrata* (OS 184). 15. Spores in hymenium: arrow shows part of loosened perisporium. 16. Spores deposited on pileipellis.

these by the relatively broad spores without perisporial loosening as seen in LM. *Galerina hygrophila*, as recently described by Arnolds (1982) from very wet to moist mossy grassland in the Netherlands, comes very close to *G. pseudomycenopsis*. Its spores apparently are slightly narrower: (5.2-)5.8-7.0(-7.2) µm broad according to Arnolds. The suprapellis of the main form of this species has gelatinous hyphae. I have observed slightly gelatinous hyphae in Svalbard material of *G. pseudomycenopsis*.

Habit and Habitats: Gregarious to subcaespitose, often forming rows, arcs or even full fairy rings. Moist to wet sites with mosses or moist mossy - grassy sites are frequently cited as habitats for *G. pseudomycenopsis*. It has often been collected close to smaller or larger lakes, along rivulets and even in running water. Other sites are littoral marshes or meadows, mire-like sites and in cracks of peaty soil. Some collections are from drier habitats. Twice it has been found in mossy habitats where *Sphagnum* was present. The most frequently noted mosses are: *Calliergon stramineum*, *Drepanocladus* spp. and *Aulacomnium palustre*. Frequent higher plants are: *Salix polaris*, *Polygonum viviparum*, *Equisetum arvense* and *E. variegatum* and *Saxifraga* spp. *Galerina pseudomycenopsis* often grows together with *G. arctica*, *G. clavata* and *Leptoglossum lobatum*. *Season*: Appears early in the agaric season, collected between 30 June and 26 August during eight different years.

Distribution: Found in all investigated parts of Svalbard. *Galerina pseudomycenopsis* is a typical subarctic, Arctic and alpine species with a circumpolar distribution. Records of *G. moelleri* and *G. riparia* Sing. from subarctic-Arctic areas probably refer to *G. pseudomycenopsis* (see Horak 1982 p. 110).

Material examined: OS 113, 114a, 119, 120, 129, 138a, 146, 147, 158, 167, 168, 174, 176, 177, 182, 185, 189, 199, 200, 202, 203, 204, 205, 208, 222, 224, 225, 229, 230, 231, 239, 241, 246, 259, 262, 266, 269, 273, 275, 276, 277, 281, 297, 299, 313, 339, 340, 342, 343, 498; leg. N. Foged 24 August 1958, 25 August 1958 (2 coll.), 26 August 1958 (TROM) - EK 12 August 1966, 14 August 1966, 15 August 1966, 16 August 1966 (2 coll.), 17 August 1966, 18 August 1966, 22 August 1966, 23 August 1966; leg. I. Kekkonen 23 August 1966; HH 26 August 1966, HH & PK 3 August 1977; SH and/or PK 17, 20, 21, 27, 48, 60, 74, 92a, 133, 140, 147 (TUR). - leg. H.L. Lövenskiold 20 July 1960; leg. B. Falkanger & P. Sunding 24 July 1960; JS 11511, 11512, 11520, 11539, 11540, 11542, 11565, 11606, 11612, 11620, 11648, 11708, 11709, 11779, 11846, 11884; leg. D. Ovstedal July 1981 (4 coll.); GG 81/81, 82/81, 83/81, 90/81, 91/81, 95/81, 96/81, 97/81, 115/81, 125/81, 150/81, 151/81, 175/81, 195/81, 230/81, 231/81, 233/81, 254/81, 267/81, 305/81, 336/81, 344/81; JM 13/84, 19/84, 30/84, 34/84, 39/84, 45/84 (O). - GAL 2625, 2661, 2666, 2670, 2697, 2729, 2730, 2731 (AL). - DL 19 July 1983, 21 July 1983, 29 July 1983 and 7 August 1983 (LY).

6. *Galerina hypnorum* (Schrank.: Fr.) Kühner, Encycl. Myc. 7: 194, 1935
Fig. 7 and Pl. V, figs. 12-14.

Four collections belong to *G. hypnorum* in the wide sense of Kühner (1935). One of them has distinctly calyptrate spores and is separated here as *G. calyptrata* Orton. Others have grown on various bryophytes and are almost identical microscopically, with less perisporial loosening than in *G. calyptrata*.

Diagnostic microscopic features: *Spores* (20/2/2): (10-)11-*11.7*-12.6 x 6-*6.7*-7.3 µm, amygdaliform, rusty brown to tawny, faintly marbled to rugulose, plage distinct, perisporium occasionally loosening, mainly in the plage area, to form small blisters (not truly calyptrate), callus/pore absent. *Cheilocystidia* variable but mainly ventricose-capitate, with clavate to globose head, some without ventral inflation, 35-46 x 7-14 x 3-6 x 5-15 µm; *dermatocystidia* restricted to stipe apex.

There are only few notes on gross morphology. It seems futile to refer the material to one or some of the many species described in the complex. However, there are distinct veil remnants as fibrils on the stipe in all collections. This should rule out *G. alpestris* Singer (1974) and *G. obscura* Smith (Smith & Singer 1964). The cystidia also are more inflated and more often capitate than described for these species. *Galerina hypsizyga* Singer (in Smith & Singer 1964) also has different cystidia, which are much more slender and nearly tibiiform (illustrated in fig. 112 in Dennis 1961). Microscopically, our material comes closest to *G. subbadipes* Huijsman and *G. rugisperma* Smith as described in Smith & Singer (1964). The former however, was originally described as a more robust species with some pleurocystidia and is considered closely related to *G. sideroides* (Huijsman 1955). *Galerina rugisperma* is described with slightly smaller spores (8-10 x 5-6 um) and with a stronger tendency to perisporial loosening. Many of these species were described from alpine habitats. All are based on scanty material (1-3 collections) and have rarely been reported later. Very probably, the descriptions do not cover the full variation of any of the species. Consequently, reference of new collections to any of them will only rarely be possible. In SEM, spores in the hymenium often were smooth or almost smooth (Pl. V, fig. 12). Those deposited on the pileus surface were more often rugulose-wrinkled (Pl. V, figs. 13-14). Probably the rugulosity is formed by a loosening of the perisporium which mainly takes place in the mature spores. The same tendency of the spores to become rugulose is seen in all the species of the subgenus.

Habit and Distribution: On *Dicranum* and other mosses. Scattered on the west coast of Spitsbergen between Longyearbyen and Raudfjorden. *Galerina hypnorum* is one of the more frequently recorded *Galerina* species from Arctic habitats, but the name has been used for many different muscicolous species. One collection in herbarium O, made by H.G. Simmons on Ellesmereland and identified as *G. hypnorum* (Rostrup 1906), turned out to be of *G. pseudomycenopsis*.

Material examined: Spitsbergen: Isfjord district: E slope of Mt. Nordenskiöldfjell above Longyearbyen 400-500 m, 16 August 1960, P. Sunding & B. Falkanger; Kongsfjorden: Gerdöya, 24 July 1960, P. Sunding & B. Falkanger (O). - Raudfjorden: E side, 6 km S of Pt. Bruce (79°30'N) 9 August 1958, OS 346 (TROM).

7. *Galerina calyptrata* Orton, Trans. Brit. Mycol. Soc. 43: 237, 1960.
Fig. 8 and Pl. V, figs. 15-16.

The only specimen attached to *Sphagnum* amongst all of the material, clearly has calyptrate spores. It agrees well with the original description of *G. calyptrata*, but there are strong reasons to include this in *G. hypnorum*. *Spores*: (10/1): 9.5-*10.7*-12.1 x 6.3-*7.0*-7.8 µm. The SEM micrographs of spores are very similar to those of *G. hypnorum*. Blisters were not readily seen in the scanning microscope.

Material examined: Spitsbergen: Isfjord district: Colesdalen near Lailadalen, moist, in *Sphagnum*, 20 July 1958, OS 184 (TROM).

8. *Galerina antheliae* Gulden, Norw. J. Bot. 27: 245, 1980.
Fig. 9 and Pl. VI, fig. 17.

The single collection of this species has characters well in accordance with Norwegian alpine material of the species.

Diagnostic microscopic features: *Spores* (10/1): 12.1-*13.4*-15 x 7.7-*8.5*-9 µm, amygdaliform, rusty brown, marbled to almost smooth, contour faintly rugulose, plage distinct, callus/pore absent. *Cheilocystidia* ventricose-capitate, 40-55 x 6-9 x 3.6-6 x 3-10.8 µm; *dermatocystidia* restricted to stipe apex.

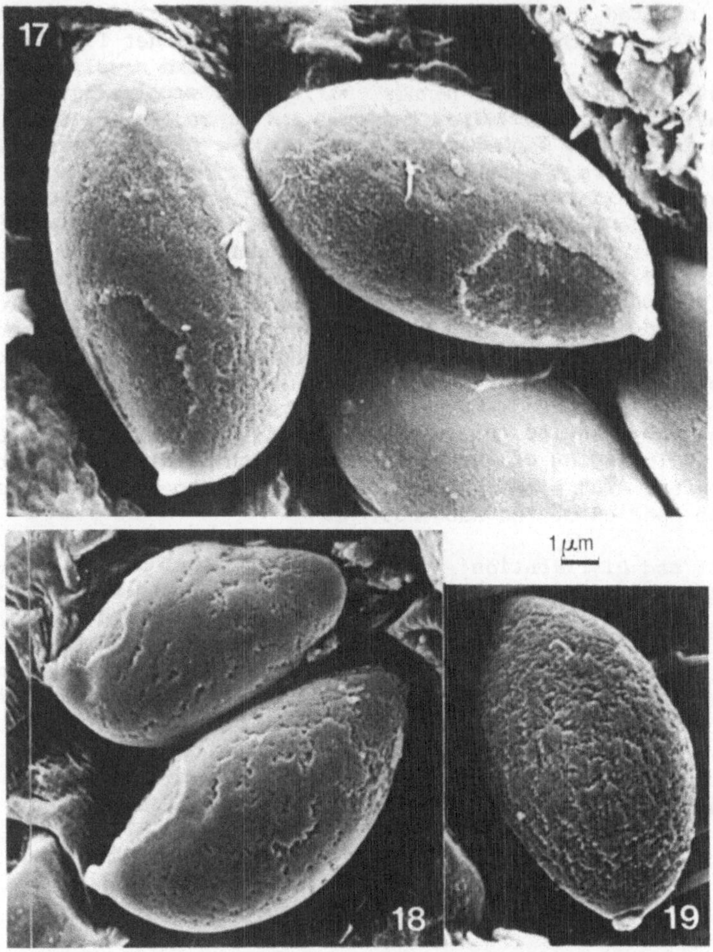

Plate VI: Fig. 17. *Galerina antheliae* (GAL 2693). Spores deposited on pileipellis. Figs. 18-19. *Galerina mniophila* (GG 260/81). Spores deposited on pileipellis.

Material examined: Spitsbergen: Longyearbyen, in wet moss in rivulet, 31 July 1983, GAL 2693 (AL).

9. *Galerina mniophila* (Lasch) Kühner, Encycl. Myc. 7: 192, 1935.
Fig. 10 and Pl. VI, figs. 18-19.

Diagnostic microscopic features: *Spores* (40/3/3): 9.8-*10.9*-12.1 x (5.8-)6.3-*6.7*-7.3 µm, amygdaliform, pale, dull brown to yellow brown, marbled to practically smooth, contour minutely rugulose, plage present, callus/pore absent. *Cheilocystidia* ventricose with cylindrical neck, only occasionally and slightly inflated or attenuated at apex, 30-53 x 6-10 x 2.5-5.5 x 3.5-7.3 µm; *dermatocystidia* restricted to stipe apex.

Galerina mniophila only rarely has been recorded from alpine and Arctic areas, but both Michelmore (1934) and Reid (1979) record it from Svalbard, the former from Edge Island, the latter from NW Spitsbergen, probably from the Woodfjord district.

The dark, somewhat olive brown pileus, frequently with paler disc, and the dull brown colour of the lamellae, characterize the material. The stipe is paler, but also with an olive cast, fibrillose from the veil and clavate. It differs from *G. pseudomniophila* in fruitbody and spore colors, and by the form of the cystidia, which is less capitate.

Habit and Distribution: Muscicolous, collected on *Dicranum* and other mosses. Kühner (1972a) records *G. mniophila* from alpine habitats in Central Europe. It was not recorded from the Finse area in Norway, but has later been found there in a low-alpine blueberry heath (GG 408/81). Dennis (1955) records it from ca. 450 m in Scotland. The material Møller (1945) described from The Faeroes, clearly belongs to another species.

Material examined: Spitsbergen: Isfjord district: Colesbukten, at Lailadalen, 20 July 1958, OS 183 (TROM). Longyearbyen, 1 August 1977, HH & PK (TUR) and 31 July 1983, on moss in wet hills down steep hillside, GAL 2684 (AL). Hotellneset - Björndalen, 27 July 1981, GG 260/81 (O), Björndalen, 3 August 1983, DL 83-68 (LY). Kong Karls Land: Svenskeöya, Kükenthalfjell, 15 August 1984 JM 36/84 (O).

10. *Galerina pseudomniophila* Kühner, Bull. Soc. Myc. Fr. 88: 152, 1972.
Fig. 11 and Pl. VII, fig. 20.

Synonym: *Galerina pumila* (Pers.: Fr.) Singer var. *subalpina* Smith, in Smith & Singer 1964.

Diagnostic microscopic features: *Spores* (10/1): 10-*10*.7-11.1 x (5.8-)6.3-*6.4*-6.5(-7.3) µm, amygdaliform, ovoid, pale, yellow brown, very faintly marbled, contour almost smooth, plage present, callus/pore absent. *Cheilocystidia* ventricose with a more or less constricted and inflated neck, often inflated at apex, 30-50 x 5-12 x 2.5-6 x 4-7 µm; *dermatocystidia* restricted to stipe apex.

The pale, yellow ochre to pale yellow colors of dried specimens, the rather pale, yellow brown, almost smooth spores and the relatively long, frequently (sub)capitate cheilocystidia distinguish this material. In the fresh condition, a darker color of the fruitbody was noted ("dark grey brown"). Kühner (1972a) characterizes the pileus color as ochre to brownish, where the brown color is "terne ou presque sale (? parfois subolivacé)", changing to cream-alutaceous upon drying. Smith (for *G. pumila* var. *subalpina*) uses the wording dull watery ochraceous tawny with an usually watery-pallid disc or the disc with an olive buff cast, fading to yellowish white (on disc first). I have little doubt that these taxa are identical, and that *G. mniophila* is the closest related (possibly even conspecific) species. For differences see *G. mniophila*.

Habit and Distribution: Muscicolous. *Galerina pseudomniophila* was previously known from alpine habitats in Central Europe and Fennoscandia, and from subalpine areas in North America and the Altai Mountains (as *G. pumila* var. *subalpina*).

Material examined: Spitsbergen: Isfjord district: E slope of Mt. Nordenskiöldfjell, 400-450 m, on *Dicranum*, 16 August 1960, P. Sunding & B. Falkanger, 2 coll. (O).

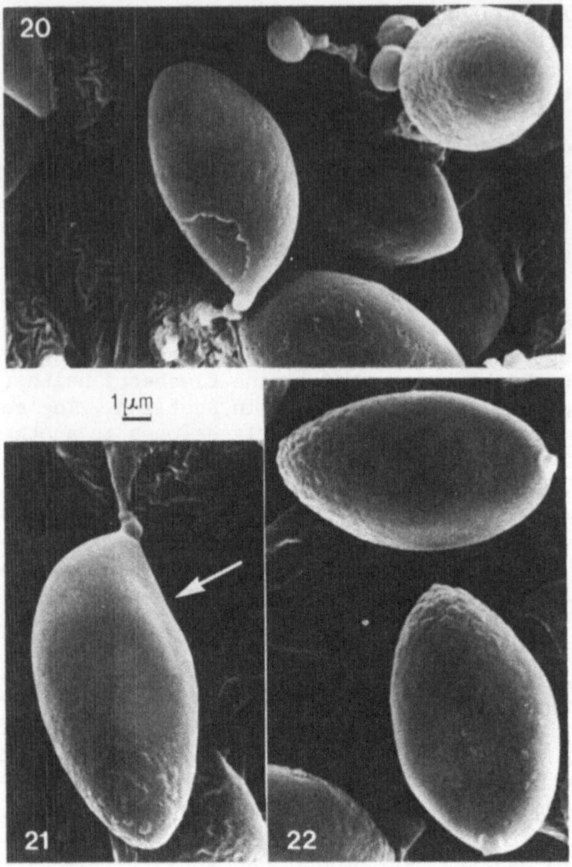

Plate VII: Fig. 20. *Galerina pseudomniophila* (Sundling & Falkanger 16.08.60). Spores in hymenium. Figs. 21-22. *Galerina pumila* (DL 31.07.83). Spores in hymenium; arrow shows the suprahilar applanation.

11. *Galerina pumila* (Pers.: Fr.) Singer, Persoonia 2: 41, 1961.
Fig. 12 and Pl. VII, figs. 21-22.

Diagnostic microscopic features: *Spores* (17/1): 10.3-*11*.7-13 x 6.8-7.*4*-8.3 µm, amygdaliform in profile view, more ovoid to ellipsoid in front view, with suprahilar applanation or slight depression but no apparent plage, or extremely faint in few spores, yellow brown, smooth or very faintly marbled, contour smooth or practically smooth, callus/pore absent. *Cheilocystidia* long protruding, cylindric or somewhat ventrally inflated, apex often beaked, lanceolate or capitate, 38-60 x 6-6.5 x 2.5-5 x 3-8.5 µm; *dermatocystidia* restricted to stipe apex.

With the SEM the spores are completely smooth except for a distinctly rough apex (Pl. VII, figs. 21-22). This is probably due to a loosening of the perisporium over the apex. The rough apex can occasionally also be seen in LM, when specially sought for. SEM micrographs also show a definite suprahilar applanation.

The characteristic cystidia and the smooth spores distinguish the species. Smith & Singer (1964) report a very minute apical pore in this

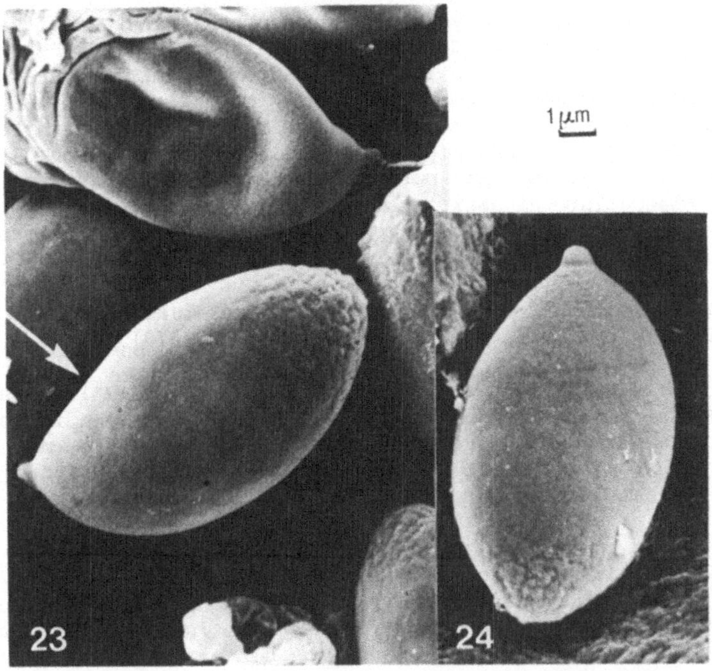

Plate VIII: Figs. 23-24. *Galerina* cf. *embolus* (GG 346/81). 23. Spores in hymenium; arrow shows suprahilar applanation. 24. Spore deposited on pileipellis.

species, which is not seen in the Svalbard material. Neither was any apical discontinuity in the spores reported by Singer (1961) or Pegler & Young (1972) in European material. *Galerina pumila* rather frequently has been reported from North Atlantic and Arctic areas, but is widespread in boreal and temperate regions as well. Reid (1979) has this species (with 9 collections) as the most frequently collected *Galerina* of the British Expedition to the Woodfjord area in 1976. But, as he indicates, the material possibly also includes *G. pseudomniophila* (= *G. pumila* var. *subalpina*). *Galerina pumila* also is known from the Subantarctic Botanical Region (Pegler et al. 1980).

Material examined: Spitsbergen: Isfjord district: N slope of Blomsterdalen near Longyearbyen, 31 July 1983, DL (LY).

12. *Galerina* cf. *embolus* (Fr.) Orton, Trans. Brit. Myc. Soc. 43: 176. 1960. Fig. 13 and Pl. VIII, figs. 23-24

Description: *Pileus* 0.6-1.1 cm, convex, becoming depressed at centre, smooth, faintly translucently striate, and dark brown (as half dry), drying to yellow brown. *Lamellae* adnate with tooth, horizontal, distant, brownish yellow with white fibrillose edges. *Stipe* 2-2.5 x 0.15-0.2 cm, yellowish at apex, brown below, apex white pruinose, minutely white silky fibrillose. *Flesh* olive brown, taste not distinctive.

Pileipellis unstratified, of radially repent, medium - to rather short-celled, inflated hyphae, up to 15 µm wide and heavily tawny brown incrusted; pileocystidia absent. *Hymenophoral trama* of filiform, medium- to long-celled, pale yellow brown to hyaline hyphae. *Basidia* 31-39 x 10-11 µm, 4-spored. *Spores* (20/1): 12-*12.7*-14 x 6.8-*7.7*-8.5 µm, amygdaliform with a suprahilar applanation or slight depression, smooth to faintly rugulose, with a tendency to perisporial loosening, plage and pore absent, dark yellow brown to rusty brown, with large internal oil drop. *Cheilocystidia* numerous, long protruding, 30-95 x 3.6-12 x 3.5-4.8 x 4.5-8.4 µm, narrowly ventricose-cylindric or subcapitate-lanceolate at apex; pleurocystidia absent. *Stipitepellis* of filiform, medium- to long-celled hyphae, evenly yellow brown or somewhat asperulate; caulocystidia in clusters at apex. *Clamps* present.

Material examined: Spitsbergen: Hotellneset near Longyearbyen, in *Polytrichum*, 29 July 1981, M. Lange & G. Gulden 346/81 (O).

Remarks: The SEM micrographs (Pl. VIII, figs. 23-24) show a very distinct apical rugulosity of the spores and this was seen in practically all spores under the scanning microscope. Judging from the micrographs there also might be a very slight ornamentation over the rest of the spore surface and a somewhat rough plage area.

Features of spores and cystidia in this collection are very similar to those found in the collection referred to *G. pumila*. However, the spores are slightly larger and slightly deeper colored. Also, there seems to be a stronger tendency of the perisporium to loosen in *G. embolus*. Macroscopically the material of *G. embolus* differs by a darker pileus color and more distant and horizontal lamellae. *Galerina embolus* as described by Orton (1960) differs somewhat from the present material, especially the spores are said to be smaller, (8-)9-11(-12) x 4.5-6(-6.5) µm. Also Pegler & Young (1972) and Bon (1977) report somewhat smaller spores.

Galerina embolus is a poorly known species. In 1872 it was recorded by Karsten from Bjornoya, the southernmost island of Svalbard. Recent finds are mainly from sand dune areas of western Europe (Bon 1977).

RESULTS AND DISCUSSION

Ecology

All the *Galerina* species recognized on Svalbard occur on or among mosses and liverworts. *Galerina pseudomycenopsis* is, however, also often found in grassland. Ohenoja (1971: 139) reports that *G. pseudomycenopsis* and "one or more *Galerina* species" in a wet, mossy plant association evidently lived on moss and killed them, and pose the question as to whether or not it could be some sort of a parasite on the moss. I have no observation of this kind, but the biology of these species needs closer examination. The important groups of lignicolous *Galerina* species, litter decomposers, such as *G. cedretorum* (Maire) Singer, and sphagnicolous forms, are practically missing on Svalbard, probably due to the scarcity of suitable substrates. A single specimen referred to *G. calyptrata* was found attached to a *Sphagnum* plant. Petersen (1977) also demonstrated a small representation of litter decomposers among the macromycetes at Godhavn in Greenland and considered this a general feature of the macromycete flora of the Arctic.

Dobbs (1942) observed that the larger fungi of Spitsbergen were "restricted to wet ground with dense vegetation, either moss-bogs or pool- or stream-marginal communities", and he believed that the general low amount of humus in the soils on Spitsbergen was the important factor. Ohenoja (1971) confirmed that the number of fruitbodies in such communities is rather high, but that the number of species is relatively low. The three very common *Galerina* species on Svalbard, *G. pseudomycenopsis*, *G. arctica* and *G. clavata*, belong to these wet communities that are often referred to as "high Arctic wet tundra". The cold climate with slow plant production and the effect of the permafrost on the water flux tend to prevent true bog and acid mire formation. More eutrophic mires with peat formation occur in some rich sites where plant production is high (Låg 1980). Of the many *Galerina* species normally found in mires and on peat, only the above mentioned species, and *G. stagnina* have been found on Svalbard.

Galerina pseudocerina seems to be a typical species of the moss heaths with *Salix* and *Dryas*, which Ohenoja found to be very rich both in agaric species and in individuals of species. The seven species of subgenus *Galerina* all are present in small numbers in the material, but seem to prefer more acid, mineral soil habitats.

The most surprising difference when compared to the Norwegian and Central European alpine *Galerina* flora, is the absence of section *Galerina* with species such as *G. vittaeformis* and *G. atkinsoniana* on Svalbard. The entire section is represented in the material by one single, tiny specimen, which by virtue of its fully cystidious stipe belongs to stirps *Vittaeformis*. However, it deviates from most species of the stirps by having almost smooth spores. Closer identification of this material was given up. Species of section *Galerina* are abundant up to mid-alpine levels in South Norway. Climate can be the limiting factor. Generally, it is assumed that the climatic conditions in the high Arctic correspond to those of the high alpine belt further south. However, only few *Galerina* species have up to now been found in the high alpine belt in Norway; i.e., *G. pseudomycenopsis*, *G. clavata* and *G. pumila*. Petersen (1977) reports several finds of *G. vittaeformis* from heath and fens with scrubs of *Betula nana* and *Salix glauca* at Godhavn, which lies close to where the high Arctic zone begins in Greenland (Lamoure et al. 1982). Ohenoja reports that pH of soils generally ranges from weakly acid to neutral on Svalbard. In my experience, *G. atkinsoniana* mainly occurs in the more oligotrophic vegetation types, but *Galerina vittaeformis* also is common in more eutrophic types (Gulden 1980).

All material herein considered was collected between 30 June and 26 August, during eight different years. Since the agaric season is very short on Svalbard, and the climatic conditions vary considerably from year to year, it is difficult to establish any seasonal aspects or preferred seasons for the different species. From my own collecting, however, it seems that an early aspect with *G. pseudomycenopsis*, *G. arctica*, *G. clavata* and *G. pseudocerina* can be recognized. Petersen reports an aestival aspect at Godhavn (69°14'N) that is restricted to wet, mossy sites, and characterized by *Galerina* spp., *Mitrula gracilis* and *Omphalina striatula*.

Distribution

The climatically most favorable parts of Svalbard are found in the middle part of Spitsbergen, around Van Mijenfjorden, Isfjorden and Kongsfjorden. South, east and north of this area the climate is considerably harsher (Rönning 1979). The bulk of the material has been collected in this central part of Spitsbergen. Considerable collecting has also taken place further north around Raudfjorden and Woodfjorden, and some few collections were made at the climatically even less favorable Kong Karls Land. All species were present in the middle part of

Spitsbergen, mainly around Longyearbyen, where also most collecting has taken place. Only the most common species on Svalbard; viz., *G. pseudomycenopsis*, *G. arctica*, *G. clavata* and *G. pseudocerina*, were collected further to the east at the inner parts of Isfjorden; i.e., Sassenfjorden, Tempelfjorden, Billefjorden, and Dicksonfjorden. These four common species and also *G. hypnorum* and *G. mniophila* also were collected further north around Raudfjorden and Woodfjorden. The four common ones (except for *G. pseudocerina*) and *G. mniophila* also were found at Kong Karls Land.

No one *Galerina* species seems to have an exclusively Arctic distribution. But some species occur more abundantly in Arctic areas than in alpine. Nezdoyminogo (1982), who studied the genus *Galerina* in polar deserts and in Arctic tundra areas of Siberia and the Siberian Islands, found six species: *G. pseudomycenopsis* (29), *G. pseudocerina* (23), *G. arctica* (10), *G. clavata* (5), and *G. pumila* (inclusive of var. *subalpina* = *G. pseudomniophila*) (3). Numbers indicate the number of collections. The first three also are among the most common on Svalbard, while they apparently are rare in alpine habitats further south. Together with *G. hypnorum* and *G. mniophila*, the six species recorded by Nezdoyminogo are also the species that occur in the most Arctic districts investigated on Svalbard. *Galerina pseudomycenopsis*, which extends down to 400 m on the Faeroes, and *G. antheliae*, which up to now was known only from the alpine belt of Norway, are the only species which can be classified as truly Arctic-alpine, since the others are also known from sites below the timberline.

The *Galerina* species found on Svalbard can be grouped in the following way:

I. Species with ecological optimum under Arctic-alpine conditions: *G. pseudomycenopsis*, *G. antheliae*, *G. arctica*, *G. pseudocerina*, and *G. pseudomniophila*.
II. Species under optimal conditions in boreal regions, but with outliers in Arctic-alpine regions: *G. stagnina*, *G. calyptrata*, and *G.* cf. *embolus*.
III. Species ± equally distributed from temperate to Arctic-alpine regions: *G. clavata*, *G. pumila*, *G. hypnorum* and *G. mniophila*.

Probably all *Galerina* species occurring on Svalbard have a northern circumpolar or at least amphi-atlantic distribution. Only *G. embolus* and the quite recently described *G. antheliae* and *G. arctica* have not yet been recognized in North America. *H. pseudomycenopsis* also is recorded from the Antarctic Continent. *Galerina hypnorum* and *G. pumila* are recorded from the Subantarctic Islands (Horak 1982).

ACKNOWLEDGEMENTS

My warmest thanks go to Mr. Ola Skifte, Tromsö, for putting at my disposal his extensively annotated collections of *Galerina* species and to Mr. J. Stordal, Gjövik, the Finnish mycologists P. Kallio, Turku, E. Ohenoja, Oulu, H. Heikkilä, Kuopio, and S. Huhtinen, Turku, to Dr. G.A. Laursen, Fairbanks, and Dr. D. Lamoure, Lyon, who also most generously lent me their *Galerina* collections. Dr. F. Roll-Hansen, Ås, is thanked for translation of a Russian text and Mrs. E. Braaten at the Electron Microscopy Laboratory for Bioscience at the University of Oslo for operating the SEM and doing most of the preparation and photographic work. I also wish to thank my husband, Eilif Dahl, and Bodil and Morten

Lange, all with long experience from the Arctic, for excellent company during my collecting on Svalbard. The Sysselmann on Svalbard is thanked for placing the old home and trapping hut of the famous trapper A. Njös at our disposal.

REFERENCES

Aleksandrova, V. D., 1980, "The Arctic and Antarctic: their division into geobotanical areas," Cambridge Univ. Press.

Arnolds, E., 1982, Ecology and coenology of macrofungi in grasslands and moist heathlands in Drenthe, the Netherlands. Part 2. Autecology. Part 3. Taxonomy, *Bibl. Myc.*, 90.

Bas, C., 1960, Notes on Agaricales-II, *Persoonia*, 1: 303-314.

Bon, M., 1977, Macromycetes de la zone Maritime Picarde (4ème supplement), *Doc. Myc.*, 7: 63-80.

Dennis, R. W. G., 1955, The large fungi in the North-West Highlands of Scotland, *Kew Bull.*, 10: 111-126.

Dennis, R. W. G., 1961, Fungi venezuelani: IV, *Kew Bull.*, 15: 67-156.

Dobbs, C. G., 1942, Note on the larger fungi of Spitsbergen, *J. Bot.*, 80: 94-102.

Eurola, S., 1968, Über die Fjeldheidevegetation in den Gebieten von Isfjorden und Hornsund in Westspitzbergen, *Aquilo, Ser. Bot.*, 7: 1-56.

Favre, J., 1955, Les champignons supérieurs de la zone alpine du Parc National suisse, *Rés. Rech. Sci. Parc Nat. Suisse V(N.F.)*, 33.

Fries, E. M., 1828, "Elenchus Fungorum," Greifswald.

Fries, E. M., 1838, "Epicrisis Systematis mycologici," Upsala & Lund.

Gulden, G., 1980, Alpine Galerinas (Basidiomycetes, Agaricales) with special reference to their occurrence in South Norway at Finse on Hardangervidda, *Norw. J. Bot.*, 27: 219-253.

Hagen, A., 1950, Notes on Arctic fungi, *Skrifter Norsk Polarinst.*, 3: 1-25.

Høiland, K., 1984, *Cortinarius* subgenus *Dermocybe*, *Opera Botanica*, 71: 1-112.

Horak, E., 1982, Agaricales in Antarctica and Subantarctica: Distribution, ecology, and taxonomy, *in:* Laursen, G. A., and Ammirati J. F., "Arctic and Alpine mycology," pp. 82-118.

Huijsman, H. S. C., 1955, Observations on agarics, *Fungus*, 25: 18-43.

Karsten, P. A., 1872, Fungi in insulis Spetsbergen et Beeren Eiland collecti, *Öfvers. Kongl. Vet.-Akad., Förhandl*, 1872 (2): 91-108.

Kobayasi, Y., Tubaki, K., and Soneda, M., 1968, Enumeration of the higher fungi, moulds and yeasts of Spitsbergen, *Bull. Nat. Sci. Mus.*, 11: 33-76, Pl. 1-4.

Kühner, R., 1935, Le genre *Galera.*, *Encycl. Myc.*, 7: 1-240.

Kühner, R., 1966, *Galerina pseudocerina* Smith et Singer, Espèce des montagnes, nouvelles pour l'Europe, *Schw. Z. Pilzk.*, 44: 92-96.

Kühner, R., 1972a, Agaricales de la zone alpine, Genre *Galerina* Earle, *Bull. Soc. Myc. Fr.*, 88: 41-118.

Kühner, R., 1972b, Agaricales de la zone alpine., Genres *Galera* Earle et *Phaeogalera* gen. nov., *Bull. Soc. Myc. Fr.*, 88: 119-153.

Lamoure, D., Lange, M., and Petersen, P. M., 1982, Agaricales found in the Godhavn area, W Greenland, *Nord. J. Bot.*, 2: 85-90.

Lange, M., 1957, Macromycetes III.-I. Greenland agaricales (pars) macromycetes caeteri. II. Ecological and plant geographical studies, *Medd. Grønl.*, 148,2: 1-125.

Lindblom, A. E., 1841, Fortechning ofver de pa Spetsbergen och Beeren Eiland anmarkta vaxter, *Bot. Not.*, 1839-40: 153-158.

Låg, J., 1980, Special peat formations in Svalbard, *Acta Agric. Scandinavica*, 30: 205-210.

Michelmore, A. P. G., 1934, Botany of the Cambridge expedition to Edge Island, S.E. Spitsbergen, in 1927. Part 1, *Kew Bull. Miscell. Inform. Appendix*, 1934: 30-39.
Møller, F. H., 1945, "Fungi of the Faeroes. Part I. Basidiomycetes," Copenhagen
Nezdoyminogo, E. L., 1982, Fungi of *Galerina* Earle gen. occurring in polar deserts and Arctic tundra of the Soviet Union, *Mik. Fitopat.*, 6: 208-211.
Ohenoja, E., 1971, The larger fungi of Svalbard and their ecology. *Rep. Kevo Subarct. Res. Stat.*, 8: 122-147.
Orton, P. D., 1960, New check list of British agarics and boleti. Part 3. Notes on genera and species in the list, *Trans. Brit. Myc. Soc.*, 43: 159-439.
Pegler, D. N., and Young, T. W. K., 1971, Basidiospore morphology in the Agaricales, *Beih. Nova Hedw.*, 35: 1-210, Pl. 1-53.
Pegler, D. N., and Young, T. W. K., 1972, Basidiospore form in the British species of *Galerina* and *Kuehneromyces*, *Kew Bull.*, 27: 483-500.
Pegler, D. N., Spooner, B.M., and Lewis Smith, R. I., 1980, Higher fungi of Antarctica, the Subantarctic zone and Falkland Islands, *Kew Bull.*, 5: 499-562.
Petersen, P. M., 1977, Investigations on the ecology and phenology of the macromycetes in the Arctic, *Medd. Gronl.*, 199,5.
Pilát, A., 1948, Velenovskyi species novae basidiomycetum, *Opera Bot. Cech.*, 6.
Pilát, A., and Nannfeldt, J. A., 1954, "Notulae ad cognitionem Hymenomycetum Lapponiae Tornensis (Sueciae)," Friesia, 5: 6-38.
Reid, D. A., 1979, Some fungi from Spitsbergen, *Rep. Kevo Subarctic Res. Stat.*, 15: 41-47.
Reid, D. A., 1984, A revision of the British species of *Naucoria* sensu lato, *Trans. Brit. mycol. Soc.*, 82: 191-237.
Rönning, O. I., 1979, "Svalbards flora. Norsk Polarinstitutt, polarhandbok" nr. 1 (2 rev. ed.).
Rostrup, E., 1906, Fungi collected by H.G. Simmons on the 2nd Norwegian Polar expedition, 1898-1902, *Rep. sec. Norw. Exp. "Fram"*, 1898-1902. No. 9.
Singer, R., 1938, De nonnullis Basidiomycetibus, *Bot. mat. Otd. spor. rast. Bot. inst. im. V. L. Komarova AN SSSR*, 10-12: 4-18.
Singer, R., 1954, VI. Fungi, *in:* Polunin, N., "The cryptogamic flora of the Arctic," *Bot. Rev.*, 20: 451-462.
Singer, R., 1961, Type studies in Basidiomycetes. X, *Persoonia*, 2: 1-62.
Singer, R., 1974, Notes on *Galerina*, *Bull. Soc. Linn. Lyon, numéro spécial*, 43: 389-405.
Skifte, O., 1979, Storsopp på Svalbard, *Ottar*, 110-112: 29-39.
Skirgiello, A., 1961, De quelques champignons supérieurs récoltes par M. Kuc au Spitsbergen 1958, *Bull. Res. Counc. Israel, sect. D: Bot.*, 10: 287-293.
Skirgiello, A., 1968, Higher fungi collected in 1958 at Hornsund, Vestspitsbergen, *Polish Spitsbergen Exped.*, 1957-1960: 113-116.
Smith, A. H., and Singer, R., 1957, The genus *Galerina*: An outline of its classification, *Sydowia*, 11: 446-453.
Smith, A. H., and Singer, R., 1964, "A monograph on the genus *Galerina* Earle.," New York & London.
Urbonas, V., and Matelis, A., 1983, Material of flora and ecology of the Agaricales of the Lithuanian SSR, *Liet. TSR Moksly Akad. darbai. C ser.*, 3: 9-24.

ASTROSPORINA IN THE ALPINE ZONE OF THE SWISS NATIONAL PARK (SNP) AND ADJACENT REGIONS

E. Horak

Herbarium, Geobotanical Institut
ETHZ, CH-8092 Zürich, Switzerland

Key words: Agaricales (Basidiomycetes), *Astrosporina*, *Inocybe*, taxonomy, ecology, arcto-alpine distribution, ectomycorrhiza, *Salix* spp., *Dryas octopetala*

ABSTRACT

Favre (1955) reported 10 species (three varieties and one form) of *Astrosporina* (= *Inocybe* p.p.) from alpine habitats in the SNP and vicinity. Subsequently this publication became of paramount importance for the identification of arcto-alpine inocyboid agarics having brown angular-nodulose spores. The critical revision of Favre's authentic collections (CHUR, now in GC) revealed however, that several taxa were misidentified by Favre or the outlined taxonomic concepts do not correspond to modern interpretation. Based upon Favre's material and additional personal topotypical specimens several taxonomical changes and corrections are proposed and discussed. At present, the following 14 species of *Astrosporina* are recorded from the alpine belt of the SNP where they are associated with dwarf willows (*Salix* spp.) and/or *Dryas octopetala* L. (* = new record for SNP):

* *A. alpigenes* Hk. sp. n.
* *A. asterospora* (Quél.) Rea (= *I. napipes* Lge. ss. Favre, misident.)
* *A. aurea* (Huijsman) Hk.
A. casimiri (Vel.) Hk.
A. concinnula (Favre) Hk.
A. egenula (Favre) Hk.
A. giacomi (Favre) Hk.
A. humilis Favre & Hk. sp. n.
A. lanuginella Schroet. ap. Cohn (= *I. decipientoides* (Peck) ss. Favre)
A. mundula Favre & Hk., sp. n. (= *I. decipiens* var. *mundula* Favre)
A. oreina (Favre) Hk.
A. praetervisa (Quél.) ss. Favre
A. pseudohiulca (Kühn. & Bours.) ss. Favre
A. taxocystis Favre & Hk., sp. n. (= *I. decipientoides* var. *taxocystis* Favre)

Table 1: Enumeration and ecological data to *Astrosporina*-species reported from arcto-alpine habitats in the Alps, N- and NE-Europe and circumpolar Subarctica

	European Alps					N-/NE-Europe circumpolar Subarctica							Arctoalpine Habitats Ecology																References
													Salix									Tundra				Scrub			
	Switzerland SNP	Austria	Germany	France	Bulgaria	Scotland	Faeroes	Norway	Iceland	Greenland	Svalbard	Alaska	spp.	arctophila	glauca	herbacea	retusa	reticulata	serpyllifolia	Dryas octopetala aggr.	Cassiope sp.	moss	lichen	sand	swamp	Rhododendron	Juniperus	Betula	
acuta								X					X														X		Gulden & Lange 1971
								X																					Gulden 1975
asterospora									X																		X		Christiansen 1941
										X																			Lange 1957
								X																			X		Gulden & Lange 1971
								X																					Gulden 1975
boltonii					X											X													Kreisel 1959
(cf. *rickenii*)								X																					Gulden & Lange 1971
								X																					Gulden 1975
		X																									X		Moser 1982
borealis										X												X							Lange 1957
												X																	Laursen & Chmielewski 1982
brevispora f.								X																					Gulden 1975
calospora										X													X						Lange 1957
								X														X							Gulden & Lange 1971
casimiri	X																								X				Favre 1955
concinnula	X															X		X											Favre 1955
decipiens	X																		X	X									Favre 1955
										X						X													Lamoure et al. 1982
									X																				Watling 1983

v. *megacystis*	X	.	.	.	.	.	.	.	.	.	.	.	.	.	.	.	.	.	.	X	.	.	.	.	.	.	.	.	Favre 1955
v. *mundulua*	X	.	.	.	.	.	.	.	.	.	.	.	.	.	.	.	X	.	.	X	.	.	.	.	.	.	.	.	Favre 1955
	.	.	.	.	.	.	.	X	.	.	.	.	.	.	.	.	.	.	.	.	.	.	.	.	.	.	.	.	Gulden & Lange 1971
	.	.	.	.	.	.	.	.	.	X	.	.	.	.	X	X	.	.	.	.	.	X	.	.	.	.	.	.	Petersen 1977
	.	.	.	X	.	.	.	.	.	.	.	.	.	.	X	.	.	.	.	.	.	.	.	.	.	.	.	.	Eynard 1977
var. ?	.	.	.	.	.	.	.	.	X	.	.	.	.	.	.	.	.	.	.	.	.	.	.	.	.	.	.	.	Christiansen 1941
decipientoides	X	.	.	.	.	.	.	.	.	.	.	.	.	.	.	.	.	.	.	X	.	.	.	.	.	.	.	.	Favre 1955
	.	.	.	.	.	.	.	.	.	.	.	X	.	.	.	.	.	.	.	.	.	X	X	.	.	.	.	.	Miller et al. 1973
	.	.	.	.	.	.	.	.	.	.	.	X	.	.	.	.	.	.	.	.	.	.	.	.	.	.	.	.	Laursen & Chmielewski 1982
v. *taxocystis*	X	.	.	.	.	.	.	.	.	.	.	.	.	.	X	.	X	.	.	X	.	.	.	.	.	.	.	.	Favre 1955
	.	.	.	.	.	.	.	.	.	X	.	.	.	.	.	.	.	.	.	.	.	X	X	.	.	.	.	.	Kobayasi et al. 1971
	.	.	.	.	.	.	.	.	.	X	.	.	.	X	X	.	.	.	.	.	.	.	.	.	.	.	.	.	Lamoure et al. 1982
egenula	X	.	.	.	.	.	.	.	.	.	.	.	.	.	X	.	.	.	.	X	.	.	.	.	.	.	.	.	Favre 1955
giacomi	X	.	.	.	.	.	.	.	.	.	.	.	.	.	X	.	.	.	.	.	.	.	.	.	.	.	.	.	Favre 1955
	.	.	.	.	X	.	.	.	.	.	.	.	.	.	X	.	.	.	.	.	.	.	.	.	.	.	.	.	Kreisel 1959
	.	.	.	.	.	.	.	.	.	X	.	.	.	.	X	.	.	.	.	.	.	.	.	.	.	.	.	.	Lamoure et al. 1982
grammata (f.)	.	.	.	.	.	.	.	X	.	.	.	.	.	.	.	.	.	.	.	.	.	.	X	.	.	.	.	.	Gulden & Lange 1971
lanuginella	.	.	.	.	.	.	.	.	.	X	.	.	.	.	X	.	.	.	.	.	.	.	.	.	.	.	.	.	Lange 1957
	.	.	.	.	.	.	.	X	.	.	.	.	.	.	.	.	.	.	.	.	.	.	.	.	X	.	X	.	Gulden & Lange 1971
	.	.	.	.	.	.	.	X	.	.	.	.	.	.	.	.	.	.	.	.	.	.	.	.	.	.	.	.	Gulden 1975
f.aff.?	.	.	.	.	.	.	.	.	.	X	.	.	.	.	.	.	.	.	.	.	.	.	X	.	.	.	.	.	Kobayasi et al. 1971
lanuginosa	.	.	.	.	.	.	.	X	.	.	.	.	.	.	.	.	.	.	.	.	.	.	.	.	.	.	.	.	Gulden & Lange 1971
	.	.	.	.	.	.	.	.	.	X	.	.	.	.	.	.	.	.	.	.	.	.	.	X	.	.	.	.	Kobayasi et al. 1971
napipes	X	.	.	.	.	.	.	.	.	.	.	.	.	.	X	.	.	.	.	.	.	.	.	.	.	.	.	.	Favre 1955
oblectabilis	.	.	.	.	.	.	.	X	.	.	.	.	.	.	.	.	.	.	.	.	.	.	X	.	X	.	.	.	Gulden & Lange 1971

Table 1 continued

	European Alps					N-/NE-Europe circumpolar Subarctica							Arctoalpine Habitats Ecology: Salix									Tundra				Scrub			References
	Switzerland SNP	Austria	Germany	France	Bulgaria	Scotland	Faeroes	Norway	Iceland	Greenland	Svalbard	Alaska	spp.	arctophila	glauca	herbacea	retusa	reticulata	serpyllifolia	Dryas octopetala aggr.	Cassiope sp.	moss	lichen	sand	swamp	Rhododendron	Juniperus	Betula	
oreina	X																	X		X									Favre 1955
			X																								X		Bresinsky et al. 1982
								X																					Gulden 1975
praetervisa	X															X	X	X											Favre 1955
										X												X							Lange 1957
								X														X	X		X				Gulden & Lange 1971
				X												X													Eynard 1977
						X										X													Watling 1981
f. *rufofusca*	X															X													Favre 1955
											X																		Ohenoja 1971
								X																					Gulden 1975
pseudohiulca	X																			X									Favre 1955
								X														X	X		X				Gulden & Lange 1971
relicina				X																									Heim 1928
rennyi										X										X							X		Lange 1957
								X																					Lange & Skifte 1967
								X														X	X		X				Gulden & Lange 1971
scabella										X										X									Rostrup 1894
f. *fulvella*				X																									Heim 1931
spp.		X																											Friedrich 1942
umboninota										X				X															Lamoure et al. 1982
umbrina							X															X							Moller 1945
trivialis cf.								X																					Gulden 1975

Illustrations and a key to the reported species of *Astrosporina* are presented. The results of the present revision emphasize not only the difficult delimitation and recognition of the arcto-alpine species of *Astrosporina*, but also the fact that the majority of their records (cf. list) require re-examination to assess their identity.

INTRODUCTION

Within the Agaricales, the taxonomy of *Astrosporina* (= species of *Inocybe* with angular-nodulose spores) is regarded unanimously as difficult. For arcto-alpine species in particular the problem of recognizing and identifying taxa is further impeded by several facts. Due to harsh climatic conditions (irradiation, drought, rain, frost, snow, etc.) most macroscopical characters of the rather small carpophores do not develop well or are distorted. However, these data (veil, color, odor) are absolutely necessary for the correct identification of the material. As a rule, dried specimens without comprehensive notes on the characteristics of fresh carpophores can not be safely identified relying only on microscopical features.

The evaluation of the literature on 34 arcto-alpine records of *Astrosporina* (Tab. 1) unfortunately demonstrate that most lack the prerequisite macroscopical information. Thus, numerous taxa cited in the relevant publications appear doubtful and need revision to confirm their identity.

The alpine species of *Astrosporina* occurring in the SNP and vicinity were collected and described by Favre (1955) with matchless scrutiny. No doubt the data submitted represent the major source of information for all papers published over the last 30 years on arcto-alpine species of *Astrosporina*. After carrying out field work for several years in Favre's "hunting grounds" I greatly felt the absence of a key to track down species necessarily gathered and taken into consideration by Favre. My effort to gain better knowledge of alpine *Astrosporina* brought me into contact with mostly unpublished original notes, drawings and paintings formerly kept in the National Park Museum Chur (CHUR) but now lodged in the Conservatoire Botanique, Geneva (GC). In the following chapters these unearthed data are worked into a compound key-description, followed by critical remarks to the 14 species now recognized in the alpine zone of the SNP.

It is safe to assume that all arcto-alpine species of *Astrosporina* enter (facultative) ectomycorrhiza and therefore play an important role in the ecology of the associated pioneer plants (*Salix* spp., *Dryas*, *Cassiope*?, *Juniperus* ?; cf. tab. 1). Unfortunately none of the mentioned species of *Astrosporina* have been successfully grown in pure culture and therefore their ecological relationships in the symbiosis host plant-fungus is deducted only from sociological observations on the collecting site. The ecological data (SNP) presented below clearly indicate, however, that the host range, in strict relationship to the pH of soil/substrate, is narrow and exclusive in the majority of taxa.

Based upon the present knowledge of 14 species of *Astrosporina*, eight *A. alpigenes* (1), *aurea* (2), *giacomi* (3), *casimiri* (5; in swamps), *taxocystis* (10), *humilis* (11), *praetervisa* (12), and *asterospora* (13), prefer snow beds with acid soil (gneiss, sandstone) supporting *Salix herbacea*; three *A. lanuginella* (4) *concinnula* (7), and *oreina* (8), are strictly connected to *Dryas* and/or calciphilous *Salix* spp. on rather subacid to neutral soil (dolomite, limestone); and for three species *A. egenula* (6), *mundula* (9), and *pseudohiulca* (14), the actual ectomycorrhizal relationships are not yet established. Records of the latter are known both on gneiss and dolomite with their respective vegetation.

If not otherwise stated the magnifications of the figures are: carpophores (nat. size), spores (x 2000), basidia, cheilo-, pleuro- and caulocystidia (x 1000), pileocutis (x 500). Personal collections are kept in ZT (Herbarium, Geobotanical Institute ETHZ, CH-8092 Zürich, Switzerland.

Key to species of *Astrosporina* in the SNP and adjacent regions

1. Spores amygdaliform to reniform, occasionally with rather blunt, indistinct angles or bulges ....cf. *Inocybe piricystis* Favre (Pl. 2,T)
cf. *Inocybe rhacodes* Favre (Pl. 2,S)

1*. Spores distinctly angular or nodulose 2

2. Base of stipe equal-cylindrical (to subclavate) 3

2*. Base of stipe distinctly bulbous to emarginate 8

3. Spores angular, knobs absent (Pl. 1,A), 7-8.5 x 4-5 µm; pileus -10 mm, hemispherical to convex, centre scurfy, fibrillose towards margin; veil pale brown, inconspicuous; lamellae dark chocolate brown or tobacco brown; stipe - 13/-2 mm, concolorous with pileus, fibrillose; odor none; cheilo-, pleuro- and caulocystidia (rare) 40-70 x 10-16 µm, fusoid, metuloid, encrusted with crystals; snow beds with *Salix herbacea-retusa*, on gneiss, 2460 m .... 1. *A. alpigenes*

3*. Spores angular with 1-4 distinct knobs or distinctly nodulose 4

4. Spores angular with 1-4 distinct knobs only (Pl. 1:B,C,D) 5

4*. Spores distinctly nodulose 7

5. Pileus golden yellow or yellow-brown, -35 mm, fibrillose, conical to umbonate, white, persistent veil remnants on margin of pileus and stipe (cortina); stipe -40/-4 mm, white to pale yellow, becoming concolorous with pileus, distinctly fibrillose; odor none; spores 8.5-10.5 x 5-7 µm (Pl. 1,B); cheilo- and pleurocystidia 40-75 x 10-22 µm, fuscid, thin-walled, crystals rare; caulocystidia none; snow beds with *Salix herbacea*, on gneiss, 2350-2420 m .... 2. *A. aurea*

5*. Pileus brown to dark brown, fibrillose, often splitting towards margin; stipe fibrillose (apex only pruinose), concolorous with pileus when young later becoming reddish tinged; caulocystidia conspicuously thin-walled, hyaline, crystals absent; odor spermatic 6

6. Cheilocystidia 45-80 x 10-25 µm, slender fusoid, mostly thin-walled, crystals absent or scattered; pileus -30 mm, hemispherical to umbonate-convex, sometimes black-brown at centre; stipe -35/-5 mm, with distinct cortinoid-fibrillose veil remnants; spores 8-11(-15) x 5-6.5 µm (Pl. 1,C); snow beds with *Salix herbacea*, 2300-2600 m, on gneiss .... 3. *A. giacomi*

6*. Cheilocystidia broadly fusoid to balloon-shaped, thick-walled, strongly encrusted with crystals, often with yellow-brown (KOH) plasmatic pigment; pileus -20 mm, umbonate to campanulate; stipe -30/-3.5 mm, veil remnants fugaceous; spores 9.5-12 x 5-6 µm (Pl. 1,D); among *Dryas* on calcareous soil, 2300-2350 m .... 4. *A. lanuginella*

7. Pileus -25 mm, hemispherical to convex, dark brown, with distinct recurved squamules; lamellae pale brown; stipe -55/-5 mm, concolorous with pileus or reddish brown, minutely squamulose, whitish at base; veil absent; odor none; spores 10-13 x 7-8.5 µm, densely covered with small hemispherical knobs (Pl. 1,E); cheilocystidia 35-60 x 10-14 µm, fusoid, often subcapitate, thin-walled (rarely metuloid), crystals scattered; pleurocystidia rare; on soil in swampy locality, 2200 m, on gneiss .... 5. *A. casimiri*

7*. Pileus -16 mm, brown to tawny, convex to subumbonata-campanulate, centre becoming depressed, smooth, squamulose towards margin; lamellae ochre-brown; stipe -20/-4 mm, pale brown to reddish brown, distinctly pruinose at apex, fibrillose towards base; veil absent; odor none; spores 7-9 x (5.5)5-7 µm, distinctly nodulose (Pl. 1,F); cheilo-, pleuro- and caulocystidia 35-65 x 8-14 µm, hyaline to pale yellow, metuloid, crystals abundant; snow beds with *Salix herbacea*, on gneiss, -2600 m or among *Dryas* and *Salix* spp. on marly schistes, -2450 m (also occurring in subalpine belt of SNP) .... 6. *A. egenula*

8. Spores with blunt angles only (Pl. 1: G,H); membrane conspicuously thick-walled (often with germ-pore); scattered clavate, hyaline, thin-walled cells among metuloid cheilo- and caulocystidia; veil remnants absent; odor none (cp. *A. mundula*) 9

8*. Spores angled-nodulose to distinctly nodulose 10

9. Pileus -13 mm, hemispherical to convex-umbonate, reddish brown with yellow tinge or golden brown, fibrillose; lamellae yellow-brown; stipe -30/-1.5 (-5) mm, pruinose, fibrillose towards base; spores 9-13 x 7-8.5 µm (Pl. 1,G); cheilo-, pleuro- and caulocystidia 45-80 x 10-20 µm, metuloid, encrusted with crystals; among *Dryas* and *Salix retusa-reticulata* on calcareous soil, 2250-2450 m, or with *Salix herbacea* on sandstone, 2650 m .... 7. *A. concinnula*

9*. Pileus -25 mm, convex to campanulate, brown, coarsely fibrillose to subsquamulose; lamellae grey-brown; stipe -30/-5 (-10) mm becoming pale brown with age, pruinose over whole length; spores 10.5-13(-15) x 6.5-8 µm (Pl. 1,H); cheilo- and pleurocystidia 45-85 x 12-26 µm, metuloid, encrusted with crystals; caulocystidia similar, but mostly thin-walled; among *Dryas* and *Salix retusa-reticulata* on calcareous soil, 2130-2500 m .... 8. *A. oreina*

10. Spores angled with one to several distinct knobs (Pl. 1: K,L); odor spermatic 11

10*. Spores distinctly nodulose 12

11. Conspicuous, white, persistent veil remnants covering centre of pileus; pileus -35 mm, convex to umbonate, brown to reddish brown; stipe -35/-5 (-9) mm, white turning pale brown or pale reddish brown with age, pruinose; spores 9-12.5(-14) x 6-8(-9) µm (Pl. 1,K); cheilo- and pleurocystidia fusoid, metuloid, encrusted with crystals; caulocystidia cylindrical, subclavate to slender fusoid, thin-walled, crystals absent; among *Dryas* on calcareous soil (in association with *Salix serpyllifolia*, *S. retusa*, *S. reticulata*, rarely also with *S. herbacea*), 2180-2550 m .... 9. *A. mundula*

11*. Conspicuous, white veil remnants on pileus absent; pileus -30 mm, conico-convex to campanulate, often splitting towards margin, hazel brown or dark brown, rarely with reddish tinge; stipe -40/-5(-7) mm, white to pale brown, marginate bulb often inconspicuous; spores 9-11(-12) x 5-7 µm ((Pl. 1,L); cheilo- and pleurocystidia broadly fusoid, metuloid, encrusted with crystals, caulocystidia oval, clavate or broadly fusoid, metuloid, 20-40 x 8-20 µm; snow beds with *Salix herbacea*, on gneiss, 2250-2650 m .... 10. *A. taxocystis*

12. Pileus hazel brown to dark brown (occasionally with red-brown tinge); stipe pale brown or concolorous with pileus (often with yellow tinge); odor none (or fruity) 13

12*. Pileus brown to dark brown, reddish tints absent; thin-walled cells among cheilo- and caulocystidia absent; odor spermatic 14

13. Lamellae yellowish to ochre; pileus -15 mm, conical to umbonate-convex; stipe -25/-2.5(-5) mm; spores 8-11.5 x 5.5-7.5 µm (Pl. 2,M;) cheilo-, pleuro- and caulocystidia 40-85 x 14-20 um, fusoid, metuloid, encrusted with crystals, hylaine; snow beds with *Salix herbacea*, on gneiss, 2400 m (type described from subalpine locality in SNP) .... 11. *A. humilis*

13*. Lamellae argillaceous-gray; pileus -45 mm, umbonate-convex to campanulate, yellowish to ochre-brown towards margin; veil absent; stipe -40/-5(-10) mm, white turning pale brown; odor none to fruity; spores 10-12 x 7-8.5 µm (Pl. 2,N); cheilo-, pleuro- and caulocystidia 45-105 x 12-25 µm, metuloid, encrusted with crystals, often with yellow-brown (KOH) plasmatic pigment, intermixed with clavate, thin-walled, hyaline cells; snow beds with *Salix herbacea* (rarely also with *S. reticulata*), on gneiss, 2180-2600 m .... 12. *A. praetervisa*

14. Spores 11-13 x 7-9.5 µm (Pl. 2,P), with prominent knobs; pileus -25 mm, conico-convex to campanulate; stipe -45/-3 (-7) mm; odor?; cheil-, pleuro- and caulocystidia often with yellow-brown (KOH) plasmatic pigment; snow beds with *Salix herbacea* and in swamps, on gneiss, 2220-2500 .... 13. *A. asterospora* (?)

14*. Spores 10-13 x 7-8.5 µm (Pl. 2,R), with low hemispherical knobs; pileus -55 m, convex to broadly campanulate, often with pale fibrillose veil remnants at centre; odor spermatic; cheilo-, pleuro-, and caulocystidia hyaline; among *Dryas* on calcareous soil (rarely also with *Salix herbacea* in snow beds on limestone-dolomite), 2400-2500 m .... 14. *A. pseudohiulca*

DESCRIPTIONS OF SPECIES AND CRITICAL NOTES

1. *Astrosporina alpigenes* Horak, sp. n., Pl. 1,A; fig. 1,A-E

Pileus -10 mm, hemisphaerico-convexus, obscure brunneus, subsquamuloso-fibrillosus, velo pallido ad margineum instructus. Lamellae emarginatae, fuscae. Stipes -13/-2 mm, cylindrico-subclavatus, pileo concolor, fibrillis brunneis dense obtectus. Odor saporque nulli. Sporae 7-8.5 x 4-5 µm, angulatae, brunneae. Cheilo- et pleurocystidia 40-70 x 10-16 µm, fusoideae, metuloideae, incrustatae. Ad terram in zona alpina (*Salix herbacea-retusa*, 2460 m), Helvetia. ZT, 2266 (Holotypus).

Pileus -10 mm, hemispherical or convex with a low umbo, subcampanulate, dark brown or tobacco brown, centre minutely scurfy or squamulose, coarsely fibrillose towards margin, dry, not striate, covered with pale subpersistent fibrils of veil. Lamellae 10-15,3(5), emarginate chocolate brown to fuscous, edges broad, concolorous. Stipe -13/-2 mm, cylindrical to subclavate, concolorous with pileus, densely covered with pale brown fibrils of veil, solid, dry, single; cortina none. Odor and taste not distinctive. Context pale brown, dark brown beneath cuticle, white in base of stipe. Spore print brown. Spores 7-8.5 x 4-5 µm, angular (like *Entoloma* spp.), brown. Basidia 30-40 x 5-7 µm, 4-spored. Cheilo- and pleurocystidia 40-70 x 10-16 µm, fusoid, metuloid, encrusted with crystals, hyaline, rarely with pale yellow-brown (KOH) plasmatic pigment. Caulocystidia rare, like cheilocystidia. Pileocutis composed of

Pl. 1 (Spores x 825): A. *Astrosporina alpigenes* (type, 2266: ZT); B. *A. aurea* (2248; ZT); C. *A. giacomi* (lectotype, Favre 124 a; GC); D. *A. lanuginella* (Favre 122 b; GC); E. *A. casimiri* (Favre 115; GC); F. *A. egenula* (lectotype, Favre 123 a; GC); G. *A. concinnula* (lectotype, Favre 117 a; GC); H. *A. oreina* (lectotype, Favre 126 a; GC); K. *A. mundula* (type, Favre 120 a; GC); L. *A. taxocystis* (type, Favre 122 a; GC).

hyaline, cylindrical hyphae, terminal cells often forked, 2-6 µm diam., cells of subcutis short-cylindrical to ovoid, encrusted with brown pigment. Clamp connections numerous.

Material examined: Switzerland, Kt. Graubünden, SNP, Val Zeznina (near Lavin), Macun, 5.IX.1983, Horak 2266 (holotype; ZT).

Ecology: On soil among *Salix herbacea-retusa*, on gneiss, 2460 m.

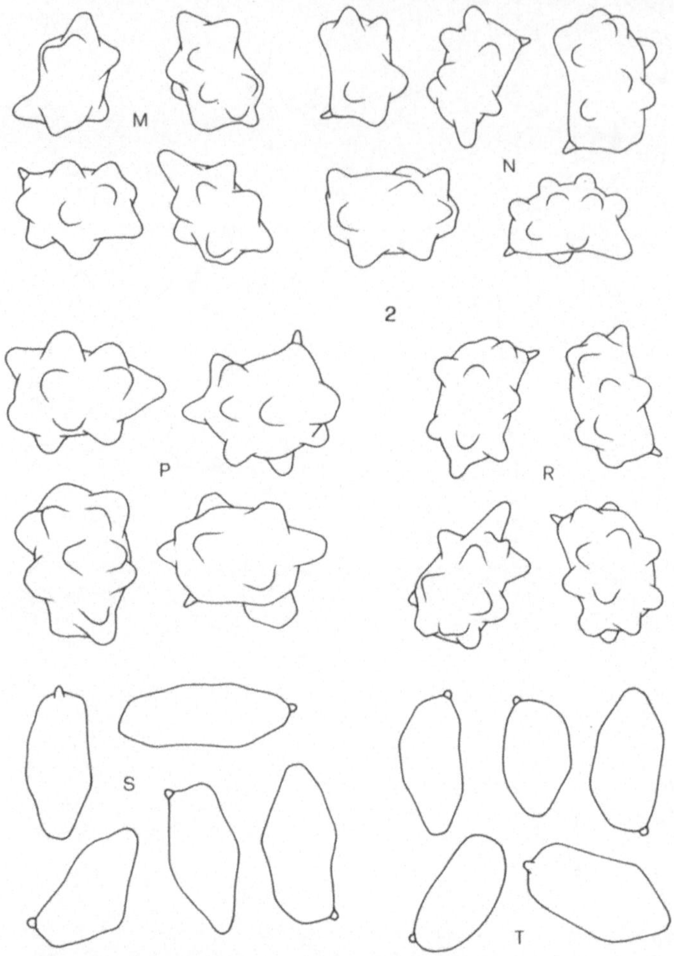

Pl. 2 (Spores x 825): M. *Astrosporina humilis* (lectotype, Favre 583; GC); N. *A. praetervisa* (Favre) 127 a; GC); P. *A. asterospora* (Favre 125 a; GC); R. *A. pseudohiulca* (Favre 131 a; GC); S. *Inocybe rhacodes* (lectotype, Favre 111; GC); T. *I. piricystis* (lectotype, Favre 109; GC).

Remarks: In the field this small, dark brown species can be readily mistaken for *Cortinarius* spp. Taxonomically *A. alpigenes* belongs to the *A. boltonii* - complex (*A. boltonii* (Heim), *A. brevispora* (Huijsman), *A. lanuginella* Schroet. ap. Cohn, *A. lanuginosa* (Bull.: Fr.), *A. proximella* (Karst.) ss. Favre (1948), *A. pseudoumbrina* (Stangl), *A. putilla* (Bres.) or *A. umbrina* (Bres.)) characterized by dark brown colours both of the pileus, stipe and lamellae, cylindrical fibrillose stipe and cortinoid veil remnants at the margin of the pileus (cf. Kühner & Boursier, 1932; Stangl 1975, 1976; Stangl & Enderle, 1983). The present species is well distinguished from the mentioned taxa, however, by the small size of the carpophores, hemispherical pileus, small angular spores (lacking knobs) and the alpine habitat in association with dwarf willows on acid soil (Stangl & Veselsky, 1974).

Comparative material examined: -*Inocybe brevispora* Huijsman: The Netherlands, Gelderland, Arnhem, Vorden, near *Pinus* sp., 24.X.1953,

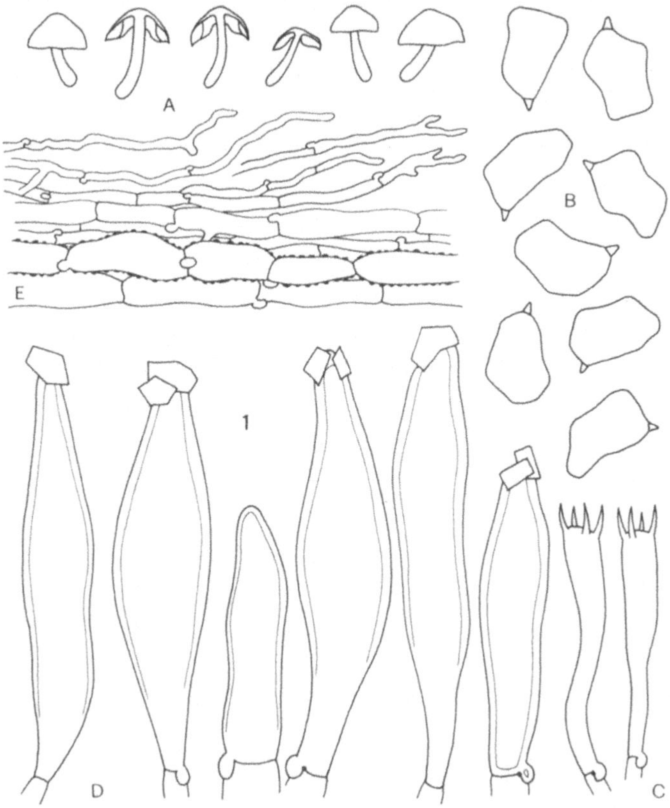

Fig. 1. A-E: *Astrosporina alpigenes* (type, 2266; GC): carpophores. B. spores; C. basidia; D. cheilo- and pleurocystidia; E. pileocutis.

Huijsman (type; L). -*Inocybe umbrina* Bresadola: Italy, Varena, in silvis abiegnis, VII.1914, Bresadola (authentic material; S). -*Inocybe umbrina* Bresadola: Italy, Varena, in silvis abiegnis, VII.1914, Bresadola (authentic material; S). *Inocybe putilla* Bresadola: Italy, sopra Arnago, estate 1883, Inzenga (type; S). -*Inocybe pseudoumbrina* Stangl: Germany, Augsburg-Berghei Grasweg in Fichtenwald, 4.VIII.1970, Stangl 654 (type; M).

2. *Astrosporina aurea* (Huijsman) Horak, comb. nov. Pl. 1,B; fig. 2,F-M
= *Inocybe aurea* Huijsman 1955, Fungus 25: 22 (basionym).

Material examined: SNP: Switzerland, Kt. Graubünden: Dischmatal, NW of Jatzhorn, 25.VIII.1963, Horak 63/105 (ZT); Flüelapass, Radönt (E of Fl. Schwarzhorn), 5.IX.1982, Horak 1718 (ZT); same locality, 3.IX.1983, Horak 2248 (ZT).

Ecology: SNP: In snow beds with *Salix herbacea*, on gneiss, 2350-2420 m.

Remarks: In the SNP the area of distribution of this striking agaric so far is restricted to the alpine zone of its western border. Macroscopically this yellow to yellow-brown *Astrosporina* resembles the

Fig. 2. F-M: *Astrosporina aurea*: F-H (2248;ZT): F. carpophores; G. spores; H. cheilo- and pleurocystidia; K (type, Huijsman 206;L): spores; L-M (63/105;ZT): L. carpophores; M. spores.

heteromorphic *Inocybe fastigiata* (Schaeff.: Fr.), the most common *Inocybe* in the Alps (Favre, 1955), and thus it might be readily overlooked. The three collections examined agree in all essential characters with the type material described by Huijsman (1955) from The Netherlands. Except for the "swollen to bulbous" stipe the Bavarian (W-Germany) record of *A. aurea* (Stangl, 1975) is also in accordance both with the type and the Swiss specimens. To my knowledge no further data are published in the current literature.

Comparative material examined: *Inocybe aurea* Huijsman: The Netherlands: Gelderland, "Groot-Hagen", in dry sandy woods of *Pinus*, 28.IX.1943, Huijsman 206 (type; L).

Fig. 3. N-T: *Astrosporina giacomi* (lectotype, Favre 124;GC): N. carpophores; P. spores; R. cheilocystidia; S. pleurocystidia; T. caulocystidia.

3. *Astrosporina giacomi* (Favre) Horak, comb. nov. - Pl. 1; fig. 3, N-T
= *Inocybe giacomi* Favre 1955, Rés. rech. scient. Parc National suisse, V: 115 (basionym)
= *Inocybe carpta* ss. Bresadola 1930, Icon. Myc. 16: 756.
= *Inocybe borealis* M. Lange 1957, Medd. Grønland 148(2): 19.

Material examined: SNP: Switzerland, Kt. Graubünden: Val S-charl, Munt Plazèr, 2.IX. 1951, Favre 124 a(lectotype;GC); Ofenpass, Buffalor, 13.VIII.1941, Favre 124 b(GC); Dischmatal, Alp Stillberg, 6.VIII.1964, Horak 64/630 (ZT): Flüelapass, Radönt (E of Fl. Schwarzhorn) 5.IX.1982, Horak 1719 (ZT).

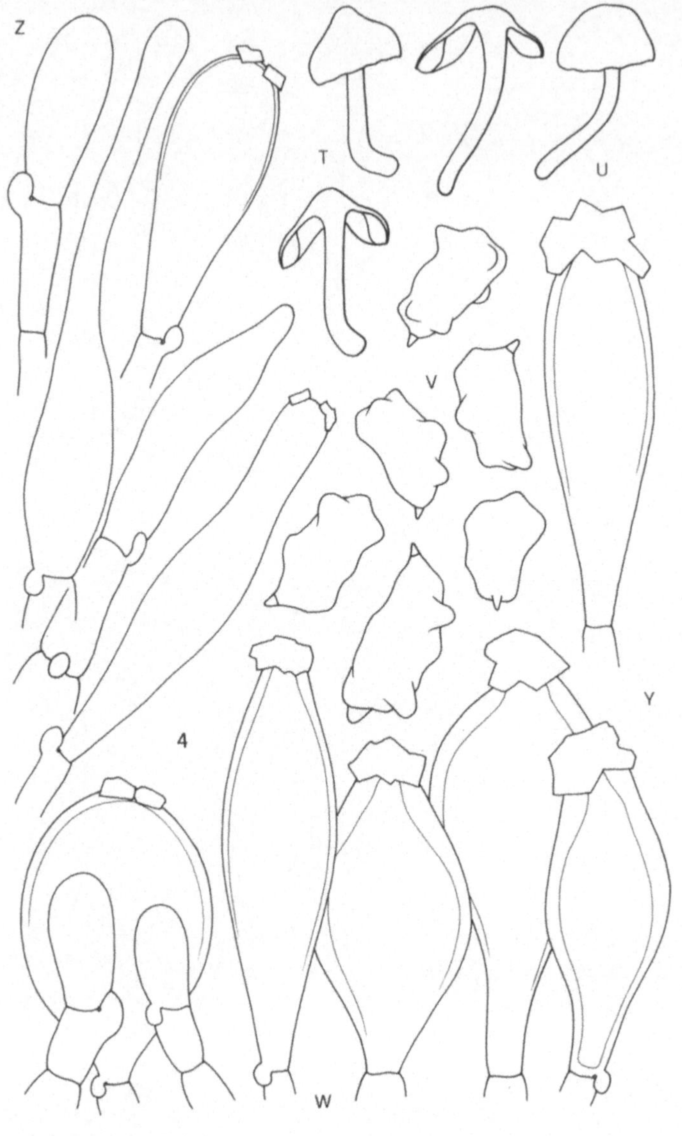

Fig. 4. T-Z: *Astrosporina lanuginella*: T. (Favre 122;GC): carpophores; U-Z (Favre 121;GC): U. carpophores (nat. size); V. spores; W. cheilocystidia; Y. pleurocystidia; Z. caulocystidia.

Ecology: SNP: In snow beds with *Salix herbacea*, on gneiss, 2300-2600 m.

Remarks: In the literature the identity of *Inocybe carpta* is much discussed (Heim, 1931; Kühner & Boursier 1932, "*I. subcarpta*", Huijsman, 1955, Alessio, 1980). To end part of the controversy and speculation Favre (1955) renamed his alpine collections of *I. carpta* ss. Bresadola as "*I. giacomi*". Subsequently Moser (1967) adopted the opinion of Kühner & Romagnesi (1953) and transferred *I. giacomi* as a synonym of *I. boltonii* Heim (1931).

After examining Favre's authentic material I support Stangl's (1976)

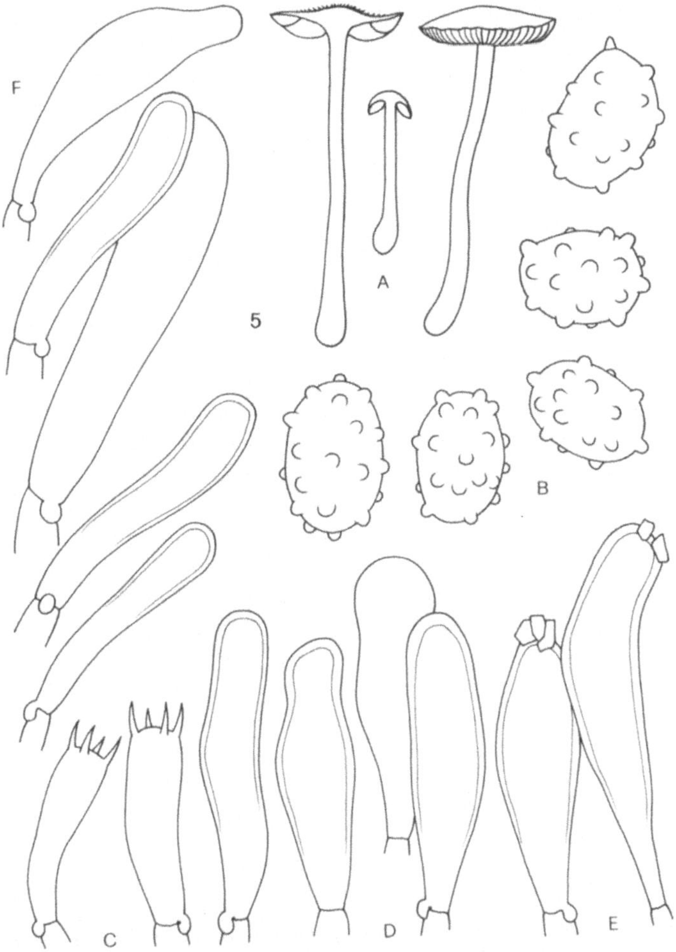

Fig. 5. A-F: *Astrosporina casimiri* (Favre 115;GC): A. carpophores; B. spores; C. basidia; D. cheilocystidia; E. pleurocystidia; F. caulocystidia.

taxonomic concept by separating *I. giacomi* Favre from *I. boltonii*. By proposing the lectotype for *I. giacomi* (Favre 124b) Stangl (l.c.) unfortunately did not take into account two facts: a) Favre's published drawings and paintings refer to 124 a (Munt Plazèr) which is herewith designated as lectotype, b) the exsiccata labelled by Favre as "*I. giacomi*" represent a mixed collection. Favre 124 c is not identical with typical *I. giacomi* but conspecific with Favre 121 a ("*I. decipientoides* var.", locality and date as in 124 c!). The two collections are not well documented in Favre's notes. It seems likely, however, that they represent *I. subcarpta* Kuhner & Boursier (1932) which frequently occurs in the subalpine conifer forests of the SNP (Favre, 1960).

The microscopical data observed on the type material of *Inocybe rennyi* (Berk. & Broome) Saccardo ("*rennei*") demonstrate that both the conspicuously elongate angular-subnodulose spores and the shape and size of the cystidia recall those of *A. giacomi*. Unfortunately the relationships of the two taxa cannot be convincingly established (Stangl,

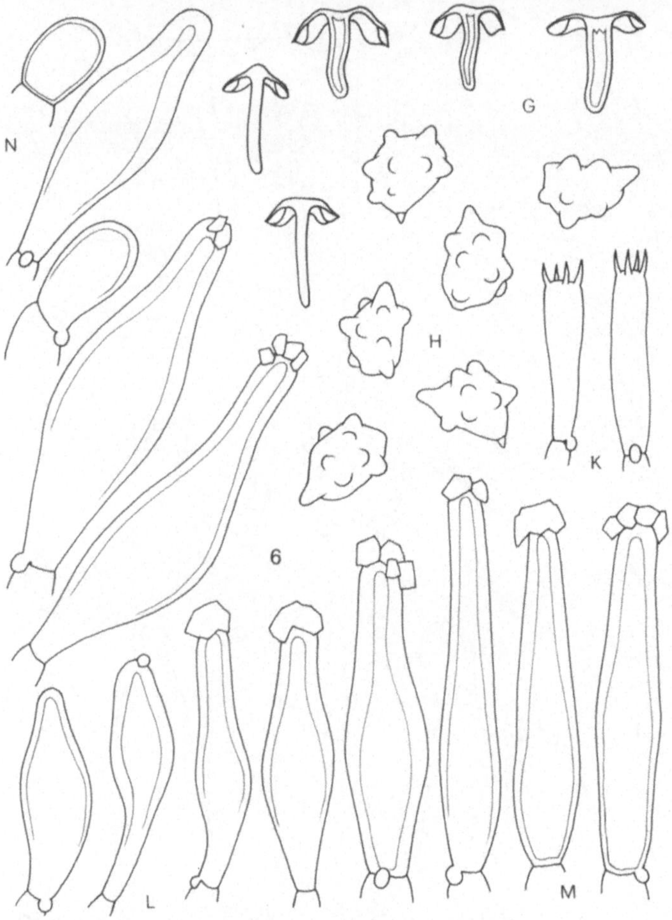

Fig. 6. G-N: *Astrosporina egenula* (lectotype, Favre 123 a;GC): G. carpophores; H. spores; K. basidia; L. cheilocystidia; M. pleurocystidia; N. caulocystidia.

1975; Alessio, 1980) because the original description of *I. rennyi* (identified once by Berkeley & Broome as "*Hebeloma*") is very poor and vague. Under these circumstances the arctoalpine records of *I. rennyi* (cf. tab. 1) appear doubtful and need careful revision.

The original description and the microcharacters found on the type collection of *Inocybe borealis* M. Lange indicate that this Greenland taxon must be considered a synonym of *A. giacomi*.

Comparative material examined: -?*Inocybe subcarpta* K. & B.: SNP, Switzerland, Val Sesvenna: Blaisch del Manaders, 26.VIII.1952, Favre 124c (as "*I. giacomi*"; GC); Blaisch del Manaders, 26.VIII.1952, Favre 121a (as "*I. decipientoides* var"; GC). -*Inocybe rennyi* (B. & Br.) Sacc.: England, J. Renny (type; K). -*Inocybe borealis* M. Lange: Greenland Sandflugtdalen, 26.VIII.1946, M. Lange 387 (type; C).

4. *Astrosporina lanuginella* Schroeter ap. Cohn 1889, Pilze Schl. 1: 577, Pl. 1,D; fig. 4,T-Z.

= *Inocybe decipiens* and *I. decipientoides* ss. Favre (1955) p.p.

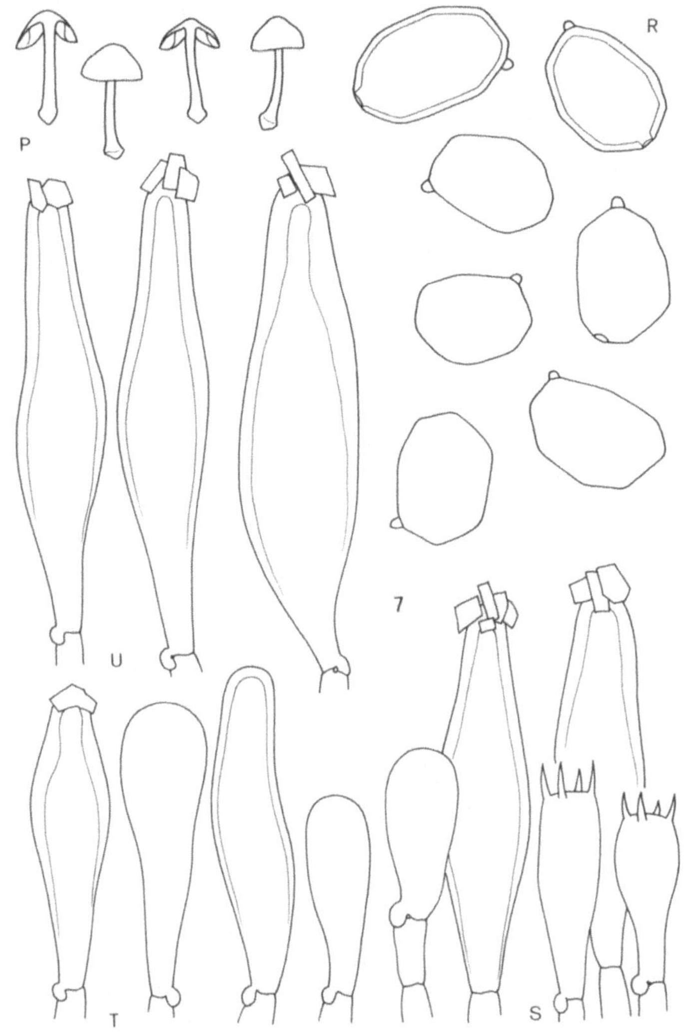

Fig. 7. P-U: *Astrosporina concinnula* (lectotype, Favre 117 a;GC): P. carpophores; R. spores; S. basidia; T. cheilocystidia; U. pleurocystidia.

Material examined: SNP: Switzerland, Kt. Graubünden: Val S-chari, Piz Mezdi, 27.VIII.1951, Favre 121 (GC); Val Sesvenna, Marangun, 29.VIII.1952, Favre 122b (as "*I. decipientoides* var. *taxocystis*"; GC); Ofenpass, W-slope of Piz d'Aint, 5.IX.1953, Favre 118 (as "*I. decipiens*"; GC).

Ecology: SNP: Among *Dryas octopetala* on calcareous soil, 2300-2350 m.

Remarks: *Astrosporina lanuginella* is another controversial species (Kühner & Boursier, 1932; Stangl, 1976; Alessio, 1980), with *Inocybe decipientoides* Peck (1907) and *I. globocystis* Velenovský (1920) as later synonyms.

In the region of the SNP *A. lanuginella* is readily confused with *A. giacomi*, because both taxa share dark brown colours on the

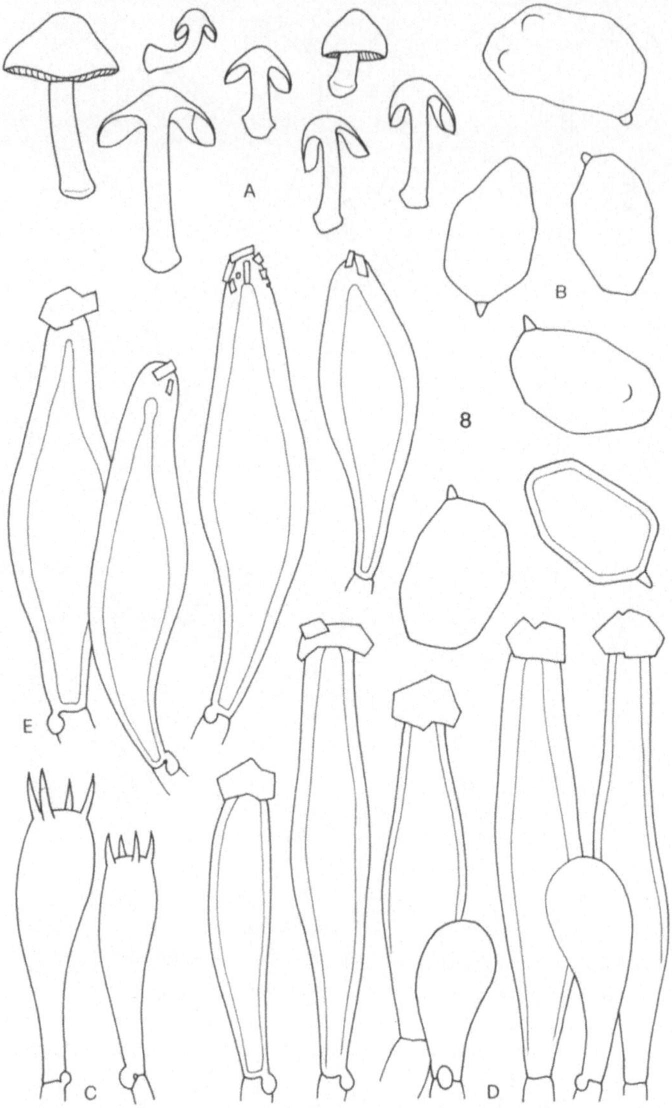

Fig. 8. A-E: *Astrosporina oreina* (lectotype, Favre 126 a;GC): A. carpophores; B. spores; C. basidia; D. cheilocystidia; E. pleurocystidia.

carpophores, subpersistent veil remnants on the margin of the pileus, and spermatic odor of the context. Ecologically, however, the two species can be separated in the field, since *A. lanuginella* is found in association with *Dryas* on calcareous soil, whereas *A. giacomi* is restricted to acid soil supporting *Salix herbacea* in snow beds. Taking these strict ecological demands into consideration the records of *A. lanuginella* from Alaska, Greenland and Norway (cf. Tab. 1) need critical reexamination to confirm their identity.

The microscopical data observed on the type collection of *Inocybe decipientoides* Peck prove that Favre's identification of several SNP collections under this name is wrong. French material, however, published

Fig. 9. F-K: *Astrosporina mundula* (type, Favre 120;GC): F. carpophores; G. spores; H. cheilo- and pleurocystidia; K. caulocystidia.

by Kühner & Boursier (1932) clearly represents the North American taxon which is well characterized by swollen fusoid cheilocystidia and metuloid pleurocystidia tapering into a long slender basal part. Taking this new information into account the proposed synonymy of *I. decipientoides* Peck and *I. lanuginella* Schroeter ap. Cohn needs further evidence.

Comparative material examined: *Inocybe decipientoides* Peck: USA, Boston, 28.VI.1906, Davis (type; NYS).

5. *Astrosporina casimiri* (Velenovský) Horak, comb. nov.

Pl. 1,E; fig. 5,A-F.

= *Inocybe casimiri* Velenovský 1920, České Houby 2: 369 (basionym)

Material examined: SNP: Switzerland, Kt. Graubünden: Val Sesvenna, Marangun, 19.VIII.1947, Favre 115 (GC). Greenland: Søndre Strømfjord, 700 m, 10.VIII.1946, Lange 267 (as "*I. calospora* Quél."; C).

Ecology: SNP: In swamp, on gneiss, 2220 m.

Remarks: Favre's material from the SNP conforms in all details with *A. casimiri* characterized by spores with numerous low hemispherical knobs, ± thin-walled hyaline fusoid cheilocystidia, and rare or absent pleurocystidia (Velenovský, 1920; Heim, 1931). The descriptions and spore-drawings published by Boursier & Kühner (1928), Stangl (1976) and Alessio (1980) evidently indicate that the material examined does not represent *A. casimiri* whose spores closely resemble those of *Inocybe salicis* Kühner (1955). In addition the reexamination of the type material of *I. leptophylla* Atkinson (1918) revealed that this North American taxon is in fact closely related if not conspecific with *I. casimiri* ss. Boursier & Kühner (1928) (cf. Stuntz, 1947). Obviously *A. casimiri* is a rare species, at least in arcto-alpine habitats. To my knowledge it also occurs on the western coast of Greenland from where Lange (1957) reported the species under the name "*Inocybe calospora* Quél.".

Comparative material examined: -*Inocybe salicis* Kühner: France, Lyon, marais des Echets, with *Salix alba*, 6.X.1944, Kühner (type, ex Herb. Kühner); The Netherlands, prov. Zeeland, Oostburg, with *Populus canadensis* in marsh, 30.VI.1980, de Meijer (ZT,683). -*Inocybe leptophylla* Atkinson: USA, N.Y., near Ithaca, 7.VIII.1902, Bradfield (type, CUP 13372, Herb. Atkinson).

6. *Astrosporina egenula* (Favre) Horak, comb. nov. Pl. 1,F; fig. 6,G-N.
= *Inocybe egenula* Favre 1955, Rés. rech. scient. Parc National suisse, V: 114 (basionym).

Material examined: SNP: Switzerland, Kt. Graubünden: God il Fuorn, 1850 m, 10.VIII.1941, Favre 581a (holotype of *I. egenula* Favre, in sched.; GC); Val S-charl, Munt Plazèr, 2.IX.1951, Favre 123a (GC); Samnaun, Alp Trida, 28.VIII.1984, Horak 2527 (ZT); Wallis: St. Nikolaus, oberhalb von Jungenalp, 2340 m, *Salix herbacea*, 25.VII.1960, Horak 2364 (ZT).

Ecology: SNP: Among moss at border of creek, on sandstone, subalpine site (holotype); in alpine zone with *Salix herbacea*, on gneiss, 2600 m, or with *Dryas octopetala* on calcareous schistes, 2420 m.

Remarks: The most distinctive features of *A. egenula* are, rather small gibbous spores, short but slender, metuloid pleurocystidia and thin-walled velar hyphae covering the pileocutis

Based on present knowledge this agaric has a wide ecological range in alpine habitats. It occurs both on dolomite/schistes and gneiss/sandstone, i.e. in association with *Dryas* or *Salix* spp.

The microscopical characters of *A. egenula* very closely resemble those of *I. nematoloma* Josserand (1959, 1974). However, the macrocharacters observed on the carpophores and the ecology separate the two taxa.

Comparative material examined: *Inocybe nematoloma* Josserand: France, Rhône, La Tour de Salvagny, Le Pré-Vieux, 21.IX.1958, Josserand (XXVI/73, holotype, ex Herb. Josserand); same locality, 19.X.1944, Josserand (XXI/75, ex Herb. Josserand).

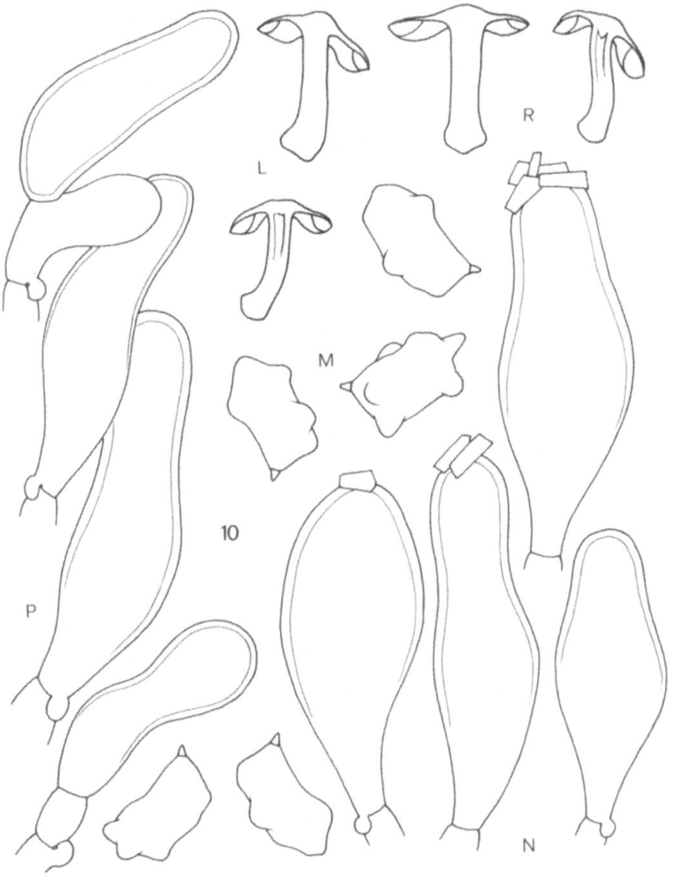

Fig. 10. L-R: *Astrosporina taxocystis*: L-P (type Favre 122 a;GC): L. carpophores; M. spores; N. cheilo- and pleurocystidia; P. caulocystidia; R. (Favre 122 c;GC): carpophores.

7. *Astrosporina concinnula* (Favre) Horak, comb. nov.

Pl. 1,G; fig. 7,P-U.

= *Inocybe concinnula* Favre 1955, Rés. rech. scient. Parc National suisse, V: 108. (basionym)

Material examined: SNP: Switzerland, Kt. Graubünden: Ofenpass, Murtaröl d'Aint, 28.VIII.1953, Favre 117a (lectotypus; GC); Ofenpass, Taunter Pizza, S of Piz d'Aint, 5.IX.1953, Favre 117b (GC) Val Tavrü, Marangun, 29.VIII.1984, Horak 2532 (ZT); Austria, Tirol, Obergurgl, Rotmoos, 2340 m, 3.IX.1965, Horak 65/270 (ZT).

Ecology: SNP: Among *Dryas octopetala* and *Salix retusa-reticulata*, on limestone-dolomite, 2250-2450 m, or with *Salix herbacea* on sandstone, 2650 m.

Remarks: Macroscopically this inconspicuous species is readily recognized by yellow-ochre lamellae and pruinose stipe with a large marginate bulb. The apex of the thick-walled spores occasionally bears a distinct pore (cf. *A. oreina*). Frequently a yellow-brown pigment (KOH)

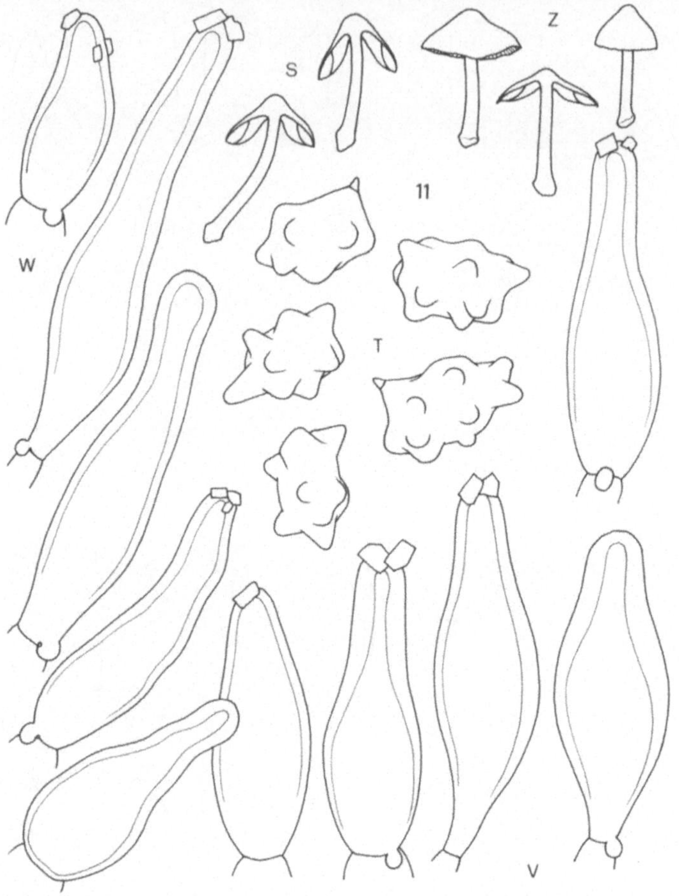

Fig. 11. S-Z: *Astrosporina humilis*: S-W (lectotype, Favre 583;GC): S. carpophores; T. spores; V. cheilo- and pleurocystidia; W. caulocystidia; Z. (64/640;ZT): carpophores.

is observed in the cell sap of the cheilo-, pleuro- and caulocystidia respectively.

The few ecological data available indicate that *A. concinnula* predominantly occurs with *Dryas* and *Salix retusa-reticulata* on calcareous soil. It also is found, however, among *Salix herbacea* on acid soil in snow beds.

8. *Astrosporina oreina* (Favre) Horak, comb. nov. Pl. 1,H; fig. 8,A-E.
= *Inocybe oreina* Favre 1955, Rés, rech. scient. Parc National suisse, V: 116, (basionym).

Material examined: SNP: Switzerland, Kt. Graubünden: Val Minger, Spadla Sura, 14.VIII.1943, Favre 126 a (lectotype; GC); Val S-charl Costainas, 8.IX.1981, Horak 1332 (ZT).

Ecology: SNP: Among *Dryas octopetala* and *Salix retusa-reticulata*, on dolomite, 2130-2500 m.

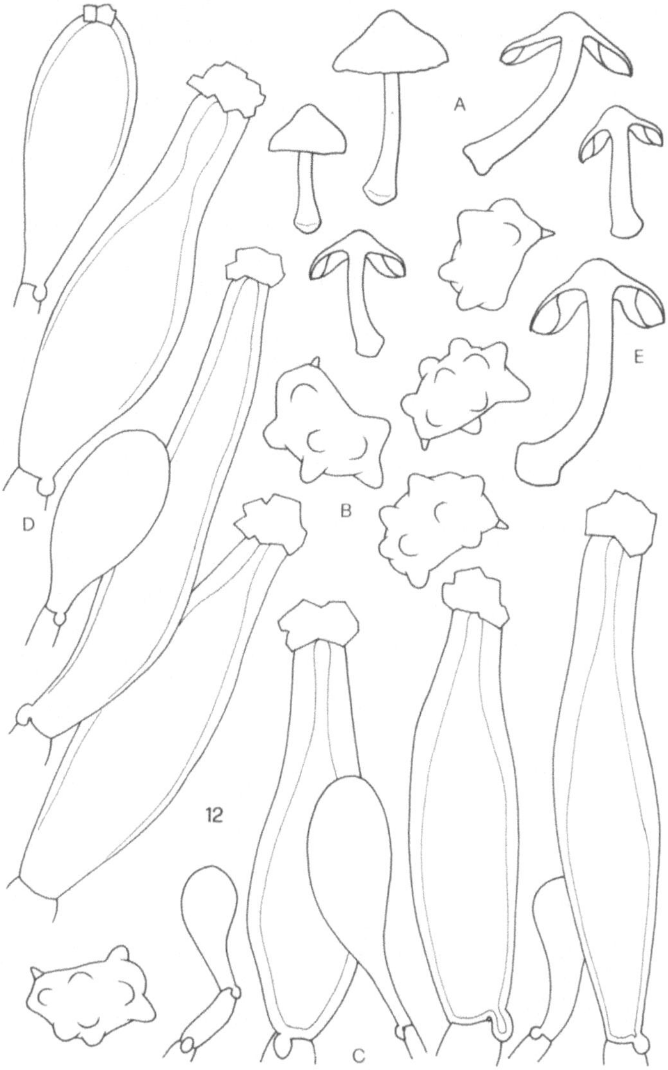

Fig. 12. A-E: *Astrosporina praetervisa*: A-D (Favre 127 a;GC): A. carpophores; B. spores; C. - cheilo- and pleurocystidia; D. caulocystidia; E. (lectotype of *I. praetervisa* f. *rufofusca* Favre 130;GC): E. carpophore.

Remarks: According to Favre (1955) *A. oreina* is closely related to *Inocybe decipiens* Bresadola (but not ss. Favre 1955: 109!, see below). In both species the coarsely fibrillose to scaly pileocutis is *not* covered by conspicuous veil remnants. The spores are angular in profile and lack knobs. The pale fibrillose stipe is pruinose at the apex only and the odor of the context is definitely not spermatic. Despite the obvious similarities, I prefer to consider the two rare and misinterpreted (by Kühner, 1933) taxa as independent species at least until further information can be drawn from fresh collections (cf. discussion concerning *I. dunensis* Orton, 1960).

Bresinsky & Schmid-Heckel (1982) report the occurrence of *A. oreina* from the Bavarian Alps (on dolomite among *Carex firma* and *Rhodothamnus chamaecistus*). The microscopical re-examination of the authentic

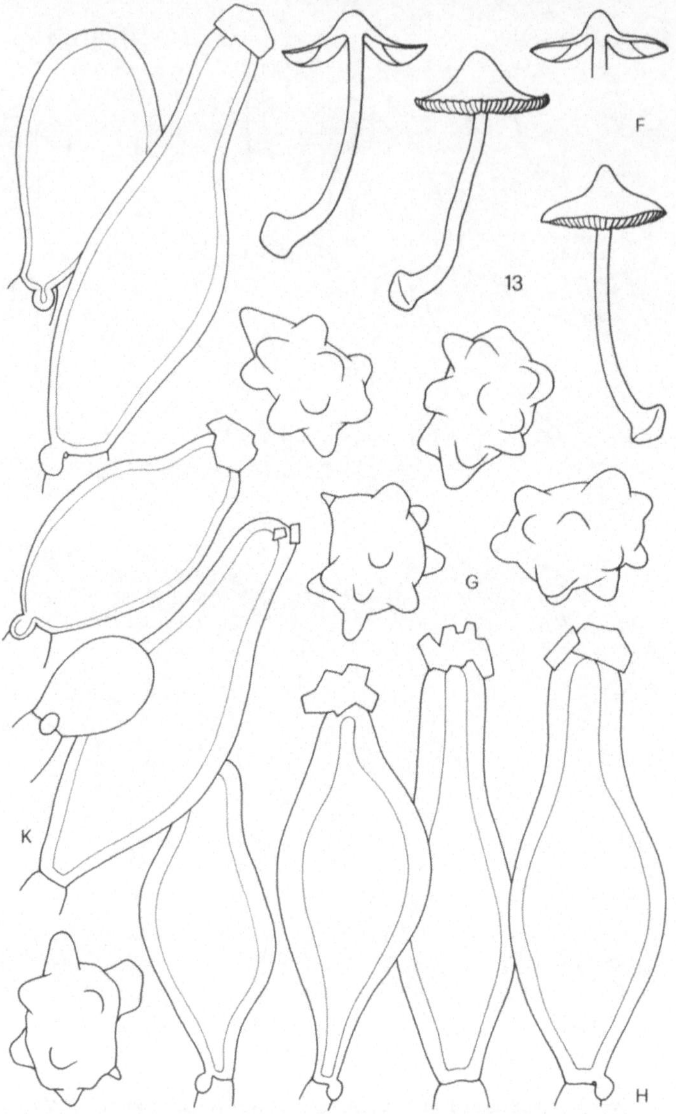

Fig. 13. F-K: *Astrosporina asterospora*: (Favre 125a;GC): F. carpophores; G. spores; H. cheilo- and pleurocystidia; K. caulocystidia.

material (822) definitely revealed, however, that the specimens do not match typical *A. oreina*. Thus the Bavarian record must be taken off the list of distribution (Table 1).

Comparative material examined: -*Inocybe decipiens* Bresadola: Italy, Villaggano, ad marginem camporum, Julio 1888, Bresadola (type; S), cf. Stangl (1979). -"*Inocybe oreina* Favre": Germany, Bavaria, Nat. Park Berchtesgaden, Kahlersberg, 2340 m, 7.VIII.1981, leg. Schmid-Heckel (822; REG).

Fig. 14. L-R: *Astrosporina pseudohiulca*: L-P (Favre 131 a;GC): L. carpophores; M. spores; N. cheilo- and pleurocystidia; P. basidium; R. caulocystidia; S. (ZT,2515): carpophores.

9. *Astrosporina mundula* Favre & Horak, sp. n.   Pl. 1,K; fig. 9, F-K.
= *Inocybe decipiens* var. *mundula* Favre 1955, Rés. rech. scient. Parc National suisse, V: 111.
= *Inocybe decipiens* Bresadola ss. Favre 1955, p.p., l.c. 109.
= *I. decipiens* var. *megacystis* Favre 1955, l.c. 111.
= *I. praetervisa* Quél. ss. Favre 1955, p.p.
= ? *I. decipiens* Bres. ss. Kühner 1933, Bull. Soc. Myc. France 49: 98.

Pileus -35(-55) mm, ex hemisphaerico-convexo subumbonato-planus, brunneus vel fuscus, striatofibrillosus, e velo albogriseo conspicue instructus. Lamellae adnato-emarginatae, griseae dein ochraceo-brunneae. Stipes -45/-1 (-16) mm, bulboso-marginatus, omnino pruinosus albus aetate pallide brunneoroseus. Odor spermaticus. Caro alba dein brunneorubescens. Sporae 10-13.5 x 6-8 µm, angulato-subnodulose. Cheilocystidia (pleurocystidia) 50-80 (-130) x 10-24 µm, fusoidea, metuloidea, incrustata. Caulocystidia cylindraceo-fusoidea, tenuitunicata (rare membrana subincrassata incrustataque instructa). Ad terram calcaream in zona alpina (*Dryas*, *Salix reticulata*, *S. retusa*, *S. serpyllifolia*), Helvetia. Favre 120a, GC (Holotypus).

Material examined: SNP: Switzerland, Kt. Graubünden, Alp Murtèr, 26.VIII.1942, Favre 120 a (type;GC). "*Inocybe decipiens* var. *mundula*": Val Nuglia, 31.VIII.1949, Favre 120 (GC); Val S-charl, Piz Mezdi, 27.VIII.1951, Favre 120 b (GC); Ofenpass, Murtaröl d'Aint, 15.VIII.1953, Favre 120 b,l (GC); Ofenpass, Munt la Schera, 9.IX.1982, Horak 1771 (ZT); Ofenpass, Munt la Bescha, Chaslot, 7.IX.1983, Horak 2271 (ZT). - "*I. decipiens*": Ofenpass, Piz d'Aint, 10.VIII.1950, Favre 118 b (GC); Val S-charl, Valbella, 14.VIII.1951, Favre 118 c (GC); Ofenpass, Murtarol d'Aint, 28.VIII.1953, Favre 118 a (GC); Ofenpass, Munt la Schera, s.d., Favre 118 b,l(GC). - "*I. decipiens* var. *megacystis*": Tarasp, Munt della Bescha, 8.IX.1945, Favre 119, lectotype of variety; GC). - "*I. praetervisa*": Val S-charl, Valbella, 27.VIII.1951, Favre 127 (GC).

Ecology. - SNP: Among *Dryas octopetala* and *Salix reticulata-retusa-serpyllifolia*, on dolomite, 2180-2550 m, rarely also in association with *Salix herbacea* in snow bed, 2400 m.

Remarks: Referring to "*Inocybe decipiens* Bres." and its new varieties Favre (1955) follows the interpretation of Kühner (1933) whose taxonomical concept obviously is not in accordance (cf. *Inocybe dunensis* Orton, 1960; Stangl, 1979) with the descriptive data and paintings published by Bresadola (1892).

The careful examination of Favre's material filed under *I. decipiens* revealed that the characters observed on the new varieties (var. *mundula*, var. *megacystis*) are not distinctive enough to warrant their separate rank.

Comparative material examined: -*Inocybe decipiens* Bresadola: Italy: Villaggano, ad marginem camporum, Julio 1888, Bresadola (type;S). -*Inocybe dunensis* Orton: England: Frehsfield, Lancs, in soc. Salicis, 9.VII.1956, Orton (type;K).

10. *Astrosporina taxocystis* Favre & Horak, sp. n. Pl. 1,L; fig. 10,L-R.
= *Inocybe decipientoides* var. *taxocystis* Favre 1955, Rés. rech. scient. Parc National suisse, V: 113.

Pileus -35 mm, e conico convexus, campanulatus vel umbonato-planus, avellaneus vel fuscus (raro castaneo tinctu), centro glabrus, fibrillo-, sorimosus marginem versus, velum nullum. Lamellae griseo-argillaceae. Stipes -35/-4 (-7) mm, bulboso-marginatus, albidus dein pallide brunneus, apicaliter pruinosus, basim verus fibrillosus. Odor spermaticus. Caro ex albido brunneola. Sporae 9-11(-12) x 5-7 µm, angulato-nodulosae. Cheilocystidia (pleurocystidia) 40-80 x 10-25 µm, fusoidea, metuloidea, incrustata. Caulocystidia 20-65 x 8-20 µm, crasse tunicata, clavata vel subfusoidea. Ad terram in zona alpina (*Salix herbacea*). Helvetia. Favre 122a; GC (Holotypus).

Material examined: SNP: Switzerland, Kt. Graubünden: Tarasp, Munt della Bescha, 8.IX.1945, Favre 122a (type; GC); Val S-charl, Munt Plazèr, 20.VIII.1952, Favre 122c; (GC); Val Sesvenna, Marangun, 5.IX.1981, Horak 1315 (ZT); Val Sesvenna, below Ils Laiets, 10.IX.1981, Horak 1351 (ZT); Flüelapass, Radönt (E of Fl. Schwarzhorn), 3.IX.1983, Horak 2249 (ZT); Flüelapass, track to Munt da Marti, 22.VIII.1984, Horak 2507 (ZT); France: Col de l'Iseran, Plan des Eaux, snow bed with *Salix herbacea*, 2650 m, 19.VIII.1982, Horak 1699 (ZT).

Ecology: SNP: In snow bed with *Salix herbacea* (rarely associated also with *S. retusa*), on gneiss, 2250-2650 m.

Remarks: In fig. 10 some of Favre's unpublished drawings clearly demonstrate (cp. Favre, fig. 100, p. 113) that the taxon "var. *taxocystis*" is not related to *Inocybe decipientoides* Peck (cf. illustrations in Stuntz, 1947), now considered to be a synonym of *A. lanuginella* Schroeter ap. Cohn.

The new species, *A. taxocystis*, is recognized by the following features, brown to dark brown pileus (occasionally with reddish tinge), absence of veil, pruinose stipe with marginate bulb at base, spermatic odor, angular-nodulose spores, metuloid clavate to short-fusoid caulocystidia.

Based upon examined material from the SNP, *A. taxocystis* is restricted to acid soils where it occurs in association with *Salix herbacea* (one record also with *S. retusa*).

11. *Astrosporina humilis* Favre & Horak, sp. nov. Pl. 2,M; fig. 11,S-Z.
= *Inocybe humilis* Favre 1960, Rés. rech. scient. Parc National suisse, VI: 480, nom. inval. - no type designated.

Latin diagnosis in Favre (1960: l.c.). Holotype: Favre 583, (GC).

Material examined: SNP: Switzerland, Kt. Graubünden: S-chanf, God God, 5.IX.1956, Favre 583, holotype; (GC); Dischmatal, Alp Stillberg, 6.VIII.1964, Horak 64/640 (ZT).

Ecology: SNP: Holotype described from subalpine zone, 1800 m, associated with *Larix*, *Picea*, *Pinus cembra*; in alpine zone with *Salix herbacea* in snow bed,on gneiss, 2300-2400 m.

Remarks: Originally described from a subalpine locality *A. humilis* (Favre, 1960) occurs in the SNP-region also above timberline in snow beds among *Salix herbacea*. Its distinguishing macroscopical characters: small carpophores, yellowish lamellae, bulbous-submarginate base of stipe, resemble those of *A. concinnula* which, however, is found exclusively in association with *Dryas* and calciphilous species of *Salix*. In case of doubt the two alpine taxa are immediately separated by the shape of the spores.

12. *Astrosporina praetervisa* (Quélet) Schroeter ap. Cohn 1889, Pilze Sch. 1: 576. Pl. 2,N; fig. 12,A-E.
= *Inocybe praetervisa* Quélet in Bresadola 1892, Fung. Trid. 1: 35 (tab. 38) (basionym)
= *I. praetervisa* f. *rufofusca* Favre 1955, Rés. rech. scient. Parc National suisse, V: 119.

Material examined: SNP: Switzerland, Kt. Graubünden: Val Tavrü, Blaisch Bella, 8.VIII.1943, Favre 127a (GC); Val Sesvenna, Ils Laiets, 6.VIII.1946, Favre 130 (lectotype of f. *rufofusca*; GC); Dischmatal, Alp Stillberg, 28.VII.1964, Horak 64/220 (ZT); Val Sesvenna, between Marangun and Ils Laiets, 10.IX.1981, Horak 1347 (ZT): Flüelapass, Radönt (E of Fl. Schwarzhorn), 11.IX.1981, Horak 1360 (ZT); same locality, 5.IX.1982, Horak 1717 (ZT); same locality, 1.IX.1983, Horak 2236 (ZT); same locality, 1.IX.1983, Horak 2241 (ZT); Val S-charl, Costainas, 8.IX.1981, Lamoure 1340 (ZT); Val Sesvenna, Ils Laiets, 8.IX.1983, Horak 2283 (ZT); Val

Tasna, between Laret and Mot da l'Hom, 27.VIII.1984, Horak 2618 (ZT); Val Sesvenna, between Ils Laiets and Marangun, 31.VIII.1984, Brunner 2545 (ZT); France: Hte. Vallé de l'Arc, Le Vallon, in snow bed with *Salix reticulata*, 2680 m, 19.VIII.1982, Horak 1697 (ZT).

Ecology: SNP: In snow beds with *Salix herbacea* (one record from the French Alps with *S. reticulata*), on gneiss, 2180-2600 m.

Remarks: The alpine collections from the SNP compare well with descriptions published in the current literature (Bresadola, l.c.; Kühner, 1933; Alessio, 1980).

In general alpine specimens have less ochre tints on the pileus whose colour ranges (depending on age and exposure) from hazel brown to reddish brown (= f. *rufofusca*; cf. topotypical records). As a rule the odor of the material studied is not distinctive (proved by Favre, 1955) but occasionally a faint fruity component can be detected on freshly cut carpophores.

13. *Astrosporina asterospora* (Quélet) Rea 1922, Brit. Basid. 213.
Pl. 2,P; fig. 13,F-K.

= *Inocybe asterospora* Quélet 1879, Bull. Soc. Bot. France 26: 50. (basionym).

= *Inocybe napipes* Lange ss. Favre 1955, l.c. 116 (misident.)

Material examined: SNP: Switzerland, Kt. Graubünden: Val Sesvenna, Ils Laiets, 20.VIII.1944, Favre 125b (GC); Val Sesvenna, Marangun, 25.VIII.1948, Favre 125a (GC).

Ecology: SNP: In swamp or in snow bed with *Salix herbacea*, on gneiss, 2220-2500 m.

Remarks: According to Favre (1955) "*Inocybe napipes* Lange" is recorded in the alpine zone of the SNP. No detailed descriptions are kept in Favre's herbarium but all microscopical data observed on the two collections clearly prove that the material is misidentified. The specimens mentioned represent *A. asterospora* and thus "*I. napipes* Lge." must be deleted from the list of alpine *Astrosporina* in the SNP.

14. *Astrosporina pseudohiulca* (Kühner & Boursier) Horak, comb. nov.
Pl. 2,R; fig. 14,L

= *Inocybe pseudohiulca* Kühner & Boursier 1933, Bull. Soc. Myc. France 49: 107 (basionym).

Material examined: SNP: Switzerland, Kt. Graubünden: Val S-charl, Mot Madlein, 12.IX.1944, Favre 131a(GC); Ofenpass, Piz d'Aint, 10.VIII.1950, Favre 131a, 1311 (GC); same locality (as "*Inocybe praetervisa*"), 10.VIII.1950, Favre 127b (GC); Samnaun, Alp Trida, 25.VIII.1984, Brunner 2515 (ZT).

Ecology: SNP: Among *Dryas octopetala* on dolomite, 2400-2450 m, rarely also in association with *Salix herbacea* (and *Dryas* ?) on lime-rich schistes, 2400 m.

Remarks: The macroscopical and microscopical characters observed on alpine collections from the SNP correspond well with the type material and specimens gathered in the subalpine forests of the SNP (Horak, 1985).

Comparative material examined: *Inocybe pseudohiulca* Kühner & Boursier: France, Rozière, 12.VIII.1927, Kühner (type, ex ex Herb. Kühner). - Switzerland, Kt. Graubünden: Schuls-Pradella, with *Picea abies*, 1170 m, 21.VIII.1983, Horak (2211; ZT).

ACKNOWLEDGEMENTS

I am grateful to the Curators in CHUR and GC for the loan of material kept in the Favre-Herbarium.

REFERENCES

Alessio, C. L., 1980, *Inocybe* in Bresadola, *Iconographia Mycologica*, 29 (2 vol.), Trento, Italy.

Boursier, J., and Kühner, R., 1928, Notes sur le genre *Inocybe*, *Bull. Soc. Myc. France*, 44: 170-189.

Bresinsky, A., and Schmid-Heckel, H., 1982, Der Lärchenporling und verschiedene Blätterpilze aus den Berchtesgadner Alpen neu für die Bundesrepublik nebst einer Liste indigener Lärchenbegleiter, *Ber. Bayr. Bot. Ges.*, 53: 47-60.

Christiansen, M. P., 1941, Studies in the larger fungi of Iceland, *Botany of Iceland*, 3(2): 195-225.

Enderle, M., and Stangl, J., 1980/1981, Beitrag zur Kenntnis der Ulmer Pilzflora: Risspilze (*Inocybe*), *Mitt. Ver, Naturw. Math. Ulm*, 31: 79-170.

Eynard, M., 1977, "Contribution à l'étude écologique des Agaricales des groupements à *Salix herbacea*," Thèse, Univ. Claude-Bernard, Lyon.

Favre, J., 1948, Les associations fongique des hauts-marais jurassiens et de quelques régions voisines, *Mat. Fl. Crypt. Suisse*, 10: 1-228.

________, 1955, Les champignons supérieurs de la zone alpine du Parc National suisse, *Rés. rech. scient. Parc Nationale suisse*, 33: 1-212.

________, 1960, Catalogue déscriptif des champignons supérieurs de la zone subalpine du Parc National suisse, *Rés. rech. scient. Parc Nationale suisse*, 42: 323-610.

Friedrich, K., 1942, Pilzökologische Untersuchungen in den Oetztaler Alpen, *Ber. Deutsch. Bot. Ges.*, 60: 218-231.

Grund, D. W., and Stuntz, D. E., 1968, Nova Scotian Inocybes. I, *Mycologia*, 60: 406-425.

Gulden, G., 1975, Mushroom inventory at Hardangervidda, autumn 1971. IBP in Norway, *Ann. Rep.*, 1974, 371-380.

________, and Lange, M., 1971, Studies in the macromycete flora of Jotunheimen, the central mountain massif of S Norway, *Norw. J. Bot.*, 18: 1-46.

Heim, R., 1928, Les champignons des Alpes, *Soc. de Biogeogr.*, 21: 1-22.

________, 1931. Le genre *Inocybe*, *Encycl. Mycol.*, 1: 1-423.

Horak, E., 1985, Oekologie der Pilzflora (Agaricales-Boletales) in 5 Pflanzengesellschaften der montan-subalpinen Stufe des Engadins (Schweiz), *Ergebn. wiss. Unters. Schw. Nationalparks*, 12(C 1 4):C 337-C 476.

Huijsman, H. S. C., 1955, Observations on agarics, *Fungus*, 25: 18-43.

Josserand, M., 1959, Notes critiques sur quelques champignons de la région lyonnaise. Nr. 6, *Bull. Soc. Myc. France*, 75: 359-404.

________, 1974, Notes critiques sur quelques champignons de la région lyonnaise. Nr. 8, *Bull. Soc. Myc. France*, 90: 231-263.

Kobayasi, Y., et al, 1971, Mycological studies of the Angmagssalik region of Greenland, *Bull. Nat. Sci. Mus. Tokyo*, 14: 1-96.

Kreisel, H., 1959, Beitrage zur Pilzflora Bulgariens, *Feddes Rep.*, 62: 34-43.

Kühner, R., 1933, Notes sur le genre *Inocybe*, *Bull. Soc. Myc. France*, 49: 81-121.

_______, 1955, Compléments à la Flore Analytique. VI. *Inocybe* goniosporés et *Inocybe* acystidiés. Espèces nouvelles ou critiques. *Bull. Soc. Myc. France*, 71: 169-201.
_______, and Boursier, J., 1932, Notes sur le genre *Inocybe*, *Bull. Soc. Myc. France*, 48: 118-161.
_______, and Romagnesi, H., 1953, "Flore analytique des champignons supérieurs," Masson, Paris, 1-556 pp.
Lamoure, D., Lange, M., and Petersen, M. P., 1982, Agaricales found in the Godhavn area, W. Greenland, *Nord. J. Bot.*, 2: 85-90.
Lange, M., 1957, Greenland Agaricales, *Medd. Grønland*, 148: 1-125.
_______, and Skifte, O., 1967, Notes on the macromycetes of northern Norway, *Acta Borealia*, 23: 1-51.
Laursen, G. A., and Chmielewski, M. A., 1982, The ecological significance of soil fungi in arctic tundra, *in:* "Arctic and alpine mycology," G. A. Laursen and J. F. Ammirati eds., ISAM I(Univ. Washington Press): 432-492.
Miller, O. K., Laursen, G. A., and Murray, B. M., 1973, Arctic and alpine agarics from Alaska and Canada, *Can. J. Bot.*, 51: 43-49.
Møller, F. H., 1945, Fungi of the Faeroes. I, Copenhagen, 1-295.
Moser, M., 1967, Die Röhrlinge und Blätterpilze, *in:* "Kl. Kryptogamenfl. II b/2," H. Gams, ed., Fischer, Stuttgart.
Ohenoja, E., 1971, The larger fungi of Svalbard and their ecology, *Rep. Kevo Subarct. Res. Stat.*, 8: 122-147.
Orton, P., 1960, New check list of British agarics and boleti, *Trans. Brit. Myc. Soc.*, 43: 159-439.
Petersen, M. P., 1977, Investigations on the ecology and phenology of the Macromycetes in the Arctic, *Medd. Grønland*, 199(5): 1-72.
Rostrup, E., 1894, Øst Grønland svampe, *Medd. Grønland*, 18: 1-39.
Stangl, J, 1975, Die eckigsporigen Risspilze (1), *Z.f. Pilzk.*, 41: 65-80.
_________, 1976, Die eckigsporigen Risspilze (2), *Z.f. Pilzk.*, 42: 15-32.
_________, 1977, Die eckigsporigen Risspilze (3), *Z.f. Pilzk.*, 43: 131-144.
_________, 1979, Die eckigsporigen Risspilze (4), *Z. f. Mykol.*, 45: 145-162.
________, and Enderle, M., 1983, Bestimmungsschlüssel für europäische eckigsporige Risspilze, *Z. Mykol.*, 49: 111-136.
________, and Veselsky, J., 1974, Beiträge zur Kenntnis seltener Inocyben. Nr. 4. *Inocybe boltonii* Heim in der Variationsbreite ihrer Formen. *Česká Mykol.*, 28: 143-150.
Stuntz, D. W., 1947, Studies in the genus *Inocybe*. I. New and noteworthy species from Washington, *Mycologia*, 39: 21-55.
Watling, R., 1977, Larger fungi from Greenland, *Astarte*, 10: 61-71.
________, 1981, Relationships between macromycetes and the development of higher plant communities, *in:* "The fungal community," Wicklow and Carroll eds., 427-458.
________, 1983, Larger cold-climate fungi, *Sydowia*, 36: 308-325.

AGARICACEAE, AMANITACEAE, BOLETACEAE, GOMPHIDIACEAE, PAXILLACEAE AND PLUTEACEAE IN GREENLAND

Henning Knudsen

Botanisk Museum
Gothersgade 130, DK-1123
Copenhagen K, Denmark

and

Torbjørn Borgen

P. O. Box 96
3940 Pamiut, Greenland

Key words: Boletales, Agaricaceae, Amanitaceae, Pluteaceae, Greenland

ABSTRACT

A review of the families Agaricaceae, Amanitaceae, Boletaceae, Gomphidiaceae, Paxillaceae and Pluteaceae in Greenland is given. Among the 29 species recorded, three are new, viz. *Amanita arctica* Bas, Knudsen & Borgen sp. nov., *A. groenlandica* Knudsen & Borgen sp. nov. and *A. mortenii* Knudsen & Borgen sp. nov. 15 others are recorded as new to Greenland including some introduced boletes which only occur with planted conifers. Notes are given on the distribution of species in related areas. The Greenland flora is relatively poor compared to similar localities at the same latitude, probably due to the more unfavourable weather and the isolated position of this island.

INTRODUCTION

This paper is part of current project undertaking a revision of the Greenland macrofungi. Earlier papers have been published by Dissing (1981), M. Lange (1980), Knudsen and Borgen (1982), Noordeloos (1984), Petersen and Korf (1982) and a series of other papers is under preparation by other mycologists.

In 1983 the authors and Jens H. Petersen made 650 collections in the southernmost part of Greenland, particularly in the Narssarssuaq area and Qinqua Valley. These two localities were revisited in 1984 by H. Knudsen and Thomas Laessøe, who made another 650 collections. The aim of these expeditions was to study the conifer plantations in these areas with special reference to parasitic and mycorrhizal fungi and to collect macrofungi in general. The expeditions were part of a larger project examining the possibilities for conifer growth in Greenland by the dendrologist Søren Ødum from the Arboretum at the Royal Agricultural and Veterinary Highschool in Copenhagen, ranger Poul Bjerge from the Upernaviarssuk Research Station in Greenland and entomologist Peter

Nielsen from the Zoological Museum in Copenhagen. The localities visited are among the most favourable for fungi in Greenland but only a few collections have previously been made in Qinqua Valley mostly in 1889 by N. Hartz whose collections were published by Rostrup (1891). In contrast the Narssarssuaq area has been visited by several mycologists, especially because one of Greenland's two airports with reqular flights from Copenhagen is located here.

Materials and methods: The revisions are mostly based on annotated, and in many cases also photographed material, collected by H. Knudsen, T. Borgen & J. H. Petersen (HK, TB & JP), H. Knudsen & T. Laessøe (HK & TL), T. Borgen (TB), P. Milan Petersen (PMP) and Morten Lange (ML).

The material was studied in the light microscope at 1200 x after revival in 2% KOH. The measurements of Amanita-spores were done in Congo Red dissolved in 10% NH_4OH.

For a general review of the vegetation of South Greenland we refer to Feilberg (1984), from which the following information has been acquired. The area around Narssarssuaq (61°10'N, 45°25'W) belongs to the subcontinental, subarctic zone with copses of *Betula pubescens* Ehrh. ssp. *tortuosa* (Ledeb.) Nyman (henceforth referred to as *Betula pubescens*), *B. glandulosa* Mich., *Salix glauca* L., *S. herbacea* L., *S. arctophila* Cockerell and *S. uva-ursi* Pursh, the first three species being particularly important forming extensive copses in the area. Other woody, but non-mycorrhizal plants include *Juniperus communis* L. ssp. *nana* (Willd.) Syme and *Sorbus groenlandica* (Schneider) A. & D. Love. *Polygonum viviparum* L. and *Dryas integrifolia* Vahl should likewise be mentioned as mycorrhizal hosts. On the southern part of the west-coast *Alnus cripsa* (Ait.) Pursh forms copses.

In Qinqua Valley (60°16'N, 44°33'W) the climate is suboceanic and subarctic. The same trees and bushes as in the Narssarssuaq area are present and here attain the largest dimensions measured in Greenland, *B. pubescens* up to 8-9 m tall and 30-40 cm in diameter at the base. The valley is 7-8 km long and is surrounded by 600-1500 m high mountains. It was visited during one week in 1983 and 1984 respectively and studies were made particularly in the southern part near Taserssuaq Lake where the largest trees are found. The season of 1983 was extremely favourable with much rain adding more than one hundred species to the Greenland list, while the season of 1984 was dry with much less fungi especially of the larger forms.

All material is deposited at the Botanical Museum in Copenhagen (C), unless otherwise mentioned. An asterisk (*) after the name indicates that the species is new to Greenland. The colour code refers to Kornerup and Wanscher (1974).

AGARICACEAE Fr.

Agaricus arvensis Schaeff.

This species was found only once, by Lange (1955). It has been reported from Iceland by Hallgrimsson (1979).

Material examined: ML 196.

A. campester L.:Fr.

This was first collected by C. Petersen in 1880 at Kagsiarssuk

(60°53'N) in Igaliko Fjord (Rostrup 1888), and later collected by Hartz in 1890 in Sydøstbugten (ca. 68°40'N), SE of Disko Island and in Cap Stewart on the East-coast near Scoresbysund (70°27'N). It has been collected by various collectors in Sdr. Strømfjord (Watling 1983, and material). We found it quite common in the Narssarssuaq area as well.

The habitat is often short grass, dry sandy or gravelly localities, grazed and fertilized by sheep, in Sdr. Strømfjord on naturally rich, loamy ground. The northernmost locality on the West coast is Godhavn (69°15'N), slightly south of the Cap Stewart collection. However, this collection as well as the other collections from the last century are lost, so they cannot be confirmed.

Agaricus campester was reported from the northernmost part of Scandinavia by Lange and Skifte (1967) and Kallio and Kankainen (1964) and from Iceland by Hallgrimsson (1979).

Material examined: H. Dissing 81.36a; HK, TB & JP 55; HK & TL 190; ML 115; 67.442; B. & L. Lange 67.30; TL & S. Elborne 84; PMP 73.140; S. Odum 28.VIII.1983.

A. fissuratus (Møll.) Møll.

Lange (1955) reported this species on the basis of a collection from Nugssuaq (70°40'N). The specimens are more coarsely scaly than in most Danish specimens, and in this character it resembles *A. tabularis* Peck. This occurs in very dry areas like steppes, prairies and semidesert and has been reported from eastern Arctic Russia by Vasilkov (1974). Both species belong to the *Arvensis* group and have neither been clearly distinquished from each other nor from *A. arvensis*.

Material examined: C. A. Jørgensen, 11.VIII.1947, Nugssuaq.

A. salicophilus M. Lange

Agaricus salicophilus was described by Lange (1955) from four collections made around Sdr. Stromfjord, a dry, continental area in Greenland. It was later collected from the same locality by P. M. Petersen, but we have not been able to find other records in the literature of this species.

Agaricus salicophilus is closely related to *A. augustus* with catenulate cystidia, but the spores are more subglobose (8-9 x 6-7 µm in the collection made by Lange). However, in PMP 73.138 the spores are more ellipsoid (7.5-9 x 5-6 µm) thus approaching the size for *A. augustus* as given e.g. by Cappelli (1984). More material is needed to elucidate the differences

Material examined: ML 132; 167 (typus); 195; 398; PMP 73.138.

A. semotus Fr.*

This species has been collected several times in recent years around Narssarssuaq, in Grønnedal, at Mestervig and on Ella Island in east Greenland, north to 72°52'. The habitat in southern Greenland is open shrub with *Betula pubescens* and further north it is dwarf shrub heath with *Dryas*, *Betula nana* and *Arctostaphylos*. It is probably identical with the species recorded as rare by Hallgrimsson (1979) from north-eastern Iceland (s. nom. "*A. rubellus*?") and the species recorded by Kallio and Kankainen (1964) from northern Scandinavia as *Psalliota rubella*. Vasilkov (1974) reports a find from the tree-tundra zone in the Ural region.

Material examined: TB 84.92; H. F. Gøtzsche 83.64; 83.79; HK, TB & JP 50; 135; 136; HK & TL 572; ML 15.VII.1975.

We have been unable to identify a number of *Agaricus* collections which lack sufficient notes or are too poorly preserved. However, a number of quite similar collections suggest a small species from the *Arvensis-group*. One collection resembles a brown scaly *campestris*-type but with cystidia. Finally, we have received reports from colleagues of a large white *Agaricus* from Jamesonland in E-Greenland which they regretfully consumed.

Lepiota clypeolaria (Bull.: Fr.) Kummer

This *Lepiota* has only been found twice, by Lange near Godhavn (Lamoure, Lange and Petersen 1982) and by the authors near Narssarssuaq. The Greenland specimens are relatively small and pale compared to collections from Denmark, but, although the complex around *L. clypeolaria* needs revision we prefer to use this name at present. *L. clypeolaria* has been reported a few times from Iceland (Hallgrimsson 1973), where the closely related *L. ventriosospora* possibly occurs as well. Lange and Skifte (1967) reported *L. clypeolaria* from Tromsø at the same northern latitude as Godhavn in Greenland and almost the same as Utsjoki in Finland, where it was reported by Kallio and Kankainen (1964). Our collection was found on a south-facing slope with an open grassy shrub of *Betula pubescens*, the same type of locality as in northern Finland.

Material examined: HK, TB & JP 119; ML 67.98.

AMANITACEAE Roze

Key to *Amanita* in Greenland and related areas

1. Stipe with annulus; pileus red with white scales .... *A. muscaria*
1. Stipe without annulus; pileus another colour 2
2. Volva developed as small scale-like fragments; stipe base bulbous 3
2. Volva saccate (sect. *Vaginatae* (Fr.) Quél.) 4
3. Pileus 2-5 cm, grey brown, paler towards the sulcate-striate margin, smooth or with small volval remnants; stipe 4-9 cm x 6-15 mm, pale greyish or brownish, slightly fibrillose; spores broadly ellipsoid, 10-12.5 x 8-10 µm (l/b mean 1.2-1.35) .... *A. friabilis*
3. Pileus up to 6 cm, white, smooth or with small volval remnants, margin sulcate; stipe up to 4 cm x 4 mm, at base up to 7 mm, white, slightly fibrillose; spores broadly ellipsoid, 11.2-13.3 x 9.4-10.8 µm (l/b mean 1.2)(not found in Greenland and not henceforth treated; description from Bas 1982) .... *A. hyperborea*
4. Pileus bright orange; stipe pale orange with prominent floccose girdles (not found in Greenland and not treated henceforth) .... *A. crocea*
4. White, cream, buff, brown, greyish or grey or olivaceous 5
5. Pileus in expanded specimens distinctly coloured grey, brown or olivaceous 7
5. Pileus in expanded expanded specimens pale, white, whitish or pale grey 6
6. Pileus white or whitish, becoming pale buff at centre, 2-8 cm; volva tall, thin and often fragmented with the upper part adhering to the stipe, whitish or with slight ochraceous tints .... *A. arctica*

6. Pileus at first grey or greyish soon becoming pale grey to whitish, sometimes with buff tint, (2-)3-5(-7) cm; volva low, usually not much higher than broad, thick and persistent, whitish below, greyish and/or ochraceous above .... *A. nivalis*
7. Volva with few scattered sphaerocysts, tough, white, slender and appressed to the stipe; pileus margin strongly sulcate, pileus brown or olivaceous brown, centre darker, with or without single volval patch .... *A. battarrae*
7. Volva with numerous sphaerocysts, grey or ochraceous, more pronounced when dried 8
8. Relatively short-stemmed, robust species, often with volval patches or scales on pileus, margin only slightly sulcate; pileus 3-12 cm,. grey, greyish, greyish brown to brown; stipe whitish with pale greyish or pale brownish floccose girdles; volva fragile and often partly disrupting, whitish below, greyish above or greyish all over .... *A. groenlandica*
8. Slender and tall species, with or without one or a few volval patches on pileus, margin strongly sulcate; pileus 2.5-6 cm, usually umbonate, yellowish, bright fulvous to olivaceous brown; stipe with or without indistinct girdles, whitish or very pale brownish; volva grey or ochraceous 9
9. Volva ochraceous; pileus fulvous without olivaceous tints .... *A. fulva*
9. Volva grey or greyish, rarely with an outer thin ochraceous layer; pileus yellowish brown to olivaceous brown .... *A. mortenii*

Amanita arctica Bas, Knudsen & Borgen sp. nov.

= *A. ? hyperborea* ss. Bas in Clémençon, 1977: 87 (based on Bas 6105). After the redescription of *A. hyperborea* by Bas (1982) the name of 6105 was changed in herb. to *A. arctica* Bas nom. prov.

In sectionem *Vaginariam* referenda. Pileus albus, in senectute medio ochraceus, 2-8 cm diam., margine sulcatus. Margines lamellarum albo-flocculosi. Stipes 4.5-13 cm altus, 0.6-2 cm crassus, omnino zonatim albo-squamosus. Volva alba, in senectute ochraceo-maculata. Sporae globulares vel subglobuares vel late ellipsoides, 9.6-13 x 9-11 µm diam. Habitat cum *Betula pubescenti, B. glandulosa, Salica glauca, S. herbacea.*

Holotypus die 15 Augusti anni 1985 in Grønnedal sub numero 85.204 a T. Borgen lectus, siccus in Museo Botanico Hauniensi (C) depositus.

Pileus hemispheric to conical then expanded to convex or applanate with umbo, 2-8 cm, at first white to ivory or pale cream becoming distinctly ochraceous buff at centre (4A3-4), slightly viscid, shining when dry, with or without a single volval patch, margin sulcate, 0.2-0.3 R. Lamellae free, white with flocculose white margin, distally slightly ventricose. Stipe cylindric or slightly enlarged towards base, 4.5-13 x 0.6-2 cm above with white floccose girdles, sometimes with a prominent fibrillose-scaly covering on a pale cream or whitish background, internal limb often highplaced; exannulate, hollow. Volva usually high and often characteristically broken leaving a fragment on the stipe, isolated from the basal volva (see Fig. 1 in Bas 1977); inside white, outside white, ochraceous spotted when old. Flesh white, without particular smell or taste. Spore print not seen.

Spores globose to subglobose or broadly ellipsoid (Q:1.0-1.2), 9.6-13(-15) x 9-11(-12.5) µm, hyaline, often with one big "oil"-droplet, nonamyloid. Basidia clavate, 60-70 x 15-18 µm, 4-spored with granulose contents. Sphaerocysts from gill-edge globose to subglobose, 25-50 µm in diam., hyaline, from outer surface of volva 20-60 µm in diam.

Fig. 1. *Amanita arctica*. Bas 6105. Photo C. Bas

This species occurs solitary or in small groups under *Betula pubescens, B. glandulosa, Salix glauca, S. herbacea or Polygonum viviparum* in dry or moist shrubs and snow-beds. It has been recorded from the Narssarssuaq area, Qinqua Valley (not preserved), Grønnedal, Frederikshåb and north to Godhavn (69°15'N). In Finland it was found at Utsjoki (Bas 6105).

Material examined: Bas 6105 (L); TB 78.85; 85.32; 85.220a (type); HK & TL 254; 517; 566; 575; ML 67.244.

A number of white *Amanitas* have been described in sect. *Vaginatae*. Some of these seem to be mere albino-forms of other species while others are species with white as their true colour. A number of the latter species have been lumped in *A. vaginata* var. *alba* by Gilbert (1941), including among other species white forms of *A. nivalis*. Likewise *A. hyperborea* (Bas 1982) should be mentioned as a white, ringless *Amanita*. In spite of this confusion we are convinced that *A. arctica* is a new species as suggested by Bas in 1977. The important characters separating it from other species in sect. *Vaginatae* are the white to buff colours, the strongly developed stipe-covering, the tall often fragmented volva and the subarctic habitat. It is difficult to scan the literature for other records due to its being mistaken for the white species, but "a pure white form (*A. alba* Gill.?)" mentioned by Jakobsson (1984) from the subarctic birch forest in Sweden most probably belongs here.

A. battarrae Boud.*

= *A. umbrinolutea* Secr. nom. illeg.

Only found three times, in Qinqua Valley in a moist bog with *Betula glandulosa, B. pubescens* and *Salix glauca* and around Narssarssuaq under the same hosts.

A. battarrae is widespread in southern and central Europe (Cetto 1980, Secretan 1833, Heim 1929, Romagnesi 1961), while it seems to be more rare or overlooked in northern Europe. It was recorded from Sweden (as less common) by Ryman and Holmåsen (1984), presumably also by Jakobsson (1984) from Hamrefjället ("a robust dark olive grey form" of *A. vaginata* in birch forest) from the southern provinces of Finland and from Iceland by Hallgrimsson (1972).

The distinguishing characters of this species as emphasized here is the zonate pileus with a dark brown to blackish centre and a paler brown or greyish brown marginal zone, the deeply sulcate margin and the white, tall, narrow, tough and persistent volva with scattered sphaerocysts in the surface layer.

The material we studied had a white, non-girdled stipe and a white volva as shown by Romagnesi (1961, Pl. 181), while Cetto (1980, Pl. 17) and Heim (1929) illustrate a distinctly coloured and girdled stipe, and Ryman and Holmasen (1984) illustrated a distinctly ochraceous volva.

Material examined: HK, TB & JP 107; 367; 393.

Furthermore we have seen material from Iceland (S. A. Elborne 132).

A. friabilis (Karst.) Bas*

This has only recorded once, by T. Borgen at Grønnedal in rich soil amongst *Salix* and *Alnus crispa*. According to Bas (1974) it is a rare but widespread species following *Alnus* and recorded with *A. viridis* up to 1900 m in the Alps.

Material examined: TB 85.135

A. fulva (Schaeff.) Pers.*

It is known from only four collections from southwestern Greenland where it grows in dry soil with *Betula glandulosa*, solitarily or in small groups. It was doubtfully recorded from Iceland by Hallgrimsson (1972) but not mentioned in his flora (Hallgrimsson 1979). *Amanita fulva* is neither mentioned from northern Norway nor from northern Finland by Kallio and Kankainen (1964). It is recognizable by the overall habit, colours and ochraceous brown volva.

Material examined: TB 79.94; 81.215; 85.259; PMP 73.538.

A. groenlandica Bas ex Knudsen & Borgen sp. nov.

= *A. groenlandica* nom. prov. Bas in Clémençon, 1977:87

Prope *A. submembranaceam* (Bon) Gröger in sectionem *Vaginariam* referenda. Pileus 3-12 cm diam., cinereus vel sepiaceus vel ambobus coloribus variegatus, margine leviter sulcatus. Margines lamellarum cinereo-sepiaceo affecti. Stipes pallide cinerascens vel fuscidulus, zonatim floccosus, 4-15 cm altus, 0.8-2(-3.3) cm crassus. Volva sub-cinerea. Sporae globulares, 9.6-12.8 µm diam.

Habitat cum *Salice glauca, S. herbacea, S. arctophila.*

Holotypus die 10 Augusti anni 1984 in valle Hospitalsdalen prope vicum groenlandicum Narssarssuaq sub numero 574 a H. Knudsen et T. Laessøe lectus, siccus in Museo Botanico Hauniensi (C) depositus.

Fig. 2. *Amanita groenlandica*. HK, TB & JP 469. Photo Jens H. Petersen.

Pileus hemispheric, then expanded to convex and broadly umbonate, finally applanate with or without a low umbo; colours ranging from pale straw to straw (3A2-3) especially when young and near margin, greyish yellow (4B5) or most commonly with deeper brown and grey colours like fulvous (5D4-7), clay buff (5C3-4), drab (5D3), hazel (5E4) or snuff brown (6E4), often darkest at centre but also uniformly coloured or with mixed grey, brown and yellowish colours, (3-)5-9(-12) cm, when young slightly viscid then dry and shiny; volval patches on pileus often present either as one or a few broad patches on pileus, or as numerous small, scale-like, whitish or pale greyish brown (5B2) to grey patches; margin weakly sulcate, 0.1-0.2 R. Lamellae free, slightly ventricose distally, white then very pale cream, edge pale grey brown, lamellulae present. Stipe slightly conical, 4-15 x 0.8-2.0 cm, at base up to 3.3 cm, ground colour whitish with pale dirty grey brown (5B2), pale greyish buff, pale greyish or pale brown (4A3), floccose girdles, becoming slightly darker when bruised, finely sulcate at apex, below (often inside volva) with a white, 2-3 mm broad, floccose limb, exannulate, hollow. Volva fragile, easily disruptable, generally whitish at base and greyish above, inside also greyish above but paler than outside, the grey colour becoming deeper when dry, sometimes with a thin outer, orange brown layer. Flesh white, pale brown under the epicutis, rather soft, without special smell or taste. Spore-print white.

Spores globose to subglobose (Q:1.0-1.1), 9.6-12.8 µm in diam., hyaline, often with one big "oil"-droplet, nonamyloid. Basidia clavate, 60-70 x 15-18 µm, 4-spored with granulose contents. Sphaerocysts from gill-edge pyriform, subglobose to globose, 15-40 µm. Sphaerocysts from outer surface of volva numerous, 20-75 µm in diam., thin-walled, globose to subglobose.

Its habit and habitat are small groups or solitary under *Salix glauca*, *S. herbacea*, *S. arctophila*, *Betula glandulosa*, *B. nana* and *B. pubescens*. The localities range from moist shrubs to dry and exposed heath and snow-beds with *Salix herbacea*. It has been found from mid-July to mid-September but most common in August. *Amanita groelandica* is the most common *Amanita* in Greenland and one of the most common and spectacular Greenland macrofungi, found from the southernmost parts north of Melville Bay (75°21'N). It has not been observed in the most continental areas of the westcoast, e.g. in the Søndre Strømfjord area (Lange 1955, as *A. vaginata*).

Material examined: T. Bernth 82.603; TB 78.6; 79.61; 80.6; 80.104; 81.76; 81.134; 81.202; 82.29; 82.46; 82.61; 82.89; 84.72; H. Dissing 81.86; J. Feilberg 3001a; 3001b; HK, TB & JP 469; 623; HK & TL 114; 574 (type); 576; ML 509; 67.103; 67.117; 67.171; 67.256; 67.257; 67.555; ML & PMP 71.2; 71.7; F. R. Petersen & M. Strandberg 112; 119; 149; PMP 64.32; 64.45; 70.105; 71.5; 28.VIII.1972 (2 coll.); 72.51; 73.158; 73.240; 73.346; 73.459; 73.581; 77.1.1; 77.1.3; 79.150; Kolderup Rosenvinge 17.VI.1888; J. Rosing 17. VIII.1980; F. Terkelsen 55.21; 55.82; 55.147.

A. groenlandica has until now passed as a form of *A. vaginata* in a broad sense, but it is more related to *A. ceciliae* (Berk. & Br.) Bas and *A. submembranacea* (Bon) Gröger. These three species constitute a group of closely related species in sect. *Vaginatae*. They are large and robust, the dominating colours are grey and brown, their volva is whitish to more or less grey and easily disrupted into patches that remain on the cap, while the volva at the base usually soon disintegrates, the stipe is floccose girdled and pale grey or pale brown and discolours more or less when bruised, the edge of the lamellae is concolorous with the pileus.

Characters distinguishing *A. ceciliae* (=*A. strangulata* Fr., *A. inaurata* Secr.) are the strongly sulcate margin, the extremely fragile, easily disappearing volva and its habitat being fertile, deciduous forests. *A. submembranacea* becomes only slightly discoloured when bruised,the cap is uniform dark grey, the margin is strongly sulcate, the volva more persistent and the habitat is acid coniferous forests. *A. groenlandica* strongly resembles *A. submembranacea*, but the margin is only slightly sulcate, the cap colour ranges from grey to brown and often mixed and it grows with *Salix* or *Betula* in subarctic-arctic areas. We admit that the separation of these last two taxa into independent species seems hazardous, but judging from the material we have seen (hundreds of collections of *A. groenlandica*, ten of *A. submembranacea*), we consider these characters as constant. No microscopical differences between the three species have been observed.

Bas (1977) recognized four types of volvas in sect. *Vaginatae*, on which we agree. The presence/absence of volval patches on the cap is determined by the composition of the volva, as well as the weather. The larger the number of sphaerocysts in the volva, the easier the volva breaks into fragments, and since the pileus is slightly viscid when young, they will adhere to it. Therefore, we regard this character as useful in the characterization of the species of sect. *Vaginatae*, not forgetting that unfavourable (dry, windy) weather during the breaking of the volva may cause it to adhere to the pileal surface even in species with a fibrillose elastic volva. Thus it may appear as numerous small polygonal scales in species with a single or few patches as in *A. nivalis*.

The species reported as *A. inaurata* Secr. from the North Slope tundra in Alaska with *Salix* and *Betula nana* by Miller, Laursen and Farr (1982) is probably *A. groenlandica*.

Fig. 3. *Amanita mortenii*. HK, TB & JP 150 (typus). Photo Jens H. Petersen

A. mortenii Knudsen & Borgen sp. nov.

In sectionem *Vaginariam* referenda. Pileus flavidus vel flavido-fulvus cinereo vel olivaceo affectus, umbone obscuriore, 2.5-5.5 cm diam., margine sulcatus. Margines lamellarum albo-flocculosi. Stipes albide pruinosa vel tomentoso-pruinosa, 7-11 cm altus, 0.7-1.1 cm crassus, non zonatus. Volva subcinerea. Sporae globulares vel subglobulares, 9.6-12.8 µm diam. Habitat cum *Betula pubescenti* et *B. glandulosa*. Holotypus die 29 Julii anni 1983 juxta arbustum Rosenvinges Plantage prope vicum groenlandicum Narssarssuaq sub numero 150 a H. Knudsen, T. Borgen, J. Petersen lectus, siccus in Museo Botanico Hauniensi (C) depositus.

Pileus at first campanulate then expanded to convex and more of less umbonate, 2.5-5.5 cm, outer part of pileus yellowish (4A4-5) or yellowish brown, at centre darker brown (5D5-6), often with greyish or olivaceous tint, young pileus whitish at extreme margin; margin strongly sulcate, 0.25-0.4 R. Lamellae free, whitish or very pale buff with concolorous, flocculose margin. Stipe cylindric to slightly enlarged at base, 7-11 x 0.7-1.1 cm, on upper half whitish pruinose but with or without indistinct girdles, whitish or very pale brown to very pale greyish brown; exannulate; hollow. Volva in the lower part appressed to the stipe, the upper part free and saccate, 3-4.5 cm high, with internal, whitish, floccose limb, whitish, below greyish (4B2) above, whitish inside; one collection with remarkable orange brown, thin outer layer overlying the grey volva. Flesh whitish, without special smell or taste. Spore print not seen.

Spores globose to subglobose (Q:1.0-1.1), 9.6-12.8 µm in diam., hyaline, often with one big "oil"-droplet, nonamyloid. Basidia clavate, 60-70 x 15-18 µm, 4-spored with granulose contents. Sphaerocysts from gill edge pyriform to subglobose, 15-30 x 10-25 µm. Sphaerocysts from outer surface of volva numerous, 25-50 µm in diam., often rather thick-walled globose, in the outer orange layer also with clavate end cells.

It has been found in open shrubs with *Betula pubescens* and/or *B. glandulosa* in rather dry soil covered with mosses and lichens (*Stereocaulon* spp., *Cladonia* spp.) around Pamiut (Frederikshåb), Grønnedal, at Rosenvinges Plantation near Narssarssuaq and in Qinqua Valley.

Material examined: HK, TB & JP 150 (type); 151; 462; PMP 73.319; TB 81.119; 84.172; 85.133.

A. mortenii has the same size and stature as *A. fulva*, but the former is more yellowish and olivaceous and the volva is grey, overlayered however, by a thin orange-brown layer in one collection.

The species is named in honour of Professor Morten Lange who made the first flora of Greenland's macrofungi.

A. muscaria (L.: Fr.) Hooker

Amanita muscaria has only been collected once, by Kolderup Rosenvinge in 1888 from Sinigtok near Julianehåb (Rostrup 1891). It is probably very rare in Greenland, since several collectors have visited some of the most favourable localities without observing it. *Amanita muscaria* has been reported much farther north on Iceland (Hallgrimsson 1979), Finland (Kallio and Kankainen 1964) and Norway (Lange and Skifte 1967).

A. nivalis Grev.*

For synonymy see Bas (1982). For a modern description based on Scottish material see Watling (1985).

Pileus hemispheric to conical, expanding to convex, sometimes with a low umbo, pale grey, greyish buff, ochraceous grey, pale buff to almost pure white usually paler with age and when exposed, often slightly darker at centre, (2-)3-5(-7) cm, with sulcate margin (0.2 R), when young slightly viscid, then dry and shiny, sometimes with one or two felted volval patches, rarely with numerous small volval warts which can be pure white, whitish or with an ochraceous tinge. Lamellae free, white then very pale cream, distally ventricose, edge whitish to pale greyish flocculose. Stipe cylindric or slightly enlarged towards base, sometimes enlarged and compressed at apex, 4-8 x 0.5-1.5 cm, white, whitish, or with a faint greyish or pinkish ground colour, with a white pruinose-floccose covering sometimes seen as indistinct longitudinal stripes, not or rarely separted into girdles, at apex often finely sulcate, at base with a fine floccose limb; exannulate, hollow. Volva white or whitish, often tinted buff of ochraceous particularly above or pale greyish at the base, or both, also sometimes spotted rust-colour at the base, inside white or very pale greyish, 2-3(-4) cm high, above with small but distinct lobes. Flesh soft, white without special smell or taste. Spore print white.

Spores globose to subglobose (Q:1.0-1.1), 9.6-12.8 µm, hyaline, often with one big "oil"-droplet, nonamyloid. Basidia clavate, 60-70 x 15-18 µm, 4-spored with granulose contents. Sphaerocysts from gill-edge globose, 15-40 µm in diam., hyaline. Surface of volva consisting of 4-6(-10) µm broad, hyaline hyphae with parallel walls and not inflated at the septae and scattered ovoid to subglobose sphaerocysts, 20-50 µm diam.

This species occurs solitary or in small groups with *Salix glauca*, *S. herbacea*, *Betula glandulosa*, *B. pubescens* and a single collection with *Polygonum viviparum*. It has been found in a number of localities from *Betula pubescens* shrubs to open heath-like vegetation with *Betula glandulosa* and *Salix glauca*, to snow-beds with *Salix herbacea*. In Greenland *A. nivalis* is found rather common from the southernmost parts to Godhavn (69°15'), from the end of July to the end of August.

Material examined: TB 78.6 p.p.; 79.58; 82.28; 82.86; 82.91; 84.30; 84.86; 84.91; 84.99; 84.101; HK, TB & JP 108; ML 67.02; F. R. Petersen & M. Strandberg 131.

Amanita nivalis was excellently described and depicted as early as 1822 by Greville from the Scottish mountains, but nevertheless, until recently it has often been confused with other white species or white forms of other species. The distinguishing characters are the small, pale, thin-fleshed pileus with a strongly sulcate margin (compared to *A. groenlandica*), the lobed white or buff, rather tough and persistent low volva, the non-girdled stipe, the relatively few sphaerocysts in the volva (compared to *A. groenlandica*, *A. vaginata*) and the arctic-alpine habitat with *Salix* and *Betula* (compared to *A. vaginata*). It is probably widespread in arctic-alpine regions but due to its strong resemblance to other species its distribution is still insufficiently known. Besides Scotland, from where it was described, it has been reported from Switzerland (Favre 1955, as *A. vaginata* f. *oreina* Favre according to Bas 1982), from Lapland (Kühner, 1972, part of *A. hyperborea* ss. Kühner), from Härjedalen in Sweden (Jakobsson 1984) and from Iceland (Hallgrimsson 1979).

Fig. 4. *Amanita nivalis*. HK, TB & JP 181. Photo Jens H. Petersen

BOLETACEAE Chevalier

Boletus subtomentosus Fr.

First reported from Greenland by Watling (1977), it was found rather commonly on our expeditions to south Greenland and recorded north to Pamiut (Frederikshåb) where it is rare (TB). *Boletus subtomentosus* prefers dry localities like open shrubs with *Betula pubescens*, *B. glandulosa* and *Salix glauca*, often with *Empetrum*, *Cladonia* and *Stereocaulon* as botton layer. It often occurs solitarily.

This is the only true *Boletus* in Greenland. It grows much further north both in Scandinavia (Kallio and Kankainen 1964, Lange and Skifte 1967) and in Iceland (Hallgrimsson 1979).

Material examined: H. Dissing 81.156; HK, TB & JP 514; HK & TL 265; ML 14.VIII.1971; TL & S. Elborne 75; PMP 73.312; TB 79.104; 82.79; 84.45.

Leccinum is one of the most important mycorrhiza-forming genera in Greenland judging from the number of carpophores observed. It is a large and difficult genus, and although efforts were made to collect and describe it extensively, the material turned out to be insufficient for a detailed treatment of the whole genus in Greenland. This was especially true for the complex around *L. scabrum*. We have therefore restricted the present treatment to easily identifiable species, as we wait for more material.

Leccinum atrostipitatum Smith & Thiers*

A close relative of the common European *L. versipelle* differing from it among other things by the beautiful orange cap (6A6) and the pale instead of greyish tubes of young specimens and possibly in some microscopical characters as well. A full description will be given when the genus has been more carefully studied.

L. atrostipitatum is common to rather common in southern Greenland found north of 62°. The habitat is shrubs with *Betula pubescens* and dry heath-like vegetation with *B. glandulosa*, solitary or in small groups.

The very firm flesh makes it and excellent, edible fungus after cooking.

Material examined: TB 78.68; 79.101; 81.118; 82.75; 82.105; N. Jakobsen 1788; HK, TB & JP 152; 639; HK & TL 262; 648; ML 596; TL & S. Elborne 38.

L. salicola Watling

Leccinum salicola was reported recently from northeastern Greenland by Watling (1983). It belongs to sect. *Leccinum* with overhanging epicutis, darkening flesh and orange to brick-coloured pileus. Only a few species within this group have been reported from arctic-alpine areas. *L. versipelle* and *L. atrostipitatum* are much larger and more robust, the flesh darkens and the stipe scales are black from the beginning. *L. arctoi* Vasilkov (1978) from eastern Arctic Siberia is described as having a "sanguineus, rubro-aurantiacus" pileus with *Arctoi alpinae* (L.) Niederzu as its mycorrhizal host, and *L. arenicola* Redhead and Watling (1979) recorded from New Brunswick in Canada has initially yellow tubes and grows with *Hudsonia tomentosa* Nutt.

A collection from Jamesonland by D. Boertmann is likewise most probably *L. salicola*, but the spores are more slender than those described by Watling.

Material examined: A. Erskine, Washburn Hut near Mestersvig, July 1979 (Mat. in E); D. Boertmann 82010.

L. rotundifoliae (Sing.) Smith, Thiers & Watling

Syn.: *L. scabrum* (Fr.) S.F. Gray ssp. *tundrae* Kallio. Reported by Lange (1957), by Watling (1977) and by Lamoure, Lange & Petersen (1982).

L. rotundifoliae is a small, firm, pale *Leccinum* with an often finely cracked pileus which grows with dwarf Betulas such as *B. nana* and *B. glandulosa*. This concept has been used by Smith and Thiers (1968) and by Lundell and Nannfeldt (in Fungi exsiccati Suecici no. 2615), while Kallio (1975) described the pileus as having a "brown, often greenish" pileus with a soft consistancy. In Singer's (1938) diagnosis, he describes the pilues as "pallide ochraceo-grisello-brunneo saepe rimoso" and later (in the key) "Kleiner, hellgefarbter Pilz". A study of Kallio's type of *L. scabrum* ssp. *tundrae* revealed no differences between this and the widely accepted concept of *L. rotundifoliae* and we, therefore consider them as synonyms.

L. rotundifoliae is rather common in Greenland, on the westcoast found north to Godhavn. On the eastcoast where collecting has been done in a much smaller scale than on the westcoast it is known from one collection from Jamesonland (71°03'N). The habitat is mostly moist boggy ground with *B. glandulosa* and *B. nana*, but it is also found in dry heath-like vegetation. From observations in the Pamiut area it seems to prefer a humid climate, occurring more abundantly near the coast than inland where it has been found up to 500 *L. rotundifoliae* has been recorded from the whole circumpolar area (Kobayasi 1967, Pomerleau 1980, Smith and Thiers 1968, Watling 1977, Gulden and Lange 1971, Ryman 1984, Kallio 1975 (s.n. *L. scabrum* ssp. *tundrae*), Vasilkov 1956). Furthermore, it has been recorded (originally) from the Altai-mountains in central Asia (Singer 1938) and from the central part of European USSR (Vasilkov 1956). It is apparently still unknown from the Alps, despite the presence of the hosts.

Material examined: T. Bernth 82.410; 82.416; D. Boertmann 82.009; TB 78-23; 80.29; 84.102; 84.193; P. Kallio (TUR-53678, type of subsp. *tundrae*); HK &TL 286; ML 356; 67.320; PMP 63.199; 64.10; 77.3; PMP & ML 71.62.

L. holopus (Rostk.) Watling var. *americanum* Smith & Thiers*

A single collection from a bog in Qinqua Valley amongst *Sphagnum* with *Salix arctophila* and *Betula glandulosa* nearby was made. It is referred to var. *americanum* due to the distinct reddening of the stipe apex when cut.

Material examined: HK, TB & JP 375.

L. variicolor Watling

This species was reported by Watling (1977) based on collections in C by P. M. Petersen. Recent collections by the authors have revealed the possible occurrence of a closely related species in Greenland which is more common than *L. variicolor*. Among the available material, we have only seen one typical collection of *L. variicolor*. Until further studies show whether they should be treated as good species or extreme forms, we restrict the occurrence of *L. variicolor* in Greenland to collection from Qinqua Valley, growing in a bog with *Betula glandulosa* and *Salix glauca*.

Material examined: HK, TB & JP 376.

Suillus brevipes (Peck) Kuntze*

Suills brevipes was found with planted *Pinus contorta* (1 m tall) near the Station Officer's house in Narssarssuaq in 1983 and also in 1984 when none of the other *Suillus*-species were found. The natural distribution is North America where it is reported as common in California (Thiers 1975), Michigan (Smith & Thiers 1971) and Québec (Pomerleau 1980), growing with 2- and 3-needled pines. The pines were grown from Alaskan seeds in Upernaviarssuk and transplanted as seedlings to Narssarssuaq. Material examined: HK, TB & JP 134; 273; HK &TL 567.

S. cf. caerulescens Smith & Thiers *

We have so far been unable to identify a collection of a *Suillus* sp. found in the plantation in Tasermiut Fjord growing under *Larix sibirica* with *S. grevillei*. The distinctve features are the tawny, red brown viscid and smooth pileus, the non-glandulose stipe with a distinct veil that adheres to the pileus margin leaving almost no trace on the stipe. The stipe is concolorous with the pileus but more mottled and the tubes are bright yellow (only young specimens seen). The flesh is yellow but becomes greenish (slightly blueing) at the lower part of the stipe. It keys out in Sect. *Boletinus* (Kalchb.) Smith & Thiers and later in the group with a smooth pileus. The closest related species seems to be *S. caerulescens*, but it is said to have "patches of agglutinated fibrillose tomentum" and it is "not associated with larch" although it may grow with *Larix*.

Larix sibirica was planted as seedlings in Upernaviarssuk from seeds collected in the western part of the Ural mountains.

Material examined: HK, TB & JP 302.

S. glandulosus (Peck) Singer* (*Fuscoboletinus glandulosus* (Peck) Pomerl. & Smith)

This fungus was found growing under *Picea glauca* in the plantation in Tasermiut Fjord. It is distributed in the U.S.A. and Canada (Quebec), the area closest to Greenland (Pomerleau 1980). The *Picea glauca* hosts were collected as seeds near Knik River of Anchorage, Alaska and raised at Upernaviarssuk near Julianehåb.

Material examined: HK, TB & JP 199.

S. grevellei (Klotzsch) Singer*

Suillus grevillei was found abundantly under planted *Larix sibirica* in Narssarssuaq, at Rosenvinge's Plantation near Narssarssuaq and in the plantations at Kugssuak in Tasermiut Fjord. Although abundant in 1983, it was not observed in the dry season at 1984. In 1985 it was also recorded from Grønnedal. The larches were raised in Upernaviarssuk from seeds collected in the western Ural mountains.

Material examined: TB 85.86; HK, TB & JP 303, 604.

GOMPHIDIACEAE Maire

Gomphidius sepentrionalis Singer*

A single collection was found under *Picea* and *Larix* at the plantation in Kugssuak in Tasermiut Fjord. *G. septentrionalis* does not occur naturally in Greenland but was introduced with the trees in the plantation. According to Singer (1949), it is found in New Brunswick and Nova Scotia.

PAXILLACEAE Maire

Paxillus filamentosus Fr.*

This species has only been recorded once, by T. Borgen on a rotten branch of *Alnus crispa* at Grønnedal. It is widespread under *Alnus* ssp. in Europe, and has been recorded to the northernmost point of Scandinavia, growing with *A. incana* (Ryman 1984).

Material examined: TB 85.169.

P. involutus (Batsch) Fr.

Paxillus involutus was first reported by Lange (1957) from the Ivigtut area. Found rather commonly by the authors in south Greenland, *P. involutus* is rare in the Pamiut-area (Frederikshåb) (TB), and has not been observed in the well investigated area around Godhavn. It was recorded recently from Iceland, somewhat further north (Hallgrimsson 1973).

The habitat is usually rather dry, heath-like vegetation with *Betula glandulosa* and *B. pubescens*. The carpohores are generally smaller than specimens from temperate areas.

Material examined: TB 79.97; 80.69; 81.123; HK, TB & JP 26; ML 520.

PLUTEACEAE Kotlaba & Pouzar

Pluteus atricapillus (Batsch) Fayod*

Pluteus atricapillus is not uncommon in the two areas visited in 1983 and 1984, occurring on old stumps and trunks of *Betula pubescens*, but, also found in *Alnus*-shrubs north of Grønnedal. Most probably restricted to favourable localites in the southwest, where *B. pubescens* attains a sufficient size to host this species. *P. atricapillus* reported recently from two localities on Iceland by Hallgrimsson (1979), from northern Scandinavia where the northern limit is far north of the Greenland limit (Kallio and Kankainen 1964, Lange and Skifte 1967) and from subalpine areas in Jotunheimen (Gulden & Lange 1971).

Material examined: TB 85.150; HK, TB & JP 13; 387; 512; HK & TL 217; 240; 520; 597.

P. romellii (Britz.)Sacc.*

Only three collections (3 specimens) of this species, from the two expeditions, in both areas growing on trunks or branches of *Betula pubescens* were made. It is not reported from northern Scandinavia or Iceland.

Material examined: HK, TB & JP 315; HK & TK 465; 571.

CONCLUSIONS

Of the 23 species here reported from Greenland (the aliens not included), *Amanita arctica, A. groenlandica, A. mortenii* and *A. nivalis* seem to be restricted to subarctic/subalpine areas. *Agaricus salicophilus* probably also belongs here, but since it is only known from the type-locality conclusions will be premature. *Leccinum rotundifoliae, L. atrostipitatum, L. salicola* and *Amanita battarrae* are boreal/mountanous species and the remaining 15 species (*Agaricus arvensis, A. campestris, A. fissuratus, A. semotus, Amanita fulva, A. muscaria, A. friabilis, Boletus subtomentosus, Leccinum holopus, L. variicolor, Lepiota clypeolaria, Paxillus filamentosus, P. involutus, Pluteus atricapillus, P. romellii*) are here considered to have their main distribution in temperate or warmer areas. The genera *Agaricus, Lepiota, Boletus, Paxillus* and *Pluteus*, are, in Greenland, mostly poorly represented as diluted versions of the sizes normal for these genera in warmer climates. *Amanita* and *Leccinum* are represented both with an element from warmer areas but also with a special subarctic/arctic element, although the *Leccinum* species so far are unnamed. In *Amanita*, 6 of the 8 species belong to sect. *Vaginatae*.

When the flora is compared to that of other areas at the same latitude, it is found that the same species occur over large areas in the whole subarctic/subalpine zone. It is also found that the northern limit of some of the species in Greenland is generally more southern than in Iceland and Scandinavia and that the number of species in Greenland generally is lower than in neigbouring areas. This could be due to the isolated position of this island and a more unfavourable climate. Species found in the neighbouring areas and belonging to the families treated here but (still) not in Greenland are, e.g. *Chalciporus piperatus, Boletus edulis, Amanita crocea* and partly *A. muscaria* (found once).

ACKNOWLEDGEMENTS

We gratefully acknowledge the financial support from the Danish Agricultural and Veterinary Research Council and M. P. Christiansen's Fund, Dr. C. Bas (Leiden) and Dr. R. Watling (Edinburgh) for valuable discussions and loans, the curators of H and TUR for loans, Mr. F. Rasmussen (Narssarssuaq) for logistical support, Dr. T. Christensen for preparing the Latin diagnoses, Dr. M. Sasa for linguistical improvements and Mr. O. Lansø and Mr. H. Gøtzsche for technical help.

REFERENCES

Bas, C., 1974, A rare but widespread *Amanita* associated with *Alnus*, *Trav. Myc. ded. R. Kühner, Bull. Soc. Linn. Lyon*, 1974: 17-23.

Bas, C., 1977, Species-concept in *Amanita* sect. *Vaginatae*, *in:* "The species concept in Hymenomycetes," H. Clémençon, ed., Vaduz, 444 pp.

Bas, C., 1982, Studies in *Amanita* - II., *Persoonia*, 11: 429-442.

Cappelli, A., 1984, "*Agaricus* L.: Fr.", Biella Giovanno, Saronno, 558 pp.

Cetto, B., 1980, "I funghi dal vero. I. Trento," 680 pp.

Dissing, H., 1982, Operculate discomycetes (Pezizales) from Greenland, *in:* "Arctic and alpine mycology," G. A. Laursen and J. F. Ammirati, eds., Univ. Washington, Seattle and London, pp. 56-81.

Favre, J., 1955, Les champignons superieurs de la zone alpine du Parc National Suisse, *Ergebn. wissens. Unters. schweiz.* National parks II 5.

Feilberg, J., 1984, A phytogeographical study of South Greenland. Vascular plants, *Medd. om Grønland, Bio-science* 15: 1-70.

Gilbert, E. J., 1941, Amanitaceae vol. II., *J. Bresadola: Iconographia Mycologica 27, suppl. I.* Mediolani, 203-427.
Greville, R. K., 1822, "Scottish cryptogamic flora. I.", Edinburgh, 60 Pl.
Gulden, G., and Lange, M., 1971, Studies in the Macromycete flora of Jotunheimen, the central mountain massif of south Norway, *Norw. J. Bot.*, 18(1): 1-46.
Hallgrimsson, H., 1972, Islenzkir hattsveppir I og II., *Act. Bot. Islandica,* 1: 73-113.
Hallgrimsson, H., 1973, Islenzkir hattsveppir III. Lepiotaceae, (Agaricaceae pro parte), Gomphidiaceae, Paxillaceae, Crepidotaceae, *Act. Bot. Islandica,* 2: 29-55.
Hallgrimsson, H., 1979, "Sveppakverid. Reykjavik, Gardyrkjufelag Islands," 158 pp.
Heim, R., 1929, Atlas pl. 29., *Bull. Soc. Myc. France* 45.
Jacobsson, S. 1984, Notes on the agarics in subalpine and alpine areas of western Härjadalen, Central Sweden, *Windahlia,* 14: 43-64.
Kallio, P., 1975, *Leccinum scabrum* (Fr.) S. F. Gray subsp. *tundrae* Kallio, a new subspecies from Lapland, *Rep. Kevo Subarctic Stat.* 12: 25-27.
Kallio, P., and Kankainen, E., 1964, Notes on the Macromycetes of Finnish Lapland and adjacent Finmark, *Ann. Univ. Turku A II 32., Rep. Kevo Subarctic Res. Stat.* 1: 178-235.
Knudsen, H., and Borgen, T., 1982, Russulaceae in Greenland, *in:* "Arctic and alpine mycology," G. A. Laursen and J. F. Ammirati, eds., Univ. Washington, Seattle and London, pp. 216-238.
Kobayasi, Y., 1967, Mycological studies of the Alaskan Arctic, *Ann. Rep. Ferm., Osaka,* 63: 1-138.
Kornerup, A., and Wanscher, J. H., 1974, "Farver i farver," Kobenhavn, 248 pp.
Kühner, R., 1972, Agaricales de la zone alpine. Amanitacées, *Ann. Sci. l'Univ. Besançon 3e Série, Botanique,* 12: 31-38.
Lamoure, D., Lange, M., and Petersen, P. Milan, 1982, Agaricales found in the Godhavn area, W Greenland, *Nord. Jour. Bot.,* 2: 85-90.
Lange, M., 1955, Macromycetes 2. Greenland Agaricles, *Medd. om Grønland,* 147(11): 1-70.
Lange, M., 1957, Macromycetes 3. I. Greenland Agaricales (pars) Macromycetes caeteri; II Ecological and plant geographical studies, *Medd. om Grønland,* 148(2): 1-125.
Lange, M., 1980, *Stropharia alpina* in Greenland, Iceland and the Faeroes, *Bot. Tidsskr.,* 75: 89-91.
Lange, M., and Skifte, O., 1967, Notes of the macromycetes of northern Norway, *Acta Bor. A,* 23: 1-51.
Miller, O. K., Laursen G. A., and Farr, D. F., 1982, Notes on Agaricales from arctic tundra in Alaska, *Mycologia,* 74: 576-591.
Noordeloos, M. E., 1984, Entolomataceae (Agaricales, Basidiomycetes) in Greenland - I. The genus *Entoloma, Persoonia,* 12: 263-306.
Petersen, P. M., and Korf, R. P., 1982, Some inoperculate Discomycetes and Plectomycetes from West Greenland, *Nord. J. Bot.,* 2: 151-154.
Pomerleau, R., 1980, "Flore des champignons au Quebec," Les editions la presse, Ottawa, 652 pp.
Redhead, S. A., and Watling, R., 1979, A new psammophilic *Leccinum, Can. J. Bot.,* 57(2): 117-119.
Romagnesi, H., 1961, "Nouvel atlas de champignons vol. 3.," Bordas, 236 Pl.
Rostrup, E., 1888, Fungi Grønlandiae. Oversigt over Grønlands Svampe, *Medd. om Grønland,* 3: 517-590.
Rostrup, E., 1891, Tillaeg til "Grønlands svampe (1888)," *Medd. om Grønland,* 3: 591-643.
Ryman, S., and Holmasen, I., 1984. "Svampar," Stockholm, 718 pp.
Secretan, L., 1833, "Mycographie suisse. I.", Geneve, 522 pp.
Singer, R., 1938, Über Lärchen-, Zirben und Birkenröhrlinge, *Schw. Zeit. f. Pilzk.,* 16(8): 123-126; 16(9): 134-137; 16(10): 148-150.

Singer, R., 1949, The genus *Gomphidius* Fries in North America, *Mycologia*, 41: 462-489.
Smith, A. H., and Thiers, H. D., 1964, "A contribution toward a monograph of North American species of *Suillus*", Ann Arbor, 116 pp.
Smith, A. H., and Thiers, H. D., 1971, "The Boletes of Michigan," Univ. Michigan Press, Ann Arbor, 428 pp.
Thiers, H. D., 1975, "California mushrooms," Hafner Press, New York, 261 pp.
Vasilkov, B. P., 1956, Berezovik - *Krombholzia scabra* (Fr.) Karst. - v SSSR., *Trudy botaniceskogo instituta im. V. L. Komarova Akademii Nauk SSSR, ser. II*, 10: 367-384.
Vasilkov, B. P., 1974, Species generis *Agaricus* in regione Arctica USSR, *Nov. Sist. Niz. Rastenij*, 11: 169-173.
Vasilkov, B. P., 1978, Species nova generis *Leccinum* S. F. Gray in parte Arctica orientis extremi, *Nov. Sist. Niz. Rastenij*, 15: 84-85.
Watling, R., 1971, A new British bolete, *Not. Roy. Bot. Gard.*, 31: 139-142.
Watling, R., 1977, Larger fungi from Greenland, *Astarte*, 10: 61-71.
Watling, R., 1983, Larger cold-climate fungi, *Sydowia*, 36: 308-325.
Watling, R., 1985, Observations of *Amanita nivalis* Greville, *Agarica*, 6: 327-335.

AGARICALES DE LA ZONE ALPINE
GENUS *CORTINARIUS* FR., SUBGENUS *TELAMONIA* (FR.) LOUD
PART III

D. Lamoure

Université Claude Bernard, Lyon I
Laboratoire de Mycologie
43 bd du 11 Novembre 1918
F 69622 Villeurbanne Cedex France

Key words: *Cortinarius*, *Telamonia*, Basidiomycetes, Agaricales, Alpine zone

ABSTRACT

Small dark coloured *Telamonia* with the smell of *Pelargonium* were collected in either the alpine zone of the Alps or in the Scandinavian mountains, and are described. One species, *Cortinarius stenospermus* sp. nov. is calciphilous or at least growing on neutro-basophilous soils together with *Dryas octopetala* and *Salix reticulata*. One variety: *Cortinarius paleiferus* Svrcek var. *brachyspermus* var. nov., is acidiphilous, and was collected in *Salix herbacea* snow-beds.

RÉSUMÉ

Description de petits *Telamonia* sombres à odeur acidule-pélargoniée, récoltés en zone alpine, dans l'arc alpin ou dans les montagnes scandinaves. Une espèce: *Cortinarius stenospermus* sp. nov., calciphile ou du moins préférant les sols neutro-basophiles, poussant parmi *Dryas octopetala* et *Salix reticulata* et une variété: *C. paleiferus* Svrcek var. *brachyspermus* var. nov., acidiphile, récoltée dans les combes à neige parmi *S. herbacea*, sont présentées comme nouvelles.

INTRODUCTION

This study of three dark-colored small *Telamonia* is a continuation of other papers on the same subject (Lamoure 1977, 1978).

TAXONOMIC PORTION

Cortinarius stenospermus sp. nov.

DIAGNOSIS: *Pileo 9-16 mm lato, obtuse convexo, haud mammoso, udo obscurius brunneo, oculo nudo glabro, aetate lucido sericeo, praeter marginem quae servat vestigia veli subtiliter araneosi et adpressi, et deinde glabrescit atque etiam transluciditate striatus fieri potest; carne tenui, praeter sub discum, obscure brunnea, sicca pallescent. Stipite 28-34 x 2,5-3 mm, aequali vel paulum imo clavato, flexuoso, obscurius brunneo, sub lamellis pallidiore; ac nudo, in media vel tertia superiore parte anulo albido, gausapato, adpresso magis minusve integro cincto, atque infra flocculis albidis, gradatim dispositis, variegato. Lamellis mediocriter stipatis, ventricosis ascendentibus, obscurioribus, primum e badiis brunneis, ante maturitatem levi reflexu e violaceo caesio vix manifesto. Odore pelargonii. Sporis 10-11 X 4,5-5 µm, a fronte obovatis, verrucosis. Typus* in Herb. D. Lamoure n° L. 63-74.

DESCRIPTION: Pileus 9-16 mm, convex, subumbonate, uniformly rather dark brown when moist: Mu. 5YR 3/4, 4/4; glabrous to the naked eye, satiny-glossy except at the margin which has adpressed, silky, velar remnants; with age, the margin also becomes glabrous, and is often translucent striate for up to 1-2 mm; flesh thin except in the center, dark brown, hygrophanous, drying paler. Stipe 28-34 x 2,5-3 mm, even, or a bit inflated at the base (x 5 mm max.) thin and flexuous; a white cottony ring girding the stipe; just in the middle or on the upper two-thirds under this more or less entire ring, the stipe is ornamented with whitish velar remnants which are evenly disposed and stand out against the dark brown cortex: Mu. 7.5YR 7/2; above the ring, just under the lamellae, the cortex is paler: Mu. 7.5 YR 5/4, 5/6, 6/4, 6/6. Lamellae subdistant, ventricose, dark colored: chocolate-brown: Mu. 5YR 4/4, 6.5YR 6/4; but when young, before the spores are ripe, they are distinctly light bluish violet. Smell: of *Pelargonium*, when intact or when cut.

Spores: 10-11 x 4,5-5 µm, narrow, ellipsoid, slender near the apex, obovate in front view; verrucose. Basidia: 4-spored. Epicutis: narrow hyphae x 4-5 µm, cell-wall incrusted with brown pigment; hypodermis typically not pseudoparenchymatic, but hyphae broader: x 10-12 µm; the incrusting brown pigment present in both the hypodermis and context. Clamp connections everywhere in the basidiocarp.

HABITAT: This species has been collected in the low and in the middle alpine zone of the W-Alps, either in *Dryas*-carpets, or in a *Salicetum retusae-reticulatae*, with *Saxifraga aizoides*, under *Salix foetida*, i.e., with calciphilous Phanerogams.

COLLECTIONS: French Alps, Vanoise: Cirque du Dard, N. 2200 m, *Dryas*, *S. reticulata*, 28-8-63 (L. 63-74); Cirque du Génépy, NW. 2300 m, *S. reticulata*, *S. foetida*, 24-8-63 (L. 63-56).

DISCUSSION

The more striking characters of *C. stenospermus* distinguishing it from other species with a smell of *Pelargonium* is that the cap is never fibrillose-scaly but with age soon quite glabrous, and that the spores are realatively large and narrow. Within our experience in the field, it seems to be calciphilous.

Cortinarius aff. *paleiferus* Svrček

DESCRIPTION: Pileus 7-18(25) mm, convex, obtusely umbonate, uniformly deep dark brown when moist: Mu. 5YR 2/4, 3/4, hygrophanous, fading to 7.5YR 3/4, 3/6, 4/6 while drying; slightly fluffy to the naked eye, under a lens one can distinguish small brownish floccules, while the margin is covered with silky brownish velar remnants. Stipe 20-26(32) x 2-4 mm, flexuous; a rather thick but not membraneous pale brown velum covers the lower part of the stipe and is limited upwards by a rather irregular, not membraneous,

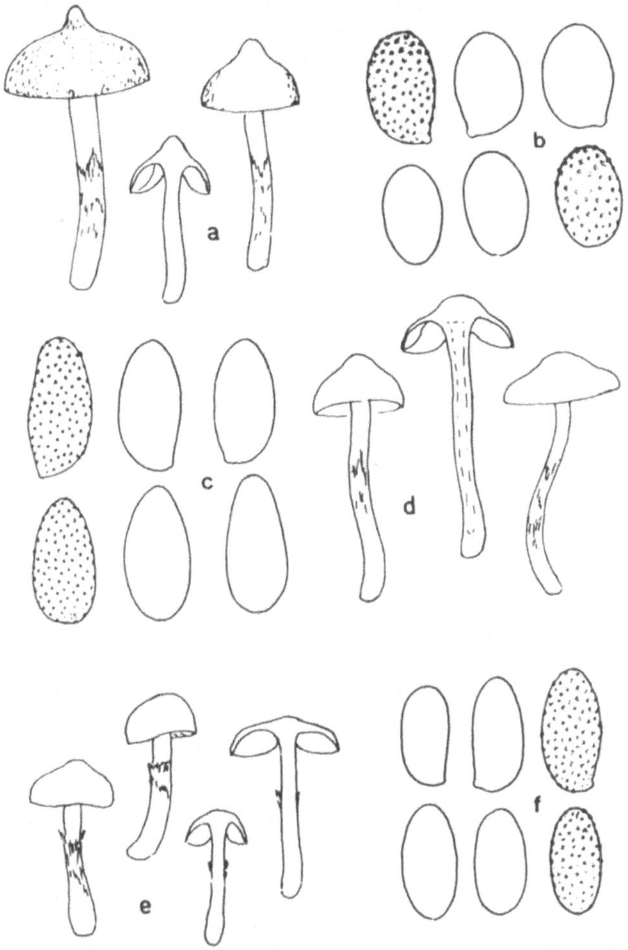

Fig. 1: a, b: *C. paleiferus* var. *brachyspermus* L. 72-64. c, d: *C. stenospermus* L. 63-74. e, f: *C.* aff. *paleiferus* L. 73-103. carpophores: x 0.8; spores: x 1650

ring-shaped zone; with age the ring-shaped zone and the velum break up, the pale brown remnants contrast sharply with the darker reddish-brown cortex. With age, the stipe becomes narrowly hollow. Basal mycelium bluish-violet tinted. Lamellae subdistant, ventricose, distinctly and undoubtful bluish-violet when young and remaining or subsisting so at the margin while the faces turn to dark chocolate-brown: about Mu. 6.5YR 4/4. Smell of *Pelargonium*, even from the outside (while the fruitbody is entire) and stronger after it has been cut in two halves.

Spores 8-9 x 4,5-5 µm, elliptical, not strongly but distinctly verrucose. Basidia 4-spored. No cheilocystidia. Anatomical studies have not been made on the fresh material. On exsiccata, one can see incrusting pigment on the walls of hyphae. Clamp connections everywhere in the basidiocarp.

HABITAT: We have collected this *Telamonia* only in the upper-alpine zone, in snow-beds.

COLLECTIONS: French Alps: Massif of L'Iseran: near the pass, NW.2750 m, between the rivlets "La Cema" and "Pisaillas", in *S. herbacea*, together with *Saxifraga oppositifolia*, 26-8-73 (L. 73-108). Scandinavian mountains: Lapland: Lullihatjarro, E. 760 m in *S. herbacea*, *Sibbaldia procumbens*, *Anthelia*, 18-8-72 (L. 72-83); Slattatjåkka, NE. 900 m in S. herbacea, 16-8-72 (L. 72-76) (basal mycelium bluish).

DISCUSSION

The collections described above agree with *C. paleiferus* by some main characters such as the smell, the darkness and the violet tint of the lamellae and violet basal mycelium; but we have never seen whitish small scales on the cap, only some pale brownish floccules. It is possible that the whitish ones become pale brown with age. Compare how we have described the pileus of other fresh collections named, *C. paleiferus* var. *brachyspermus* Lamoure, a variety differing mainly by broader spores.

Cortinarius paleiferus Svrček var. *brachyspermus* var. nov.

DIAGNOSIS: *A typo differt sporibus latioribus. Typus in Herb. D. Lamoure n° L. 72-64.*

DESCRIPTION:: Pileus 16-24 mm, convex, distinctly umbonate even when the margin spreads itself; strongly whitish floccose-scaly at first, then pale-brown with small tufts standing out from the dark brown background: Mu. 5YR 2/4, 3/4, fading only slightly when drying: to Mu. 7.5YR 4/4, 4/6; margin always pale: the remnants of the velum forming a rather thick tomentous silky zone, and in some cases, these remnants break up into whitish-pale brownish fragments. Stipe 18-30(43) x 3-5 mm, equal, trimmed with rather thick whitish girding bands after the telamonic velum, which is at first entire breaks up; the whitish contrast sharply with the dark brown or slightly reddish-brown cortex: Mu. 5YR 4/6, 5/4, paler to the base: about Mu. 10YR 5/4. Under the lamellae, the naked cortex appears pale under a silky adpressed fibrillum, and in some cases has a violet tint. Lamellae subdistant, rather dark chocolate-brown: Mu. 7.5YR 3/6, 3.5/6, 4/6 pale violet when young. Smell to *Pelargonium*, rather stark from the outside, and after having cut the cap in two halves.

Spores 6-9 x 6-6,5(7) um, so really short and broad, strongly ornate with warts rather protruding in profile, more or less flowing together in front view. Basidia 4-spored. No well-defined cheilocystidia, but here and there some sterile hairs mingle with basidia. Epicutis: hyphae 3-5 um, with brown-incrusted walls. Hypodermis: hyphae broader: x 15-20 um, with shorter cells: 30-40 um. Gill-trama: very dark hyphae with incrusted walls.

HABITAT: We have collected var. *brachyspermus* in the middle and upper alpine zones of Scandinavian mountains.

COLLECTIONS: Scandinavian mountains: Hardanger: Olavsbuhaii, SE. 1250 m, *S. herbacea*, 6-8-72 (L. 72-26). Lapland: Slattatjåkka, NE. 1100 m, *S. herbacea*, *Sibbaldia procumbens*, *Polytrichum norvegicum*, 14-8-72 (L. 72-64); Katterjaure, southern to the Lake, 700 m, *S. herbacea*, *S. lapponum*, *Rubus chamaemorus*, *Sphagnum*, 15-8-74 (L. 74-80) ibid., wet snow-bed with *S. herbacea*, 20-8-72 (L. 72-91).

DISCUSSION

At Vanoise in the French Alps, only one collection is referred to *C. paleiferus* var. *brachyspermus*, but as all the characters don't agree with those of the northern collections described above, we prefer put it aside for the moment. The fruit-bodies were frozen, and this fact may account for the fact that the smell was not exactly that one of *Pelargonium*, but more like *Melissa*, "*Citronella*" (deviant metabolism ?. Other data: concerning the ecology prompt caution: the fruit-bodies were not growing between *Salix herbacea*, but in a wet place irrigated with water melting from a glacier, under *Salix foetida*, in *S. retusa* + *S. reticulata* carpets, together with *Saxifraga aizoides*, i.e. with calciphilous Phanerogams. Another discordant character was the spore dimensions. The measurments were within the ranges for *C. paleiferus* var. *brachyspermus*, but the spores appear not to be so broad when drawn.

REFERENCES

Lamoure, D., 1977, Agaricales de la zone alpine. Genus *Cortinarius* Fr., subgenus *Telamonia* (Fr.) Loud. - lere partie, *Trav. sci. Parc Natl. Vanoise* 8: 115-146.

_________, 1978, Agaricales de la zone alpine. Genus *Cortinarius* Fr., subgenus *Telamonia* (Fr.) Loud. - 2eme partie, *Trav. sci. Parc Natl. Vanoise* 9: 77-101.

ARCTIC GASTEROMYCETES. THE GENUS *BOVISTA* IN GREENLAND AND SVALBARD

Morten Lange

Institut for Sporeplanter, University of Copenhagen
Ø. Farimagsgade 2 D, DK-1353 København K, Denmark

Key words: Gasteromycetes, Greenland, Svalbard

ABSTRACT

Divisions of the arctic zone and variations in climate and soil are briefly outlined. The genus *Bovista* is redefined by largely excluding sect. *Globaria*. *Bovista tomentosa* and *B. limosa* both reach the high arctic zone where they are confined to alkaline soils and subcontinental-continental climate. *Bovista nigrescens* is known from subarctic tracts where it grows in grassy pastures. Distribution of the species in other arctic, subarctic and alpine areas are indicated.

INTRODUCTION

The present study is the first in a series dealing with the occurrence and distribution of arctic Gasteromycetes, mainly based on material from Greenland and Svalbard.

Since the monographic treatment of Greenland gasteromycetes was published (M. Lange 1948), several authors have dealt with gasteromycetes from arctic and alpine tracts without, however, intending a more coherent presentation. In some groups there is a strong need for a clear delimitation of the taxa, and the distributional patterns of the species are not worked out. It is hoped that the materials now available for study will make it possible to contribute to these two needs.

CLIMATE AND SOIL

Climatic and edafic factors are the main factors determining the distributions of the species. Hence, it is necessary here to include a brief account of climate and soil conditions in Greenland and Svalbard. The chapter also shall define the terms used for describing the distributional patterns.

The limit of the arctic is generally drawn by the 10° July isotherm which is quite often the limit of forest vegetation. The adjacent part of the northern boreal zone may, however, also be without true forest, especially in oceanic regions. In line with the classification of alpine zones, marginal parts of the northern boreal zone are often termed subarctic. Greenland is in the arctic region, except for some inland stretches south of 62°N, on the South West Coast. These localities have a

Table 1. Main climatic characteristics of stations in Greenland and Svalbard.

For comparison the table also includes records from - mostly - subarctic Iceland, Faeroes, North Scandinavia, and from Barrow, Alaska. - The columns indicate: Number of months with positive average temp. - Average temp. of warmest month - Sum of average temps of + months - Difference between average temp of warmest and coldest month. - Last column gives average yearly precipitation. - Data compiled from various sources with different observation periods, hence not fully comparable.

	+ months	max T°C	+months T°C	max/min diff. °C	precip. mm
Greenland					
W. Coast					
Prins Christians Sund 60°10'N	6	6.2	22.9	12.5	2354
Julianehåb 60°45'N	7	8.2	33.6	12.3	871
Narssarssuaq 61°10'N	6.5	10.2	39.1	18.2	585
Godthåb 64°10'N	5	6.6	20.7	14.3	757
Sdr. Strømfjord 67°N	5	10.5	34.5	29.0	165
Godhavn 69°15'N	4	7.0	20.4	19.0	465
Jakobshavn 69°15'N	4	8.1	23.2	22.1	261
Upernavik 72°45'N	4	4.8	12.1	25.5	245
Dundas 76°30'N	3	3.9	9.0	27.9	129
E. Coast					
Angmagssalik 65°40'N	5	6.3	20.5	15.3	960
Mestersvig 72°15'N	3	6.0	13.2	28.0	-
Daneborg 72°45'N	2	0.6	1.2	27.6	-
Svalbard					
Longyear 78°15'N	4	6.5	14.0	21.5	240
Isfj. Radio 78°05'N	4	4.5	11.5	16.6	378
Iceland					
Vestmannaeyr S.Coast 63°25'N	12	10.2	63.8	9.5	1367
Stykkisholmur W.Coast 65°05'N	10	10.1	50.8	11.4	691
Reykjahlid N.Coast-inl. 65°40'N	6	10.2	40.8	14.3	394
Teigarhorn E.Coast 64°40'N	11	9.7	53.8	10.2	1163
Faroes					
Thorshavn 62°N	12	11.1	85.2	7.4	1434
Sandur 61°50'N	12	11.4	76.7	7.6	1181
North Scandinavia					
Tromsø W. Coast 69°40'N	7	11.9	48.2	15.9	1052
Karasjok Inland 69°30'N	6	17.1	55.1	38.0	374
Vardø N. Coast 70°15'N	6	9.2	36.1	14.4	497
Barrow, Alaska 71°20'N	3	4.3	9.0	31.0	105

woodslike climax vegetation of *Betula tortuosa* with high scrub of *Salix glauca* and some stands of *Alnus viridis*.

In the low arctic zone, extending to about 70-71°N, the climax vegetation is a scrub of *Salix glauca*, *Betula nana* or *B. glandulosa*. In the high arctic zone north of this limit, dwarfish *Salix* spp. are among the dominants. Svalbard is in the high arctic zone. The terminology used to designate these zones is not very well established while their definition is almost generally accepted. The terms used above are preferred here because of their accordance with corresponding terms for the alpine zonations to which there is a clear parallel, at least for an overall consideration.

The temperatures for some typical Greenland stations are given in Table 1. The total variation is naturally much broader than these figures show. There are great differences between coastal and inland stations, especially south of 66-67° N, and the inland stations show very striking microclimatic gradients. For the development of the fungi it is noteworthy that the northern stations may have a very long frost-free period, up to 70 days as far north as Brøndlund Fjord at 72°20°N. This is due to the midnight sun effect.

There are many large differences in the distribution of the annual precipitation (Tabl. 1). It is high in SW and SW Greenland coastal localities while even extremely low figures are recorded at the head of the deep fjords. Also, the entire high arctic zone 11 of 70-71° has very low precipitation. Relatively high on the coasts, Svalbard has a fairly high precipitation, compared with other stations at this latitude. The picture is further complicated by the duration of the snow cover. With low precipitation the snow cover is fairly thin and it may disappear in early spring from exposed localities, thus adding to the assimilation period for the green plants and probably also the growth period of the fungi. Protected and north exposed sites have long persisting snow beds.

Precipitation and temperature define the oceanic and continental climate types. Typical oceanic subarctic and low arctic stations have an annual precipitation of greater than 1000 mm and a difference between average temperature of warmest and coldest month of less than 12°. For typical continental stations in some zones the corresponding figures are less than 200 mm, and greater than 25°. The annual (or summer) precipitation value, which is the main factor when defining the oceanic zones, decreases with decreasing summer temperature, and corresponding lower rate of evaporation. From Table 1 intermediate figures characterize most Greenland stations which then range from suboceanic to subcontinental. Most high arctic inland stations have a continental climate. The Svalbard archipelago, however, is largely with a coastal, mildly subcontinental climate.

Soil characters play a similarly important role for the distributions of fungi. Gneissic rock is dominant in Greenland as it is in North Scandinavia. There also are gneissic zones in Svalbard. Basaltic rock, forming the brito-arctic shield forms a belt across Greenland from Scoresbysund to Nugssuaq, Disco and Svartenhuk. The same shield also forms the Faroes and E and NW Iceland. Smaller basaltic intrusions occur, for example, at Narssarsuaq in South Greenland. Various sediments, mostly alkaline, are found at Mestersvig in NE Greenland and also are typical for northernmost Greenland, and the Isfjord region in Svalbard from where the collections studied here originate. The typical soil, of course, mostly reflect the nature of the bedrock, although often modified by a fairly thick peat layer. The Sdr. Stromfjord area has a special status with aeolic, mostly basic loess sediments.

MATERIALS AND METHODS

The main source for the present study are collections made by the author from Greenland and Svalbard. Collections from several west coast stations in 1946, published in 1948, have been revised and are supplemented with new W. Coast collections from 1967, 1978, and 1982.

Also, revised are collections from Mestersvig, E. Greenland, 1975, (published in 1976). Several other botanists have gathered specimens of Gasteromycetes which are deposited in the Copenhagen Botanical Museum (C). They have been at my disposal and add further to the total covering of the vast area under consideration. Noteworthy contributions are by Henning Knudsen and T. Borgen from S.W. Greenland, P. Milan Petersen from various W. Greenland stations and by Henry Dissing from E. Greenland. B. Fredskild has collected in both E. and W. Greenland stations and G.S. Mogensen has provided specimens from extreme northern localities.

Most of the Svalbard collections were gathered by the author in the Isfjord region in 1981 and 1982. They are supplemented with a few collections from the Botanical Museum in Oslo (O). The number of records indicated for the species refer to specimens seen by the author.
Lycoperdaceae

Bovista Pers.

Typified by *Bovista plumbea* Pers., the genus is primarily characterized by a smooth, rather thick, two-layered exoperidium, a durable endoperidium with well defined opening and homogeneous gleba. The capillitium is floccose, of discrete elements. The main stem thick, with dichotomous branching. Spores pedicellate. Whole fruit body detached when mature. Undoubtedly related are some species with fruit bodies remaining fixed to the substrate, sometimes also with compact subgleba, but with peridium, capillitium and spores typical for the genus. From the boreo-arctic zones *Bovista cretacea*, *B. paludosa*, and *B. tomentosa* belong here. *Bovista limosa* deviates in its Geastrum-like stoma and in its capillitium of transitional type (Kreisel 1967).

Excluded from the genus are species with spores and capillitium as in *Bovista* but with irregular dehiscence of exoperidium (*Calbovista*) or with cellular subgleba, distinct diaphragm, and exoperidium of connivent spines (*Bovistella*).

Excluded from the genus, referred to *Lycoperdon*, are species with all or most of the following characters: Exoperidium breaking up into spines when ripe, subgleba cellular with pseudocolumella, capillitium simple, elastic or brittle, with subseptal and dichotomous branching and simple pores, spores with pedicel detached.

Kreisel (1964, 1967) proposed a much broader concept of *Bovista*. As defined above the genus largely corresponds to subgenus *Bovista sensu* Kreisel. His subgenus *Globaria* (Quél.) Kreisel includes species such as *Lycoperdon polymorphum* Vitt. and *L. pussillum* Batsch, which share a large number of features with typical *Lycoperdon* species, while having in fact no positively expressed character in common with the type of the genus *Bovista*.

In fact the circumscription of *Bovista* proposed by Kreisel (1967) lets the separation from *Lycoperdon* rest on two correlated characters: absence of pseudocolumella and a non-cellular structure of the subgleba when present. Both characters also are found in species which undoubtedly belong in *Lycoperdon*, and by their "negative" definition they are hardly of any clear significance. It is further noteworthy that probably all species which are here referred to *Lycoperdon* have an exoperidium which at least in part breaks up into minute spines, due to a dominant presence of short inflated cells.

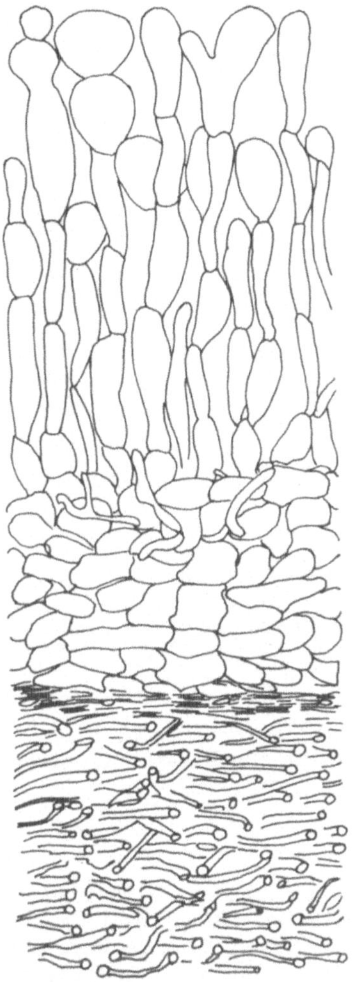

Fig. 1. *Bovista nigrescens* (H.D. 81.102). Section of peridium (subschematic).

From subg. *Globaria, B. limosa* is retained in *Bovista*. Its peridial and spore characters are as in *Bovista*, while the capillitium is of a transitional type. A further study may retain other species of this subgenus, which certainly is not a good taxonomic unit.

The three Greenland species enumerated below are all characterized and described in the monographic treatment by Lange (1948) and Kreisel (1967). Hence, full and formal descriptions are not included here. The notes are restricted mostly to peridium anatomy, completing the previous observations.

Bovista nigrescens Pers. per Pers.

18 collections from Greenland.

The material studied seems in all ways typical. However, apparently there are no records of the peridial structure of this common European

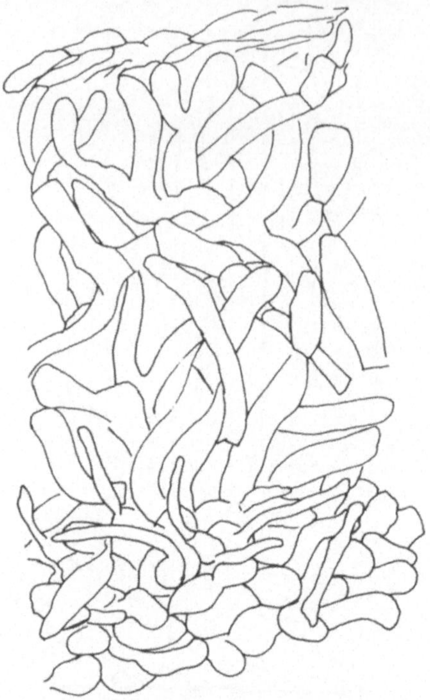

Fig. 2. *Bovista nigrescens* (H.D. 81.102). Section of exostratum and outer endostratum (subschematic).

species. The exoperidium is white and almost smooth, the ripe endoperidium robust, dark brown. Sections from fully developed, unripe specimens (Fig. 1) show the exoperidium to have two main layers, an exostratum 400 µm thick, of radiating or loosely interwoven hyphal chains, elements ± inflated, 5-20 µm broad, and an endostratum a 300-400 µm thick, of large-celled pseudoparenchyma cells 10-30 µm broad, upper part of this layer with some irregularly branched cyanophilic elements with refractive content, protruding into exostratum. Endoperidium a 200 µm thick dense plechtenchyma, hyphae non-septate, unbranched, rather thick-walled, 2-3 um broad, distinctly cyanophilic.

Distribution: All the collections are from S.W. Greenland. Northermost stations around 61°11'N.

Bovista nigrescens is rather common in subarctic, subcontinental to suboceanic Greenland, especially in the medieval Norse settlement areas at Erik Raudes Fjord, confined to pastures grazed by sheep. This European species may well have been introduced with Norse settlers.

Also in Iceland, *Bovista nigrescens* seems to prefer grazed pastures (Hallgrimsson 1963). There are only a few records from the Faroes (cp. Møller 1958). It is a frequent species in the lowland in northernmost Norway (Eckblad 1971). In Central Europe it is mostly recorded from subalpine and low alpine meadows (Kreisel 1967, Gross et al. 1975), however, it is also found in truly alpine vegetation (Favre 1955).

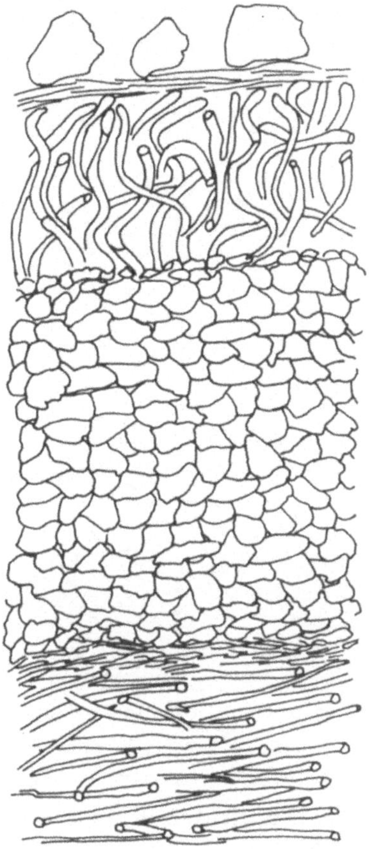

Fig. 3. *Bovista tomentosa* (Odum 28.8.83). Section of peridium with Loess particles on top (subschematic).

Its distribution in arctic and subarctic Siberia is not too well documented. In the North American continent, it is replaced by the closely related *Bovista pila* Berk. & Curt., while in tropical mountain zones, *Bovista fusca* Lév. is a close relative.

Bovista tomentosa (Vitt.) Quél.

16 collections from Greenland, 1 from Svalbard.

The material includes a few collections of unripe specimens, which makes a study of the peridial configuration and structure possible. The surface of the peridium is even, in places somewhat cracking into patches. It has a finely tomentose appearance, due to incrusting, very fine soil particles in combination with a minutely fibrillose texture. In half ripe specimens, the exoperidium can be detached in flakes.

The anatomy of the peridium (Fig. 2) differs somewhat from the description given by Kreisel (1967). The exostratum is duplex. There is an outer layer 20-40 µm thick of narrow subparallel hyphae, covering a 150 µm thick layer of interwoven, branched, septate hyphae, 2-4 µm broad. The endostratum is pseudoparenchymatic, 300 µm thick, the elements 10-25 µm broad. No cyanophilic elements are observed. The endoperidium is a

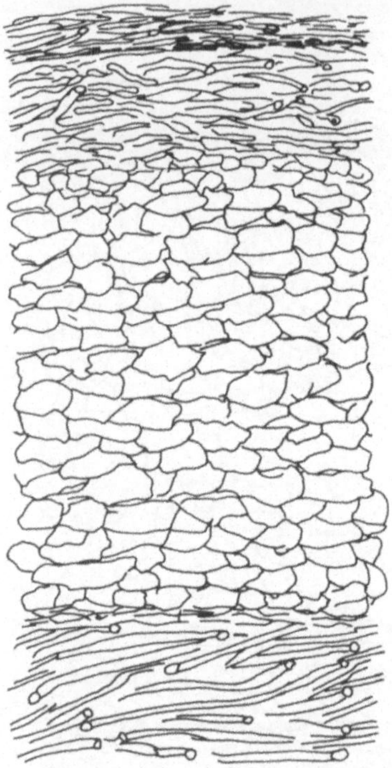

Fig. 4. *Bovista limosa* (H.D. 82.216). Section of peridium (subschematic).

100 µm thick layer of cyanophilic, thick-walled, subparallel-interwoven, unbranched, non-septate hyphae.

Other microscopic characters are as indicated by Kreisel (1967), with rather little variation. There is hardly any difference in the capillitium elements in various parts of the gleba. The species has been compared with *Bovista paludosa* Lév., which has an almost identical peridial anatomy.

Distribution: This species was previously recorded from Sdr. Strømfjord 67°N, and from Ikorfat 70°46'N, 53°07'W in W. Greenland (M. Lange 1948). New W. Greenland material includes several collections from Narssarsuaq (61°N). In E. Greenland there is a previous record from Mestersvig (71°N; M. Lange 1976), and several recent collections from the same area. A collection from Nordl. Mudderbugt 82°28'N, 21°11'W leg. G.S. Mogensen, places the species as a member of the very selective flora of the high arctic desert. All stations registered are from localities with alkaline loess or with alkaline mesozoic or tertiary formations. The labels often indicate sunny south slopes.

I have made a single collection of *Bovista tomentosa* in Hotelness, Longyearby (78°14'N, 15°30'W) growing on coal silt. It seems to be rare in Svalbard, not found by me on the calcareous soils in Sassendal, where I have made extensive collections of other gasteromycetes.

Bovista tomentosa is known from several records in N. Iceland (Hallgrimsson 1963, and own observations). It is noted as a rare species in continental mountain tracts in N. Norway (Eckblad 1971). It has a

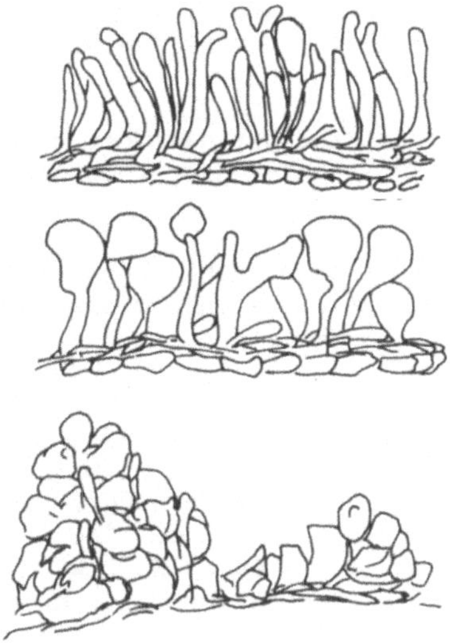

Fig. 5. *Bovista limosa* (H.D. 82.216). Sections of exostratum. All from same specimen (subschematic).

widely scattered distribution in Europe, here mostly confined to continental, calcareous tracts. Favre (1955) records it as frequent on limestone in high alpine zones.

Bovista limosa Rostr.

17 collections from Greenland.

Macroscopically, the geastrum-like stoma is a unique character, clearly observed in almost all suitable specimens. The surface of the peridium is rather variable, as is often the case in arctic material of Lycoperdaceae. It may be almost smooth and glabrous, but in most specimens it forms very low, irregular polygons, with or without a small tuft in the middle. When ripening the peridium becomes minutely felty. It may peel off in flakes or shrink to leave small tufts, which are irregular or sometimes even more or less stellate, on the ripe peridium.

Sections of the peridium (Fig. 3) show a 20-60 µm thick outer layer of cyanophilic, densely interwoven-subparallel, septate and branching hyphae 2-3 µm broad. Under this is a slightly less compact, and hardly cyanophilic 30-40 µm thick layer. In places the outer layer disappears, and the second layer develops to become more loosely interwoven, or gives rise to erect hyphae or more often to a mass of hyphae with inflated elements up to 20 um broad. This layer may become greater than 100 µm thick. The configuration and variation in the peridial surface originate from the development of these two layers. The endostratum is pseudoparenchymatic, 300-400 µm thick, elements up to 40 µm broad, smaller in outermost zone. The inner peridium is 50-70 µm thick, of cyanophilic thick-walled, unbranched and non-septate, tangentially arranged, 2-3 µm broad hyphae. The capillitium varies considerably. It belongs in the transitional type described by Kreisel (1967). Dichotomous branching is typical, but subseptal branching is frequent in the peripheral gleba. Truly discrete elements are rarely seen. The dimensions of the main stem varies in different collections, from 6 to 12 µm. In some mounts a few scattered, large and simple pores are observed. The spores are always finely warty, the pedicel is very short, 3-8 µm, in some mounts some pedicels have broken off.

The above description of the peridium corresponds in the main to data presented by H. Lohwag (1933). His interpretation of the development is, however, less convincing.

There are good reasons to place this species on the borderline between *Lycoperdon* and *Bovista*. The characters of the peridium point, however, mostly to *Bovista*, as do the spore characters. The peculiar stoma is hardly enough to validate a specific genus.

Distribution; The type collection is from high arctic E. Greenland, Scoresby Sund region 70°N. There are, however, several recent records as far North as Ella Ø 72°50'N 25°W at 400 m. The northernmost locality in W. Greenland is Angertussup qaqai, 68°33'N 52°W. There are records from S. Strømfjord (M. Lange 1948), and several recent collections from Narssarsuaq 61°N.

All Greenland records are from more or less alkaline soil, on localities similar to those where *Bovista tomentosa* is found. Both seem to prefer a continental or subcontinental climate. I have, however, not been able to find it on the many suitable habitats on Svalbard.

The species is also on record from similar localities in North Scandinavia (Eckblad 1971) and from Swedish Lappland (Fries, 1921). Eckblad (1971) also reports finds from Southern Norway. Kreisel (1967) quotes several records from various European localities, and there is a recent record from Spain (Calonge & Demoulin, 1975). The description of the Spanish specimens, and probably also of the other Central European collections, indicate the absence of the characteristic stoma. It may well be a separate taxonomic unit. Kreisel (1967) records several small Bovista-species which are close to *B. limosa*.

ACKNOWLEDGEMENTS

The collectors mentioned above have been extremely cooperative in paying attention to the Gasteromycetes, bringing home many and well preserved collections. - Bodil Lange has assisted in the field work on our joint travels in Greenland and Svalbard.

REFERENCES

Bowerman, C. A., and Groves, J. W., 1962, Notes on fungi from Northern Canada. V. Gasteromycetes, *Can. J. Bot.*, 40: 239-254.

Calonge, F. D., and Demoulin, V., 1975, Les Gasteromycetes d'Espagne, *Bull. Soc. Myc. Fr.*, 91: 247-292.

Eckblad, F. -E., 1971, The Gasteromycetes of Finmark (Northernmost Norway), *Astarte*, 4: 7-21.

Favre, J., 1955, Les champignons supérieurs de la zone alpine du Parc National Suisse, *Ergeb. Wiss. Untersuch. Schweiz. Nationalparkes 5 (N.F.)*, 212 pp.

Fries, T. C. E., 1921, Sveriges Gasteromyceter, *Arkiv. f. Botanik*, 17,9: 1-63.

Gross, G., Runge, A., and Winterhoff, W., 1980, Bauchpilze (Gasteromycetes s.l.) in der Bundesrepublik und Westberlin, *Beih. Zeitschr. Mycol.*, 2: 1-220.

Hallgrimsson, H., 1963, Eldsveppir (Íslenskir belgsveppir II), *Náttúrufraedingurinn*, 33: 138-147.

Kreisel, H., 1964, Vorlaufige Übersicht der Gattung *Bovista* Dill. ex Pers., *Feddes Repertorium,* 69: 196-211.

__________, 1967, Taxonomisch Pflanzengeographische Monographie der Gattung *Bovista, Beih. Nova Hedwigia,* 25: 1-244.

Lange, M., 1948, Macromycetes Part I. The Gasteromycetes of Greenland, *Meddr. om Grønl.,* 147,4: 1-32.

__________, 1976, Some Gasteromycetes from North East Greenland, *Kew Bull.,* 31: 635-638.

Lohwag, H., 1933, Mykologische Studien VIII. *Bovista echinella* Pat. und *Lycoperdon velatum* Vitt., *Beih. Bot. Centralbl.,* 51: 269-286.

Møller, F. H., 1958, "Fungi of the Faroes II," Copenhagen, 286 pp.

Ulvinen, T., 1969, Über einige *Bovista*- und *Lycoperdon*- Arten in Finland, *Aquilo, Ser. Bot.,* 8: 25-41.

HYGROPHORACEAE FROM ARCTIC AND ALPINE TUNDRA IN ALASKA

Gary A. Laursen

Agricultural and Forestry Experiment Station
University of Alaska
Fairbanks, Alaska 99701 U.S.A.

Joe F. Ammirati

Department of Botany, KB-15, University of Washington
Seattle, Washington 98195 U.S.A.

and

David F. Farr

Mycological Laboratory, Rm. 313, Bldg. 011A, SEA BARC-West
Beltsville, Maryland 20705 U.S.A.

Key words: *Hygrocybe*, *Hygrophorus*, tundra, Alaska

ABSTRACT

Nine taxa in the Hygrophoraceae, including five informally designated variants, are described and discussed from Alaskan Arctic and alpine tundra. These include, *Hygrocybe citrinopallida*, *H.* aff. *citrinopallida*, *H. coccineocrenata*, *H.* aff. *coccineocrenata*, *H. conica*, *H.* aff. *lilacina*, *Hygrophorus* aff. *chrysodon*, *H.* aff. *eburneus* and *H. melizeus*.

INTRODUCTION

Relatively few studies have been conducted on the Hygrophoraceae in Alaska. Hesler and Smith (1963) did not include Alaskan specimens in their treatment, "North American species of *Hygrophorus*" *sensu lato*. Reports of Hygrophoraceae from Alaska include, Saccardo, Peck and Trelease (1904), Cash (1953), Kobayasi et al. (1967), Miller et al. (1973), Miller and Laursen (1978), Miller (1972), Laursen and Ammirati (1982b), Laursen and Chmielewski (1982), Miller (1982a, 1982b), and Miller et al. (1982). See Table 1 for a list of taxa. Further information on Alaskan Hygrophori may be available from checklists of Alaskan fungi based on the herbarium of V. L. Wells (deceased) and P. E. Kempton, Anchorage, Alaska. However, these checklists have not been published to our knowledge.

TABLE 1. Hygrophoraceae reported from Alaska

Hygrocybe miniata [Scop. ex Fr.] Karsten (Kobayasi et al. 1967)
Hygrophorus cantharellus Fr. (Cash 1953)
Hygrophorus ceraceus Fr. (Cash 1953)
**Hygrophorus chrysodon* (Fr.) Fr. (Miller 1982a, 1982b)
**Hygrophorus citrinopallidus* Smith and Hesler (Kobayasi et al. 1967; Laursen and Chmielewski, 1982; Miller et al. 1982)
**Hygrophorus* aff. *citrinopallidus* Smith and Hesler (Laursen and Chmielewski 1982)
**Hygrophorus conicus* (Fr.) Fr. (Cash 1953; Laursen and Chmielewski 1982; Miller 1982a)
**Hygrophorus laetus* (Fr.) Fr. (Miller 1982b)
*?*Hygrophorus lilacina* (Laest.) M. Lange (Kobayasi et al. 1967)
Hygrophorus limacinus Fr. (Cash 1953; Saccardo et al. 1904)
**Hygrophorus melizeus* Fr. (Miller 1982b)
**Hygrophorus miniatus* (Fr.) Fr. (Miller 1972, 1982a; Miller et al. 1982)
**Hygrophorus vitellinus* Fr. (Laursen and Ammirati 1982b)
**Hygrophorus vitellinus* Fr. *sensu* Möller (Kobayasi et al. 1967)

*Reports known to be from Arctic and/or alpine tundra sites

In general, the Hygrophoraceae, as well as other agarics that occur in Alaska are poorly known. This is evident from papers dealing with agarics in the Alaskan Arctic and alpine tundra. See, for example, Ammirati and Laursen (1982), Laursen and Ammirati (1982a & b), Miller (1982) and Gulden (1983). No doubt, new taxa await description. However, the most important studies conducted involve a careful evaluation of the morphological and anatomical characteristics of taxa, particularly in relation to their habitat(s) and locations. Also, critical to determining correct species identifications and distributions of fungi, other than and including Hygrophori, are establishing clear relationships, or lack thereof, among agarics that inhabit Alaskan Arctic and alpine tundra sites. In addition, the occurrence of agarics on tundra sites in Alaska has to be related to those occurring in the northern coniferous forest (boreal, taiga, etc.) in Alaska and Canada. Finally, circumpolar distributions have to be given careful consideration. This must be done on a broad scale for all ecosystems in the United States and Canada, particularly in relation to other northern regions of the world.

The *Hygrocybe citrinopallida* complex exhibits various kinds of taxonomic problems one encounters with Alaskan species. *Hygrocybe citrinopallida* was originally described from Mt. Rainier National Park by Smith and Hesler (1954). However, populations of this species in North American alpine habitats have not been studied in terms of variation in taxonomic characteristics. *Hygrocybe citrinopallida* in the Alaskan Arctic tundra exists in two color forms; populations that are basically yellow and populations that develop olive to gray coloration on the disc in addition to the basic yellow color. Compare *Hygrocybe cintrinopallida* and *H.* aff. *citrinopallida* here and see *Hygrophorus citrinopallidus* Smith and Hesler (Miller et al. 1982). Other significant differences between the original description by Smith and Hesler and our Arctic collections are the generally larger basidiospores, the occurrence of 4-, 2- and 1-spored basidia, and variation in basidiospore shape; characteristics that are probably interrelated to some extent.

Taxa in the European literature that are similar to *Hygrocybe citrinopallida* include, *Hygrocybe glutinipes*, *H. vitellina*, *H. ceracea* and *H. citrina*. Do one or more of these taxa overlap in taxonomic characteristics, habitat, etc. with *Hygrocybe citrinopallida*? It would seem they do share several characteristics with this taxon if one uses the original descriptions as a basis of comparison. For example, *Hygrocybe vitellina* may be the same as *Hygrocybe citrinopallida*. If *Hygrocybe citrinopallida* is related to or the same as an European taxon, how do populations of these taxa vary in their anatomical and morphological characteristics on the separate continents? Does the variation seen in the specimens from the Alaskan Arctic tundra fit within the circumscription of an existing taxon? Do these specimens represent one or more new taxa?

In this paper we have chosen not to designate new names for Alaskan Arctic and alpine specimens that do not fit descriptions of published taxa. These taxa which are in the *Hygrocybe coccineocrenata*, *Hygrocybe lilacina*, *Hygrophorus chrysodon* and *H. eburneus* complexes, simply are not well enough documented or understood to make decisions as to whether or not they are in fact new.

MATERIALS AND METHODS

Specimens are described and identified as completely as possible, followed by comments on their taxonomy. Collection numbers are those of one or more collectors designated as follows: Orson K. Miller, Jr. (OKM), Gary A. Laursen (GAL), Joe F. Ammirati (JFA) and David F. Farr (DFF). All collections are deposited at the Herbarium of the University of Alaska, Fairbanks (ALA) except for those of Orson K. Miller which are in the Herbarium at the Virignia Polytechnic Institute and State University (VPI).

Basidiospores and tissues of the basidiocarps were studied from dried material revived in a moist chamber and mounted in 3% aqueous potassium hydroxide. Basidiospore shapes and measurements were made from lamellae. The apiculus was not included in the length measurements of basidiospores. Color standards include Ridgway (1912), for which the color names are capitalized, e.g., Ochraceous Tawny, and Kornerup and Wanscher (1967) indicated as follows, burnt sienna (7D8).

Detailed habitat descriptions for Driftwood Camp are provided in Ammirati and Laursen (1982) and Laursen and Ammirati (1982a). Dalton Highway site descriptions can be found in Brown and Kreig (1983). Murray (1978) provides an overview of the general vegetation and floristics for locations in northern Alaska. See also Miller (1982a, 1982b) and Laursen and Chmielewski (1982) for general comments on the subarctic and Arctic tundra in Alaska.

DESCRIPTIONS OF TAXA

Hygrocybe citrinopallida (Smith and Hesler) Kobayasi apud Kobayasi et al.
Fig. 1

Pileus 12–20 mm wide, convex to plane with the disc somewhat depressed; edge slightly incurved to somewhat decurved, opaque to striate, even sulcate-undulate when more expanded; surface smooth but appearing somewhat fibrillose when faded, moist, lubricous, viscid, hygrophanous on margin, color bright yellow to yellow or light yellow (3A8–5), disc at times pallid or dull, margin fading with loss of moisture to white. *Lamellae* distant, decurrent to long decurrent, color light yellow to yellow (3A7–5) or bright yellow. *Stipe* 12–55 mm long, 1–3 mm thick, tapering toward base; surface smooth, viscid when fresh, color pale yellow or light yellow to yellow (3A6–3) or bright yellow, base often white, fading to white overall with loss of moisture, or at times grayish to dingy light buff; hollow.

Figures 1-6. Fig. 1 *Hygrocybe citrinopallida* GAL/JFA 1782; Fig... 2. *H.* aff. *citrinopallida* GAL/JFA/DFF 2154; Fig. 3. *H.* aff.*citrinopallida* GAL 1374; Fig. 4. *H.* aff. *citrinopallida* OKM 10987; Fig. 5. *Hygrocybe coccineocrenata* GAL/JFA 2273; and Fig. 6. *H. coccineocrenata* OKM 15758.

Basidiospores 7.5-10 (-12) x (4.5-) 5-6 µm, in profile view usually ovate and somewhat inequilateral, some ± strangulate. *Basidia* normally

4-spored, also 1- and 2-spored, sterigmata often large. *Hymenial cystidia* absent. *Pileus cuticle* an ixocutis; hyphae narrow, mostly 3.5-9 µm wide, interwoven, loosely arranged, at times tending to be upturned as in an ixotrichoderma.

Habit and Habitat: Scattered to gregarious; in grass and moss covered peat soil of high center polygons, in grassy troughs, in mosses and sedges, in moss under *Salix* sp. and *Betula nana* and in *Polytrichum* sp. under *Vaccinium vitis-idaea*; early to late August.

Material Examined: GAL 1165, 1238, 1371, 1375 and 1376 and GAL/JFA 1782, USIBP Tundra Site, NARL, Barrow; GAL/JFA 1727 and 2287, Driftwood Camp, Delong Mountains; GAL/JFA/DFF 3266, Atqasuk (Meade River Camp).

Discussion: The above collections are the most typical we have seen for *Hygrocybe citrinopallida* in the Alaskan Arctic tundra; that is, they do not have the gray to olive coloration characteristic of many of the Arctic collections (see below). The disc of some pilei, however, do tend to be somewhat dull or pallid, for example, in the Atqasuk collection. Basidiospore measurements are in the range for *H. citrinopallida*, but in general tend to be somewhat broader and longer from some specimens. This could be because these specimens produce a larger percentage of basidia with two sterigmata, or even one sterigma. By way of comparison, the number of sterigmata per basidium in specimens studied from collection GAL/JFA/DFF 3266 was usually four; only occasionally were two sterigmata seen. The basidiospore measurements were normally 7.5-8.5 x 5-5.5 µm. In collection GAL/JFA 1727 the number of sterigmata per basidium was often four, but two sterigmata or one sterigma per basidium were fairly common. This seems to have produced a greater range in basidiospore measurements and generally large basidiospores, 8.5-10(-12) x 4.5-6 µm. Also, the shape was not so consistently ovate and inequilateral.

Hygrocybe vitellina (Fr.) Karsten is similar to our material but has smaller basidiospores (5-8 x 3-4 µm, Moser, 1983; 6-8 x 4.5-5 µm, Orton, 1960). Lange (1955) gave basidiospore measurements of 7.5-9 x 4-5 µm and shape subcylindric, for specimens he named *Hygrophorus vitellinus* from Greenland. Our collection GAL/JFA/DFF 3266 has very similar basidiospore measurements but no subcylindric basidiospores. *Hygrocybe citrina* (Rea) lange *sensu* Lange (Moser, 1983) may be the fungus we describe here but further study of a wide range of materials is needed to determine this possibility. *Hygrocybe glutinipes* and *H. ceracea*, as defined in Moser (1983) and *Hygrophorus nitidus* B. & C. also appear to be related to this group.

Hygrocybe aff. *citrinopallida* Figs. 2-4

Pileus 6-8 mm wide, convex or convex with depressed disc; edge incurved becoming straight, even or undulate to crenate; margin opaque or striate; surface viscid fresh, more or less hygrophanous, color bright yellow (4A6) to yellow (3A8-6, 3B8-6; Primuline Yellow, Wax Yellow or Strontian Yellow) overall at first, or yellow on the margin and gray to grayish brown on the disc, in age the general color may be duller yellow (Citron Yellow) and the disc may be olive-gray to dark olive, generally the gray to olive colors develop more strongly in age; flesh to 3 mm thick, more or less concolorous with surface; taste mild; pleasant; odor none or fungoid. *Lamellae* short to long decurrent, thick, alternating with short lammulae, color yellow to light yellow (3A7-5) to somewhat duller yellow (Colonial Buff to Deep Colonial Buff). *Stipe* 10-45 mm long, 1-3.5 mm thick, tapering toward base; surface smooth, viscid when fresh, color bright yellow (4A6) to yellow (Wax Yellow) or paler yellow

(Primrose Yellow) below and yellow buff (Ivory Yellow) at apex, grayish or tinted with Olive Buff in age, apex at times darkening like the pileus disc; hollow.

Basidiospores 7.5-11 (-13) x (4.2-) 4.5-6 µm, in profile view broadly to narrowly ovate or ovate-elliptic, inequilateral, sometimes constricted (slightly strangulate). *Basidia* 4-spored, some 2- or 1-spored. *Hymenial cystidia* usually absent, at times cystidia-like cells (clavate with sterigma-like extension) present in hymenium. *Pileus cuticle* an ixocutis; hyphae narrow, mostly 3.5-9 µm wide, interwoven, appressed to loosely arranged, at times upturned.

Habit and Habitat: Scattered to gregarious, in moist moss or in moist moss and grass on higher polygon tops, and in grassy polygon troughs; early to late August.

Material Examined: GAL 1240, 1373 and 1374, GAL/JFA/DFF 2154 and OKM/GAL 10557, 10572, 10950, 10951, 10987, 10988, 10990, 10991, 10992, 11016, 11017, 11031, 11035, 11132, 11136 and 11185, USIBP Site, NARL, Barrow; OKM/GAL 11106 and 11156, Atqasuk (Meade River Camp).

Discussion: The specimens described above are closely related to *Hygrocybe citrinopallida*. However, they differ in that they tend to have gray to olive colors, especially on the central depressed disc of the pileus, and on the stipe base as well. This was originally pointed out in Miller et al. 1982. In the description by Hesler and Smith (1963) there is no indication that this was the case for any of the collections from high elevations on Mt. Rainier in Washington state. Also, we have Arctic collections (see above) that do not develop these colors or at most only slightly so. The reason for the development of the olive to gray colors is not known. It may be environmental or perhaps characteristic of only some populations of the *H. citrinopallida* complex. It would be advisable to check other Arctic locations for similar specimens; all of the collections listed here are from Barrow or Atqasuk (Meade River Camp). Also, alpine collections from several areas should be studied in reference to the gray to olive color development.

Basidiospore measurements tend to be somewhat larger for this variant of *H. citrinopallida* than for the more commonly found, yellow colored Arctic specimens described above. Basidiospore shape is somewhat variable as well, the general outline of the basidiospores varies from ovate to narrowly ovate or ovate-elliptic in profile view. They are inequilateral but some are constricted or strangulate. The combinations of basidiospore size and shape present in a single specimen yield basidiospores that range from short and relatively broad to long and narrow.

Normally cystidia-like cells are absent from the hymenium, but in some instances certain clavate elements have a sterigma-like elongation from the apex of the cell. For example, these elements are fairly common in the hymenium of OKM 10951 and for all practical purposes look like cystidia. They usually are within the hymenium, rather than protruding beyond it, and do not appear to have special contents. Since basidia with a single sterigma, usually off to one side rather than at the apex of the basidium, are present in the hymenium, they may be nothing more than basidia which have produced a single, apically located sterigma.

Hygrocybe coccineocrenata (Orton) Moser Figs. 5-6

Pileus 7-20 mm wide, obtusely conic to convex or plano-convex, some expanded pilei shallowly depressed centrally; edge narrowly incurved to

Figures 7-12. Fig. 7. *Hygrocybe* aff. *coccineocrenata* GAL/JFA/DFF 3103; Fig. 8. *Hygrocybe conica* GAL/JFA/DFF 3077; Fig. 9. *H. conica* GAL/JFA 1762; Fig. 10. *Hygrocybe* aff. *lilacina* GAL/JFA 2327; Fig. 11. *Hygrophorus* aff. *chrysodon* GAL/JFA/DFF 3033; Fig. 12. *Hygrophorus melizeus* GAL/JFA/DFF 2993.

decurved, dentate to even; margin opaque or slightly striate; surface moist, matted fibrillose to minutely scaly, color red to orange-red to copper red or burnt orange (7C8; Russet, Kaiser Brown, Vinaceous-Rufous, Burnt Sienna, Mars Orange) at times becoming more brownish orange (6B7-8, 6C8) or light orange (5A4), center usually darker than the margin, some felty fibrils brownish yellow (5C8); context up to 2 mm thick in disc, soft, buff in color; odor not distinctive; taste acidulous. *Lamellae* broadly adnate to sinuate or adnexed with or without a decurrent line, or short decurrent, distant to subdistant, color cream (4A3) to yellow (Antimony Yellow, Maize Yellow) to pale orange-yellow or light orange-yellow (4A4-6), reddish orange color of pileus apparent between lamellae in some specimens; two tiers of lamellulae. *Stipe* 17-28 mm long, 2-6 mm thick, equal or tapering downward; surface smooth, glabrous, moist, color orange (5A7; Mars Orange, Orange Rufous) grayish orange (6B6), light orange (6B7; Capucine Orange) orange-yellow (Light Orange-Yellow) or yellow (5A6, 4A5-7), base similarly colored or pale yellow (4A2) to white; hollow or partly stuffed.

Basidiospores 7.5-12.5 x (5-) 5.5-7.5(-8) μm, in profile view elliptic, oblong-elliptic or slightly obovate, some slightly strangulate. *Basidia* 4-spored, rarely 2-spored. *Hymenial cystidia* absent. *Pileus cuticle* a cutis, with some distinct fascicles of hyphae; hyphae up to 11(-18) μm wide, terminal cells at times cystidium-like.

Habit and Habitat: Scattered to gregarious; in moist, mossy peat soil, on raised rims of low centered polygons under *Betula glandulosa* and *Vaccinium vitis-idaea* or in *Pogonatum alpinum* near *Vaccinium vitis-idaea*, and *Dryas integrefolia* in Arctic tundra, in alpine tundra under *Vaccinium vitis-idaea*; late August.

Material Examined: GAL/JFA 2273, Driftwood Camp, Delong Mountains; GAL/JFA/DFF 3161, Steese Highway, mile 110; OKM/GAL 11157, Meade River Camp (Atqasuk); OKM 15758, Eagle Summit south of circle.

Discussion: These collections have many features in common with *H. coccineocrenata*, i.e. minute scales on the center of the pileus, general color, and basidiospore size. A common characteristic of dried pilei is for the centers to be brownish tinted and minutely scaly (*sub lenta*). There also are some differences. The stature of the basidiocarps is somewhat reminiscent of *Hygrocybe miniata* (Fr.) Kummer, particularly since the lamellae are not typically decurrent. In addition, the shape of the basidiospores is variable, with some being centrally constricted, and there are at least some 2-spored basidia present. Except for the more slender stature (which may be due to growth in *Sphagnum* moss beds) *Hygrocybe turunda* var. *sphagnophila* (Peck) Bon is similar to our material. Also, the description allows for variable lamella attachment, "at first adnate, remaining so or becoming deeply decurrent" (Hesler and Smith, 1963). If one considers *Hygrocybe coccineocrenata* and *Hygrocybe turunda* var. *sphagnophila* to be the same taxon, as suggested by Orton (1960), then these Arctic and alpine specimens would fit this expanded concept of *H. coccineocrenata*. For the time being we are using this approach. However, this group of Hygrocybes is in need of careful study in North America and further investigations could alter our currently position. Collection OKM/GAL 11157 was placed in *H. miniatus* (Fr.) Fr. by Miller et al. (1982).

Hygrocybe aff. *coccineocrenata* (Orton) Moser Fig. 7

Pileus up to about 10 mm wide, obtusely conic to convex, disc rounded or slightly depressed, margin decurved to nearly straight expanded, faintly striate, the very edge usually narrowly incurved and slightly crenate; surface moist to dry, somewhat hygrophanous, slightly appressed fibrillose, color dark reddish brown (9E7), almost a purple reddish brown, margin red (9B7) to orange-red (8B7) becoming orange (6A7) to reddish orange-yellow (4A6); flesh concolorous with surface. *Lamellae* thick, distant, adnate to adnexed or notched, color yellow to yellow-orange or in some parts tinted dull reddish orange (7B6) to dull red or grayish red (8B6). *Stipe* 7-20 mm long, 1-3 mm thick, equal, often curved; surface moist to dry, smooth, color about like pileus surface at apex, grading from red to yellow downward, base light yellow (4A5) to light orange (5A6); hollow.

Basidiospores (7.5-) 9.5-13.0 x (4.5-) 5-6 (-7.5) µm, narrowly elliptic, broadly elliptic, slightly ovate-elliptic, lacrimoid or obovate, some centrally constricted. *Basidia* 4-spored or less commonly 2-spored. *Hymenial cystidia* absent. *Pileus cuticle* a cutis with hyphae tending to form fascicles; hyphae up to 10.5(-14) µm wide, some end-cells cystidium-like.

Habit and Habitat: Scattered to gregarious in mosses and lichens or on moist to wet peat soil, higher plant associates *Ledum palustre* subsp. *decumbens* and *Arctostaphylos alpina*; late August.

Material Examined: GAL/JFA/DFF 3103 and 3116, Dalton Highway, mile 247.

Discussion: The larger basidiospores and strong red coloration of the basidiocarps place this collection in the *H. coccineocrenata* complex. The pileus surface is not described as "minutely fibrillose scaly" but only as slightly appressed fribrillose. In our color photographs of collection GAL/JFA/DFF 3103, small appressed patches can be seen in places on the pilei disc. The damp, cool weather may have prevented the small scales from forming on the pileus surface. However, the disc of dried pileus are not minutely scaly *sub lenta*, as in our *H. coccineocrenata* above, indicating that the non-fibrillose scaly pileus may be a typical feature. If this were so, the collections may be closer to the *Hygrocybe substrangulata* (Orton) Moser complex. Another feature which does not conform to the description of *H. coccineocrenata* in Orton (1960) is the attachment of the lamellae, being adnate to notched in our material, rather than decurrent.

Hygrocybe conica (Scop.: Fr.) Kummer Figs. 8-9

Pileus 11-44 mm wide, conic, campanulate, obtusely conic or convex, umbonate, the umbo rounded, subacute or acute; edge straight or narrowly decurved; surface viscid to moist, opaque or slightly striate on the margin, usually radially streaked, color somewhat variable, entirely light yellow or more commonly yellow-orange (4A6) to orange-yellow or orange, rarely slightly reddish orange at first, soon developing grayish or gray brown to blackish stains and discolorations, then grayish yellow (4B7, 4C8) to brownish orange (7C7) or with light brown (7D8) to olive-brown (4F4) tones, eventually black on some areas. *Lamellae* nearly free, thick, distant, color pale yellow (2A2) to light yellow, becoming grayish yellow (2B4) to olive-yellow (3D4) or more grayish, finally stained black in places; edges entire to serrulate. *Stipe* 25-70 mm long, 2-11 mm thick, equal or tapering slightly toward apex, at times somewhat compressed and ridged; surface moist, color yellowish white (1A2), pale yellow (1A3), light yellow (1A4-5), greenish yellow (1A6-7), yellowish orange (4A7) or light orange (5A7), developing grayish tones or becoming olive-brown (4D6, 4E6, 4F6), eventually staining black; context eventually staining black.

Basidiospores (8-) 9.5-12.5 (-14, rarely -17) x (4.5-) 5.5-6.5 (-8.5) µm, in profile view elliptic, oblong-elliptic or obovate, often centrally constricted. *Basidia* 4-spored, also 2- or rarely 1-spored. *Hymenial cystidia* present or absent, if present usually on lamella edges, typically not well differentiated and not protruding much beyond the hymenium. *Pileus cuticle* an ixocutis; hyphae narrow, 2-7 µm wide, interwoven, usually forming a thin layer.

Habit and Habitat: Scattered to gregarious; in gravel and soil on airplane landing strip, in sedge meadow, in rocky stream bed near *Salix alaxensis* or in wet, moss covered peat; late August.

Material Examine: GAL/JFA 1600, 1623, 1762 and 2236, Driftwood Camp, Delong Mountains; GAL/JFA/DFF 3077 Dalton Highway, mile 247.

Discussion: *Hygrocybe conica* occupies a variety of habitats in the Arctic and also occurs in the Arctic/alpine tundra of Atigun Pass. Pileus color varies from yellow to orange or slightly reddish orange, often being a mixture of these colors, and the lamellae are pale to light yellow. Basidiocarp size varies somewhat but is within the range given by Orton (1960), except for a somewhat thicker and stouter stipe in some specimens. To date we have not found basidiocarps which combine red to orange-red, bluntly conical, pilei and more robust stipes, as is described for *Hygrocybe nigrescens* (Quél.) Kühner. The basidia in the specimens studied are mainly 4-spored, with some 2-spored or even 1-spored basidia present but usually not common. Basidiospore size covers a broad range, even from a single basidiocarp and reflects to some degree the presence of 1-, 2- and 4-spored basidia.

Hygrocybe aff. *lilacina* (Laest.) Moser Fig. 10

Pileus margin decurved, striate; surface glabrous, moist to slightly viscid, color light orange-yellow. *Lamellae* short decurrent, subdistant to distant, yellowish white to light yellow, more yellow near pileus edge. *Stipe* smooth, moist to slightly viscid, light pinkish lilac to more lilac at the base, apex tinted orange yellow.

Basidiospores 8.5-11.0 x 5.5-7.5 (-8.5) µm, in profile view elliptic to broadly elliptic or more ovate. *Basidia* (1-, 2-) 4-spored. *Hymenial cystidia* absent. *Pileus cuticle* a cutis; hyphae narrow, mostly 3-8 µm wide, interwoven, tending to be repent, not particularly gelatinized.

Habit and Habitat: Scattered in mosses and lichens; late August.

Material Examined: GAL/JFA 2327, Driftwood Camp, DeLong Mountains.

Discussion: Only one collection of the *Hygrocybe lilacina* group has been made by us from Alaskan tundra. The description above comes in part from a color slide of fresh material and therefore, is not as accurate as actual notes on fresh specimens. Kobayasi et al. (1967) reported specimens from Barrow under the name *Hygrophorus lilacinus* (Laest.) M. Lange. Their material differs from ours in that the pilei in their specimens had greenish or purplish tints and the lamellae at times had purplish tints. They reported the basidiospore shape as ovoid and gave measurements of 6-7.5 x 4-6 µm. Basidiospore measurements from our collection are larger than these measurements and also larger than those reported by Lange (1955), 7.5-9.5 x 5-6.2 µm, for *Hygrophorus violeipes* Lange (= *Hygrocybe lilacina*) and Lange and Skifte (1967) 7.5-9 x (4.2-) 4.5-5 µm, for *Hygrophorus lilacinus*.

Hygrophorus aff. *chrysodon* (Batsch: Fr.) Fr. Fig. 11

Pileus 15-45 mm wide, convex to plano-convex, uplifted in age, edge narrowly inrolled, tomentose; surface viscid fresh, opaque when fresh, color cream (4A3) to light yellow (4A4) on disc, whitish to cream or light yellow (4A6) on margin and often grayish yellow (4B4) to sordid brownish in age; context white; odor chalk-like. *Lamellae* thick, anastomosing in places, distant, slightly decurrent, adnate or sinuate with a decurrent tooth, color whitish to cream or pale yellow. *Stipe* 17-35 mm long, 7-12 mm thick, equal or tapering downward; surface viscid when fresh, apex furfuraceous, fibrillose downward, basic color whitish to pale yellow or light yellow, surface fibrils often yellow-orange-brown, minute points on apex yellow-brown.

Basidiospores 8-10(-11) x 4-5.5 µm, in profile view elliptic to narrowly elliptic. *Basidia* 4-spored. *Hymenial cystidia* absent. *Pileus cuticle* an ixocutis; hyphae narrow, mostly 2-6(-9) µm wide, somewhat branched, tending to be repent.

Habit and Habitat: Solitary to gregarious in wet, mossy peat; late August.

Material Examined: GAL/JFA/DFF 3033, Dalton Highway, mile 272, Jade Mt., Toolik Lake.

Discussion: This collection seems most closely related to *Hygrophorus chrysodon* and actually may be that species. It differs from *H. chrysodon* in that it lacks distinctly decurrent lamellae and has a rather short, stout stature. The fresh specimens, particularly the old ones, are more colored (discolored) than normally would be expected. Dried specimens, especially more mature ones, are brownish and the most expanded pileus appears to have had a thin, watery, striate edge. Fresh specimens gave a yellow (4A6) reaction when 5% KOH was applied to the pileus surface and stipe apex. The most convincing evidence for placing the material in *H. chrysodon* are the minute golden brown tomentose patches or points on the pileus edge and stipe apex of dried specimens.

Miller (1982a) first reported *Hygrophorus chrysodon* from Alaska, stating that it is "commonly found in subalpine tundra", "usually associated with *Betula nana* and *Dryas octopetala*", "among mosses with only the top of the pileus visible". He further reported that the species occurs in the taiga, not surprisingly since *H. chrysodon* is widespread, but Miller did not find it in Arctic tundra north of the Brooks Range. This collection, if *H. chrysodon*, would extend the range of this species some twenty miles north of Atigun Pass, onto the north slope of the Brooks Range.

Hygrophorus aff. *eburneus* (Bull.: Fr.) Fr.

Pileus 36 mm wide, somewhat plano-convex expanded; surface moist to viscid, margin striate, color white to buff. *Lamellae* distant, decurrent, color white. *Stipe* 75 mm long, apex 9 mm thick, tapering downward; apex furfuraceous, longitudinally fibrillose below, color white above, white to pale yellow below.

Basidiospores 7.5-10 (-11) x 4.5-6.5 µm, in profile view ovate to elliptic or at times lacrimoid or somewhat strangulate. *Basidia* 4-spored. *Hymenial cystidia* absent. *Pileus cuticle* an ixocutis or an ixotrichodermium in places; hyphae narrow, mostly 3-5 µm wide, more or less interwoven, somewhat branched.

Habit and Habitat: Solitary, on moss under *Betula nana*; late August.

Material Examined: GAL/JFA/DFF 3160, Steese Highway, mile 110.

Discussion: Only one specimen was found, as described above, and the material was mature. Therefore, we do not have much information to go on at this point in time. If the stipe surface was dry or only slightly viscid rather than viscid, that would remove this specimen from the *H. eburneus* complex. The strongly striate pileus also is somewhat curious for this taxon.

Hygrophorus melizeus (Batsch: Fr.) Fr. Fig. 12

Pileus 20-70 mm wide, convex to plano-convex, edge narrowly inrolled; surface glutinous to viscid becoming dry, edge tomentose, color white or yellowish white (4A2) to pale yellow (4A3) on the margin, pale orange (5A3) on the disc; context solid, white; odor of plastic. *Lamellae* thick, short decurrent to decurrent, subdistant to distant, color white with pinkish tones or cream to pale yellow with salmon pink tinge, darkening or brownish tinted (5D5 to 5E5) in age. *Stipe* 40-55 mm long, 5-14 mm thick, equal or tapering downward; surface fibrillose to furfuraceous on the apex, below glutinous to viscid or dry, color white to pale yellow, staining or changing to yellow-brown or light orange-brown where handled or bruised; context white to buff.

Basidiospores 7-9 (-10) x 4-5.5 (-6) µm, in profile view elliptic. *Basidia* 4-spored. *Hymenial cystidia* absent. *Pileus cuticle* an ixotrichodermium but with some interwoven hyphae; hyphae narrow, mostly 3-6 µm wide, somewhat intertwined, somewhat branched.

Habit and Habitat: Solitary to densely gregarious, with *Betula nana* or *Vaccinium vitis-idaea* and *Betula nana*, on slopes or in wet peat; mid- to late August.

Material Examined: GAL/JFA/DFF 2873, Dalton Highway, mile 316; GAL/JFA/DFF 2993 and 3036, Dalton Highway, mile 272.

Discussion: These specimens fit the concept of *Hygrophorus melizeus* in Moser (1983) fairly well. Also, they are in good agreement with the description in Miller (1982a) except that our material did not have an odor that was "aromatic and pleasant." The collections described by Miller (1982a) are from Polychrome Pass, Mt. McKinley National Park, collected in alpine tundra in deep moss among *Betula nana* and *Salix* spp. Our collections represent the first report of this species in the Arctic tundra north of the Brooks Range.

ACKNOWLEDGEMENTS

The authors gratefully acknowledge the financial support for this research provided under contracts N00014-77-C-0003 and N00014-77-C-0162 made to the University of Alaska by the Office of Naval Research and the Graduate School Research Fund, University of Washington. Logistics and field camp efforts were supported by the Naval Arctic Research Laboratory, the Alaska Department of Fish and Game, in particular Mr. Harry Reynolds, and Mr. Jim Rood, Kotzebue, Alaska.

Special thanks to Professor Orson K. Miller, Jr., Virginia Polytechnic Institute and State University for the loan of Alaskan Hygrophoraceae specimens.

REFERENCES

Ammirati, J. F., and Laursen, G. A., 1982, Cortinarii in Alaskan arctic tundra, *in:* "Arctic and Alpine Mycology, The First International Symposium on Arcto-Alpine Mycology," G. A. Laursen and J. F. Ammirati, eds., University of Washington Press, Seattle, pp. 282-315.

Brown, J., and Kreig, R. A., eds., 1983, "Elliot and Dalton Highways, Fox to Prudhoe Bay, Alaska (Guidebook to Permafrost and Related Features)," Fourth International Conference on Permafrost, July 18-22, 1983. University of Alaska, Fairbanks, Alaska.

Cash, E. K., 1953, A check list of Alaskan fungi, *Plant Dis. Reptr., Suppl.,* 219, Feb. 15.

Gulden, G., 1983, Studies in *Lepista* (Fr.) W.G. Smith. Section *Lepista* (Basidiomycotina), Agaricales, *Sydowia, Ann. Myc. Ser. II,* 36: 59-73.

Hesler, L. R., and Smith, A. H., 1963, "North American species of *Hygrophorus*," University of Tennessee Press, Knoxville.

Kobayasi, Y., Hiratsuka, H., Aoshima, K., Korf, R. P., Soneda, M., Tubaki, K., and Sugiyama, J., 1967, Mycological studies of the Alaskan arctic, *Ann. Rept. Inst. Ferm.,* Osaka No. 3, 138 pp.

Konerup, A., and Wanscher, J. H., 1967, "Methuen Handbook of Color," Methuen & Co. Ltd., London.

Lange, M., 1955, Macromycetes Part II. Greenland Agaricales, *Medd. Grøn.,* 147: 1-69, illus.

Lange, M., and Skifte, O., 1967, Notes on the Marcomycetes of northern Norway, *Acta Borealia, A. Scientia,* 23: 1-51.

Laursen, G. A., and Ammirati, J. F., 1982a, Lactarii in Alaskan Arctic tundra, *in:* "Arctic and Alpine Mycology, The First International Symposium on Arcto-Alpine Mycology," G. A. Laursen and J. F. Ammirati, eds., University of Washington Press, Seattle, pp. 216-244.

Laursen, G. A., and Ammirati, J. F., 1982b, The FISAM in retrospect, *in:* "Arctic and Alpine Mycology, The First International Symposium on Arcto-Alpine Mycology," G. A. Laursen and J. F. Ammirati, eds., University of Washington Press, Seattle, pp. 532-544.

Laursen, G. A., and Chmielewski, M. A., 1982, The ecological significance of soil fungi in Arctic tundra, *in:* "Arctic and Alpine Mycology, The First International Symposium on Arcto-Alpine Mycology," G. A. Laursen and J. F. Ammirati, eds., University of Washington Press, Seattle, pp. 432-492.

Miller, O. K., Jr., 1972, "Mushrooms of North America," E. P. Dutton and Co., Inc., New York, N.Y.

Miller, O. K., Jr., 1982a, Higher fungi in Alaskan subarctic tundra and taiga plant communities, *in:* "Arctic and Alpine Mycology, The First International Symposium on Arcto-Alpine Mycology," G. A. Laursen and J. F. Ammirati, eds., University of Washington Press, Seattle, pp. 123-149.

Miller, O. K., Jr., 1982b, Mycorrhizae, mycorrhizal fungi and fungal biomass in subalpine tundra at Eagle Summit, Alaska, *Holarctic Ecology,* 5: 125-134.

Miller, O. K., Laursen, G. A., and Murray, B. M., 1973, Arctic and alpine agarics from Alaska and Canada, *Can. J. Bot.,* 51: 43-49.

Miller, O. K., and Laursen, G. A., 1978, Ecto- and endomycorrhizae of arctic plants at Barrow, Alaska, *in:* "Vegetation and Production Ecology of an Alaskan Arctic Tundra, Ecological Studies 29," L. L. Tieszen, ed., Springer-Verlag, New York, N.Y., pp. 229-237.

Miller, O. K., Jr., Laursen, G. A., and Farr, D. F., 1982, Notes on Agaricales from Arctic tundra in Alaska, *Mycologia,* 74: 576-591.

Moser, M., 1983, "Keys to Agarics and Boleti," R. Phillips, ed., Gustav Fischer Verlag, Stuttgart.

Murray, D. F., 1978, Vegetation, floristics and phytogeography of Northern Alaska, *in:* "Vegetation and Production Ecology of an Alaska Arctic Tundra, Ecological Studies 29," L. L. Tieszen ed., Springer-Verlag, New York, N.Y., pp. 19-33.

Orton, P. D., 1960, New check list of British agarics and boleti. Part III. Notes on genera and species in the list, *Trans. Brit. Mycol. Soc.*, 43: 246-271.

Ridgway, R., 1912, "Color Standards and Color Nomenclature," Published by the author, Washington, D.C.

Saccardo, P. A., Peck, C. H., and Trelease, W., 1904, The fungi of Alaska. Harriman Alaska Expedition, "Cryptogamic Botany. V," pp. 44-49.

Smith, A. H., and Hesler, L. R., 1954, Additional North American Hygrophori. *Sydowia, Ann. Myc. Ser. II*, 8: 304-333.

HIGHER FUNGI IN TUNDRA AND SUBALPINE TUNDRA FROM THE YUKON TERRITORY AND ALASKA

Orson K. Miller, Jr.

Biology Department, Virginia Polytechnic Institute and State University, Blacksburg, VA 24061, U.S.A.

Key words: *Cortinarius*, *Inocybe*, nunatak, *Salix*, Arctic, alpine

ABSTRACT

Descriptions and illustrations are provided for six species of *Inocybe* and one *Cortinarius* collected in tundra and subalpine tundra mainly from the St. Elias Mountains, the Yukon Territory and Alaska. A new combination, *Inocybe boltoni* ssp. *giacomi*, is proposed.

INTRODUCTION

The majority of the fungi reported in this paper were collected in the St. Elias Mountains during the Icefield Ranges Research Project conducted jointly by the Arctic Institute of North America and the American Geographical Society. These mountains are located in northwestern British Columbia, the southwestern Yukon Territory and extend north to the White River in Alaska. Some of the fungi have been reported in prior publications which have been assembled in Volumes 2 and 4 of the Icefield Ranges Research Project, Scientific Results (Bushnell & Ragle, Eds., 1970 & 1974). However, other taxa from remote stations within the Icefields have not been studied and are reported here.

One of these areas is a nunatak located in the Kaskawulsh Glacier above the Slims River which originates at the foot of the glacier. The Kaskawulsh Nunatak is an extensive bedrock ridge which protrudes through the ice of the Kaskawulsh Glacier (Murray & Murray, 1969). The nunatak is about 24 kilometers upglacier from the glacier terminus and it divides the upper Kaskawulsh Glacier into the north and central arms. The nunatak is a low, vegetated knoll only a few miles above the firn limit. According to Murray & Murray (1969) it is separated on each side from the adjacent mountain slopes by 1.5 to 2 kilometers of ice and it lies at 1768 m (5799 ft) elevation with a vegetated surface about 400–500 m long.

The nunatak has 61 species of Angiosperms of which 41 species are dicotyledonous plants and 20 species are monocotyledonous plants. The latter group is composed mostly of grasses and sedges. The dicots are distributed among 9 families. *Salix* is the only ectomycorrhizal genus (Miller, 1982) and is represented by four species including *S. brachycarpa* Nutt. ssp. *niphoclada* (Rydb.) Argus, *S. polaris* Wahl., *S. reticulata* L. and *S. rotundifolia* Trautv. There are no other candidates listed by Murray (1978) on the nunatak which are known ectomycorrhizal higher plant hosts. This snowfree site is well above the regional snowline and it provides a present day opportunity to observe those fungi which can grow and act as decomposers or ectomycorrhizal symbionts under severe conditions (Murray, 1981). Such arctic refugia provide us with a clue as to survival of both plant host and fungus associate during continental glaciation. Lastly, the putative mycorrhizal associates provide us with a knowledge of the fungus species and higher plants which are able to initially colonize totally isolated land recently freed from ice. In addition, tundra vegetation is usually composed of an intermixed series of different ectomycorrhizal plants. Here we have only four species of *Salix* with which the putative ectomycorrhizal fungi can be associated.

A second area is the Chitistone Pass (1774 m) at the headweaters of the White River located at the northern end of the St. Elias Mountains. It is similar in many ways to the Kaskawulsh Nunatak vegetation but it does have nearby icefree tundra vegetation in the Chitistone (Skolai) region. This has allowed the development of a somewhat more diverse plant assemblage. Scott (1974) includes this area in the middle alpine zone (1446-2215 m). A wet meadow tundra in the pass is dominated by *Carex bigelowii* Torr. and *Petasites frigidus* (L.) Fr. with *Arctagrostis latifolia* (R. Br.) Griseb. in the wet swales. The somewhat drier sites just above the valley bottom support *Dryas octopetala* L., *Salix arctica* Pull., *S. polaris* and *S. reticulata*. These four species are the only ones that form ectomycorrhizae with higher fungi (Miller, 1982). *Cassiope tetragona* (L.) D. Don and *Vaccinium uliginosum* L. ssp. *microphyllum* occupy exposed ridges and the steeper valley sides. They form endomycorrhizae of the ericoid type as described and illustrated by Miller (1982). More protected, dry sites are occupied by *Hierochloe alpina* (Sw.) R. & S. and *Trisetum spicatum* (L.) Richt. (Scott, 1974). The area investigated for fungi directly in the Pass has about 45 species of Angiosperms. The direct westerly exposure and the resulting high winds, coupled with late melting snow, make this a somewhat more severe site than that of the Kaskawulsh Nunatak.

All collections are housed in the herbarium at Virginia Polytechnic Institute and State University, Blacksburg, Virginia (VPI) unless otherwise indicated. Ridgway (1912) colors begin with capitals and are enclosed in quotation marks. Kornerup and Wanscher (1967) color designations are indicated by page, row, and number (p7A-B3).

Cortinarius cf. *inconspicuus* Favre — Figs. 1-3

Pileus 4-12 mm broad, convex to broadly convex or even plane in age, dry, deep red brown, "Natal Brown" to "Bone Brown", margin obscurely translucent, striate. Lamellae ascending, adnexed, broad, distant, brown, "Army Brown". Stipe 10-15 mm long, 1.5-2 mm broad, equal, dry, glabrous with some sparse, basal fibrils, light brown, "Fawn Color".

Spores (7-)8.5-11 X 5-6.5 µm, broadly elliptical, thin-walled with small but distinct separate warts, brown in KOH and Melzer's solution. Basidia 23-33 X 7.5-11 µm clavate to nearly ventricose, thin-walled, hyaline, four-spored. Pleurocystidia none. Cheilocystidia 18-37 X 3.5-11 µm, clavate to narrowly clavate, thin-walled, with dark reddish-brown contents in KOH, infrequent and part of an oleiferous system. Pileocutis a thin, irregular trichodermium of thin-walled pileocystidia (13-)25-64 X 5-9 µm, cylindric, narrowly clavate to clavate, often with a basal clamp connection, sparse to frequent, arising from a dense layer of deep rusty brown in KOH and Melzer's solution, often incrusted hyphae, 4-11.5 µm

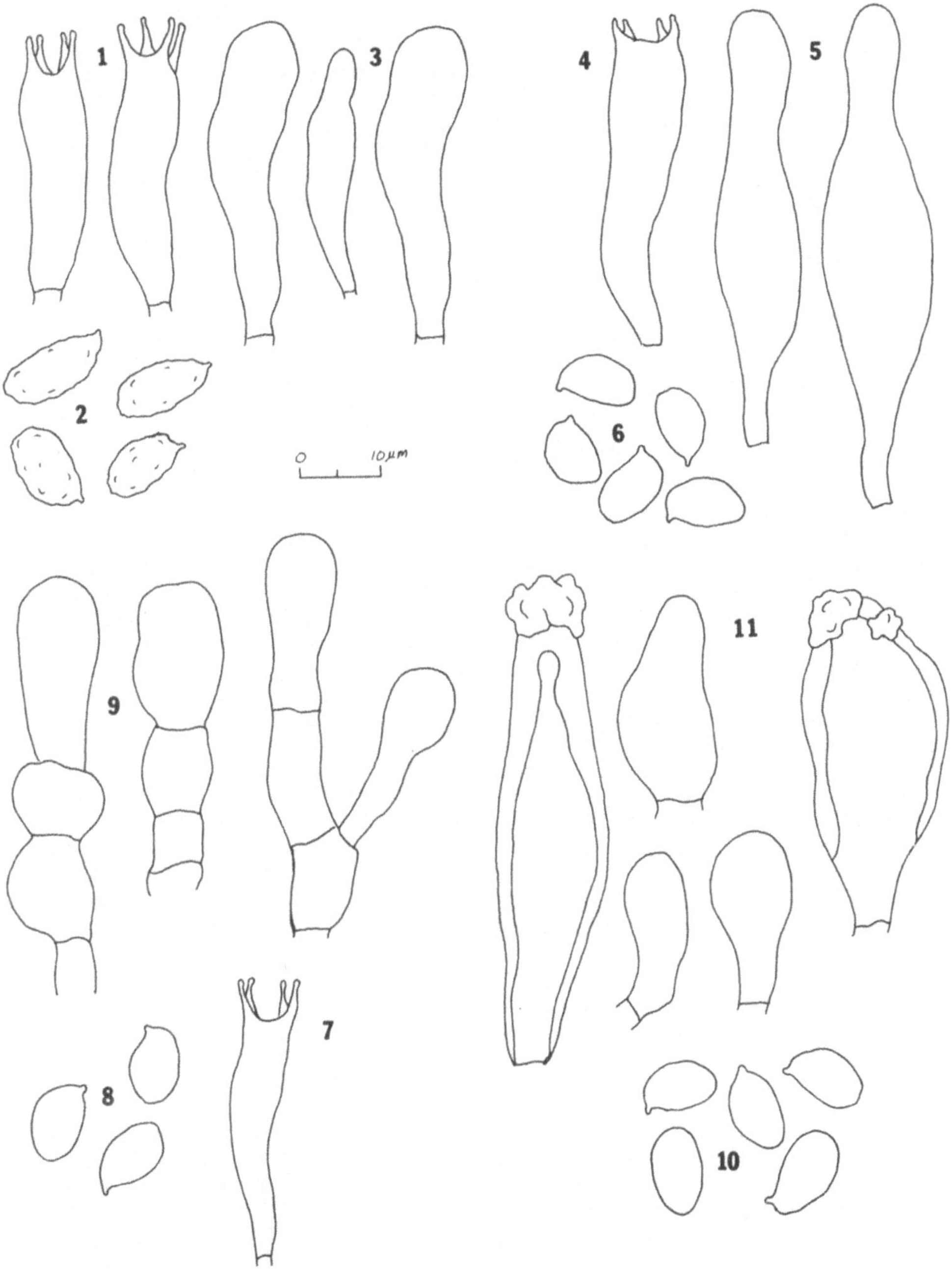

Figs. 1-3. *Cortinarius inconspicuus*. Fig. 1. Basidia. Fig. 2. Basidiospores. Fig. 3. Cheilocystidia. Figs. 4-6. *Inocybe bongardii*. Fig. 4. Basidium. Fig. 5. Cheilocystidia. Fig. 6. Basidiospores. Figs. 7-9. *Inocybe dulcamara* f. *pygmaea*. Fig. 7. Basidium. Fig. 8. Basidiospores. Fig. 9. Cheilocystidia. Figs. 10-11. *Inocybe gausapata*. Fig. 10. Basidiospores. Fig. 11. Cheilocystidia.

diam., somewhat thick-walled. Pileotrama of very broad to ventricose, thin-walled hyphae (8-)20-32 µm diam. with a reddish brown hue but hyaline individually. Lamellar trama of broad, often inflated hyphae 8-18(-40) µm diam., loosely parallel to irregular with reddish brown walls and scattered incrustations. Clamp connections common in all tissues.

HABIT, HABITAT and DISTRIBUTION: Several together, sometimes nearly caespitose, on the ground among lichens and mosses under dwarf willows (*Salix rotundifolia* noted) in alpine tundra. Fruiting occurs in late July and August.

MATERIAL EXAMINED: CANADA: Yukon Territory; David & Barbara Murray OKM 5910, OKM 5918, OKM 5926 (VPI).

OBSERVATIONS: This minute species has been recorded only in tundra. It is distinctive by virtue of its small size, warted spores, hyaline pileocystidia, lack of pleurocystidia, infrequent deep red-brown caulocystidiaa, very large tramal cells and numerous clamp connections. Although it is close to *Galerina*, the spores have no plage and the characteristic cheilocystidia found in *Galerina* are lacking. Infrequent cystidia with reddish-brown contents are present and resemble contorted basidioles. One could easily conclude that it does not have cheilocystidia depending on the interpretation of these contorted cells with brown contents. They are described for the Section *Submelinoideae* Sing. of *Naucoria*. In this section the only taxon which resembles our material is *N. submelinoides* Lge. (Moser 1983 p. 341). The only other possibility is a *Hydrocybe* (*Cortinarius*) close to *C. inconspicuus* Favre but our material has somewhat larger spores and does not quite fit a specific taxon described and illustrated by Favre (1955). There is also no evidence of the cortinus veil or fibrillose annular zone usually noted in the subgenus *Hydrocybe* but Favre does indicate that there are a few species with no annulus. It is found in Section *Telemoniae* in Moser (1983, p. 417) among the species without a white annulus on the stipe. Lange (1957) reports "a form of the *C. uraceus*-group, close to *C. inconspicuus* Favre, but with larger spores". This is also the situation with the taxon described above which may be a closely related species from arctic tundra. Additional study of fresh material is needed to fully resolve the placement of this taxon.

Inocybe bongardii (Weinm.) Quél. Figs. 4-6

Pileus (12-)18-25 mm broad, conic to convex-umbonate, dry, matted fibrils, yellow brown, "Apricot Buff" to orange brown over the umbo "Cinnamon-Rufous" in age, fine reddish fibrils densely arranged over the center becoming well separated over the margin. Context firm, white. Lamellae adnate, subdistant with one or two tiers of lamellulae between each lamella, broad, "Cinnamon-Rufous". Stipe 18-28 mm long, 3-5 mm thick, equal, dry, covered at first with downy white fibrils darkening to "Apricot Buff" in age and downy white pubescence over the base. Veil a dense, white cortina soon separated sometimes leaving a scant annular zone near the apex.

Spores 8.5-11 X 5-6 µm almost oblong to reniform, slightly thick-walled, brown in 3% KOH. Basidia 28-36 X 8-10 µm clavate, thin-walled, four-spored, hyaline in KOH. Cheilocystidia infrequent, 23-40 X 6-9 µm fusiform to clavate, thin-walled, hyaline. Cuticle a trichodermium of regular tufts of dark brown cystidial hyphae 5-10 µm diam with basal clamp connections. Pileotrama of interwoven hyphae, 8-12 µm, thin-walled, hyaline or light brown in KOH.

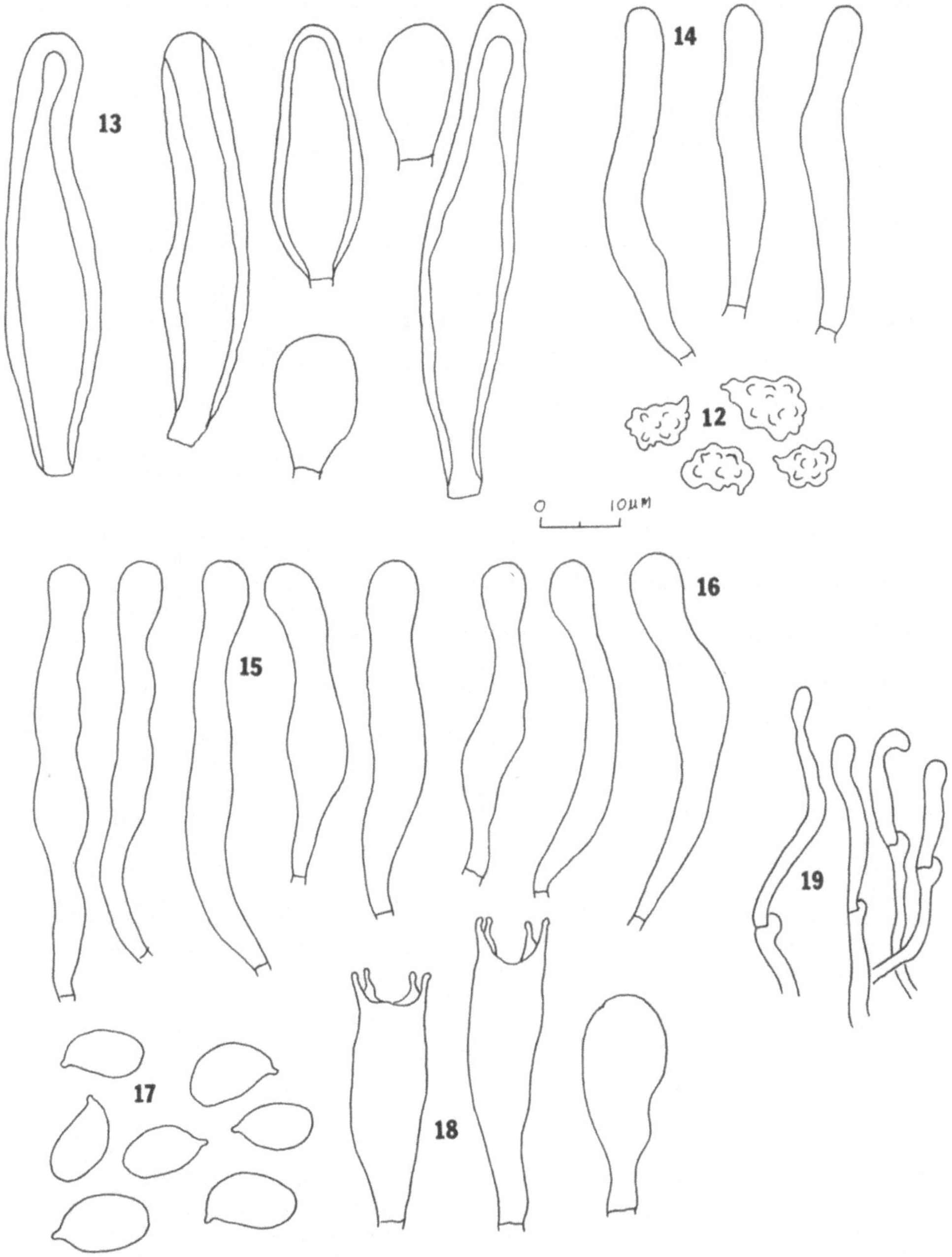

Figs. 12-14. *Inocybe rufo-alba*. Fig. 12. Basidiospores. Fig. 13. Cheilocystidia. Fig. 14. Pileocystidia. Figs. 15-19. *Inocybe tenuicistidiata*. Fig. 15. Cheilocystidia. Fig. 16. Pleurocystidia. Fig. 17. Basidiospores. Fig. 18. Basidia. Fig. 19. Hyphae of cuticle.

HABIT, HABITAT AND DISTRIBUTION: Several together in deep moss in alpine tundra, Polychrome Pass, McKinley National Park, Alaska. Fruiting in early August.

MATERIAL EXAMINED: CANADA: Yukon Territory; Kaskawulsh Nunatak OKM 5889 (VPI). USA: Alaska, McKinley Nat. Park, OKM 15434 (VPI). SWITZERLAND: Ramosch; Horak 4.IX, 1963. (ZT).

OBSERVATIONS: There are only four or five species of *Inocybe* with no metuloids. These species have scattered to numerous thin-walled cheilocystidia. The characteristics of *I. bongardii* are those of our taxon in the sense of Alessio.

Inocybe dulcamara (Fr.) Kummer f. *pygmaea* Favre Figs. 7-9

Pileus 8-15(-35) mm broad, convex, dry, brown, "Sayal Brown" covered with conspicuous nearly appressed to slightly raised squamules whcih are brown tipped, "Snuff Brown", in age densely fibrillose with squamulose only over the disc. Lamellae adnate to broadly adnate, close to subdistant, brown, "Mikado Brown". Stipe 20-40 mm long, 2-3(-5) mm broad, dry, covered with appressed, downy fibrils mostly "Sayal Brown" and scattered, raised, dark brown, "Snuff Brown" squamules, context soft and hollow in age. Veil fibrous, separates early and leaves no annulus.

Spores 9-11 X 5.5-7 µm, elliptical, smooth, thin-walled, light yellow brown in KOH and Melzer's solution. Basidia 25-35 X 6-7.5 µm clavate, thin-walled, four-spored. Cheilocystidia 17-28 X 8-16 µm, thin-walled, ovoid, pyriform, clavate, often in chains of several cells, numerous, forming a sterile gill edge. Pleurocystidia absent or occasional only near the gill edge. Cuticle a loose trichodermium of rusty brown, erect hyphae in KOH with 5-16 µm diam., thin-walled, with some incrustations, often in fascicles with interspersed cystidial end-cells. Pileotrama beneath cellular cuticle, 5-17 µm diam., thin-walled, hyaline to light yellowish in KOH becoming interwoven, hyphal-like near the gill trama which is composed of subparallel hyphae 5-11 µm diam. and also hyaline with yellowish walls. Clamp connections frequent in all tissues.

HABIT, HABITAT and DISTRIBUTION: Several close together on the ground among bush willows (*Salix* spp.) or at higher elevations among *Dryas* or dwarf willow in subalpine tundra. Fruiting in late June at low elevations to early or mid july at higher elevations in the Donjek Mountains, Kluane Game Preserve, Yukon Territory, Canada.

MATERIAL EXAMINED: CANADA: Yukon Territory; OKM 5365, 5496, 5514, 5518, 5604, 5605, 5606, 5731, 5878, 5884, 5885, 5888 (VPI).

OBSERVATIONS: *Inocybe dulcamara* is regarded as "a very rare species in the United States" according to Stuntz (1947). I have yet to encounter it during extensive field work in Virginia, Idaho and Montana. However, we have many collections from the Yukon and Alaska where it is common above timberline among species of *Salix* and *Dryas* in arctic and subalpine tundra. At lower elevations it has been noted under or near aspen (*Populus tremuloides* Michx.) and white spruce (*Picea glauca* (Moench) Voss). However, the ever present dwarf and shrubby species of *Salix* and species of *Dryas* could easily be the main ectomycorrhizal hosts. *Inocybe dulcamara* has also been found in pioneering situations on the Kaskawulsh Nunatak in the Donjek Mountains, Yukon Territory, Canada in the Chitistone (Skolai) Pass on the Yukon-Alaska border which was covered by mountain glaciers in the recent past. It has been reported by Malloch (1973) from arctic and boreal sites in Alberta, Ontario, and the Northwest Territories in Canada. The description by Malloch suggests that he may have sampled some of the f. *pygmaea* along with the v. *dulcamara*.

The distinctive field characters include small brown squamules over the cap surface, smooth spores and ventricose to pyriform cheilocystidia.

The illustrations in Alessio (1980, Tab. 1) and form *pygmaea* Favre (Alessio 1980, Tab. 2,1) well illustrate the different populations which I have encountered. The latter taxon is described here since it has never been recognized in North America and it agrees well with the treatment by Favre (1955).

Inocybe gausapata Kühner Figs. 10-11
= *Inocybe flocculosa* (Berk.) Sacc. Sensu Lange

Pileus 7-30 mm broad, conic, campanulate to convex-umbonate, dry, light brown, "Clay Color", with radially arranged squamules or fascicles of hairs which appear lighter than the ground color, margin in age conspicuously rimose. Lamellae nearly free, close, light brown, "Pale Cinnamon Pink" to nearly "Avellaneous". Stipe 16-30 X 2-4 mm broad, nearly equal or enlarging toward the base, glabrous, even, dry, very pale brown to nearly white.

Spores (8-)10-11(-13) X 5.5-6(-7) µm, subreniform in profile, oblong-elliptical in face view, dingy yellow-brown with a single oil body in KOH. Basidia 28-34 X 8.5-9.5 µm, clavate, thin-walled, four-spored, hyaline in KOH. Cheilo- and pleurocystidia similar, 42-70 X 12-19 µm, clavate to broadly fusiform, thick-walled, hyaline in KOH frequently with apical incrustations. Cuticle a trichodermium, hyphae with incrusted walls 6.5-14 µm diam., often in loose fascicles with many clamp connections. Pileotrama of nearly cellular elements, 6-24 µm diam., hyaline to light yellowish in KOH. Lamellar trama of loosely parallel hyphae 6-13 µm diam., hyaline, thin-walled in KOH.

HABIT, HABITAT AND DISTRIBUTION: Single or in troops on the ground under deciduous trees or decumbent shrubs (including *Salix*) in tundra plant communities. Fruiting in July in the Yukon Territory and Alaska.

MATERIAL EXAMINED: CANADA: Yukon Territory: OKM 5422, 5423, 5427, 5428 (VPI), Kaskawulsh Nunatak OKM 5881 (VPI). USA: Chitistone (Skolai) Pass; OKM 5729, 5788 (VPI).

OBSERVATIONS: The minute, light squamules on the darker brown ground color, white stipe without a basal bulb and large spores combine to make this a distinctive species. It is found associated with deciduous angiosperms and in tundra it is found near species of *Dryas* and under species of *Salix* (Favre, 1969). It is illustrated by Phillips (1981), Alessio (1980, Tab. 53).

Inocybe rufo-alba Pat. et Doassans sensu Lange Figs. 12-14

Pileus 5-10 mm broad, convex with a distinct umbo, dry, radially fibrillose, brown to rusty brown, "Sayal Brown" to "Cinnamon Rufous". Lamellae adnate, distant, thick, brown, "Mikado Brown". Stipe 10-20 mm long, 1-1.5 mm broad, nearly equal, glabrous with minute fibrils visible under a lens, light brown, "Sayal Brown", with a downy, white pubescence over the slightly enlarged base.

Spores 6.5-8.5 X 5-6.5(-7.5) µm, ovoid to short elliptical, tuberculate warted, light brown in KOH. Basidia clavate, thin-walled, hyaline, four-spored. Cheilocystidia fusiform to broadly cylindrical, thick-walled, 30-60 X 11-13 µm diam., intermixed with ovoid, clavate often fluted thin-walled cells 12-20 X 8.5-12 µm diam., hyaline in KOH. Pleurocystidia numerous but all similar to the thick-walled cheilocystidia. Pileocuticle a deep reddish-brown trichodermium with tightly bound fascicles or scattered, thin-walled pileocystidia, 40-50 X

5-6 µm diam., cylindric arising from a basal layer of hyphae 3.5-7 µm diam. in KOH. Pileotrama of dense, interwoven, reddish brown hyphae, 2.5-11 µm diam. almost the same color as the cuticle. Lamellar trama of parallel, thin-walled hyphae, 3.5-7 µm, yellowish to yellow-brown in KOH. Clamp connections present in all tissues.

HABIT, HABITAT AND DISTRIBUTION: Fruiting in clusters in moss and *Cladonia* near shrubby willows and scattered conifers on a lateral moraine in an area glaciated prior to 1936 at the foot of Mendenhall Glacier, Tongass National Forest, Alaska in early August (Aug. 4, 1967).

MATERIAL EXAMINED: USA: Alaska; Tongass National Forest OKM 6003 (VPI).

OBSERVATIONS: This is a minute species with a distinct umbo and odd shaped tuberculate warted spores. The illustration by Alessio (1980, tab. 92) fits our material very well. *Inocybe egenola* Favre (1955) is very close and found in alpine tundra, however, it is lighter in color and not as distinctively umbonate as our taxon and the spores are not as distinctly tuberculate and variable. I favor, therefore, referring our material to *I. rufo-alba*. This group of Inocybes needs to be intensively studied.

Inocybe tenuicystidiata Horak & Stangl Figs. 15-19
= *Inocybe leptocystis* Atk. var. *ambigua* Favre

Pileus 10-20 mm broad, convex, glabrous, moist, light red-brown. Lamellae adnate, close, light brown, "Avellaneous". Stipe 15-25 mm long, 2-5 mm broad, glabrous, dry, nearly white. Odor not distinctive.

Spores 9-11 X 5.5-6.5 µm, broadly elliptical, thin-walled, smooth, light brown. Basidia 24-30 X 8-11 µm, clavate, thin-walled, four-spored with long sterigmata (3.5-7 µm), hyaline. Cheilocystidia 37-50(-72) X 5-8.5 µm, cylindric to narrowly clavate, thin-walled, hyaline with a basal clamp connection. Pleurocystidia similar in shape and size, less numerous.Cuticle a loose mixocutis of hyphae 2.5-7 µm diam., thin-walled, hyaline with clamp connections which arise from a subcutis of densely interwoven, incrusted hyphae 3-9 µm diam., hyaline in KOH. Pileotrama of hyphae 4-12 µm diam., nearly cellular, thin-walled, hyaline to pale yellow in KOH.

HABIT, HABITAT AND DISTRIBUTION: Several together on the ground in alpine tundra on the Kaskawulsh Nunatak located in the upper Kaskawulsh Glacier, St. Elias Mountains, Yukon Territory. Fruting in late July.

MATERIAL EXAMINED: CANADA: Yukon Territory; Kaskawulsh Nunatak, David & Barbara Murray OKM 5875, 5885, 5886, 5893, 5902, 5925 (VPI). WEST GERMANY: Sigmaringen; Stangl 25, IX, 1974 (Holotype) (ZT).

OBSERVATIONS: The material from tundra in the Yukon is small in stature with the uniformly long, cylindric, thin-walled cystidia pictured by Favre (1955, Fig. 80) and described by Horak & Stangl (1980). No other tundra species of *Inocybe* has this combination of characteristics combined with large, elliptical, smooth spores.

Inocybe boltoni Heim ssp. *giacomi* (Favre) comb. et stat. nov. Figs. 20-22
= *Inocybe giacomi* Favre, 1955: 115.

Pileus 18-22 mm broad, conic at first to campanulate with a prominent

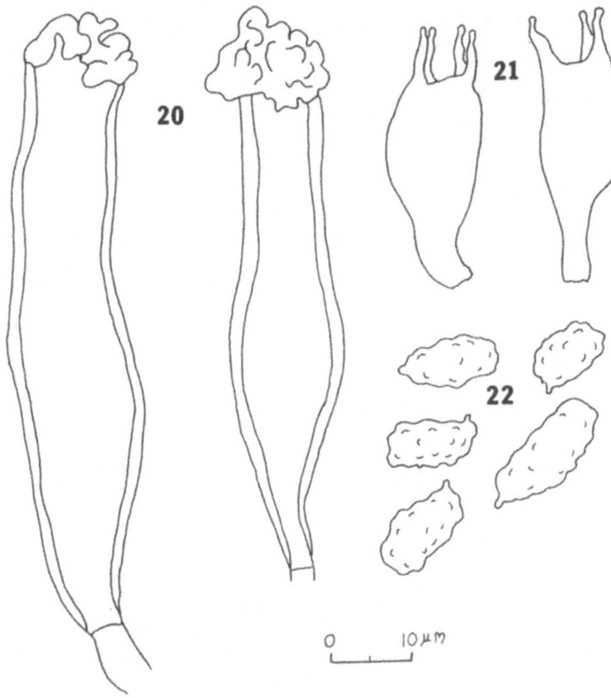

Figs. 20-22. *Inocybe boltoni* ssp. *giacomi*. Fig. 20. Cheilocystidia. Fig. 21. Basidia. Fig. 22. Basidiospores.

umbo, dry with a radially striate, matted, brown surface (p7E6-7) often almost corded, revealing a light brown (p7A-B-3) ground tissue beneath. Lamellae adnate, subdistant with alternating lamellulae, light brown (p7D5-6). Stipe 18-26 mm long, 4-5 mm broad, tapering somewhat toward the apex, dry, light brown with scattered loose lighter fibrils on the surface. No sign of a veil or veil remains. Odor not distinctive.

Spores (9-)10-15.5 X 5-6.5 µm diam., oblong to elliptical angular with irregular low warts, slightly thickened wall, yellow brown in KOH and Melzer's solution. Basidia 24-30 X 10-13 µm diam., broadly clavate, thin-walled, hyaline with sterigmata 5-7 µm long. Pleurocystidia and cheilocystidia 60-80 X 12-18 µm diam., narrowly clavate to fusiform with a rounded apex, thick-walled (1.25 um), often incrusted at the apex, hyaline with yellow walls in KOH.

Cuticle a dense rusty brown trichodermium in KOH composed of hyphae 3.5-6(9-8.5) µm diam. Pileotrama of large often ventricose elements 4.5-15.5 µm diam., deep brown in KOH and distinct from the cuticle. Lamellar trama of parallel thin-walled hyphae 3.2-14 µm diam., nearly hyaline. Clamp connections seen in all tissues.

HABIT, HABITAT AND DISTRIBUTION: Scattered among *Salix rotundifolia* on bluffs south of Barrow Village on the Arctic North Slope of Alaska. Fruiting in late August.

MATERIAL EXAMINED: USA: Alaska; Barrow, OKM 11913 (VPI).

OBSERVATIONS: A strikingly similar *Inocybe* is described by Favre

(1955) on Mount Plazer at 2600 m under mats of *Salix herbacea*. Trimbasch (1978) has rejected the complete synonymy of *I. giacomi* Favre with *I. boltoni* Heim and instead considers the former a variety of the latter. Alessio (1980, Tab. 98, lower left) illustrates a fungus considered to be part of the phenotypic variation of *I. boltoni* which resembles my taxon. The dark brown pileus color, dark lamellae, narrowly fusiform cystidia, and short stipe set this fungus apart from *I. boltoni* and I, therefore, prefer to recognize the taxon as a subspecies. When the population is studied further, it, may in fact, warrant recognition as a separate species as proposed by Favre.

REFERENCES

Alessio, C. L., 1980, "Iconographia Mycologica 29: Suppl. III, *Inocybe*," Scuola Grafica Salesiana, Trento.

Atkinson, C. F., 1918, Some New Species of *Inocybe*, *Am. J. Bot.*, 5: 210-218.

Bushnell, V. C., and Ragle, R. H., 1970, "Icefield Ranges Research Project Scientific Results. Vol. 2," Am. Geographical Society & Arctic Institute of North America, 138 p.

Favre, J., 1955, "Les champignons superieurs de la zone alpine du Parc National Suisse," Ludin AG., Liestal, 212 pp., illus.

________, 1960, Catalogue descriptif des la zone alpine du Parc National Suisse, *Ergeb. Wis Untersuch. Schweiz. National Parks*, 6: 326-610.

Grund, D. W., and Stuntz, D. E., 1968, Nova Scotian Inocybes I, *Mycologia*, 60: 406-425.

________, 1983, Nova Scotian Inocybes VII, *Mycologia*, 75: 257-270.

Horak, E, 1979, *Astrosporina* (Agaricales) in Indomalaya and Australasia, *Persoonia*, 10: 157-205.

________, 1980, *Inocybe* (Agaricales) in Indomalaya and Australasia, *Persoonia*, 11: 1-37.

Horak, E., and Stangl, J., 1980, Notizen zur Taxonomie und Verbreitung von *Inocybe leptocystis* Atk., *Sydowia*, 38: 145-151.

Lange, M., 1955, Macromycetes Part II, Greenland Agaricales, *Medd. Grøn.*, 145: 1-69.

________, 1957, Macromycetes Part III, Greenland Agaricales, *Medd. Grøn.*, 148: 1-125.

Malloch, D., 1973, "Fungi Canadenses No. 3. *Inocybe dulcamara*," Agric. Can., Ottawa.

Miller, O. K., Jr., 1982, Ectomycorrhizae in the Agaricales and Gasteromycetes, *Can. J. Bot.*, 61: 910-916.

Moser, M., 1983 (Eng. Translation), "Key to Agarics and Boletes 4th Ed.," Roger Phillips, 15a Eccleston Sq., London S.W. 1: Gustav Fischer Verlag, Stuttgart, 535 p.

Murray, D. F., 1971, Comments on the Flora of the Steele Glacier Region, Yukon Territory, *in:* "Expedition Yukon," Marnie Fisher, ed., Thomas Nelson & Sons Ltd., Canada.

________, 1978, "Ecological Studies on a Nunatak in the Kaskawulsh Glacier, Yukon Territory," Unpublished Report to the Arctic Institute of North America, 36 p.

Murray, D. F., and Murray, B. M., 1969, Notes on Mammals in Alpine Areas of the Northern St. Elias Mountains, Yukon Territory and Alaska, *The Can. Field-Naturalist*, 83: 331-338.

Phillips, R., 1981, "Mushrooms and Other Fungi of Great Britain and Europe," Pan Books, Ltd., London, 288 p.

Scott, R., 1974, Alpine Plant Communities of the Southeastern Wrangell Mountains, Alaska, p. 283-306, *in:* "Bushnell & Marcus, Ed. Icefield Ranges Research Project Scientific Results Vol. 4: 385 pp. Am. Geographical Soc. and Arctic Inst. of N. Am".

Stuntz, D. E., 1947, Studies in the Genus *Inocybe* I. New and Noteworthy Species from Washington, *Mycologia*, 39: 21-55.
Trimbach, J., 1978, Materiel pour une check list des Alpes Maritimes, *in:* "Documents Mycologiques," 29.

FUNGI (AGARICALES, RUSSULALES) FROM THE ALPINE ZONE OF YELLOWSTONE NATIONAL PARK AND THE BEARTOOTH MOUNTAINS WITH SPECIAL EMPHASIS ON *CORTINARIUS*

Meinhard Moser

Institut für Mikrobiologie
Universität Innsbruck
Sternwartestrasse 15
A-6020 Innsbruck, Austria

and

Kent H. McKnight

Plant Protection Institute, Agricultural Research Service
U.S. Department of Agriculture
Beltsville, Maryland 20705

Key words: alpine fungi, *Cortinarius*, *Myxacium*, *Lactarius*, new species, Rocky Mountains, *Telamonia*, SEM

ABSTRACT

A first series of observations on the fungus flora of the alpine zone in mountains of Wyoming and Montana is presented here. Special attention is given to species of the genus *Cortinarius*. Thirteen species of this genus are discussed or described, 5 of them are new: *Cortinarius absarokensis* sp. nov., *C. fuscoflexipes* sp. nov., *C. mucronatus* sp. nov., *C. rufoanuliferus* sp. nov., and *C. vulpicolor* sp. nov. Several of the other species treated are new for North America, including *Lactarius nanus* Favre.

INTRODUCTION

The importance of alpine ecosystems as part of the high elevation watersheds and pasturelands has been appreciated for a long time. As interest in alpine ecosystems steadily increases, interest in alpine fungi increases also. Thus, in the summer of 1983, several visits were made to the alpine zone on Mt. Washburn in Yellowstone National Park and to several sites on the high plateau in the region of Beartooth Pass along the common border of Wyoming and Montana. Both localities can be reached with comparative ease. The summit of Mt. Washburn is accessible by a well maintained footpath which was formerly an auto road. U.S. Highway 212 winds along the high plateau of the Beartooth Mts. above the treeline in an alpine area approximately six miles wide and 12 miles long.

The alpine crest of the Beartooth Range represents the eroded surface of an uplifted fault block, overthrust into the Big Horn Basin to the east (Spencer, 1959). It has a northwest-southeast orientation. The exposed precambrian granitic rocks are cut by both basaltic and acid-porphyry dikes. In their detailed report on the alpine vegetation of the Beartooth Plateau in relation to cryopedogenic processes and patterns, Johnson and Billings (1962) described four main vegetation types which intergrade along environmental gradients: (1) *Geum* turf, (2) *Deschampsia* meadow, (3) *Carex* bog, and (4) *Salix* thicket. Within each vegetation type community composition varies greatly in response to diverse microhabitats.

The Washburn Range is located in North Central Yellowstone Nat'l Park. Mount Washburn is the remnant of an eroded, glaciated volcanic cone with exposed andesitic breccias and conglomerates of the Eocene Langford Formation.

Our interest, during this brief study, was primarily in collecting species of *Cortinarius*, especially Sect. *Telamonia*, which we found mostly in association with two species of dwarf *Salix*. We know of no reports of macrofungi from the alpine zone of these areas, although Johnson and Billings (1962) reported two parasitic fungi. The Beartooth Pass area and Mt. Washburn were visited in the first and fourth weeks of August. No macrofungi were found there during the first week.

Most fungi collected were associated with an undetermined alpine species of *Salix*, which seems to be the most favorable phanerogamic mycorrhizal partner in this habitat. Fewer species were found associated with *Salix reticulata* L. and one was found under *Salix planifolia* Pursh.

As expected, most of the fungus species readily identified from study areas of the alpine zone are known from similar high mountain habitats in Europe, but a significant number of them seem to be new for North America. Five taxa are described here as new, however, one of these was collected earlier in Europe. Nine of the 13 species of *Cortinarius* treated here are recorded for both continents. A very conspicous fungus, which occurred frequently but at only one of the collecting sites, is *Lactarius nanus* Favre.

Herbarium designations where exsiccati are located follow Holmgren et al. (1981). Names of colors in the species descriptions are indicated as follows:

Expo. = Cailleux et Taylor (1912), R. = Ridgway (1912), Mu. = Munsell (1966), and Meth. = Kornerup and Wanscher (1967).

ENTOLOMATACEAE

Entoloma (Fr.) Kummer

Entoloma clypeatum (L.: Fr.) Kummer var. *alpicola* Favre

This taxon was very frequent in the Beartooth Pass area of Montana and Wyoming at about 3,300 m elevation during the last week of August, 1983. The American material agrees fully with European collections.

Collections studied: 83/329, Beartooth Mts., Shoshone Nat'l. Forest, Park Co., Wyo., 22.VIII.1983, IB; 83082402, Beartooth Mts., Absaroka Wilderness, Custer Nat'l. Forest, Carbon Co., Mont., 24.VIII.1983, BPI.

CORTINARIACEAE

Inocybe Fr.

Many species of *Inocybe* were observed but not collected, an exception being *Inocybe calamistrata* (Fr.) Gill. This species was observed at several localities in the Beartooth Pass area. One collection was preserved: 83/356, north of Beartooth Pass, Absaroka Wilderness, Custer Nat'l Forest, Carbon Co., Montana, ca. 3,200 m, among dwarf *Salix* spp., 24.VIII.1983, IB.

Cortinarius Fr.

Subgen. *Myxacium* (Fr.) Loud.

Only two species of *Myxacium* were found associated with dwarf or low shrubby species of *Salix* in alpine zones of Yellowstone Park and the adjacent Beartooth Mts. Both species also occur in Europe. One is the widely distributed and fairly variable *Cortinarius favrei* Moser ex Henderson; the other is a new species.

Cortinarius favrei Moser ex Henderson

The type variety was observed both on Mt. Washburn, Yellowstone Nat'l. Park, Wyo., at about 3,050 m among plants of *Salix* sp. and in the Beartooth Mountains just north of the south summit of Beartooth Pass at several places between 3,200 and 3,300 m with *Salix planifolia* and *S. reticulata*.

There are no macro- or microscopic differences between the American and European material.

Spores 11.6-14.3 X 5.5-7.7 µm (means 12.87 [s = 0.82] X 7.33 [s = 0.54]), Q = 1.77 in face view, 11-16 X 6.1-9.4 µm (means 13.24 [s = 1,32] X 7.26 [s = 0.71]), Q = 1.84 in side view.

Collections studied: 83/388, 27.VIII.1983, Mt. Washburn, Absaroka Range, Yellowstone, Nat'l. Park, Wyo., lB; 83082410, Beartooth Pass, Beartooth Mts., Shoshone Nat'l. Forest, Park Co., Wyo., BPI.

This species was reported from arctic tundra in Alaska by Kobayashi et al. (1967) as *C. alpinus* Boud. and by Ammirati and Laursen (1982). We know of no other reports from alpine tundra in North America.

Cortinarius favrei Moser ex Henderson forma *pallida* Moser et McKnight, fm. nov.

Differt a typo in colore pilei pallide ochraceo, interdum fere eburneo, dein pallide brunneo.

Holotypus IB 83/338, Beartooth Pass, Beartooth Mts., Shoshone Nat'l. Forest, Park Co., Wyo., sub Salice, 22.VIII.1983, isotypus BPI.

In the Beartooth Pass area a comparatively pale form of the species was observed several times which, except for its more pallid colors, agrees in aspect and in microscopic characters with typical *C. favrei*.

Pileus 2.0-5.5 cm across, hemispherical to campanulate or convex, often irregularly wavy, glutinous, color fairly pale ochraceous, sometimes nearly ivory to pale brownish (Expo. 77 K, Mu. 2.5Y 8/4-8/6) to sometimes pale orange brown (Expo. 65 L, rarely 59 M), sometimes with brown spots (Expo. 57 N) on paler background when old, always much paler than typical *C. favrei* even in very young specimens. Occasionally becoming rimose. *Lamellae* very pale lilac (Meth. 19A2) in very young, often unopened carpophores, but soon becoming pale grayish brown (Expo. 54 C), finally rust colored (Expo. 56 E), adnate to broadly but shallowly emarginate, edge somewhat eroded, 3-8 mm wide (=2-3 X thickness of pileus context), crowded, L = 45-48, l = 3, 15-16 per cm near margin. *Stipe* 2-4 cm X 6-10 mm, cylindric or slightly clavate, rarely tapering downward, apex dingy whitish (Expo. 71 K - 75 K), pure white below the white cortina, glutinous, base turning yellowish to brownish, sold; *context* dingy whitish underneath cuticle, brownish in stipe base. Taste mild; smell not distinctive.

Microscopic characters: *Spores* elongate amygdaline, 11-16.5 X 6.6-8.0 µm (means 13.36 [s = 0.96] X 7.53 [s = 0.43]), Q = 1.78 in face view, 12.1-15.4 X 6.6-8.9 µm (means 13.5 [s = 0.99] X 7.39 [s = 0.51]), Q = 1.83 in side view, strongly verrucose, wrinkled under SEM. *Basidia* 4-spored, 50-55 X 12-13(14) µm, with basal clamp connection, sometimes with brown content (phaeobasidia). No *cheilocystidia*. The gelatinous layer on cap 150-170 µm thick, hyphae ± straight and periclinal, 2-3.5 µm thick, often enlarged at the septae, clamp connections present. Epicutis hyphae 7-10 um, brown, with conspicuous, irregular, ± granular incrustations, subcutis of hyphae 10-15 um, mixed with some narrower (5-8 µm), strongly pigmented ones. *Pileus trama* ± irregular, hyaline, hyphae 10-15-20 µm thick. *Lamellular trama* ± subparallel, hyphae 7-8 µm, hyaline. Cortina of clamped hyphae 2-4 µm in diam.

Habitat: several localities north of Beartooth pass, Beartooth Mts., Shoshone Nat'l. Forest, Park Co., Wyo., ca. 3,300 m, among *Salix* sp. and *S. reticulata*. *Collections studied*: 83/338, 83/338b, 22.VIII.1983, IB; 83082205, 83082207, 22.VIII.1983, BPI.

Cortinarius absarokensis Moser et McKnight, sp. nov. Figs. 1a, b, 2, 6a.

Pileo (3)-5-11-(13) cm lato, semiorbiculari margine involuto, dein convexo usque applanato, plus minusve flavobrunneo usque fulvo, saepe pallidiore, saepe maculato-variegato, glutinoso, lamellis primo pallide griseis, dein argillaceis usque subfuscis, denique pallid flavo-brunneis, acie serrulato, subconfertis, 5-11 mm latis, stipite 3-6.5 cm longo, 10-22 mm crasso, cylindrico sive, basin versus attenuato albido, glutinoso, basi flavescente vel brunnescente, carne albida, in cortice brunneolo-variegata. Sapore miti, odore nullo vel debiliter fruticoso. Sporis elongate-amygdaliformibus, 12.1-16.5 x 6.6-8.8 µm valde verrucosis, basidiis 4-(vel 2-) sporigeris, 40-50 x 11-12(13) µm, absque cheilocystidiis.

Habitatio sub Salice planifolia. Holotypus IB 83/352, Absaroka Wilderness, Beartooth Pass, Montana, 24.VIII.1983, isotypi BPI, YELLO.

Pileus (3)5-11(13) cm, hemispheric with incurved margin when young, then convex, finally applanate, with narrowly overhanging cuticle (1-2 mm), center often somewhat depressed, yellow brown to fulvous (Expo. 68 D, 56 D) but often much paler even in young carpophores (Expo. 77 K 89 K, 60 L, 65 L, 56 B), and then center only yellow brown (Expo. 58 E), often with spots of these colors on paler ground, often finely radially wrinkled, not translucent-striate, glutinous. *Lamellae* at first pale grayish (Expo. 61 B), then argillaceous to gray brown (Mu. 2.5Y 7/4), finally light yellow brown (Expo. 65 M), edge often somewhat paler, strongly serrulate, adnate to shallowly emarginate, moderately crowded, L = 60, l = 1-3, 10

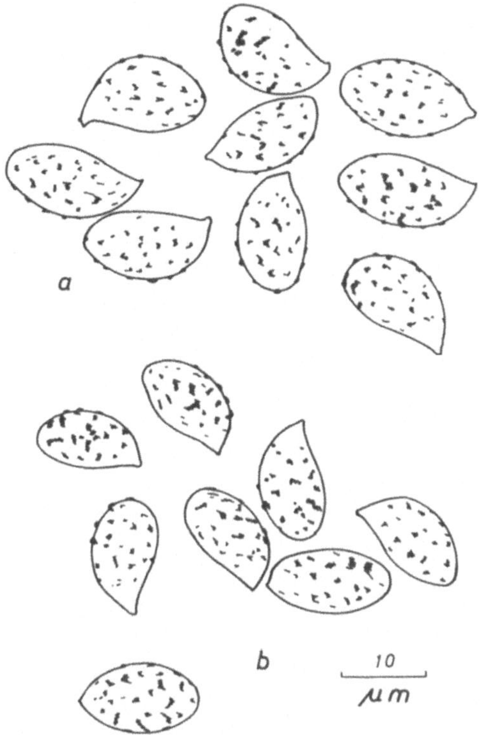

Fig. 1. Spores of *Cortinarius absarokensis*: a. Coll. 83/352 (type), b. Coll. 73/41 (Obergurgl, Tyrol).

per cm at pileus margin, 5-11 mm broad (=3 X thickness of pileus context), sometimes unequally wide. *Stipe* 3.0-6.5 cm X 10-22 mm, cylindric or tapering toward base, sometimes also tapering upward from cortina, sordid whitish to slightly brownish above cortina, pure white and glutinous below, sometimes becoming somewhat floccose and wrinkled when old, base becoming yellowish to brownish. *Context* firm, whitish except slightly brownish just beneath cuticle, sometimes also cortex of stipe streaked brownish. *Taste* mild, odorless or with slightly fruity smell.

Microscopic characters: *Spores* elongate almond-shaped, 12.1-16.5 X 7.2-8.8 µm (means 13.58 [s = 0.90] X 7.68 [s = 0.40]), Q = 1.77 in face view, 12.1-15.4 X 6.6-8.8 µm (means 13.05 [s = 0.69] X 7.53 [s = 0.41]), Q = 1.73 in side view, strongly verrucose, wrinkled under SEM. *Basidia* 4-(or 2-) spored, clavate, 40-50 X 11-12(13) µm, with basal clamp connection. No cheilocystidia or only insignificant basidium-like clavate cells. Gelatinous layer on pileus 120-180 µm thick, of narrow hyphae (2-3.5 µm) with clamp connections, $\pm$ parallel to surface, somewhat wavy, partiallly collapsing. *Epicutis* with a layer of hyaline hyphae (4-8 µm thick) and a thick layer of brown hyphae with lumpy encrusted wall, *subcutis* mixed with hyphae up to 20 µm thick with brown pigmentation. *Pileus trama* hyaline, irregular, hyphae 10-20 µm diam. Clamp connections present but not on all septae. Hyphae of glutinous veil from stipe 2-4(5) µm diam., with clamp connections. *Context of stipe* of strongly parallel hyphae, 8-20 µm thick.

Habitat: under *Salix planifolia*, ca. 3,200 m, frequent and in large quantities, Abasaroka Wilderness north of Beartooth Pass, Beartooth Mts., Carbon Co., Montana, U.S.A. (coll. 83/352, 24.VIII.1983, IB, BPI,

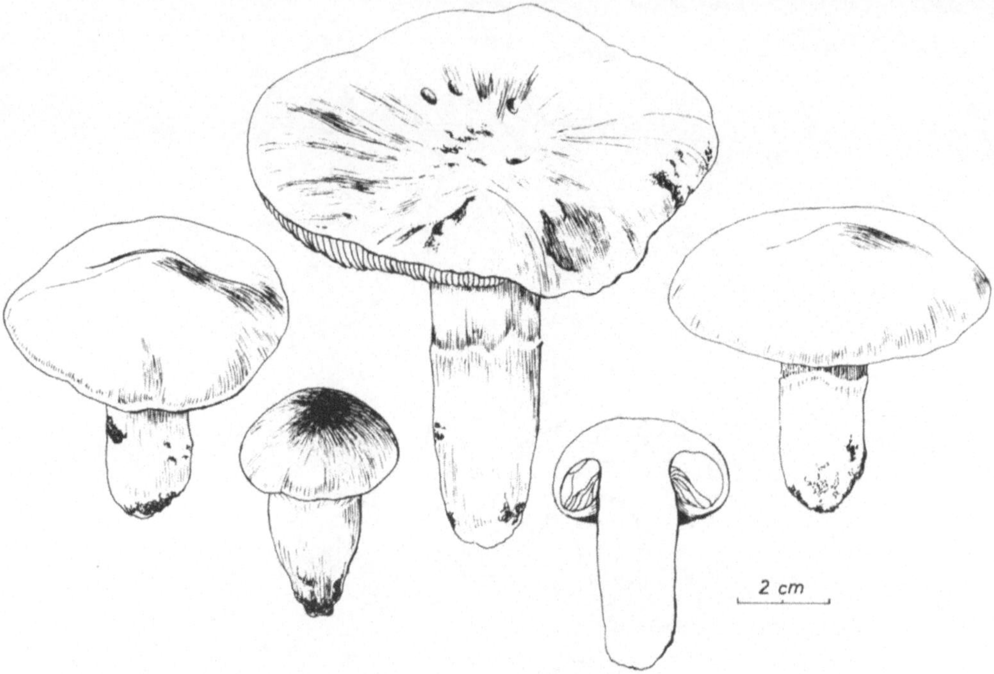

Fig. 2. Carpophores of *C. absarokensis*: Coll. 83/352 (type).

YELLO); and under *Salix* sp. near the embankment of a mountain stream, Obergurgl Tyrol, Austria, 2,000 m (coll. 73/41, 19.VIII.1973, IB).

Comments: This fungus was found by one of us (M.M.) in Tyrol about 10 years ago but there was only one specimen and the species remained undeterminable. The rich collections made now in the Beartooth mountains allowed study of the species in its full range of variation. Microscopically it is very similar to *C. favrei*, the spore shape and size being about the same. However, the habitat, size and habitat of *C. absarokensis* are quite different from that of *C. favrei*. Where as *C. favrei* is found with alpine dwarf willows in the strict sense (*S. reticulata*, *S. retusa*, *S. herbacea* etc.) the above described species is found with (and usually under) *Salix* bushes which are, in the case of *Salix planifolia*, 20-30 cm high, and in the case of the Tyrolean record 1.5-2.0 m. *Cortinarius alpinus* Boud. has basidiocarps the size of *C. favrei* but with significantly larger spores and is a very rare species. The microscopic characters place our fungus in Sect. *Myxacium*.

Subgen. *Telamonia* (Fr.) Loud.

(a) Species with yellow veil.

Cortinarius chrysomallus Lamoure Figs. 3a, b, 5a, b, 6b

A yellow-veiled *Telamonia* belongs to the most common species in the alpine zone both in the Beartooth Pass area and on Mt. Washburn. Some of our collections correspond exactly to the original description of *Cortinarius chrysomallus* Lamoure. Others appeared to differ enough in field characters that we suspected two species. Microscopic characters are too much alike in all collections, however, to admit two species. The North American material agrees well with European material of *C. chrysomallus*.

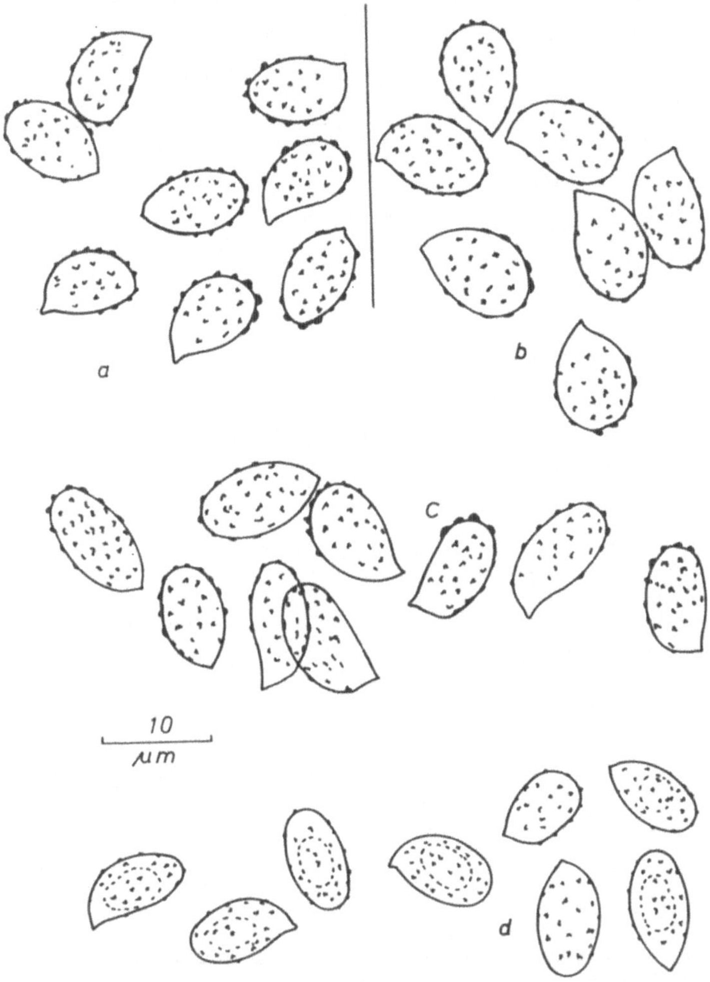

Fig. 3. Spores of: a. *Cortinarius chrysomallus*, Coll. 83/330; b. *C. chrysomallus*, Coll. 83/331; c. *Cortinarius fuscoflexipes*, Coll. 83/384; d. *Cortinarius sp.* cf. *C. stenospermus*, Coll. 83/353.

The original description emphasizes the presence of a bluish color in the basal mycelium, but admits considerable variation in intensity of the blue color. This character was observed by one of us (M.M.) only once on European material in a collection from Lappland. None of the collections from the Alps, which we have seen in the field, showed bluish colors. Several American collections from both Mt. Washburn and Beartooth Pass showed the very typical bluish basal mycelium, which on one covered the lower third of the stipe. On the other extreme, we have a considerable number of collections showing no trace of blue even on young carpophores. These usually have a longer, more slender stipe.

Lamoure (1977) states that this species seems not strictly bound to alpine dwarf *Salix*, although it appears to be predominantly an alpine species. She mentions one record under *Salix arbuscula* and *S. caesia*. We have one collection from Falls Campground on Brooks Lake Creek, Shoshone Nat'l. Forest, Fremont Co., Wyoming, where it grew in a

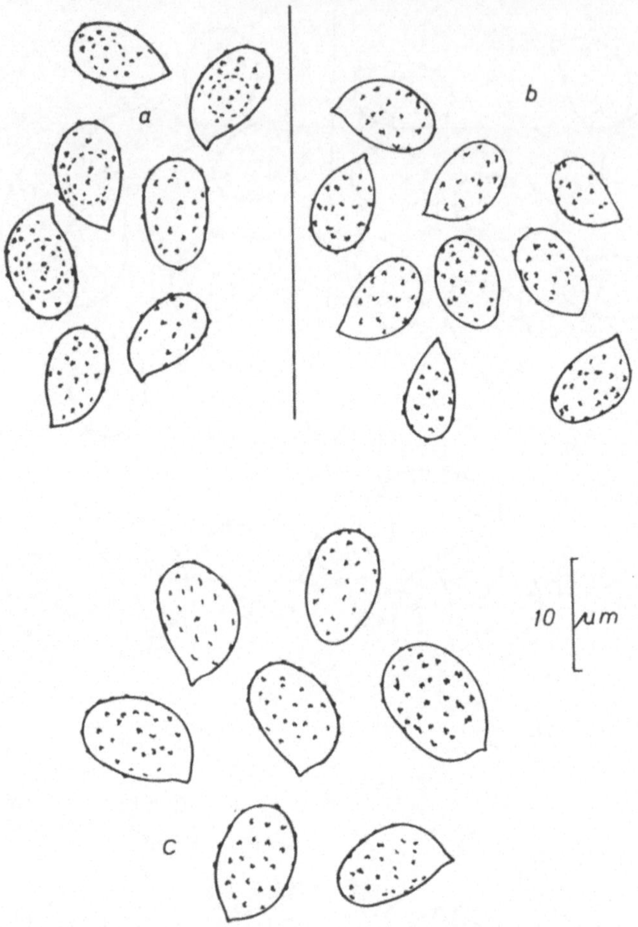

Fig. 4. Spores of: a. *Cortinarius mucronatus*, Coll. 83/342; b. *C. vulpicolor*, Coll. 83/355; c. *Cortinarius rufoanuliferus*, Coll. 83/386.

subalpine forest under *Picea engelmannii* and an unnamed species of *Salix*.

Collections studied: 83/330, IB,BPI,83/330b, IB, 83/331, 83/331b, 83/331c, IB, BPI, 83/331d, 83/331e, 83/331f, 83/331g, 83/331h, IB, Beartooth Pass, Shoshone Nat'l. Forest, Park Co., Wyo., ca. 3,300 m, 22.VIII.1983, 83/362, 83/363, IB, BPI, Absaroka Wilderness north of Beartooth Pass, Carbon Co., Mont., ca. 3,050 m, 24.VIII.1983; 83/382, 83/399b, IB, 83/399, IB, BPI, Mt. Washburn, Yellowstone Nat'l. Park, Wyo., ca. 3,050 m, 27.VIII.1983; 83/148, Brooks Creek Falls Campground, Shoshone Nat'l. Forest, Fremont Co., Wyo., ca. 2,800 m, 28.VII.1983.
(b) Dark brown to grayish brown species with white veil.

Cortinarius fuscoflexipes Moser et McKnight sp. nov. Figs. 3c, 5c, 6c.

Pileo 0.8-2.8 cm lato, acute vel obtuse umbonato, rariore semiorbiculari, hygrophano, numquam striato, udo obscure nigrobrunneo, sicco obscure rufofusco, in iuventute margine a velo albo obtecto, lamellis pallide fuscis, adnatis usque subemarginatis, 3-5 mm latis,

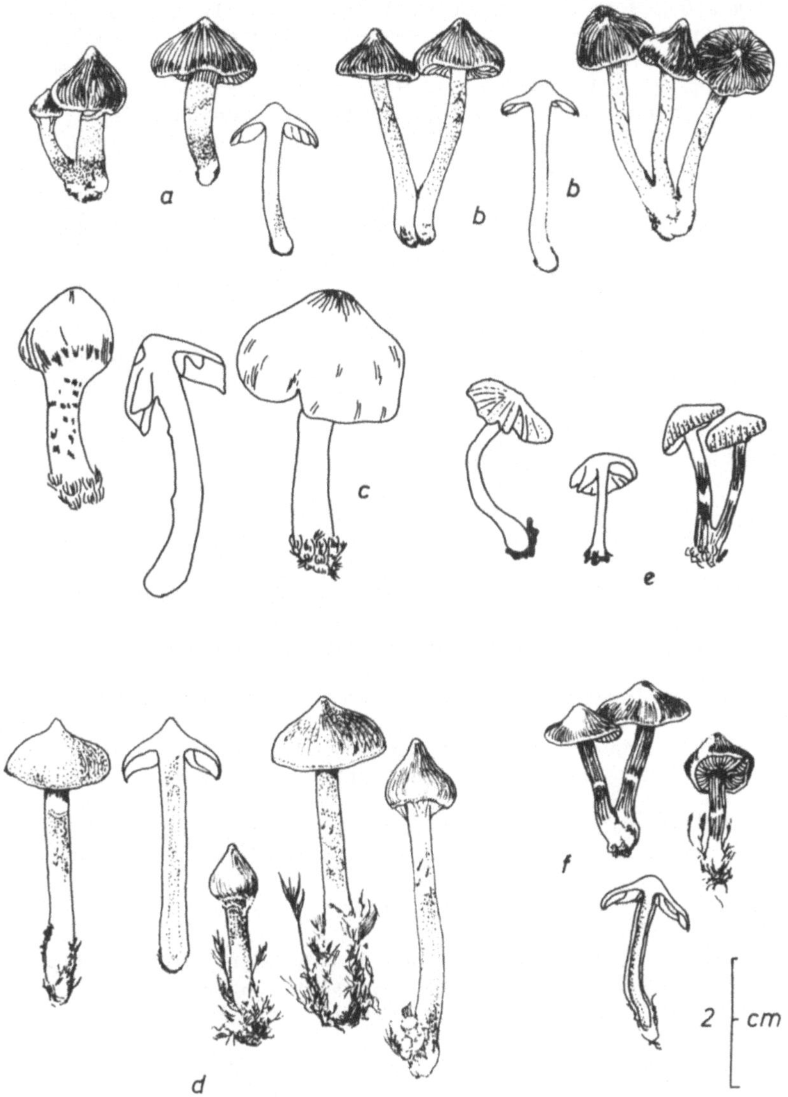

Fig. 5. Carpophores of: a. *Cortinarius chrysomallus*, Coll. 83/330; b. *C. chrysomallus* 83/331; c. *Cortinarius fuscoflexipes*, Coll. 83/384; d. *Cortinarius mucronatus*, Coll. 83/342; e. *Cortinarius vulpicolor*, Coll. 83/355; f. *Cortinarius rufoanuliferus*, 83/332.

subconfertis, stipite 2-4 cm X 2-6 mm, basin versus leviter incrassato, albo fibrilloso in fundo umbrino, in parte inferiore cum anulo albo irregulari et infra anulum floccis albis praedito, carne argillaceo-fusca, odore nullo, sapore miti.

Sporis elongatis, 7-10.6(11) X (4.8)5.3-6.2 µm, verrucosis, basidiis 4-sporigeris (raro bisporigeris) 30-34 X 9 µm.

Habitatio sub Salice sp., in vicintate Piceae engelmanii, 3,050 m altitudine, Mt. Washburn, Wyo.

Holotypus IB 83/384, 27.VIII.1983, leg. M. Moser, isotypus BPI.

Pileus 0.8-2.8 cm across, obtusely to ± acutely conical, rarely ± semiglobose, hygrophanous, not translucent-striate, dark blackish brown when humid (Expo. 70 T, 30 S-R), dark grayish brown with a slight red-brown tinge when dry, (Expo. 42 H to near 42 F), margin covered by a white veil in young carpophores (often visible even in older carpophores). Odor not distinctive, taste mild. *Lamellae* dark grayish brown (Expo. near 54 F to near 49 R, between Mu. 7.5YR 5/6 and 4/6), edge entire to slightly eroded, adnate to slightly emarginate, moderately crowded, L = 25-38, l = 3-7, 19 per cm at the margin, 3-5 mm broad (=4-5 X thickness of pileus context). *Stipe* 2-4 X 0.2-0.6 cm with slightly enlarged base, surface fibrillose with fine, silky white fibrils covering dark umber brown ground color (near Expo. 69 R) or slightly paler, with ± well-developed, irregular white ring in lower part and often white floccose below the ring; becoming hollow. *Context* argillaceous grayish-brown (Expo. 70 L), with umber brown variegation in cortex and upper part of stipe.

No reaction under ultraviolet light.

Microscopic characters: *Spores* ellipsoid to elongate, 7.0-10.6 X 5.3-6.2 µm (means 9.28 [s = 0.86] X 5.81 [s = 0.36]), Q = 1.6 in face view, 7.9-10.6 X 4.8-6.2 µm (means 9.42 [s = 0.71] X 5.5 [s = 0.39]), Q = 1.72 in side view, verrucose, strongly so in apical region, with irregular, interconnected warts under SEM. *Basidia* 4-, rarely 2- (or 1-) spored, 30-34 X 9 µm, with basal clamps. No *cheilocystidia*. Epicutis of hyphae 8-10 µm thick with fulvous walls, subcutis of thick, often subcellular hyphae up to 15(-20) µm thick, in the upper part of the trama the hyphae 8-15 µm with yellowish walls, more or less regular, in the lower part irregular to pseudoparenchymatic. *Cortina* of hyaline hyphae 3-4 µm diam. Clamp connections present on cortina and in trama.

Habitat: Among *Salix* sp., not far from *Picea engelmannii* at timber line, 3,050 m, Mt. Washburn, Yellowstone Nat'l. Park, Wyo.

Collection studied: 83/384, IB, BPI.

This species belongs to the group around *Cortinarius flexipes* Fr., from which it differs by its habitat, smaller, darker colored carpophores, and larger spores. *Cortinarius fuscoflexipes* has smaller spores than *Cortinarius casimiri* (Vel.) Huijs.

Cortinarius (Telamonia) sp.
(cf. *stenospermus* Lamoure ined.) Fig. 3d.

Our collection 83/353 (IB, BPI) comes close to or may even be identical with a species Lamoure intends to publish under the name *Cortinarius stenospermus*. Our collection agrees macroscopically fairly well, but the spores are somewhat shorter and slightly broader. We give a provisional description of our collection.

Pileus semiglobose to conic-campanulate, 0.6-1.5 cm across, hygrophanous, dark umber with slight red-brown admixture, (Expo. 50 S, 69 S), not translucent-striate, drying out in radial streaks, becoming red brown in dry condiiton, (Expo. 45 P, 47 P - 57 N), margin paler from the veil, glabrous, only the very margin occasionally with fine fibrils and squamules (visible under a hand lens). *Lamellae* pale grayish brown (Expo. 53 N-P), in age dark grayish brown (Expo. 55 R - 49 R), adnate to emarginate, crowded, L = 20, l = 1-(3), 22 per cm at pileus margin, 2 mm broad (=2-3 X thickness of context). *Stipe* pale brown (Expo. 67 N, 7 M), somewhat fibrillose, darker toward base (near Expo. 62 D), with

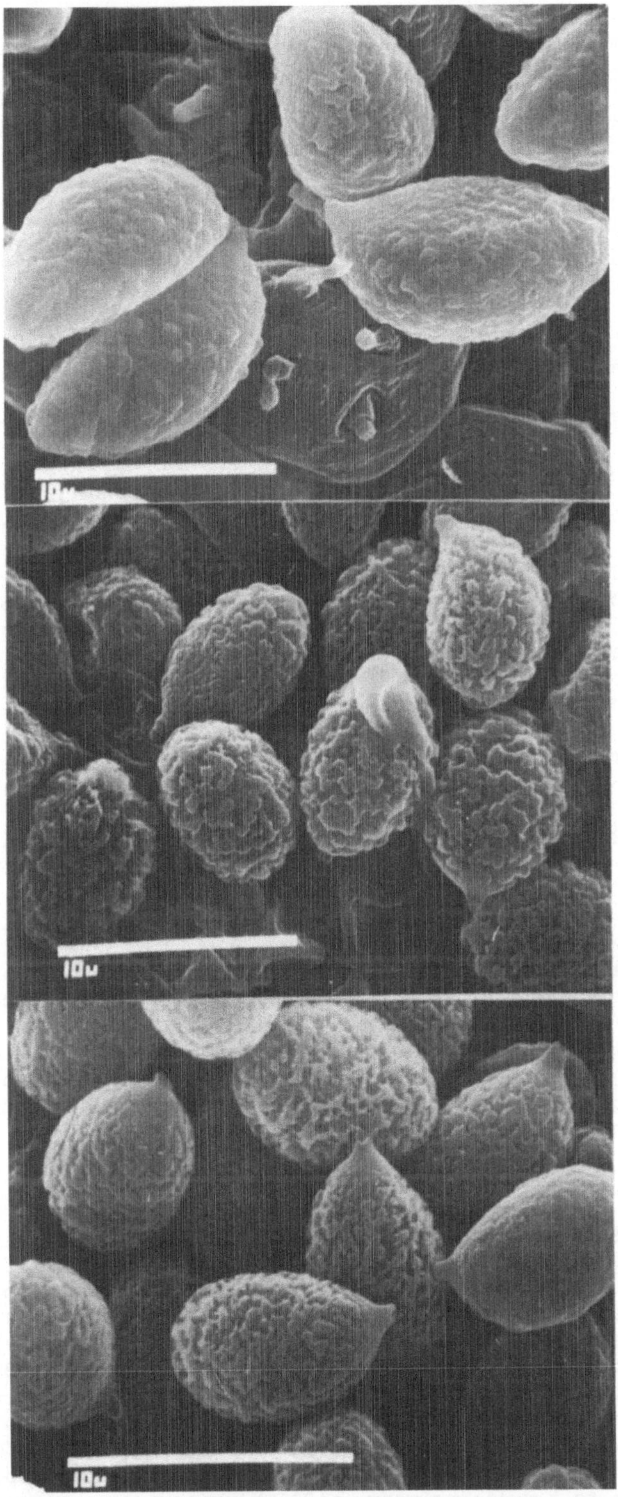

Fig. 6. Scanning electron micrographs of spores of: a. *Cortinarius absarokensis*, Coll. 83/352; b. *Cortinarius chrysomallus*, Coll. 83/331; c. *Cortinarius fuscoflexipes*, Coll. 83/384.

median, white, irregular belt and sometimes with white floccose scales and fibrils below. Context argillaceous in central part of pileus and stipe, dark brown at pileus margin and in stipe cortex (Expo. 62 H, 64 D). Odor weakly of *Pelargonium* leaves (sometimes practically odorless); taste mild. Spores in coll. 38/353: 7.4-9.2 X 4.4-5.7 um (means 8.33 [s = 0.44] X 5.42 [s = 0.39]), Q = 1.65 in face view, 7.4-9.7 X 4.4-5.7 um (means 8.68 [s = 0.58] X 4.93 [s = 0.45]), Q = 1.77 in side view, in coll. 83/353b: 8-10(11) X 5-6 um (Lamoure, 1977, gives 10-11 X 4.5-5.0 um). A remarkable feature of our collections is the strong brown, lumpy encrustations on all hyphae in the pileus and the presence of brown, amorphous intercellular pigment accumulations.

Habitat: Among *Salix* sp., ca. 3,200 m, Beartooth Pass area, Beartooth Mts., Shoshone Nat'l. Forest, Park Co., Wyo. 24.VIII.1983.

This species belongs to the group around *C. paleaceus*.

Cortinarius hemitrichus Pers.: Fr. fm. *improcerus* Favre

Several of our collections from the Beartooth Pass area belong to the group around *C. hemitrichus* and seem to fit best the description of *C. hemitrichus* fm. *improcerus* Favre. Most typical is our collection 83/333 (IB, BPI) from Beartooth Pass, Beartooth Mountains, Wyo., among *Salix* sp. at about 3,300 m elevation. The specimens in this collection has a ± acute to obtusely conic pileus of chocolate brown ground color densely covered by white squamules. The stipe (1.5-3.5 cm X 3-4 mm) was covered by the white floccose and peronate veil up to the middle. Spore measurements are 8.5-10(11) X 5.5-6(6.5) µm, and they are moderately verrucose. Thus some spores in the print are slightly larger than reported both by Favre (1955) and Lamoure (1977).

Coll. 83/334 (IB, BPI) from the same area agrees macroscopically well with 83/333, but the spores have weaker sculpturing, being only slightly roughened.

Two other collections from the same area, 83/335 and 83/335b, (both IB, BPI) have similarly sculptured spores of the same size as 83/333, but pilei have only a slight covering of white hairs with no squamules. Spores of 83/335b measure 8.3-10.6 X 4.8-5.7 µm (means 9.41 [s = 0.49] X 5.31 [s = 0.18]). However, in these collections all of the carpophores were older. The collections may belong to the above taxon but in the absence of specimens of all ages in prime condition we are unable to venture a final decision.

Cortinarius albonigrellus Favre

Collections 83/336 and 83/337 (both IB, BPI) from Beartooth Pass area, Wyo., among plants of *Salix* sp. at about 3,300 m fit fairly well in *Cortinarius albonigrellus* Favre. We have noted some minor differences but these seem not justify a separate taxon: (1) Both Favre (1955) and Lamoure (1977) report pileus size up to 1.8 cm. In collection 83/336 the pilei were from 1.0 to 2.1 cm. diam., in coll. 83/337 from 1.0 to 2.7 cm. (2) Favre (1955) and Lamoure (1977) indicate large, often appendiculate patches of veil at the pileus margin. In both of our collections we observed a characteristic line at a distance of about. 1-1.5 mm distance inward from the pileus margin with small patches of veil, similar to a quilting seam. (3) The spores are relatively slender. Favre reports 8-10 X 4-5 µm; Lamoure (1977) gives (7)8-9(10) X (4)4.5-5.0 µm. Our measurements indicate slightly broader spores: 8.8-12.3 X 5.3-7.0 µm (means 10.15 [s = 0.75] X 5.86 [s = 0.57]), Q = 1.74

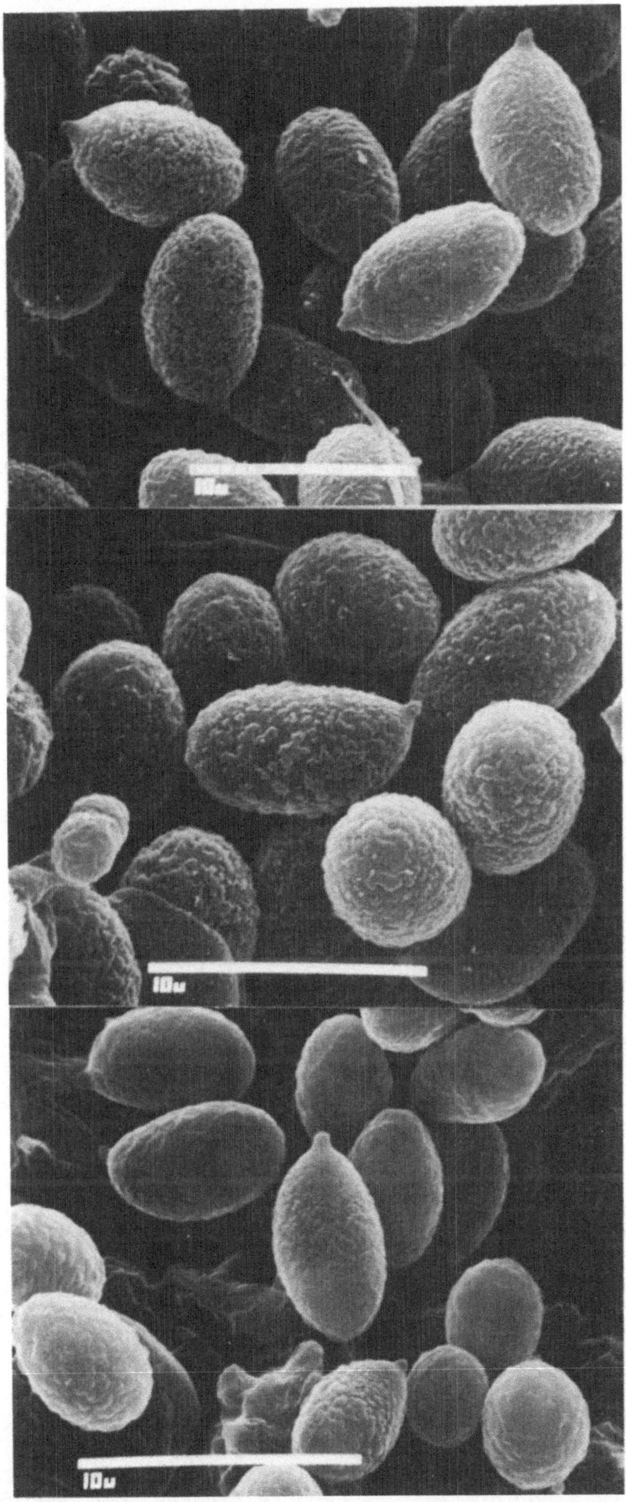

Fig. 7. Scanning electron micrographs of spores of: a. *Cortinarius mucronatus*, Coll. 83/342; b. *Cortinarius vulpicolor*, Coll. 83/355; c. *Cortinarius rufoanuliferus*, Coll. 83/387.

in face view, 8.8-11.0 X 5.3-6.6 µm (means 9.87 [s = 0.65] X 5.82 [s = 0.15]), Q = 1.70 in side view. Not all have the characteristic cylindric shape as figured by the above authors.

Cortinarius mucronatus Moser et McKnight sp. nov. Figs. 4a, 5d, 7a.

Pileo 1,2-2,2 cm lato, acute conico campanulato, dein convexo acute umbonatoque, hygrophano, nec pellucido striato, obscure rufobrunneo, primo margine e vel albo obtecto, dein nudo, sericeo, sicco flavobrunneo, lamellis subfuscis, dein flavo-ferruginascentibus, acie serrulato, subemarginatis, subconfertis, stipite aequali, 2.5-4.55 cm X 3-5 mm, pallide griseo-brunneo, e velo albo peronato anulatoque, carne pallide argillacea, odore subnullo vel subtiliter raphanoideo, sapore miti. Sporis ellipticis usque subcylindraceis, verrucosis, (8.2)8.5-10(11) X (4.5)5-6 µm, basidiis 4-(2-) sporigeris, 30-34 X 8-9 µm.

Habitatio inter Salice sp., in locis alpinis, Holotypus IB 83/342, Beartooth Pass, Beartooth Mts., Wyo., 22.VIII.1983, leg. M. Moser, isotypus BPI.

Pileus 1.2-2.2 cm across, at first acutely conic-campanulate, then expanding and more conic-convex, finally convex with an acute umbo, hygrophanous, not translucent-striate, dark red-brown when humid (Expo. 53 P), radially fibrillose, silky, never squamulose, margin covered by a white veil in young carpophoes, yellow brown in dry condition (Expo. 59 M), margin often retaining a darker color for a long time, center remaining ± red-brown (Expo. 57 P-40 R), in age the margin dark brown (Expo. 69 M), in young carpophores margin sometimes slightly plicate, splitting radially in old pilei. *Lamellae* grayish brown (R. = XXIX Tawny-Olive, or slightly paler) later yellowish rust-brown (R. = XV Ochraceous Tawny), edges eroded, indistinctly emarginate, slightly crowded, L = 20, l = 3, 18-19 per cm at the margin, 2-3(4) mm broad (=5-6 X thickness of pileus context). *Stipe* 2.5-4.5 cm X 3-5 mm, ± cylindric, pallid grayish brown (Expo. 70 L), lower part peronately covered by white veil, disappearing in the lower part or dissolving into floccose scales, leaving a ring-like belt at the upper end, stuffed at first, becoming hollow.

Microscopic characters: *Spores* verrucose, elliptic to subcylindric, 8.3-11.0 X 4.8-5.7 µm (means 9.37 [s = 0.66] X 5.40 [s = 0.22]), Q = 1.74 in face view, 7.9-11.8 X 4.8-6.2 µm (means 9.5 [s = 0.69] X 5.36 [s = 0.29]), Q = 1.78 in side view. *Basidia* 4-(2-) spored, 30-34 X 8-9 µm. *Pileus surface* with hyaline hyphae, 4-7(8) µm thick (veil remnants), epicutis hyphae 8-12 µm, with brown, slightly encrusted walls, subcutis not much differentiated but hyphae thicker, up to 15 µm. *Pileus trama* less pigmented, hyphae in the upper part ± subparallel 10-14 µm thick, irregularly pseudoparenchymatic below. Hyphae of the veil 5-7(12) µm, of the cortina 4-5 µm, with clamp connections.

Habitat: among plants of *Salix* sp., at about 3,300 m, Beartooth Pass area, Beartooth Mts., Shoshone Nat'l. Forest, Park Co., Wyo. *Collections studied*: 83/342, IB and BPI, 83/342b, IB, both 22.VIII.1983.

Comments: The species has the habit of certain forms of *C. paleaceus* Fr., but has a completely smooth pileus, no smell of *Pelargonium* leaves and is somewhat smaller. We considered a possible identity with *C. albonigrellus*, as Lamoure figures some specimens under that name with a similar habit (somewhat in contradiction to her description). However, the typical habit of *C. albonigrellus* is different and so is the coloration. These two species may be related.

(c) Reddish brown species with white veil.

Cortinarius pauperculus Favre

Collection 83/354 (IB, BPI) found growing between plants of *Salix* sp. on the plain north of Beartooth pass (3,200 m), on 24.VIII.1983 agrees well with *C. pauperculus* Favre. The pileus was rarely umbonate as figured by Favre (1960) but agreed well with the figures given by Lamoure (1978). The striking features are dark umber brown color of the humid pileus which turns reddish brown, (more rarely yellowish brown) from the center outward on drying. The white veil is scanty on the stipe, covering the basal third and forming a ± incomplete ring on some carpophores.

Spores in our collection are elongate, measuring 7.9-9.7 X 4.8-5.8 µm (means 8.64 [s = 0.50] X 5.31 [s = 0.32]), Q = 1.63.

Cortinarius vulpicolor Moser et McKnight sp. nov. Figs. 4b, 5e, 7b.

Pileo 0.8-1.8 cm lato, semiorbiculari usque convexo, raro leviter umbonato, hygrophano, udo ad hemidiametrum pilei vel utra pellucido striato, udo obscure rubro-brunneo, sicco palliodiore, fulvobrunneo, subtiliter fibrilloso, lamellis ferrugineis, subdistantibus, 1.5-4.0 mm latis, stipite 1.0-2.5 cm X 1-2 mm, rubro-brunneo, fibrillis albidis fugacibus, interdum subanuliformibus praeditis, caro brunneo, sapore miti, odore subnullo. Sporis late ellipsoideis, 9.3-11 X 6.3-8.0 µm, verrucosis, basidiis 4-sporigeris, 32-36 X 8-9 µm.

Habitatio inter Salice sp., Holotypus IB 83/355, Beartooth Pass, Mont., 3,200 m, 24.IV.1983, leg. M. Moser, isotypus BPI.

Pileus 0.6-1.8 cm across, mostly semiglobose to convex, rarely slightly umbonate, hygrophanous, translucent-striate inward from margin 1/2 to 2/3 of radius when moist, dark reddish brown (Expo. 47 P) on disc and striae, paler (Expo. 45 P) between striae, yellowish rust brown (Expo. 60 P) when dry, finely matted fibrillose, margin sometimes slightly crenulate. *Lamellae* rust brown (Expo. 45 P-R), slightly emarginate, edge entire to slightly wavy, subdistant, L = 15, l = 1-3, 13 per cm at margin, 1.5-4.0 mm broad (= 6 X thickness of context). *Stipe* 1.0-2.5 X 0.1-0.2 cm, vivid reddish brown to rusty brown, apex paler (Expo. 45 P), darker downward (47 P - 45 R), with fugacious whitish fibrils sometimes forming a fugacious ring about midway. *Context* dark brown (Expo. 47 R-S) in moist pileus, paler (45 R) in stipe. Smell indistinct, taste mild.

Microscopic characters: *Spores* broadly elliptic, 8-11 X 5.4-7.0 µm (means 9.57 [s = 0.75] X 5.98 [s = 0.39]), Q = 1.6 in face view, 8-11 X 5.3-8.5 µm (means 9.54 [s = 0.77] X 6.09 [s = 0.67]), Q = 1.58 in side view, verrucose. *Basidia* 4-spored, 32-36 X 8-9 µm. Hyaline hyphae (remnants of universal veil) on surface of *pileus cuticle* 5-9 µm diam., with clamp connections; epicutis of ± parallel hyphae, 8-14 µm diam., with strong brown, lumpy encrustations; subcutis of hyphae 7-20 µm thick, with lumpy encrustations less pronounced than in epicutis; *pileus trama* paler but also with lumpy encrustations, hyphae subparallel in upper part, irregular in deeper layers. Stipe hyphae 10-15 µm diam., with brown lumpy encrustations. Hyphae of veil from stipe surface 6-9 µm, hyaline, with clamps.

Habitat: Among *Salix* sp., on plain north of Beartooth Pass, 3,200 m, Absaroka Wilderness, Custer Nat'l. Forest, Carbon Co., Montana.

Collections studied: Holotype, 83/355, IB, 24.VIII.1983, leg. M. Moser, isotype, BPI.

The species is close to *Cortinarius galerinoides* Lamoure, which, however, is smaller, less striate and has smaller spores. *Cortinarius rufostriatus* Favre as figured and described by that author has a more conic pileus, larger size, and larger spores. *Cortinarius minutalis* Lamoure is a species with a more strongly developed veil and comparatively shorter and stouter stipe.

A macroscopically very similar collection from the same area (IB 83/355b) has somewhat smaller and more slender spores (8.5-10.0 X 4.6-6.0 µm) but has the same type of characteristic pigment incrustation in all parts of the carpophore except the veil. Thus we place it here with some reservation.

Cortinarius rufoanuliferus Moser et McKnight sp. nov. Figs. 4c, 5f, 7c.

Pileo 0,7-2,2 cm lato, obtuse, rarior acute conico, dein campanulato usque convexo, hygrophano, nec pellucido striato, udo rubrobrunneo usque ferrugineo, sicco flavobrunneo disco obscuriore, glabro, interdum areolato et in squamis inutis diffracto, lamellis ferrugineis, adnatis usque emarginatis, subconfertis, stipite (1)-1,5-3 cm X 1-3 mm, rufo-brunneo usque fulvo, velo albo peronate velato, evanescente, carne flavobrunnea, sicco argillacea, odore nullo, sapore miti. Sporis (6,8)7-9(10) X 4-5(6) um, basidiis 4-(2-) sporigeris, 30-33 X 7-9 µm.

Habitatio inter Salice sp. in zona alpina, ca. 3,050 m, Mt. Washburn, Yellowstone Nat'l. Park., Wyo., Holotypus 83/386, IB, 27.VII.1983, leg. M. Moser, isotypus BPI.

Pileus 0.7-2.2 cm across, obtusely (rarely acutely) conic, campanulate to convex and slightly umbonate, hygrophanous, not translucent-striate, bright red brown, rust brown, (Expo. 57 P - 59 P, 45 P - 40 N), darker and more dingy in older carpophores, (Expo. 55 P, 47 P) yellow brown when dry (Expo. 59 M, 58 C, 57 N), the center darker (38 E, 39 R) or remaining more reddish brown, smooth, in older specimens sometimes breaking up and becoming areolate to finely squamulose. *Lamellae* rust brown, (Expo. 45 P-R), darkening in older carpophores (to Expo. 47 R), adnate to emarginate, slightly crowded, L = 28-35, l = 1(2), 10-17 per cm at pileus margin, 1-3 mm broad, some ± ventricose, ((3)4-5 X thickness of context), edge entire to slightly eroded. *Stipe* (1)1.5-3.0 cm X 1-3 mm, red-brown to fulvous brown, (Expo. 59 P, 45 M to 57 N), becoming darker downward (Expo. 40 N), the basal part, sometimes up to the middle, peronately covered by the white veil, which may partly disappear on older carpophores but leaves mostly a white, often submembranaceous belt. Context dingy yellowish brown (Expo. 57 P, 57 N), argillaceous (69 M) in dry condition. No particular smell, taste mild. No reaction under UV.

Microscopic characters: *spores* 6.6-8.3 X 4.4-5.7 µm, (means 7.53 [s = 0.48] X 4.96 [s = 0.45]), Q = 1.69 in face view, 6.6-8.8 X 4.4-5.7 µm, (means 7.49 [s = 0.52] X 5.08 [s = 0.46]), Q = 1.48 in side view, elliptic to pip-shaped, moderately verrucose to slightly roughened, *basidia* 4-(rarely 2-) spored, 30-33 X 7-9(10) µm, lamellar edge with some vesiculose, rarely somewhat fusoid sterile cells of basidium size; cuticle on surface with some hyaline hyphae (4-6 µm thick) from veil remnants, epicutis of (5)6-10 µm thick hyphae with yellowish brown wall, sometimes with brown, lumpy encrustations, subcutis of thick, often subcellular hyphae (up to 20 µm and more), walls yellowish brown but not encrusted; *pileus trama* ± hyaline, subcellular to pseudoparenchymatic; Cortina hyphae 3-4(5) µm, hyaline, with clamp connections.

Habitat: Among *Salix* sp., at and above timber line. Coll. 83/386 and 83/387 Mt. Washburn, Yellowstone Nat'l. Park., Wyo., at about 3,050 m, 27.VIII.1983, leg. M. Moser. Coll. 83/332 Beartooth Pass, Beartooth Mt., Shoshone Nat'l. Forest, Park Co., Wyo., ca. 3,200 m, 22.VIII.1983. All collections in IB and BPI.

Comments: This species has some resemblance to *C. vulpicolor*. It differs however by the stronger developed veil, the ± persistent, nearly membranous ring, the lack of translucent striations on the pileus, by smaller spores, the less intensely colored context and the scarce hyphal incrustations which are found here only in hyphae of the epicutis.

(d) Species with veil inconspicuous or lacking.

Cortinarius tenebricus Favre

Three of our collections fit quite well with Favre's(1955) description of *C. tenebricus*; 83/381 from Mt. Washburn, Yellowstone Nat'l. Park, Wyo., at ca, 3,050 m, among *Salix* sp.; 83/368 from the plain north of Beartooth Pass, Park Co., Wyo., ca. 3,200 m, among *Salix* sp., 24.VIII.1983; 83/361 same area as 83/368, 24.VIII.1983. All collections in IB and BPI.

Favre (1955) reports spore measurements of 8.0-10.0 X 5.0-6.5 µm and also the indications of Lamoure (1978) fall in the same range. Our collections 83/381 and 83/368 show the range of variability, e.g. 83/368 has spores measuring 7.9-10.6 X 4.4-6.2 µm (means 9.17 [s = 0.75] X 5.46 [s = 0.36]), Q = 1.69 in face view, 7.9-9.7 X 4.4-6.2 µm (means 8.7 [s = 0.58] X 5.4 [s = 0.41]), Q = 1.62 in side view. Coll. 83/361 has slightly smaller spores: 7.4-9.7 X 4.8-5.7 µm (means 8.54 [s = 0.60] X 5.27 [s = 1.62]), Q = 1.62 in face view, and 7.9-10.1 X 4.8-5.7 µm (means 8.74 [s = 0.51] X 5.33 [s = 0.21]), Q = 1.64 in side view. These dimensions place them in the lower range of variability, which might be expected for the species. This could be regarded as a small spored variety which is macroscopically hardly recognizable. As we have only one collection to date with spores so small, we are uncertain of the constancy of this character and decline to describe it as a distinct taxon at this time.

In collection 83/368 (IB) slight bluish tints were observed in some younger carpophores both at the stipe apex and base. These tints were ephemeral and no other differences were observed in comparison with 83/381.

Cortinarius uraceus Fr.

This fungus was found on Mt. Washburn, Yellowstone Nat'l. Park, Wyo. at an altitude of ca. 3,050 m (83/383, 27.VIII.1983, leg. M. Moser, IB and BPI). It grew among *Salix* sp., but not far from the last spruce trees (*Picea engelmannii*). It is not a typical alpine fungus as indicated by Favre (1960), who reported it from spruce forest near timberline (1,850-2,000 m). Although our collection was found among dwarf *Salix*, its true mycorrhizal host may well be *Picea*. Due to the high altitude, the carpophores were relatively small (pilei 0.8-3.0 cm diam., stipes 1.5-3.5 X 0.2-0.7 cm). Otherwise the collection is in full agreement with European material from both Sweden and Central Europe.

Spores are ellipsoid, strongly verrucose with warts connected by irregular ridges, partly visible even under the light microscope, and measure 7.4-9.7 X 5.3-7.0 µm (means 8.25 [s = 0.53] X 6.23 [s = 0.35]), Q = 1.33 in face view, 7.0-9.7 X 5.3-6.6 µm (means 8.12 [s = 0.46] X 5.96 [s = 0.40]), Q = 1.37 in side view.

We understand this species in the sense of J.E. Lange, Favre, (Ricken sub. nom. *H. rubricosus*), not in the sense of Kühner (= *C. rigidipes* Moser). Fries indicates the species from coniferous forests ("in pinetis") and calls the stipe "e fusco nigricante, apice olivascente." This fits our fungus fairly well. *Cortinarius rigidipes* Moser has a paler greenish to olivaceous, non-blackening stipe and occurs in deciduous forests (*Fagus*, *Quercus*, *Carpinus*) or frequently in open places near shrubby plants such as *Corylus*.

RUSSULACEAE

Lactarius (DC: Fr.) S.F. Gray

Only one species of this genus has been found in association with dwarf *Salix* in the area investigated.

Lactarius nanus Favre

This species is well known from the European Alps and from the mountains of Scandinavia but we find no previousy reports of it from North America. It was collected in the Beartooth Pass area, where it was found frequently and abundantly growing with *Salix reticulata* and *Salix* sp. at elevations ca. 3,200-3,300 m. We did not find it on Mt. Washburn. *Collections studied*: 83/339, 22.VIII.1983, IB; 83082204, BPI.

Russula Pers.: S.F. Gray

Russula nana Killerm.

Coll. 83082403, BPI, Beartooth Summit, Shoshone Nat'l. Forest, Park Co., Wyo. Common on moist soil among dwarf *Salix*. Observed frequently in the study area. Our collections fit the interpretation of this species given by Knudsen and Borgen (1982). Another collection, 83082206, BPI, agrees in microscopic characters and macroscopically except for a complete lack of acrid taste. The species has been reported from continental North America under dwarf *Salix arctica*, *S. rotundifolia* and *Betula nana* in arctic and subarctic tundra of Alaska and Canada as *Russula emetica* Schaeff.: Fr. var. *alpestris* Boud. by Miller (1982) and Miller et al. (1973). We follow the interpretation of this species given by Knudsen and Borgen (1982).

Russula queletii Fr. in Quél.

Coll. 27.VIII.1983, IB, Mt. Washburn, Yellowstone Nat'l. Park, Wyo., among dwarf *Salix* but not far from the last *Picea engelmannii* at timber line (3,050 m) and certainly associated with the spruce.

ACKNOWLEDGMENTS

We gratefully acknowledge our debt to the following: the University of Wyoming and the National Park Service for the use of the facilities of the Biological Science Research Center at Moran, Wyo., and particularly to Dr. Kenneth L. Diem, Director; to the National Park Service Offices of Research for permission to collect and study in Yellowstone and Grand Teton National Parks and for the use of their facilities at Mammoth Hot Springs, and Lamar Ranger Station. Special thanks is due to N.P.S. scientists Dr. Donald Despain, Dr. Mary Meaghar, and Dr. Robert Wood. Dr. Fred Meyer, U.S. National Arboretum, ARS helped with identification of the dwarf willows. Scanning electron micrographs were prepared by James Plaskowitz; drawings were prepared by Vera B. McKnight and Mag. Regine Finkernagel.

REFERENCES

Ammirati, J. F., and Laursen, G. A., 1982, Cortinarii in Alaskan arctic tundra, *in:* "Arctic and alpine mycology," G. A. Laursen and J. F. Ammirati, eds., University of Washington Press, Seattle, 559 p.

Cailleux, A., and Taylor, G., 1912, "Code expolaire. (resp. Cailleux, Code des coleurs des sols)." Boubée et Cie, ed.

Favre, J., 1955, Les champignons supérieurs de la zone alpine du Parc National Suisse, *Rés. rech. sci. entr. Parc Nat. suisse*, 5: 1-212, Pl. I-XI.

Favre, J., 1960, Catalogue descriptif des champignons supérieurs de la zone subalpine du Parc National suisse, *Rés. rech. sci. entr. Parc Nat. suisse*, 6: 322-610, Pl. I-VIII.

Holmgren, P. K., Keuken, W., and Schofield, E. K., 1981, "Index Herbariorum, Pt. I., Herbaria of the World. 7th edition," F. A. Stafleu, ed., Reg. Veget. 106: 1-452.

Johnson, P. L., and Billings, W. D., 1962, The alpine vegetation in relation to cryopedogenic processes and patterns, *Ecol. Monog.*, 32: 105-133.

Knudsen, H., and Borgen, T., 1982, Russulaceae in Greenland, *in:* "Arctic and alpine mycology," G. A. Laursen and J. F. Ammirati, eds., University of Washington Press., 559 p.

Kobayashi, Y., Hiratsuka, N., Korf, R. P., Tubaki, K., Aoshima, K., Soneda M., Sugiyama, J., 1967, Mycological Studies of the Alaskan Arctic, *Ann. Rep. Inst. Fermentation, Osaka*, 3: 1-138.

Kornerup, A., and Wanscher, J. H., 1967, "Methuen Handbook of Colour," Methuen Publ. Co., N.Y., 243 p.

Lamoure, D., 1977, Agaricales de la zone alpine. Genre *Cortinarius*, sous-genre *Telamonia*. Part 1, *Trav. Sci. Parc Nat. Vanoise*, 8: 115-146.

______, 1978, Agaricales de la zone alpine. Genre *Cortinarius*, sous-genre *Telamonia*. Part 2, *Trav. Sci. Parc Nat. Vanoise*, 8: 77-101.

Miller, O. K., Jr., Laursen, G. A., and Murray, B. M., 1973, Arctic and alpine agarics from Alaska and Canada, *Can. J. Bot.*, 51: 43-49.

Miller, O. K., Jr., 1982, Higher fungi in Alaskan subarctic tundra and taiga plant communities, *in:* "Arctic and alpine mycology," G. A. Laursen and J. F. Ammirati, eds., Univ. of Washington Press, Seattle, 559 p.

Munsell, H. H., 1966, "Munsell Book of Color," Munsell Color Company, Baltimore, Maryland.

Ridgway, R., 1912, "Color Standards and Color Nomenclature," Published by the author, Washington, D.C., Pl. I-LIII, 43 p.

Spencer, H., 1959, Geologic evolution of the Beartooth Mountains. Montana and Wyoming. Pt. 2. Fracture patterns, *Geol. Soc. Am. Bull.*, 70: 467-508.

LICHENIZED AGARICS: TAXONOMIC AND NOMENCLATURAL RIDDLES

S.A. Redhead

Biosystematics Research Centre,
Agriculture Canada, Ottawa, Ontario, Canada, K1A 0C6

and

Th. W. Kuyper
Rijksherbarium, Schelpenkade 6,
2313 ZT Leiden, The Netherlands

Key words: Basidiolichens, *Botrydina*, *Coriscium*, *Omphalina*, *Phytoconis*, nomenclature, typification, taxonomy, thallus-identification

ABSTRACT

The fully lichenized omphaloid agarics common in arctic and alpine regions of the world represent natural genera distinct from their non-lichenized allies in *Omphalina*. Based on thallus morphology these lichens may be distinguished from other basidiomycetous lichens. *Semiomphalina leptoglossoides* (Corner) Redhead develops a reduced basidiome lacking lamellae. Its globular thalli are typical of primitive *Botrydina* species. *Botrydina viridis* (Ach.) Redhead & Kuyper develops a complex foliose thalli (*Coriscium*) from simpler thalli initially resembling the globular thalli of more primitive *Botrydina*. *Botrydina lobata* sp. nov. develops an intermediate form of thallus. The morphological differentiation of the thallus and the basidiome belies suggestions that the lichenized taxa have only recently evolved and are unstable. The reliability of pigmentation differences at the generic level as proposed by Singer without corroborating morphological or ecological data is questioned and rejected in light of the theory of evolution proposed for these basidiolichens.

Byssus botryoides L., the basionym for the type of the generic name *Botrydina* Bréb. is lectotypified by a Dillenian illustration for which original material is available. It is shown to represent the commonest lichenized species which is often labelled *Omphalina ericetorum*. The generic name *Phytoconis* is rejected for this species.

Botrydina is recognized as a separate agaric genus from *Omphalina* Quél., lectotypified by *O. epichysium* (Pers.: Fr.) Quél. The typification of the latter generic name is reviewed in view of the fact that "*O. ericetorum*" was suggested to be the type of the name by some. Contrary to its apparent isolation from the primarily bryophilous Omphalinas, the lignicolous *O. epichysium* is also linked with bryophytes and algae in non-lichenized associations on surface films. These facultative fungus-alga and bryophyte associations exhibited by *Omphalina* species logically serve as precursors to the development of lichenization, hence the close alliance between *Omphalina* and *Botrydina*.

New combinations proposed include: *Botrydina botryoides* (L.) c.n., *Botrydina chromacea* (Clel.) c.n., *Botrydina luteovitellina* (Pilát & Nannf.) c.n., *Botrydina velutina* (Quél.) c.n., and *Botrydina viridis* (Ach.) c.n. Two new species, *B. aurantiaca* and *B. lobata*, are described from alpine habitats in South America.

Based on type studies *Merulius turfosus* Pers., *Omphalina fulvopallens* Orton, and *Omphalia sphagnophila* Peck are considered to be synonyms of *B. botryoides* (L.) Redhead & Kuyper.

Nomenclaturally, *Agaricus ericetorum* Pers. and *A. umbelliferus* Fr. are superfluous names for *A. pseudoandrosaceus* Bull. The latter is lectotypified and placed in synonymy with *B. botryoides*. The name *Agaricus umbelliferus* L., lectotypified by a Micheli plate, and a possible synonym of *Mycena capillaris* (Schum.: Fr.) Kumm., is rejected as type of *Omphalina*.

Lichenization is believed to be an adaptation for survival in arctic and alpine habitats.

A key to the thalli of *Botrydina* species is given.

INTRODUCTION

Lichenized agarics are primarily an arctic-alpine group of species forming conspicuous elements in the mushroom flora of the far north. Bigelow (1970) spoke of *Omphalina ericetorum* (Fr.: Fr.) M. Lange as probably the most common arctic agaric. It is the intent of the present contribution to show that the taxonomy and nomenclature of this small group of species is badly confused, and to offer an analysis of the nomenclatural problems guided by international rules, the ICBN (Voss et al., 1983). It will also be argued on taxonomic grounds that these fungi represent distinct genera, recognition of which has been suppressed by the overpowering image of the lamellate basidiome.

A generation ago Gams (1962) discovered that some agarics were lichenized, but even today many mycologists remain reluctant to fully accept the fact that some agarics are obligately lichenized. Doubts have clouded taxonomic decisions (generic and species distinctions) as well as the nomenclature pertaining to these conspicuous, arctic and alpine species. Evidence of the lichenization of certain omphaloid Tricholomataceae has accumulated. All authors since Gams (1962) with the exception of Watling & Richardson (1971) have found the basidiome commonly labelled "*Omphalina ericetorum*" to be associated with a lichen thallus (Acton, 1909) to which the name *Botrydina vulgaris* Bréb. is commonly applied. However, even Watling & Richardson's collections were later found to be associated with the lichen thallus (Watling, pers. comm. 1983). Heikkilä & Kallio (1966, 1969), leaders in the study of arctic basidiolichens, found the association to be constant over three continents. Similarly, Bigelow (1970) found the association to be constant in numerous North American collections.

Complicating these observations on *O. ericetorum* has been the discovery that a number of other omphaloid species are constantly associated with a *Botrydina* thallus. Basidiomes commonly labelled *Omphalina velutina* Quél. (or *O. grisella* (P. Karst.) Moser) and *Omphalina luteovitellina* (Pilat & Nannf.) M. Lange have been shown to always form near thalli collectively labelled *Botrydina vulgaris* (Heikkilä & Kallio, 1966, 1969). The name *Botrydina* has also been applied to a host of other algal-basidiomycete ball-like lichen thalli (Oberwinkler, 1970). Furthermore, another omphaloid species, *Omphalina hudsoniana* (Jennings) H. Bigelow, is linked to a second lichen thallus which has been named *Coriscium viride* (Ach.) Wainio (Gams, 1962; Heikkilä & Kallia, 1966; Bigelow, 1970; Oberwinkler, 1970), and not to a traditional *Botrydina*. Some of the basidiomycetes forming the *Botrydina*-like thalli form clamp-connections on hyphae connected to their algal balls, thus clearly establishing the basidiomycetous nature of the lichen. The omphaloid agarics lack clamp-connections and therefore the lichen mycobiont is not immediately recognizable as a basidiomycete. Henssen & Kowallik (1976) and Oberwinkler (1984) have reported the presence of dolipore septa in the hyphae of *Coriscium viride* and dolipore septa have been demonstrated for the mycobiont of *Botrydina* thalli associated with *O. ericetorum* in transmission electron microscope preparations by Boissiere (1980), Honegger & Brunner (1981) and Oberwinkler (1984). Lamoure (1968) has shown the mycobiont of some *Botrydina* to be dikaryotic. Redhead has also observed the fusion of basidiospores with the hyphae forming the *Botrydina* under *O. ericetorum* (Fig. 21) and links between *Botrydina* thalli splashed up on lamellae of *O. ericetorum* with the hyphae of the basidiomes (Figs. 8, 9). These observations provide solid evidence that the omphaloid agarics and the mycobionts of the associated lichens represent morphs of the same species.

LICHEN NOMENCLATURE

Nomenclatural implications of the above conclusion have not been fully appreciated by agaricologists. The older lichen names have priority over the names based on the omphaloid teleomorphs (ICBN, Art. 59.1). Thus the oldest validly published name for what has been called *Omphalina hudsoniana* (Jennings) H. Bigelow is *Endocarpon viride* Acharius (1810).

The disposition of the name *Botrydina vulgaris* Brébisson (1839) is a more complex issue. Brébisson (1839: 36) not only introduced a new generic name but also a new species epithet for *Byssus botryoides* Linnaeus (1753). Although Linnaeus (1753) gave only the briefest description "*Byssus botryoides saturate virens,*" he referred to both the Ray (1724) publication edited and enlarged by Dillenius, and Dillenius's (1741) own work, tab. I, fig. 5, where it was stated that the species was collected in Hampstead heath, London, England. Dillenian material still exists in Oxford (OXF) and this material labelled *Historia muscorum* tab. I, fig. 5 has been examined by J.R. Laundon (pers. comm. 1983) who confirmed that it is indeed what commonly has been called *Botrydina vulgaris*. Although Drouet & Daily (1956, p. 145) have stated that the type of *Byssus botryoides* is the specimen in the Linnaean herbarium in London (LINN) labelled "*Byssus botryoides*" there is no indication that Linnaeus consulted this specimen prior to 1753. Therefore, it is at best a neotype. The Linnaean description is essentially that of Dillenius (Ray 1724, Dillenius 1741) and the fig. 5 referred therein is available as lectotype and therefore has priority (ICBN Art. 7.5, 8). Interpretation of the plate is nearly impossible but the specimen examined by Laundon clarifies its disposition.

These relevations immediately raise the question as to which species known under a traditional agaric or hymenomycete name the name

Byssus botryoides has priority. Singer (1970, 1975a) has been of the opinion that the link cannot be made but he did not offer an analysis of the lichen or trace specimens. Oberwinkler's (1970) synopsis of the basidiolichens is the most complete overview of the known forms. Among the genera forming ball-shaped thalli only those assignable to omphaloid or clavarioid species produce clampless sheaths and subtending hyphae. Furthermore, only the omphaloid species produce totally enclosing sheaths leaving no gaps between the sheathing cells forming the cortex of the ball-shaped thalli (Figs. 8-10, 12, 15-19, 34). Thus, the lichen thallus of the omphaloid species can be recognized even when the basidiomes are absent. *Multiclavula mucida* (Pers.: Fr.) R. Petersen, which can be eliminated by the presence of clamp-connections, but which appears to form totally enclosing sheaths in Oberwinkler's illustration of the lichen thallus (Abb. 10), forms a more open type of sheath in specimens examined by the senior author (DAOM). The lichen thalli of the omphaloid agarics are characterized by prismatic, cortical cells abutting adjoining cells on all sides, and by secondary septa often occurring at oblique angles (Figs. 12, 15-19, 34). The only non-lamellate species known to produce this type of lichen thallus is *Semiomphalina leptoglossoides* (Corner) Redhead (1984) and that species probably evolved from other *Botrydina* ancestors much as *Arrhenia* Fr. was derived from omphaloid ancestors.

Byssus botryoides forms lichen thalli (Figs. 10, 12) characteristic for the omphaloid agarics and distinct from other hymenomycetes. Only two species of omphaloid agarics, *O. ericetorum* and *O. velutina*, among those forming the *Botrydina* type of thalli, are known to occur in southern England (Watling, 1981) near London, and could have been collected by Dillenius. All others are restricted to alpine areas or are otherwise geographically distant. Morphologically the thalli are virtually identical but comparison of the subtending hyphae linking the individual ball-thalli has revealed that the two are distinguishable. The commonest species, *Omphalina ericetorum*, forms hyphae in the range 3-4 µm diam. (Figs. 20-22); whereas the less common *O. velutina* forms them in the range of 2-3 µm diam. (cf. Figs. 39, 41 and Poelt & Jülich 1969). Through the courtesy of J.R. Laundon (British Museum), slides prepared from the Dillenian lectotype were examined, and the subtending hyphae measured to be 3-4(-5) µm diam. (Fig. 13), confirming that *Byssus botryoides* = *O. ericetorum*.

Which then in the correct species name for the fungus and what is an appropriate or correct genera disposition? The name *Omphalina ericetorum* has a long and convoluted history which involves the name *Agaricus umbelliferus* Linnaeus (1753), published simultaneously with *Byssus botryoides*. The name *A. umbelliferus* is implicated in the typification of the generic name *Omphalina* Quélet (1886) (see Lange, 1981; Redhead & Weresub, 1978) over which the generic name *Botrydina* Brébisson (1839) would have priority if combined. Thus, the history of *O. ericetorum* and the typification of *Omphalina* need to be reappraised. A second question to be addressed is the status of the generic name *Botrydina* listed as an illegitimate synonym of *Phytoconis* Bory de St. Vincent.

PHYTOCONIS

The genus *Phytoconis* Bory de St. Vincent (1797) is listed as an earlier name for *Botrydina* Bréb. in *Index Nominum Genericorum* (Farr et al. 1979). The basis of this statement was the proposed lectotypification of *Phytoconis* by Drouet & Daily (1956) by *P. botryoides* (L.) Bory and the proposed lectotypification of *Byssus* L. by Ross & Irvine (1967) by *B. cryptarum* L. when the latter authors recommended the formal rejection of *Byssus*. The recommendation by Ross & Irvine has not been accepted and is not listed in the International Code of Botanical Nomenclature. Furthermore, *Byssus* had been earlier typified [lectotypified by definition] by Fries (1825: 309), "...*ex qua Byssus Jolithus generis typus*...," a typification having priority over that of Ross & Irvine (ICBN, Art. 8). In addition, when *Phytoconis* was first circumscribed it included *B. jolithus* L., type of the earlier name *Byssus* and hence is automatically typified by *B. jolithus* not *B. botryoides* (ICBN, Art. 7.11). *Byssus jolithus* is now considered to be a *Trentepohlia* species under the conserved generic name *Trentepohlia* C.F.P. Martius.

Drouet & Daily (1956) listed seven additional generic synonyms published earlier than *Botrydina*. However, none are available, as they are either substitutes for *Phytoconis* and hence typified by *B. jolithus*, later homonyms of other names, or were earlier typified by other species (Farr et al. 1979). We conclude that *Botrydina* Bréb. is the earliest legitimate name available for the lichenized omphaloid agarics.

OMPHALINA

Lange (1981) disagreed with Redhead & Weresub's (1978) unravelling of the nomenclatural and taxonomic problems which plague the name *Omphalina* Quél. At the center of the controversy is the circumscription of the species selected as type for the generic name by Singer & Smith (1946), "*O. umbellifera* (L. ex Fr.) Quél." Lange's opinion (p. 692) that "... the species is well known and characteristic for the genus..." ignores two major controversies.

The first is a problem in taxonomy. Singer & Smith (1946) jointly tpyified the generic name *Omphalina* and then independently proceeded to apply the name to two mutually exclusive groups (Singer, 1975a, b; Smith, 1973). The generic name *Omphalina* Quél. continues to be a source of confusion, used for two taxonomically mutually exclusive groups (Bigelow,1982; Clémençon, 1982; Kühner, 1980; Lange 1981). The second controversy is nomenclatural and involves the named species selected as type. Quélet (1886a) adopted a Linnaean species epithet "umbellifera, Linn." which Redhead & Weresub (1978) contended was not applicable to the well known lichenized species Lange and others would prefer.

Rogers (1950) has noted that *Omphalina* Quélet (1886a) was not an avowed substitute for any earlier name; notably Quélet's (1872) earlier use of the name *Omphalia* which he applied differently in 1886. He did not list any generic synonyms or refer to any other publications in the protologue except for the individual species included. Strictly speaking, the name *Omphalina* Quél. is not a *nomen novum* (ICBN, Art. 7.9). There has been considerable controversy over the typification of *Omphalia* (Pers.) S.F. Gray (1821) or various other homonyms but all are predated by *Omphalea* Linnaeus (1759) *nom. cons.* which is taken to be an orthographic variant (ICBN, Art. 64.2) of *Omphalia* (Donk, 1962). Illegitimate names (ICBN, Art. 64.1) are to be rejected and replaced by legitimate ones (ICBN, Art. 72.1). Thus ICBN, Art. 7.11 which establishes automatic typification for names which *ought* to have been adopted is irrelevant; in retrospect the name *Omphalia* ought *not* to have been adopted by Quélet.

These facts unequivocably establish *Omphalina* Quélet as a new independent generic name (see ICBN, Art. 33. Note 1) which must be typified on the basis of the 1886 protologue *alone*. Typification of the earlier *Omphalia* homonyms is nomenclaturally irrelevant and dwelling on them seems to be counterproductive.

The earliest known typification proposal was by Earle (1909) based on the American Code of Botanical Nomenclature (1907). Following the American Code, Earle chose *Omphalina hydrogramma* (Bull. ex Fr.) Quél. as type; it being the first illustrated species in Quélet's treatment. Currently, this selection may be superseded (ICBN, Art. 8).

The next typification was by Singer & Smith (1946) who chose *Omphalina umbellifera* (L. ex Fr.) Quél. They did not use a mechanical system of selection but carefully weighed the options available. Donk (1949) concurred with their choice, citing it as *Agaricus umbelliferus* L. ex Fr. Succeeding authors had more or less accepted this typification in name except that different interpretations of which species was chosen led to divergent usage of the name *Omphalina*. Singer (1975a) has argued that the type of *Omphalina* is *O. umbellifera* (L. ex Fr.) sensu Quél., which he believes to be different from the true *O. umbellifera*, as interpreted by most authors; hence the split usage of *Omphalina*. While arguments based on species concepts might have been debatable prior to the Sydney congress, such debate has been terminated by the new wording of ICBN, Art. 10.1 and 10.2 which states that the type of a genus is the type of a species name, and furthermore, included species names have priority. The nomenclatural history of the selected species must be traced and its type determined.

Quélet (1886a) treated the epithet *umbellifera* in his listing of *Omphalina* species as "*umbellifera*", Linn. Fl. Dan. t. 1015 A., (*pseudoandrosacea* Bull., t. 276)." Until the recent changes in starting point dates this epithet has been interpreted as based on the basionym *Agaricus umbelliferus* L. ex Fr. (1828); treated by current authors as a later synonym of *Agaricus ericetorum* Pers. ex Fr. (1821). It is clear that Quélet closely followed the taxonomy and nomenclature of Fries, and therefore, this presumption was logical even though there is no reference to Fries in the treatment of this species. The *Flora danica* illustration cited by Quélet (Vahl, 1790, tab. 1015A; see Figs. 3-4) is, as Lange (1981) noted, a reasonable likeness of the basidiolichen he considers as type of *Omphalina*. However, nomenclaturally the Linnaean epithet "*umbelliferus*" is traced to Linnaeus (1753: 1175) where it was simultaneously validated, along with the name *Byssus botryoides*, as *Agaricus umbelliferus*. It has been noted that in light of our present knowledge of the large number of agarics in Sweden we can never be certain to which species Linnaeus intended the name to apply (see Lange, 1981; Redhead & Weresub, 1978). It appears certain that Linnaeus included more than one species in his concept. The description from his earlier work (Linnaeus, 1745) which he repeated in 1753 by way of synonymy reads in part "... *capitulo turbinato* ...," while the cited description by Micheli (1729) reads "... *pileo hemisphaerico*...,". Traditionally, *A. umbelliferus* has been taken to be synonymous with *Agaricus ericetorum* of Fries (1821) which is the same as *Byssus botryoides* and which inhabits peaty soils, well decayed wood surfaces, and moss beds in arctic tundra, alpine sites, boreal forest, and temperate coastal coniferous rain forests (Bigelow, 1970; Heikkilä & Kallio, 1966, 1969). Although it may be presumed that Linnaeus included more than one species in his concept as suggested by the synonymized descriptions, his description of the habitat, minimally altered in different publications, "*inter semiputrida dejecta folia sylvarum*" (Linnaeus, 1745), "*inter folia congesta, semiputrida*" (Linnaeus, 1753), leaves little doubt that the species did not include *A. ericetorum* of Fries (1821). Accumulated leaf litter is not a suitable substrate for the lichen as it shades the lichenized photosynthetic thallus and inhibits colonization by the basidiolichen. Accumulated leaf litter is not a substrate for the species Fries knew as *A. ericetorum*.

Fig. 1. Bulliard (1786) pl. 276, *Agaricus pseudoandrosaceus*, central figure (lectotype). Fig. 2. Bulliard (1786) pl. 276, *A. pseudoandrosaceus*, entire plate showing excluded dark species (arrow). Figs. 3 & 4. Vahl (1790) Tab. 1015', *Agaricus niveus*. Fig. 5. Micheli (1729) Tab. 80, fig. 11. Fig. 6. Micheli (1729) Tab. 73, fig. 6. Fig. 7. Micheli (1729) Tab. 74, fig. 5.

Lange (1981: 697) erred when he stated that the text of Linnaeus did not exclude *A. umbelliferus* of Fries. Any typification of *Agaricus umbelliferus* L. by a basidiolichen which does not inhabit leaves is in *serious* conflict with the protologue. As an alternative, the selection of Micheli's (1729) Tabula 80, Fig. 11 which was constantly cited by Linnaeus (1745, 1753, 1755, 1764) allows for an educated disposition of the name (Fig. 5). It should also be noted that Linnaeus from the beginning (Linnaeus, 1737) had in mind a fungus with a long stipe, quite unlike that typical for the lichen, and he clearly stated that he was following Micheli's system of classification for agarics. Micheli illustrated a long stalked species with a conspicuously umbrella-like pileus (Fig. 5); hence the species epithet, *umbelliferus*. If Linnaeus had intended an omphaloid species among mosses he would have cited Micheli's omphaloid species (Fig. 6) figured as Tab. 73, Fig. 6, and as

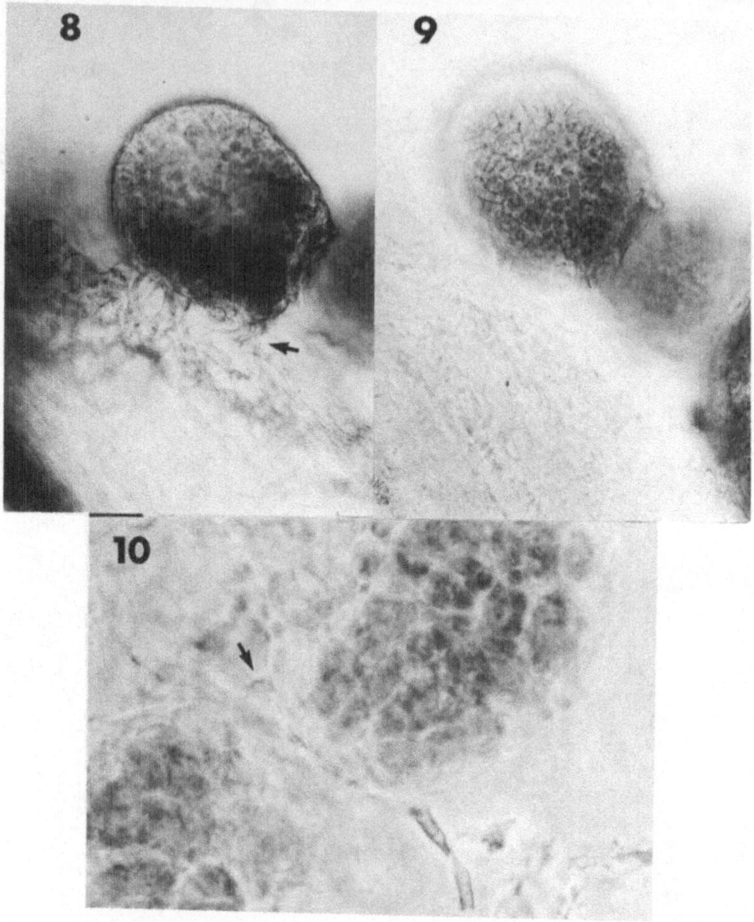

Figs. 8 & 9. *Botrydina botryoides* lichenized granule on lamellar edge showing hyphal connections (arrow), DAOM 189776, ca. 250X. Fig. 10. *Byssus botryoides* L. from slide in British Museum, lichenized granule and subtending hypha (arrow), ca. 990X.

discussed on p. 146. By the citation of a nonomphaloid species on leaves it is interpreted by us that he clearly indicated a different intent.

Fries (1821) cited Micheli's figure (Tab. 80, Fig. 11) under the name *Agaricus capillaris* Schum.: Fr. and judging from Micheli's figure, Fries was probably correct. Therefore, Micheli's (1729) Tab. 80, Fig. 11, is here selected as lectotype of *Agaricus umbelliferus* Linnaeus (1753). The correct name is *A. capillaris*, the name adopted in a sanctioned work (Fries 1821) for *Fungi caeteri* (ICBN, Art. 13).

Fries (1821) listed *Agaricus umbelliferus* L. as a synonym of *A. ericetorum* Pers. Considering that he cited part of the protologue elsewhere (i.e. Micheli, Tab. 80, Fig. 11, under *A. capillaris*) and that the excluded portion has now been selected as type, Fries' citation of *A. umbelliferus* can only be considered as *pro parte*. Nomenclaturally *Agaricus umbelliferus* L. is not a synonym of *A. ericetorum* Pers.: Fr. (ICBN, Art. 63.2). Fries (1821) accepted Persoon's (1796) species name *A. ericetorum*, cited Persoon's (1801) later treatment, and recognized Persoon as the authority for the name in the index to Volume 1 of

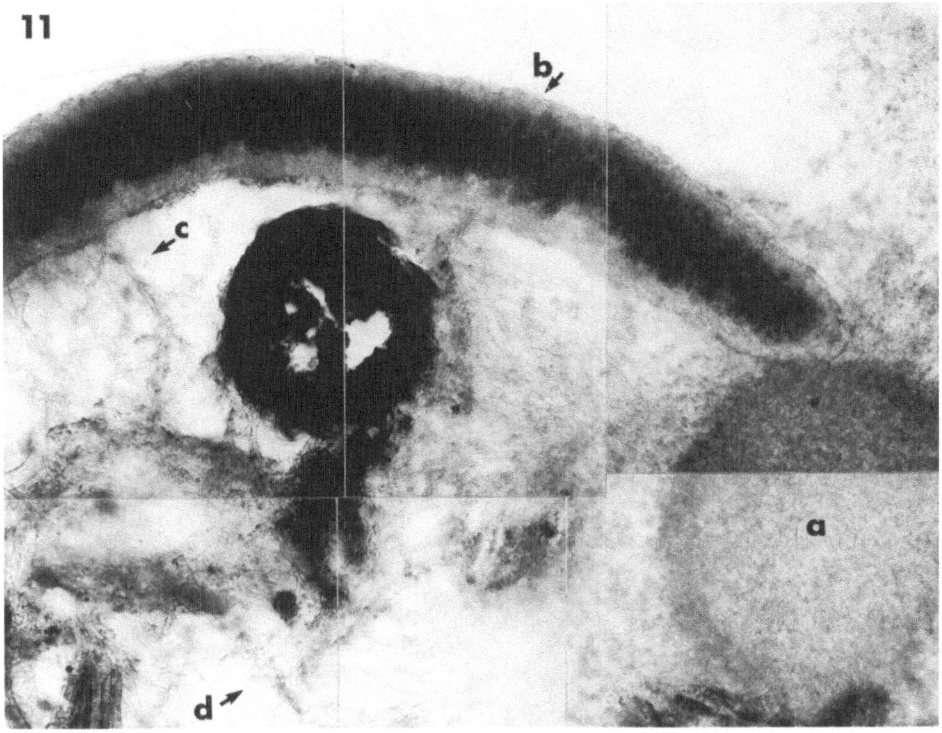

Fig. 11. *Botrydina viridis*, a. base of basidiome stipe, b. lichenized thallus, c. thick-walled central anchoring tuft of hyphae, d. thinner-walled hyphae further from thallus, DAOM 189780, ca. 185X.

Systema mycologicum (Fries 1821) and in the general index to both the *Systema Mycologicum* and *Elenchus Fungorum* (Fries 1832) thus excluding none of Persoon's protologue. Nomenclaturally, there is no reason to recognize an *Agaricus ericetorum* Persoon and an *Agaricus ericetorum* Fries (ICBN, Art. 63.1, 63.2). Nomenclaturally, Persoon's name was a superfluous name for *Agaricus pseudoandrosaceus* Bulliard (1786, pl. 276), which he cited as a synonym (ICBN, Art. 63.1, 63.2), and therefore, *A. ericetorum* is automatically typified by the type of *A. pseudoandrosaceus* (ICBN, Art. 7.11). Bulliard & Ventenat (1809) correctly listed Persoon's *A. ericetorum* as a synonym of *A. pseudoandrosaceus*. Fries (1821) cited Bulliard's *A. pseudoandrosaceus*, Pl. 276, in his synonymy of *A. ericetorum*. As noted in an earlier paper (Redhead & Weresub, 1978), Bulliard almost certainly included more than one species, as we understand them in modern taxonomy, in his concept in 1786. Bulliard (1786) also cited Micheli (1729, Tab. 74, Fig. 5), a marasmioid fungus (Fig. 7)), and a link between the name *Agaricus pseudoandrosaceus* and *Agaricus androsaceus* L. (= *Marasmius androsaceus* (L.: Fr.) Fr.). The dominant form of Bulliard's Pl. 276 was whitish. His illustration (Figs. 1, 2) shows a central moss cushion bearing 14 whitish basidiomes. Two whitish basidiomes are depicted in section to the left, and only 2 greyish basidiome are depicted to the right. The protologue (Bulliard 1786, Pl. 276) states "il varie du blanc au gris cendré, il est quelquefois d'un blanc jaunâtre." Bulliard &

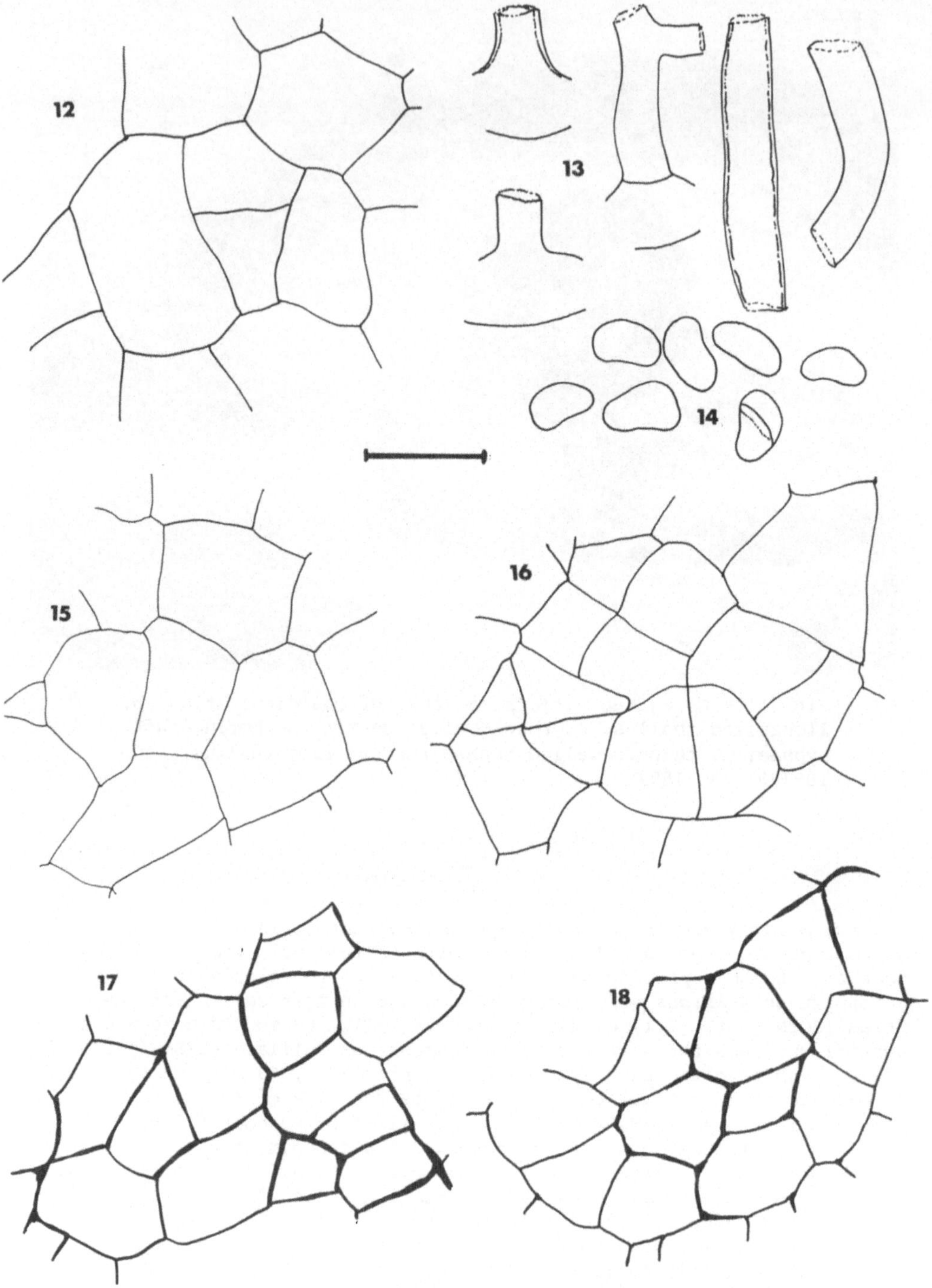

Fig. 12. Surface cells of lichenized granule from *Byssus botryoides* (BM). Fig. 13. Subtending hyphae from previous collection. Fig. 14. Algal cells from previous collection. Fig. 15. Surface cells of lichenized granule, DAOM 190482, *Botrydina botryoides*. Fig. 16. As in 15, DAOM 190483, *B. botryoides*. Fig. 17. As in 15, DAOM 189782, *B. luteovitellina*. Fig. 18. As in 15, DAOM 190533, *B. luteovitellina*. Scale bar = 10 um.

Ventenat (1809: 539) later described the colours as "*modo niveus, modo substramineus, interdum fuligineo-cinereus*" or "blanc ou de couleur jaune-paille, et quelquefois d'un gris-bistré." Logically the names *A. pseudoandrosaceus* and *A. ericetorum* should be applied to the whitish elements of the protologue and not the greyish element despite the fact that the reverse was Persoon's dominant element. Bulliard's illustration of the central species on a moss cushion (Pl. 276) is recognizable as one of the forms currently considered to be *A. ericetorum* of Fries (1821). We therefore select the central figures as lectotype. This typification pre-empts M. Lange's (1981) proposed neotypification of *Agaricus ericetorum* of Fries 1821 (ICBN, Art. 8) which nomenclaturally is automatically typified by the type of the earlier *A. ericetorum* Pers., that in turn is automatically typified by the type of *A. pseudoandrosaceus* Bull. It may be argued by some that a name in a sanctioned work such as *Systema mycologicum* is not automatically typified by earlier names but in this case *the species is a lichen* and such names are *not sanctioned* (ICBN, Art. 13.1(d)). Thus Persoon proposed a new name for *A. pseudoandrosaceus* Bull. which is illegitimate.

The name *Omphalina pseudoandrosacea* has been applied correctly in recent years by Møller (1945), Poelt & Oberwinkler (1964) and Watling (1977) although usually in a restricted sense. The name *Byssus botryoides* Linnaeus (1753), however, has priority for this organism.

Fries (1828) proposed that *Agaricus umbelliferus* L. was the correct name for *A. ericetorum* as he treated it in 1821. However, in 1821 he had excluded a part of the protologue, the lectotype of *A. umbelliferus*, considering it to represent *A. capillaris*. Therefore, the name *Agaricus umbelliferus* in Fries (1828) is to be attributed to Fries alone (ICBN, Art. 63.2) as an unsanctioned, illegitimate, superfluous name for *A. pseudoandrosaceus* Bull., a lichen.

With two competing but historically linked names, *A. umbelliferus* L. (1753) and *A. umbelliferus* Fr. (1828) recognized, a problem of resolution is created when the typification of *Omphalina* by Singer & Smith (1946) is examined. Redhead & Weresub (1978) had earlier suggested that because of the citation of Linnaeus by Quélet, that the chosen type was *A. umbelliferus* of Linnaeus. If this were the position adopted, *Omphalina* might become a taxonomic synonym of *Mycena* (Pers.) Roussel (1806) where *Agaricus capillaris* is placed. However, historically all treatments of the name *A. umbelliferus* after 1828 have been in the sense of Fries even though the authority has often been attributed to Linnaeus. In addition, Quélet treated *Mycena capillaris* (Schum.: Fr.) Kumm. as a distinct species and that is the correct and sanctioned name for *A. umbelliferus* L. when the genus *Agaricus* is used. Also, Quélet (1886a) listed Bulliard's *Agaricus pseudoandrosaceus* as a synonym of *Omphalina umbellifera*. Therefore, it may be argued that Quélet's (1886) citation of Linnaeus was a bibliographic error (ICBN, Art. 33.2, 63.2) which could be corrected to read *A. umbelliferus* Fr. 1828 in which case *Omphalina* Quél. would be a synonym of *Botrydina*. Alternatively, because there is this doubt as to what was selected as type, Singer & Smith's (1946) lectotypification might be considered as insufficient for lectotypification purposes and not valid.

The current wording of ICBN, Art. 10 states that "The type of a name of a genus... is the type of a name of a species... For purposes of designation or citation of a type, the species name alone suffices...," and "if... one or more species names is definitely included, the type must be chosen from among the types of these names." Quélet used Linnaeus' species name and therefore the type of *Agaricus umbelliferus* L. would be

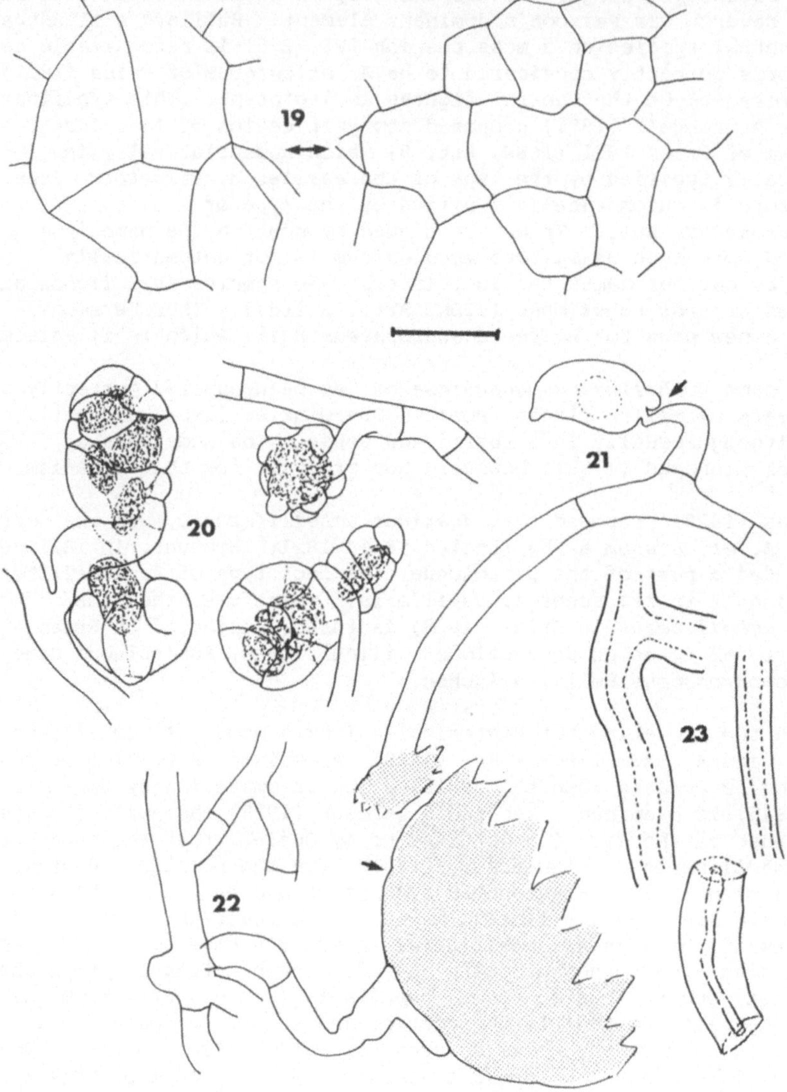

Fig. 19. Surface cells of lichenized granules from Fungi exsiccati Suecici 1752 = *Botrydina velutina* (DAOM). Fig. 20. Early developmental stages of lichenized granules of *Botrydina botryoides* showing algal cells in encapsulation process, DAOM 174850. Fig. 21. Hyphal connection (arrow) between basidiospore and hypha subtending lichenized granule (shaded undetailed portion), DAOM 174850. Fig. 22. Hyphae subtending lichenized granule (undetailed shaded portion), DAOM 174850. Fig. 23. Subtending thick-walled hyphae from thallus of *Botrydina luteovitellina*, DAOM 28451. Scale bar = 10 um for Figs. 19, 23, 12 um for Figs. 20-22.

considered type of *Omphalina* Quél. This is clearly a case where the designated type is in serious conflict with the protologue (ICBN, Art. 8) and better species are available. Quélet treated the species depicted in Micheli (Tab. 80, Fig. 11) in another genus as *Mycena capillaris* as noted above. The choice of *Omphalina umbellifera* as lectotype for *Omphalina* is most definitely an unacceptable typification.

Redhead & Weresub (1978) selected *Omphalina epichysium* (Pers.: Fr.) Quél. as lectotype of *Omphalina* Quél., partly on the basis of arguments no longer valid. However, *O. epichysium* is a suitable species, not only for the reasons given by Redhead & Weresub but also because it is a well known species and it removes the ambiguity inherent with the selection of *O. umbellifera*. This typification has been accepted by Kühner (1980) and Clémençon (1982) and is accepted here.

TAXONOMY

Traditionally, lichenologists have maintained *Botrydina vulgaris* Bréb. (= *Byssus botryides* L.) and *Coriscium viride* (Ach.) Wainio in two separate genera based on gross morphological differences (cf. Figs. 8 and 9 to 11). Traditionally, agaricologists have not distinguished between the lichenized and non-lichenized species of omphaloid Tricholomataceae generically. All have been treated in *Omphalina*, or, on the basis of pigmentation differences in the basidiomes, both *B. vulgaris* and *C. viride* have been treated in the genus *Gerronema* Sing., and a second *Botrydina* species, *Omphalina velutina*, was treated in *Omphalina*. Thus, there is a contradiction between the two systems of classification.

Singer (1975a: 17-18) has discussed lichenized agarics and concluded that distinct genera for the lichenized species were not warranted because they were closely allied to non-lichenized species and because the lichenized status of some related species was suspect. Another factor which influenced Singer (1970) was the apparent instability of the form of the lichen thalli over the range of the mycobiont. He reported that *Gerronema luteovitellina* (Pilát & Nannfeldt) Singer formed a *Coriscium* thallus in South America and a *Botrydina* thallus in Arctic regions, while *G. hudsonianum* (Jennings) Singer, formed a *Botrydina* thallus in South America and a *Coriscium* thallus in Arctic regions.

We find the features of the various thalli to be remarkably constant, even more so than that of the basidiomes. Singer's puzzling South American finds represent two unnamed species (see below) which differ from their northern counterparts in both thallus form and subtle basidiome features. No switching of thallus forms occurs as reported.

As for the question of obligate lichenization of the few well known species, there seems little doubt. The species commonly labelled *O. ericetorum*, *O. hudsoniana*, *O. luteovitellina* and *O. velutina* have been shown to be constantly lichenized by repeated observations over large geographic areas. They are difficult or impossible to culture on artificial media (Heikkilä & Kallio, 1966; Lamoure, 1982). The lichen thalli are characteristic for these agarics and distinguishable from other basidiolichens. The thalli are composed of highly evolved tissues relative to other tissues formed by basidiomycetes.

In addition, the phycobiont of *O. ericetorum* has been shown to be *Coccomyxa icmadophilae* Jaag, the same as that of the wood inhabiting ascolichen *Icmadophila ericetorum* (L.) Zahlbr. (Honegger & Brunner 1981). *O. ericetorum* is as much a lichen as is *I. ericetorum*.

Reports that species such as *Myxomphalia maura* (Fr.) Hora (James 1965, Hawksworth et al. 1980) and *Omphalina griseopallida* (Desm.) Quél. (Hawksworth 1972, Hawksworth et al. 1980) are associated with a *Botrydina* in at least one collection do not qualify as proof that they form lichen thallus. Similar accidental "associations" have been reported for other species when they fruit in the vicinity of lichenized species (Heikkilä & Kallio 1966). Given the fact that considerable morphological complexity is exhibited by these agarics in their anamorphs and that it has irreversibly modified their ecological status and mode of nourishment, the fully lichenized species represent better defined genera than are many genera now proposed for basidiomycetes.

Defining the generic limits, though, is not without difficulties. *Semiomphalina leptoglossoides* is a basidiolichen, closely related to the other lichenized omphaloid species. However, the senior author (Redhead, 1984) recognized it as a distinct genus based on gross morphological differences in the basidiomes and not the thalli. *Coriscium viride* differs from allied lichenized species by the gross morphological differentiation of the thallus and not by differences in the basidiomes. Other species such as *Omphalina velutina* differ from other *Botrydina*-formers in pigmentation, a feature Singer earlier emphasized as a generic distinction. Are the important features, the basidiome morphology, the thallus morphology, and the pigmentation of the basidiome, all of equal taxonomic value? It is difficult to decide *a priori*, particularly from the biased view of traditional agaric systematics. A phylogenetic analysis of omphaloid Tricholomataceae (Kuyper, 1986) offers a better understanding of the relative importance of the various characters, and leads to a more rational classification. It becomes evident, then, that *Omphalina velutina* and *O. ericetorum* (= *B. botryoides*) cannot be classified among different genera. *O. hudsoniana* (= *Coriscium viride*) differs in thallus morphology in a rather dramatic way, but considering the resemblance of young *Coriscium* thalli with those of *Botrydina*, the existence of two species with thalli intermediate between the two (see below), and the nearly identical morphology of the basidiomes, we prefer to regard it also as congeneric. Doubts Singer (1975: 789-790) raised concerning the identity of the basionym *Endocarpon viride* Ach. are unfounded. Mr. Heino Vanska (H) in a letter to Dr. I. Brodo (CANL) confirmed that a specimen of *Endocarpon viride* exists in the Acharian herbarium (H-ACH no. 869) and it represents what is currently to be *Coriscium viride*.

On the other hand, we regard *Semiomphalina* as a distinct genus. The level of differentiation exhibited by the basidiomes seems sufficient for generic recognition, just as in *Arrhenia* Fr. (Redhead, 1984).

Obviously this classification brings to question the generic distinction between the non-lichenized omphaloid species of the Tricholomataceae such as *Omphalina* with incrusting pigments, and those with non-incrusting pigments such as *Gerronema*, as defined by Singer (1975). Considering how many genera of basidiomycetes contain species which display diverse pigmentation involving two major types of pigmentation, generic distinction based on this feature now seems to be tenuous unless it is correlated with other features (Kuyper, 1986). At present we recognize a distinct genus, *Omphalina* Quél., typified by *O. epichysium*, for the non-lichenized omphaloid Tricholomataceae. The species in this genus are loosely associated primarily with bryophytes, and sometimes with algae, but do not form morphologically differentiated thalli, and generally are not obligately symbiotic with any particular host species. *Omphalina epichysium*, although usually noted to be lignicolous, is, in all specimens examined, associated with bryophytic and algal films on the surface of old logs it inhabits. Furthermore, the mycelium is not restricted to the woody substrate but also ramifies through this film of photosynthetic organisms, coiling around rhizoids and protonema of some of the bryophytes and some types of algal cells. Thus, we believe the species to be a facultative symbiont supplementing its woody diet by either the leached nutrients or dead cells of the associated photosynthetic cryptogams. It is evident that just such loose associations as exhibited by the Omphalinas would be precursors to the formation of the lichen *Botrydina*. *Omphalina* and *Botrydina* are closely allied as indicated by Singer (1975) but a distinct hiatus is present. As proposed for the evolution of *Arrhenia* (Redhead, 1984), the evolution of lichenization in this group was probably promoted by the rigours of the harsh arctic and alpine environments. The following species are recognized in the lichenized genera discussed above:

Botrydina aurantiaca sp. nov. (see following text)

Botrydina botryoides (L.) comb. nov.
- = *Byssus botryoides* Linnaeus, Species Plantarum: 1169. 1753 (basionym).
- = *Botrydina vulgaris* Brébisson, Mém. Soc. Acad. Agricol Indust. & Instruct. Falaise: 36. 1839 [illegitimate, Art. 63].
- = *Agaricus pseudoandrosaceus* Bulliard, Herbier de la France, Pl. 276. 1786.
- = *Agaricus ericetorum* Persoon, Obs. mycol. 1: 50. 1796. [illegitimate, Art. 63].
- = *Agaricus nothus* J.F. Gmelin, Caroli a Linné Systema Vegetabilium 2: 1423. 1796. [illegitimate, Art. 63].
- = *Agaricus umbelliferus* Fries, Elenchus Fungorum 1: 22. 1828 [illegitimate, Art. 63].
- = *Agaricus valgus* Holmskjold, Beata ruris otia Fungis danicis 2: 62. 1799.
- = *Merulius turfosus* Persoon, Mycologia europaea 2: 26. 1825.
- = *Omphalia luteola* Peck, Bull. Torr. Bot. Club 23: 411. 1896.
- = *Omphalia sphagnophila* Peck in Saccardo, Peck & Trelease, Fungi of Alaska, Harriman Alaska Exped. 5: 47. 1904.
- = *Omphalina fulvopallens* Orton, Notes Roy. Bot. Gard. Edinburgh 41: 605. 1984.

Botrydina chromacea (Clel.) comb. nov.
- =*Omphalia chromacea* Cleland, Toadstools and mushrooms and other larger fungi of South Australia 1: 86. 1934.

Botrydina lobata sp. nov. (see following text)

Botrydina luteovitellina (Pilát & Nannf.) comb. nov.
- = *Omphalia luteovitellina* Pilát & Nannfeldt, Friesia 5: 22. 1954 (basionym).

Botrydina velutina (Quél.) comb. nov.
- = *Omphalia velutina* Quélet, C. R. Ass. franç. Av. Sci. (Grenoble, 1885) 14: 445. 1886 (basionym).
- = *Omphalia grisella* P. Karsten, Medd. Soc. Fauna Flora Fennica 16: 92. 1890 [as "(Weinm.?) n. sp."].
- = *Omphalina oreades* Singer, Pap. Mich. Acad. Sci. Arts & Lett. 32: 123. 1946.
- = *Clitocybe albimontana* H. Bigelow, Rhodora 68: 178. 1966.

Botrydina viridis (Ach.) comb. nov.

= *Endocarpon viride* Acharius, Lichenogr. Univers.: 300. 1810 (basionym).

= *Verrucaria laetevirens* Borrer in Hooker, Suppl. Engl. Bot. 1: Tab. 2658. 1830 [illegitimate, Art. 63].

= *Hygrophorus hudsonianus* Jennings, Mem. Carnegie Mus. pt. III (Botany) 12: 2. 1936.

= *Agaricus alpinus* Britzelmayr, Ber. naturw. Ver. Agusburg 30: 13. 1890.

= *Omphalia luteolilacina* Favre, Champ. sup. Parc Nat. Suisse: 199. 1955.

Semiomphalina leptoglossoides (Corner) Redhead, Can. J. Bot. 62: 886. 1984.

= *Pseudocraterellus leptoglossoides* Corner, Cantharelloid Fungi: 161. 1966.

NEW SPECIES

Botrydina aurantiaca Redhead & Kuyper sp.nov. Figs. 72-74

Pileus 10-15 mm diam. convexus glabratus striatus ex parte aurantiacus umbilicatus ad centrum. Lamellae salmoneo-aurantiacae distantes vel subdistantes arcuatae. Stipes 9-19 x 1-2 mm, aurantiacus glabratus. Basidia clavata (uni-) bi- (tri-) vel tetraspora 29-37 x 6.5-8.8 µm. Basidiosporae 8-10 x 5.5-6.9 µm lato-ellipsoideae vel subglobosae pallido-aurantiacae. Hyphae fibulis nullis. Thallus globosus 40-135 µm diam.

Typus: Colombia: Cundinamarca, Paramo Chisaca-Sumapas, 3000-4000 um alt., May 8, 1968, R. Singer B-7019 (F). Other collection: Same location and date, Singer B-7024 (F).

Singer (1970) treated this species under the name *Gerronema hudsonianum* although its basidiomes resemble most closely those of *B. luteovitellina*. However, the spore size, reported by Singer 5.3-8 x 5.8(-7) µm, and by S.A.R., 8-10 x 5.5-6.9 µm (Fig. 32), is much broader than that of *B. luteovitellina* (cf. Heikkilä & Kallio 1966, Bigelow 1970). In addition, the size of the thallus granules vary mainly in the range of 40-135 µm diam., while the subtending hyphae are 2.5-4 µm diam. (Fig. 33) with refringent slightly thickened walls, but not as greatly thickened as in *B. luteovitellina* (see key). Thus, there are a number of differences which indicate to us that *B. aurantiaca* is a distinct species. Its geographically allied companion species is clearly another new species.

Botrydina lobata Redhead & Kuyper sp. nov. Figs. 30-21

Pileus 10-14 mm diam. convexus glabratus striatus ex parte luteus umbilicatus vel planus ad centrum. Lamellae luteae distantes adnatae vel arcuatae. Stipes 10-15 x 1.5-2(-2.5) mm, niveus vel luteus ex parte glabratus. Basidia clavata tetraspora 25-40 x 5.5-7 µm. Basidiosporae 7-9.5 x 4.2-5 µm, cylindraceae vel ellipsoideae inamyloideae hyalinae. Hyphae fibulis nullis. Thallus lobatus convexus 2-3 mm diam. 225-270 µm crass. griseo-viridis.

Typus: Colombia: Cundinamarica, Paramo Chisaca-Sumapas, 3000-4000 m alt., May 8, 1968, R. Singer B-7023 (F). Other collection: Venezuela: Merida, Paramo de Mucubaji, 3600 m alt., Sept. 1977, leg. Sleumer (F, determined as *G. luteovitellinum* by Singer).

Singer (1970) recorded this species from South America as *Gerronema luteovitellina*. He regarded this species and the former as Arctic species which had reversed their lichen thalli formation (?or association) in the South American alpine populations. However, the thallus form is unlike that formed by any Arctic species. It is lobed from an early stage (Fig. 31), convex, consisting of a number of subglobose algal-containing internal units 25-75 µm diam., bounded by a thin cortical layer 12-25 µm deep. The thalli are a dull greyish colour with concolorous margins. *Botrydina viridis* thalli are concave, disc-shaped with an even, thickened, whiter margins, which only becomes crisped or somewhat lobed and concave in extra large growth forms, and are a brighter green colour. The upper cortex is thicker than the lower in *B. viridis*, unlike *B. lobata*. All other *Botrydina* thalli except *B. chromacea* consist of clustered, globose thalli. Heikkila in an annotation with Singer-B-7023 also noted that the thalli differed from *B. viridis* thalli. The thalli of *B. lobata* and *B. chromacea* are intermediate in organization between *B. viridis* thalli and other *Botrydina* thallus types, hence our transfer of *Coriscium viridis* to *Botrydina*.

The basidiomes of *B. lobata* bear a greater resemblance to those of *B. viridis* than to *B. luteovitellina* despite Singer's statements regarding the collections cited above. *Botrydina viridis* typically has a whitish stipe as does *B. lobata* unlike *B. luteovitellina*. Both *B. viridis* and *B. luteovitellina* exhibit colour variations for their stipes from yellow to orange over their geographic ranges (Tables 1 & 2). Singer reported the spores to be 5-8 x 3-4 µm, most frequently 6-7 x 3-3.5 µm, whereas, S.A.R. found them to be 7-9.5 x 4.2-5 µm (Fig. 30), within the range of *B. viridis* (cf. Favre 1955). The essentially glabrous stipe of *B. lobata* is a distinguishing feature from typical *B. viridis*.

Botrydina chromacea was brought to our attention by Dr. V. Demoulin, who collected material in Australia. The basidiomes are chrome colored and omphaloid (Cleland, 1934). The thalli in Demoulin's collection are 200-900 µm in diam. and 100-150 µm thick. They resemble flattened thalli of *B. botryoides*. In mass they form an areolate green crust on silty soil. The basidiomes are typically clampless, with 2- (3-) 4-spored basidia, and nonamyloid spores. The disc-shaped to slightly angular thalli differ from those of *B. lobata* by the lack of discrete elongated or branched lobes, and the thinner cortical layer. In cross section thalli of *B. lobata* are elliptical to semicircular in outline with the convex surface up. Cross sections of the thalli of *B. chromacea* show a number of irregular wedge-shaped segments with convex outer surfaces facing up, bound by a very thin cortical layer.

NOTES ON SYNONYMY

The synonymy of *Botrydina botryoides* in part follows from the above discussion of the Linnaean, Bulliardian, Persoonian, and Friesian epithets. *Omphalia luteola* is accepted as a synonym on the basis of Bigelow's (1970) paper.

The lectotype of *Merulius turfosus* (herb. L) was studied by the junior author. Lichen thalli of *B. botryoides* were found at the base of the stipe. Singer (1962) already concluded that this species was identical with *Omphalina ericetorum* (= *B. botryoides*).

Omphalia sphagnophila was described by Peck with the stated reservation that it might only be a variety of *O. umbellifera* (= *B. botryoides*). Murrill (1916) in fact placed it in synonymy with this species, but Bigelow (1970) reported the occurrence of clamp-connections in the basidiomes of the type collection and, therefore, maintained it as a distinct species. The type of O. sphagnophila (NYS) was re-examined by S.A.R. Tissues revived well (Figs. 24-25) and compared well with other materials (Figs. 27-29). All hyphae and basidia were simple-septate. Sterigmata were also seen. Lichen thalli of *Botrydina botryoides* were also present with Peck's type. Thus *O. sphagnophila* is placed in synonymy with *B. botryoides*.

A specimen sent as type of *Agaricus sphagnicola* Berkeley from Kew labelled "*Agaricus sphagnicola* - Charnwood Forest June 22, 1839" is in fact *B. botryoides* (see Fig. 26). The spores were 8-10 x 5-6.5 µm, nonamyloid, and smooth. The hyphae of the basidiomes lacked clamp connections and incrusting pigments. Typical thalli were present. As the specimen was stamped "M.C. Cooke 1885" it is possible that this was the collection illustrated by Cooke (1881-1891). However, there is no guarantee that it represents the original *A. sphagnicola* based on a specimen collected June 21, 1827, Chartley Moss, Staffs, on *Sphagnum acutifolium* (Berkeley 1836). Currently Berkeley's species is recognized as a distinct *Omphalina*, *O. sphagnicola* (Berk.) Moser, in Europe. Based on the uncertainty of the status of the specimen sent as type, the current and traditional application of the name *O. sphagnicola* is maintained.

Orton (1984) recently attempted to distinguish two taxa, *Omphalina ericetorum* and *O. fulvopallens* on the basis of biology, the former lichenized, the later nonlichenized, and also on pigmentation variations. However, examination of the type of *O. fulvopallens* (Orton 5105, E) by S.A.R. (Figs. 35-37) revealed that it is associated with *Botrydina botryoides* thalli contrary to Orton's findings. There is no reason to believe the type basidiomes do not represent the mycobiont of these thalli, which leaves only the pigmentation differences noted by Orton.

TABLE 1

Botrydina viridis: colours when fresh

DAOM	Pileus	Lamellae	Stipe
190526	yellow	yellow	white
190506	yellow	-	white
190527	yellow	-	white
190512	yellow	yellow	white
190504	yellow	-	white
190508	yellow	-	white
190515	yellow	yellow	white
190514	yellow	-	white
189781	-	-	white
190523	yellow	yellow	white
190525	-	-	white
190520	yellow	yellow	white
190524	yellow	yellow	white
94376	orange	-	-
21536	dull orange	light orange	white
28422	orange	orange yellow	cream
191220	pale luteous	pale luteous	whitish with faint lilac tint
117597	pale yellow	-	-
117612	pale orange yellow	-	-
189780	pale luteous	pale luteous	whitish with pale luteous apex

These we believe are too variable to represent specific characters as noted by other authors. In the allied *B. viridis* discussed below there is variation in pigmentation over its geographic range (Table 1).

Contrary to the conclusions of Bresinsky & Stangl (1974), we consider *Agaricus alpinus* Britz. not a synonym of *B. luteovitellina* but of *B. viridis*. Britzelmay's (1890, 1893, 1898) descriptions indicating that the stipe was white, leave little doubt as to the identity of the species. As noted by Heikkilä & Kallio (1966) and shown in our table 1, *B. viridis* is clearly distinguished from *B. luteovitellina* by the whitish stipe of the former and the yellow to orange stipe of the latter. Besides *B. viridis* has been collected from the same region where Britzelmayr found his *A. alpinus*, whereas *B. luteovitellina* is not known from recent finds.

TABLE 2

Botrydina luteovitellina: colours when fresh

DAOM	Pileus	Lamellae	Stipe
96377	orange	-	-
96378	orange	-	-
190528	-	-	orange
190533	pale yellow	pale yellow	pale yellow
189782	-	-	yellow
21743	orange yellow	yellow	orange yellow
21806	bright yellow	bright yellow	bright yellow
21858	orange yellow	-	-
28451	orange	orange	orange
26686	orange yellow	orange yellow	orange yellow

Both *B. luteovitellina* and *B. viridis* show considerably more variation in pigmentation than attributed to them in the literature (Tables 1 & 2). Western North American collections of *B. viridis* cited generally had yellow pilei while eastern North American collections often had orangish pilei. Lilac tints are not recorded from the stipes of North American collections (cf. Bigelow 1970) but are known from European collections. Similarly, *B. luteovitellina* ranges from orange to yellow. Because of this form of variation in well characterized species, we do not consider similar subtle differences in the *B. botryoides* taxon to be significant for distinguishing species.

The synonymy of *B. velutina* is less than clear. Singer & Clémençon (1972) adopted the name *Agaricus rusticus* Fries (1838) for the lichenized *B. velutina*, and Clémençon (1983) recognized two lichenized taxa, *Omphalina rustica* (Fr.) Quél. and *O. grisella* (Karsten) Moser (as "(Weinm.) Moser") in this lichenized group with 2-spored basidia and dark, incrusting basidiome pigments, and one with 4-spored basidia, *O. pararustica* Clemençon. Bigelow (1982) discussed another name, *Clitocybe albimontana*, differentiating a taxon under this name from *B. velutina*. He did this on the basis of colour differences, citing original descriptions. Although old original descriptions are useful for determining application of names, they do not reflect the full range of variation exhibited by species and we do not think they substitute for a more complete understanding of the species which has developed over time. Favre (1955), using the name *Omphalia grisella* Karst., described and illustrated alpine forms in Switzerland exhibiting colours similar to the North American *C. albimontana*. A Favre collection was examined by S.A.R. and found to be associated with *Botrydina velutina* thalli.

Bigelow (1983) did not cite Favre's work and gave spore lengths of *C. albimontana* to be 7.5-10 µm long, based on his earlier study (spores 7.5-10 x 3-4 µm, Bigelow & Barr, 1966) but the name *C. albimontana* is based on *Omphalina oreades* Singer (1946), which had spores reported to be 6.8-8.2 x 3-7 µm. Obviously some variability is admitted. Similarly, Singer & Clémençon (1972) when adopting the name *Omphalina rustica* for a lichen reported the spores to be 8-10.3(-11) x 4-5.5 µm, even though citing Favre's publication where spores were reported to be 6-7 x 3.5-4 µm. Later, when Clémençon (1982) differentiated between two taxa, *O. rustica* and *O. velutina* on the basis of spore size, 8-11 x 4-4.5 µm versus 6-8 x 3-4 µm he did not address past discrepancies. We found spores 6.5-8.3 x 4-4.7 µm (Figs. 38, 40). It is apparent to us that the *Botrydina velutina* complex requires a critical revision. We would not be surprised to find all of these taxa to be conspecific.

Although it is tempting to adopt the name *Agaricus rusticus* for the taxon we treat as *B. velutina* because Fries (1838) stated that Persoon's illustration (1786, tab. 4, fig. 12) was optimal, we are convinced the name is not applicable. The lichenized species has relatively broad, decurrent but not profoundly decurrent lamellae, and a velutinous stipe. Fries (1838) treated *A. rusticus* in subgeneric group *Pyxidati* (p. 122) characterized by subnarrow lamellae. He also described the stipe as glabrous and the lamellae as grey. At the same time he continued to treat a grey form of *A. umbelliferus* in that species in subgeneric group *Umbelliferi* (p. 124) characterized by broad lamellae and by implication, pale whitish lamellae. More explicitly, he (Fries 1874) p. 161 described a taxon we consider to be *B. velutina* as a form of *A. umbelliferus*, "*Exstat quoque alia varietas grisea pileo stipiteque velutinis in terra deusta*" while continuing to distinguish *A. rusticus* which must be another taxon (cf. comments by Gulden & Lange 1971). Fries (1821) had earlier described the species as variety β *velutinus* which he knew from living specimens. His variety B (Fries 1821) seems also to be this taxon. It is noteworthy that Fries recognized that variety B was associated with algae.

The name "*Omphalina grisella* (Weinm.) Moser" has been used but it is difficult to interpret its basionym. Weinmann (1836) described a *form* of *A. umbelliferus* Linn. as, "b. grisellus. Pileus ...." but the purported name was not italicized as he did for varieties, and clearly it was not his coinage. Fries (1821) treated a form similarly, "b. grisellus." Karsten (1879) first treated it formally as a variety with the name "var. *grisella* Weinm." of *Omphalia umbellifera*. He later elevated it to species level (Karsten 1889) as "*Omphalia grisella* (Weinm.?) Karst. n.sp." Thus, at the variety level it has priority from 1879, but not at the species level until 1889. Quélet (1886b) in the interval published the name *Omphalia velutina* which has priority at the species level.

The name *Agaricus sagittula* R. A. Hedwig (1802) was listed as a synonym of *A. ericetorum* b. grisellus by Fries (1821) and needs to be considered as an earlier name for *B. velutina* if combined. However, as implied by its epithet, Hedwig's fungus had an obconical, arrow-head like appearance, as well as forked lamellae, and white radicating hairs at the base. In addition, it was said to occur where the preceding species, *Agaricus niveus* Hedwig, occurred, i.e. on walls, naked earth, and among grasses. The habitat and habit do not correspond well with that of *B. velutina* but do correspond remarkably well with that described and illustrated by Boudier (1905-1910: pl. 71, p. 36) under the misapplied name *Cantharellus helvelloides* (Bull.) Quél. The fungus described by Boudier had an obconical form, forked lamellae, white radicating hair-like

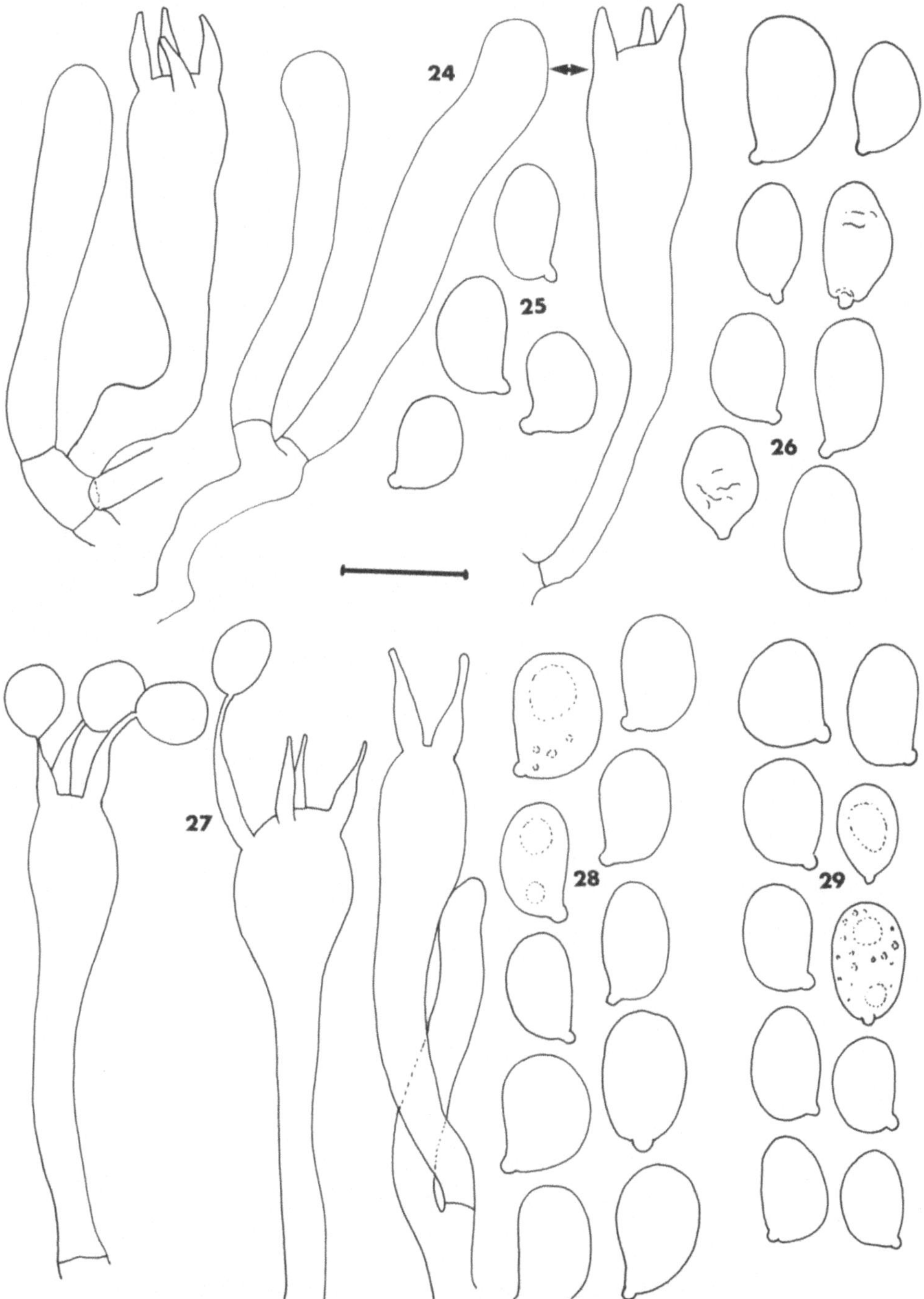

Fig. 24. Basidia from type of *Omphalia sphagnophila* (NYS). Fig. 25. Basidiospores from previous collection. Fig. 26. Basidiospores from purported type of *Agaricus sphagnicola* (K). Fig. 27. Basidia from DAOM 174850, *Botrydina botryoides*. Fig. 28. Basidiospores from previous collection. Fig. 29. Basidiospores from DAOM 174851, *B. botryoides*. Scale bar = 10 um.

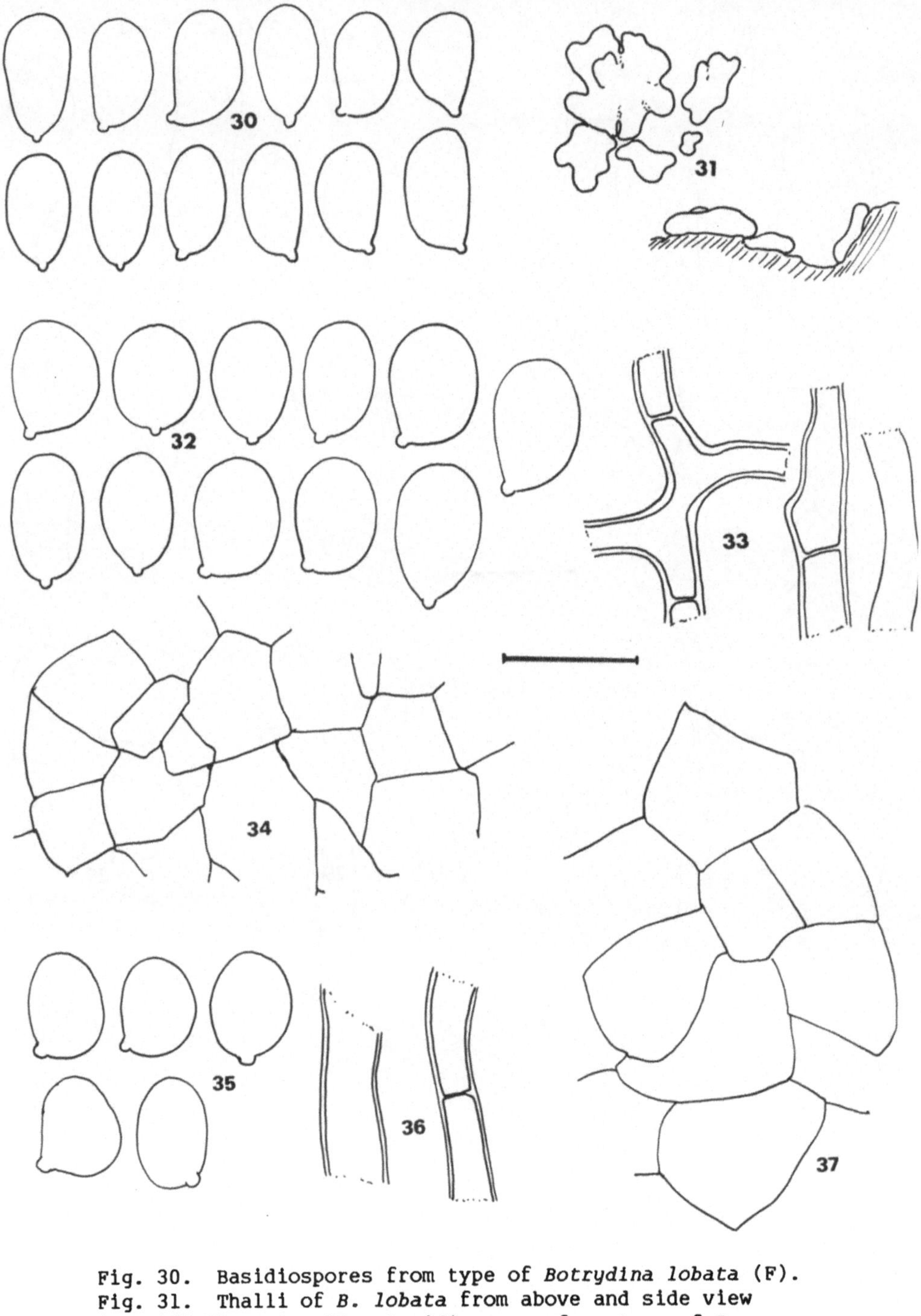

Fig. 30. Basidiospores from type of *Botrydina lobata* (F). Fig. 31. Thalli of *B. lobata* from above and side view (magnified). Fig. 32. Basidiospores from type of *B. aurantiaca* (F). Fig. 33. hyphae subtending granules of thalli of *B. aurantiaca*. Fig. 34. Surface cells of thallus of *B. aurantiaca*. Fig. 35. Basidiospores from type of *Omphalina fulvopallens* (E). Fig. 36. Hyphae subtending granules of thalli with type of *O. fulvopallens*. Fig. 37. Surface cells of thallus from type of *O. fulvopallens*. Scale bar = 10 µm.

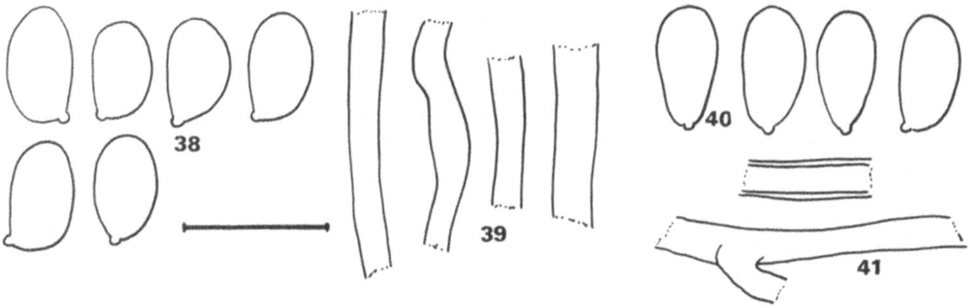

Fig. 38. Basidiospores of *Botrydina velutina*, Favre coll. (G). Fig. 39. Hyphae subtending granules of thallus of *B. velutina*, Favre coll. (G). Fig. 40. Basidiospores of *B. velutina*, Fayod coll. (G). Fig. 41. Hyphae subtending granules of thallus of *B. velutina*, Fayod coll. (G). Scale bar = 10 µm.

mycelial strands, and occurred on walls with small moss cushions. Hora (1960) published the name *Omphalina rickenii* Hora for a small fungus resembling *A. sagittula*, said to grow on walls, sandy soil and among grasses. In the New Check List of British Agarics and Boleti, Dennis, Orton & Hora (1960) listed *C. helvelloides* sensu Boudier as *O. rickenii*. We believe *A. sagittula* is an earlier name but hesitate to transfer it to a modern generic placement until more microscopic details are studied.

KEY TO BOTRYDINA THALLI

A. Thallus foliose, squamulose, or disc-shaped .... B
A. Thallus composed of globose granules .... D
   B. Thallus concave when young, disc shaped; margins whitish, thickened, except in age; Arctic and alpine; North America and Eurasia .... *B. viridis*
   B. Thalli compressed; margins concolorous, not thickened; alpine; northern South America or Australian C
C. Thalli discretely lobed, 2-3 mm broad; lobes elongated, sometimes branched; alpine in northern South America .... *B. lobata*
C. Thalli disc-shaped to slightly angular, 200-900 µm broad, forming an areolate crustose surface when crowded; alpine in SE Australia .... *B. chromacea*
   D. Hyphae subtending lichenized granules 2-3 µm diam. .... *B. velutina*
   D. Hyphae subtending lichenized granules (2.5-)3-4(-5) µm diam. .... E
E. Subtending hyphae mainly thick-walled (1-1.5(-2.8) µm thick); granules rarely over 100 µm diam.; cortical cells with slightly thickened walls; Arctic and alpine areas; North America and Eurasia .... *B. luteovitellina*
E. Subtending hyphae thin-walled to slightly thick-walled (0.1-0.5 µm thick), granules often over 100 µm diam. (thalli of *Semiomphalina* resemble the following two) .... F
   F. Granules with a wide range of sizes up to 300 µm diam.; subtending hyphae 3-4(-5) µm diam.; Arctic, alpine and boreal; North America and Eurasia .... *B. botryoides*
   F. Granules mainly 40-135 µm diam.; subtending hyphae 2.5-4 µm diam.; alpine; northern South America .... *B. aurantiaca*

ACKNOWLEDGEMENTS

Mr. J.R. Laundon (BM) and Dr. I. Brodo (CANL) provided invaluable assistance by way of loans, literature and discussion. Dr. V. Demoulin (LG) kindly provided collections, in particular pointing out the lichenized *B. chromacea*. Dr. J. Ginns (DAOM) especially assisted with numerous Yukon collections and constructive review. Thanks are also given to Parks Canada for collecting permits for Gros Morne, Glacier, Pacific Rim, and Riding Mt. Natl. Parks, and the personnel at Naikoon Prov. Park. The loan of types by Drs. D.A. Reid (K), R. Watling (E),, O. Monthoux (G), R. Singer (F), and J. Haines (NYS) was greatly appreciated.

REFERENCES

Acharius, E., 1810, "Lichenographia universalis," Gottingen.

Acton, E., 1909, *Botrydina vulgaris*, Brébisson, a primitive lichen, *Ann. Bot.*, 23: 579-585.

American Code of Botanical Nomenclature, 1907, *Bull. Torrey Bot. Club*, 34: 167-178.

Berkeley, M. J., 1836, Fungi. Vol. 5, Pt. 2, *in* "The English Flora," J. E. Smith, ed., London.

Bigelow, H. E., 1959, Notes on fungi from northern Canada. IV. Tricholomataceae, *Can. J. Bot.*, 37: 769-779.

Bigelow, H. E., 1970, *Omphalina* in North America, *Mycologia*, 62: 1-32.

Bigelow, H. E., 1982, North American species of *Clitocybe*, Part I, *Beih. Nova Hedw.*, 72: 1-213.

Bigelow, H. E., 1983, Some clampless species of *Clitocybe*, *Cryptogamie (Mycol.)*, 4: 93-98.

Bigelow, H. E., and Barr, M. E., 1966, Contribution to the fungus flora of northeastern North America. IV, *Rhodora*, 68: 175-191.

Boissiere, J. -C., 1980, Un vrai Basidiolichen européen: l'*Omphalina umbellifera* (L. ex Fr.) Quél. Étude ultrastructurale, *Cryptogamie (Bryol.-Lich.)*, 1: 143-149.

Bory de St. Vincent, J. B. G. M., 1797, "Mémoire sur les genres *Conferva* et *Byssus*, du chevalier O. Linné," L. Cavazza, Bordeaux.

Boudier, J. L. E., 1905-1910, "Icones mycologicae ou inconographie des champignons de France principalement Discomycètes avec texte descriptif," P. Klincksieck, Paris.

de Brébisson, M., 1839, De quelques nouveaux genres d'Algues, *Mém. Soc. Acad., Agri., Indust. & Instruct. (Falaise 1839)*,: 34-37.

Bresinsky, A., and Stangl, J., 1974, Beiträge zur Revision M. Britzelmayrs "Hymenomyceten aus Sudbayern" 12, *Z. Pilzk.*, 40: 69-104.

Britzelmayr, M., 1890, "Hymenomyceten aus Sudbayern. Teil VI. Boleti, Cortinarii, Dermini, Hydnei, Hyporhodii, Leucospori, Melanospori," Berlin.

______, 1893, Materialien zur Beschreibung der Hymenomyceten, *Bot. Centralbl.*, 54: 33-40.

______, 1898, Revision der Diagnosen zur den von Britzelmayr aufgestellen Hymenomyceten-Arten, *Bot. Centralbl.*, 73: 129-148.

Bulliard, J. B. F., 1786, "Herbier de la France. Vol. 6." Fasc. 65-72. Pls. 256-288, Paris.

Bulliard, J. B. F., and Ventenat, E. P., 1809, "Histoire des champignons de la France II(1)," p. 369-372, 509-540.

Cleland, J. B., 1934, "Toadstools and mushrooms and other fungi of South Australia. Part 1," Britsh Science Guild (South Australia Branch), Adelaide.

Clémençon, H., 1982, Kompendium der Blätterpilze Europäische Omphalinoide Tricholomataceae, *Z. Mykol.*, 48: 195-237.

Code of Botanical Nomenclature, 1904, *Bull. Torrey Bot. Club*, 31: 249-290.

Cooke, M. C., 1881-1891, "Illustrations of British Fungi (Hymenomycetes)," Williams and Norgate, London.
Dennis, R. W. G., Orton, P.D., and Hora, F.B., 1960, New Check List of British Agarics and Boleti, *Trans. Brit. Mycol. Soc.*, 43 (Suppl.): 1-225.
Dillenius, J. J., 1741, "Historia muscorum... ," Theatro Sheldoniano, Oxford.
Donk, M. A., 1949, New and revised nomina generica conservanda proposed for Basidiomycetes Fungi, *Bull. Bot. Gard. Buitenzorg III*, 18: 83-168.
Donk, M. A., 1962, The generic names proposed for Agaricaceae, *Beih. Nova Hedw.*, 5: 1-320.
Drouet, F., and Daily, W. A., 1956, Revision of the coccoid myxophyceae, *Butler Univ. Bot. Stud.*, 12: 1-218.
Earle, F. S., 1909, The genera of the North American gill fungi, *Bull. New York Bot. Gard.*, 5: 373-451.
Farr, E. R., Leussink, J. A., and Stafleu, F. A., 1979, "Index Nominum Genericorum (Plantarum)," Bohn, Scheltema and Holkema, Utrecht.
Favre, J., 1955, Les champignons supérieure de la zone alpine du Parc National Suisse, *Rés. recher. sci. entrepr. Parc National Suisse.*, 5(N.F.): 1-212.
Fries, E. M., 1821, "Systema mycologicum, I," Lund.
Fries, E. M., 1825, "Systema Orbis Vegetabilis. Pars I. Plantae Homonemeae," Typographia Academica, Lund.
Fries, E. M., 1828, "Elenchus fungorum I & II," Griefswald.
Fries, E. M., 1832, "Systema mycologicum III," Griefswald.
Fries, E. M., 1836-1838, "Epicrissis Systematis Mycologici seu Synopsis Hymenomycetum," Typographia Acad., Upsaliae.
Fries, E. M., 1874, "Hymenomycetes Europaei sive Epicriseos Systematis Mycologici. Ed. Altera," Berling, Uppsala.
Gams, H., 1962, Die Halbflechten *Botrydina* und *Coriscium* als Basidio-lichen, *Österr. Bot. Zeit.*, 109: 376-380.
Gray, S. F., 1821, "A natural arrangement of British plants I," London.
Gulden, G., and Lange, M., 1971, Studies in the macromycete flora of Jotunheimen, the central mountain massif of south Norway, *Norw. J. Bot.*, 18: 1-46.
Hawksworth, D. L., 1972, The natural history of Slapton Ley nature reserve. IV. Lichens, *Fld. Stud.*, 4: 535-578.
Hawksworth, D. L., James, P.W., and Coppins, B.J., 1980, Checklist of British lichen-forming, lichenicolous and allied fungi, *Lichenologist*, 12: 1-115.
Hedwig, R. A., 1802, "Observationum botanicorum fasciculus primus," Ex officina Hirschfelida, Lipsiae.
Heikkilä, H., and Kallio, P., 1966, On the problems of subarctic basidiolichens, I, *Ann. Univ. Turku, Ser. A., II Biol. - Geograph.*, 36: 48-74.
Heikkilä, H., and Kallio., P., 1969, On the problem of subarctic basidiolichen, II., *Ann. Univer. Turku, Ser. A., II. Biol.-Geograph.*, 40: 90-97.
Henssen, A., and Kowallik., K., 1976, A note on the mycobiont of *Coriscium viride* (Ach.) Vain., *Lichenologist*, 8: 197.
Honegger, R., and Brunner, U., 1981, Sporopollenin in the cell walls of *Coccomyxa* and *Myrmecia* phycobionts of various lichens: an ultrastructural and chemical investigation, *Can. J. Bot.*, 59: 2713-2734.
Hora, F. B., 1960, New check list of British Agarics and Boleti. Part IV. Validations; new species and critical notes, *Trans. Brit. Mycol. Soc.*, 43: 440-459.
James, P. W., 1965, A new check-list of British lichens, *Lichenologist*, 3: 95-153.

Karsten, P. A., 1879, Rysslands, Finlands och den Skandinaviska halföns Hattsvampar. I. Skifsvampar., *Bidrag til Kännedom af Finlands Natur och Folk*, 32: I-XXVII; 1-571.
Karsten, P. A., 1889, Symbolae ad Mycologiam Fennicam., *Meddel. af Soc. pro Fauna et Flora Fenn.*, 16: 84-106.
Kühner, R., 1980, Les Hyménomycètes agaricoides, *Numéro spécial. Bull. Soc.Linné. Lyon 49e Année*, 1-1027.
Kuyper, T. W., 1986, Generic delimitation in European Omphalinoid Tricholomataceae., in La Famiglia delle Tricholomataceae, *Atti del Centro Studi per la Flora Mediterranea (Borgo Val di Taro, Italy)* 6: 83-104.
Lamoure, D., 1960, Preuve caryologique que le Basidiomycète *Omphalina ericetorum* (Pers. ex Fr.) M. Lange peut être le mycobionte du lichen *Botrydina vulgaris* Bréb., *C.R. Acad. Sci. Paris D*, 266: 2339-2340.
Lamoure, D. 1982, Alpine and circumpolar *Omphalina* species p. 201-215, in: "*Arctic and Alpine Mycology. The first international symposium on Arcto-Alpine Mycology*," G. A. Laursen, and J. F. Ammirati, eds., Univ. Wash. Press, Seattle.
Lange, M., 1981, Typification and delimitation of *Omphalina* Quél., *Nord. J. Bot.*, 1: 691-696.
Linnaeus, C., 1737, "Flora Lapponica," Amsterdam.
Linnaeus, C., 1745, "Flora Suecica," Stockholm.
Linnaeus, C., 1753, "Species plantarum," Stockholm.
Linnaeus, C., 1755, "Flora Suecica. 2 ed.," Stockholm.
Linnaeus, C., 1759, "Systema Naturae. 10th ed. vol. 2," Stockholm.
Linnaeus, C., 1762, "Species Plantarum. 2nd ed.," Stockholm.
Micheli, P. A., 1729, "Nova Plantarum Genera," Florence.
MØller, F. H., 1945, "Fungi of the Faeroes. I. Basidiomycetes," Copenhagen.
Murrill, W. A., 1916, *Omphalina* Quél., *N. Amer. Flora*, 9(5): 344-352.
Oberwinkler, F., 1970, Die Gattungen der Basidiolichenen, *Dtsch. Bot. Ges., Neue Folge*, 4: 139-169.
Oberwinkler, F., 1984, Fungus-alga interactions in basidiolichens, *Beih. Nova Hedw.*, 79: 739-774.
Orton, P. D., 1984, Notes on British Agarics: VIII, *Notes Roy. Bot. Gard. Edinburgh*, 41: 565-624.
Persoon, C. H., 1796, "Observationes mycologicae. I," Leipzig.
Persoon, C. H., 1801, "Synopsis methodica fungorum," Gottingen.
Poelt, J. 1975, Basidienflechten eine in den Alpen lange übersehence Pflanzengruppe, *Jahrbuch d. Vereins.*, (1975): 81-92.
Poelt, J., and Oberwinkler, F., 1964, Zur Kenntnis der flechtenbilden Blätterpilze der Gattung *Omphalina*, *Österr. Bot. Zeit.* 111: 393-401.
Quélet, L., 1872, Les champignons du Jura et des Vosges, *Mem. Soc. Emul. Montbel. II*, 5: 43-332.
Quélet, L., 1886a, "Enchiridion fungorum in Europa media," Lyon.
Quélet, L., 1886b, Quelques espèces critiques ou nouvelles de la Flore Mycologique de France, *C.R. Ass. franç. Av. Sci. (Grenoble 1885)*, 14(2): 444-453.
Ray, J., 1724, "Synopsis methodica stirpium britannicarum. 3rd ed.," J. Dillenius, ed., Guilielmi and Joannis Innys, London.
Redhead, S. A., 1984, *Arrhenia* and *Rimbachia*, expanded generic concepts and a reevaluation of *Leptoglossum* with emphasis on muscicolous North American taxa, *Can. J. Bot.*, 62: 865-892.
Redhead, S. A., and Weresub. L. K., 1978, On *Omphalia* and *Omphalina*, *Mycologia*, 70: 556-568.
Rogers, D. P., 1950, Nomina conservanda proposita and nomina confusa - fungi, *Supplement. Farlowia*, 4: 15-43.
Ross, R,. and Irvine. L. M., 1967, The typification of the genus *Byssus* L. (1753), *Taxon*, 16: 184-186.

Singer, R., 1946, New and interesting species of basidiomycetes. II, *Pap. Mich. Acad. Sci. Arts & Lett.*, 32: 103-150.
Singer, R., 1962, Type studies on Basidiomycetes X, *Persoonia* 10: 1-62.
Singer, R., 1970, Omphalinae (Clitocybeae-Tricholomataceae, Basidiomycetes), *Flora Neotropica*, 3: 1-81.
Singer, R., "1975, The Agaricales in modern taxonomy. 3rd ed." J. Cramer, Vaduz.
Singer, R., and Clemençon, H., 1972, Notes on some leucosporous and rhodosporus European Agarics, *Nova Hedwigia*, 23: 305-351.
Singer, R., and Smith, A. H., 1946, Proposals concerning the nomenclature of the gill fungi including a list of proposed lectotypes and genera conservanda, *Mycologia*, 38: 240-299.
Smith, A. H., 1973, Agaricales and related secotoid gasteromycetes. Chap. 23. p. 421-450, *in:* "The Fungi. An advanced treatise. IVB," G. C. Ainsworth, F. K. Sparrow and A. S. Sussman, eds., Academic Press, New York.
Vahl, M., 1790, "Flora Danica. Vol. 6," Fasc. 18, Pl. 961-1020, Kobenhavn.
Voss, E.G. et al, 1983. International code of botanical nomenclature adopted by the thirteenth international botanical congress, Sydney, August 1981, *Regnum vegetabile*, 11: 1-472.
Wainio, E. A., 1890, "Étude sur la classification naturelle et la morphologie des lichens du Brésil," Helsingfors.
Watling, R., 1977, Larger fungi from Greenland, *Astarte*, 10: 61-71.
Watling, R., 1981, Lichenicolous agarics, *Bull. British Lich. Soc.*, 49: 28-37.
Watling, R., and Richardson M. J., 1971, The agarics of St. Kilda, *Trans. Bot. Soc. Edinburgh*, 41: 165-187.
Weinmann, C. A., 1836, "Hymeno- et Gastero- mycetes hucusque in Imperio Rossico Observator," Acad. Imp. Sci., Petropoli.

COLLECTIONS EXAMINED (excluding types)

BOTRYDINA BOTRYOIDES

Austria: Tirol, Patscherkogel near Innsbruck, Sept. 22, 1981, Kuyper 1814 (L 980.30.803). Belgium: Prov. Luxembourg, Eupen, Hestreux, Sept. 13, 1981, Schreurs 624 (L 981.70.235) and Sept. 14, 1981, Schreurs 625 ((L 981.70.254). Canada: British Columbia: Burnaby, April 17, 1968, G.F. Crossley (DAOM 128468); Fort St. James, June 22, 1940, T. McCabe 196 (DAOM 114849); Glacier Natl. Park, Illecillewaet R., Sept. 12, 1980, S.A. Redhead 3671 (DAOM 190407); Gulf Is., Galiano I., April 10, 1983, S.A.R. 5042 (DAOM 188204); Kootenay Natl. Park, Aug. 30, 1977, H.M.E. Schalkwyk (DAOM 170691); Monashee Mts. area, W. of Cherryville, Sept. 25, 1980, S.A.R. 4067 (DAOM 184115); Queen Charlotte Is.: Graham I.: S. Naikoon Prov. Park, Geikie Cr., Sept. 17, 1982, S.A.R. 4337 (DAOM 190537), Phantom Cr., 1000' elev., 9 km E. Shields Bay, Sept. 19, 1982, S.A.R. 4414 (DAOM 190406), 7 km E. Chown R. mouth, Sept. 19, 1982, S.A.R. 4327 (DAOM 185571); Surrey, Oct. 10, 1973, S.A.R. #AU5 (DAOM 185748); Vancouver, Mar. 15, 1982, R.G. Thorn (DAOM 189882); Vancouver I.: Cowichan L., May 15, 1976, W.G. Ziller (DAOM 109813), Jordan R., May 11, 1968, D. Evans (DAOM 128542), Oct. 4, 1979, S.A.R. 3335 (DAOM 190404), Pacific Rim Natl. Park, Long beach sect., Sept. 29, 1979, S.A.R. 3248 (DAOM 175324), 56 km W. Port Alberni, Sept. 30, 1979, S.A.R. 3256 (DAOM 190405), Saanich, March 27, 1942, G.A. Hardy (DAOM 11168). Manitoba: E. of Churchill, June 23, 1948, J. Gillett & W. Cody 1787 (DAOM 21527); Gillam, June 13, 1950, W.B. Schofield 958 (DAOM 25862); Riding Mt. Natl. Park, Aug. 27, 1979, S.A.R. 3093 (DAOM 176592). Newfoundland: I. of Newfoundland: L'Anse aux Meadows, Sept. 22, 1983, S.A.R. 4922 (DAOM 189777), Gros Morne Natl. Park, Sept. 18, 1983, S.A.R. 4785 (DAOM 189776), St. Anthony, June 4 and 7, and July

15, 1951, D.B.O. Savile & J. Vaillancourt 1729, 1744, 2051 (DAOM 28851, 27225, 27223): Labrador; Crater L., Aug. 11, 1954, J.M. Gillett 8962 (DAOM 45035), Goose Bay, Aug. 10, 1949, W.B. Schofield 697 (DAOM 27517). Northwest Territories: Baffin I.: Inugsuin Fiord, July 26, 1967, J.A. Parmelee & J.R. Seaborn (DAOM 190532), Icy Arm, Clyde Inlet, Aug. 5, 1950, V.C. Wynne-Edwards (DAOM 27901), July 30, 1950, V.C. W-E. (DAOM 27735), July 1965, G.A. Petrie (DAOM 110483), Frobischer Bay, July 10, 1948, H.A. Senn & J.A. Calder 3876 (DAOM 21526); Baker L., Aug. 18, 1974, E. & M. Ohenoja (DAOM 155306); Chesterfield Inlet, July 19, 1950, D.B.O. Savile 1038 (DAOM 26688); Ellef Ringnes I., SW. of Isachsen, July 16 and 28, 1960, D.B.O.S. 4258, 4315 (DAOM 75022, 75021); Inuvik, June 24, 1963, G.W. Scotter 2816 (DAOM 109791); Lower Hay R., July 17, 1951, W.H. Lewis 750 (DAOM 27865); Repulse Bay, Aug. 4, 1974, E. & M. Ohenoja (DAOM 159710). Ontario: Algonquin Prov. Park, June 12, 1983, R.G.T. (DAOM 189883); Pukaskwa Natl. Park, July 26, 1982, R.G.T. (DAOM 189884). Quebec: Great Whale River, June 23 and Sept. 5, 1949, D.B.O.S. (DAOM 21742, 21846); Parc Laurentide, Grand Jardin (47°38'N, 70°52'W), Aug. 15, 1981, S.A.R. et al. (DAOM 180811); Réserve Chibougamau: 15.75 km N. of Buchart, Aug. 22, 29 and 29, 1976, S.A.R. 1992, 2101, 2102 (DAOM 174850, 174851, 174853), Lac Harquail, Aug. 21, 1976, S.A.R. 1961 (DAOM 174849). Yukon Territory: S. Canol Rd., km 55, 61°45'N, 133°05'W, Aug. 2, 1980, J. Ginns 5327 (DAOM 190535); Dempster Hwy. km 148 (65°04'N, 138°09'W), July 2, 1982, J.G. 6476 (DAOM 190482), km 447 (66°55'N, 136°17'W), July 4, 1982, J.G. 6590 (DAOM 190483); Granville, 1949, W.W. Judd (DAOM 21741); Jensen Flats, July 24, 1949, L.G. Billard & J.A. Calder 3918 (DAOM 21745); Klondike Hwy., km 326, 61°51'N, 136°06'W, Aug. 10, 1980, J.G. 5798 (DAOM 190534); Nisling R., 2700', 61°54.4'N, 137°52'W, July 24, 1980, J.G. 4945 (DAOM 190536); Ogilvie-Wernecke Mts.: 5500', 64°22'N, 137°59'W, July 2, 1984, J. Ginns 8035 & W.J. Cody (DAOM 190505), 4050', 64°45'N, 138°07'W, July 4, 1984, J.G. 8091b & W.J.C. (DAOM 190531), 16 mi. SW Chapman L., 64°40'N, 138°28'W, July 5, 1984, J.G. 8097 & W.J.C. (DAOM 190511), Cassair Dome, 4300', 64°15.5'N, 140°22'W, July 7, 1984, J.G. 8128 & W.J.C. (DAOM 190513), 5500', 64°19'N, 137°34'W, July 14, 1984, J.G. 8234 & W.J.C. (DAOM 190517); Reindeer Mt., SW of Dawson, 4800', 63°37'N, 139°22'W, July 16, 1984, J.G. 8263, 8264, 8337, 8338 (DAOM 190518, 190519, 190521, 190522); Richardson Mts., 68°35'N, 137°20'W, July 12, 1982, and 67°44.5'N, 137°28'W, July 15, 1982, W.J. Cody & J.G. 6846, 6904 (DAOM 190484, 190481). Great Britain: Scotland: Fort William, Ben Nevis, Sept. 15, 1983, Kuyper 2377 (L 983.236.134); Perthshire, Davan near Kindrogan, Sept. 23, 1983, Kuyper 2417 (L 983.236.179), Cairnwell, Sept. 27, 1983, Kuyper 2446 (L 983.236.29) and Kuyper 2447 (L 983.236.146), Dell Lodge, Sept. 7, 1981, Noordeloos 1451 (L 982.330.115) and Noordeloos 1452 (L 982.330.100). Federal Republic of Germany: Nordreinland-Westfalen, Bentheim, Bentheimerwald, May 1, 1960, Barkman (L 960.110.715). The Netherlands: Noord Brabant, Asten, Grote Peel, Aug. 21, 1960, Jansen (L 960.216.686); Utrecht, Lage Vuursche, Sept. 28, 1958, Daams (L 959.16.039); Drente, Lieveren, Brunnerveen, May 9, 1954, Koster (L 954.017.091); Drenthe, Dwingeloo, Sept. 20, 1952, Huijsman (L 954.320.184) and Oct. 1965, van der Eb (L 964.296.113). Switzerland: Jura, Chaux d'Abel, May 29, 1969, Huijsman (L 966.14.456).

BOTRYDINA CHROMACEA

Australia: New South Wales: Northern Budawang Range, Morton Natl. Park, Wog Wog Creek, 2 km from Korra hill, 700-750 m alt., Aug. 12, 1981, V. Demoulin 6154 (LG).

BOTRYDINA LUTEOVITELLINA

Canada: Manitoba: Fort Churchill, June 22-25, 1956, H.A. Crum 6640 (DAOM 54141). Newfoundland: St. Anthony, June 14, 1951, D.B.O. Savile & J. Vaillancourt 1783 (DAOM 28451). Northwest Territories: Axel Heiberg I., Aug. 15, 1967, M. Kuc F32 (DAOM 124714); Baffin I.: Icy Arm Inlet, July 1965, G.A. Petrie (DAOM 190530), Sunneshine Fiord, July 18, 1967, J.A. Parmelee & J.R. Seaborn (DAOM 117576); Chesterfield Inlet, July 22, 1950, D.B.O. Savile 1079 & C.T. Watts (DAOM 26686); Ellef Ringnes I., Isachsen, Aug. 16, 1961, C.R. Harington 530 (DAOM 91524); Melville I., Bridport Inlet, July 12, 1961, J.S. Tener & C.R. Harington 401 (DAOM 91523); Southampton I., 7 mi. W. Boas R., June 23, 1952, G. Cooch 4 (DAOM 38599); Tuktoyaktuk, July 2, 1963, J.A. Parmelee 2604 (DAOM 96377). Quebec: Great Whale R., June 21, 1949, D.N. Jenkins (DAOM 21743), June 23, 1949, D.B.O.S. (DAOM 21858). Yukon Territory: Dawson, June 26, 1949, L.G. Billiard & J.A. Calder 3370 (DAOM 21806); Dempster Hwy., km 94, 64°40'N, 138°24'W, July 1, 1982, J. Ginns 6446 (DAOM 190533); Ogilvie-Wernecke Mts., 4050', 64°45'N, 138°07'W, July 4, 1984, J.G. 8091a & W.J. Cody (DAOM 190509); Richardson Mts.: 66°22'N, 135°49'W, July 6, 1982, J.G. 6752, 6753 & W.J.C. (DAOM 189782, 190529), 67°10.5'N, 136°17'W, July 9, 1982, J.G. 6794b & W.J.C. (DAOM 190528); Shingle Pt., 68°56'N, 137°12'W, July 9, 1963, J.A. Parmelee 2698 (DAOM 96378). Great Britain: Scotland, Perthshire, Cairnwell, Sept. 17, 1983, T.W. Kuyper 2448 (L 983.236.194).

BOTRYDINA VELUTINA

Federal Republic of Germany: Reinland-Pfaly, Dreibach, Oct. 4, 1979, T. Kuyper 1351 (L 982.217.357). Italy: Piémont, Sept. 25, 1887, V. Fayod (as *Omphalia grisea*, G). Norway: Spitsbergen, Isflorden, Grumantdalen, July 24, 1979, Brand 8062 (L 980.78.275). Sweden: Småland, Femsjö Parish, Sept. 21, 1943, S. Lundell (Fungi Exsiccati Suecici 1752 as *Omphalia anthodia*). Switzerland: Kanton Bern, Steingletscher, Sept. 21, 1984, Kuyper 2566 (L 984.185.022); Pente sud du Mot dal Gajer, 2650 m alt., Aug. 17, 1948, J. Favre (as *Omphalia grisella*, G).

BOTRYDINA VIRIDIS

Austria: Tirol, Rosskogel, Sept. 7, 1982, Schreurs 712 (L 982.38.475). Canada: Island of Newfoundland, Gros Morne Natl. Park, summit of Gros Morne, Sept. 18, 1983, J. Ginns & S.A. Redhead 4790 (DAOM 189780); St. Anthony, June 20, 1951, D.B.O. Savile 1867 & J. Vaillancourt (DAOM 28422). Quebec: Baie James, T. N.-Q., Ile du sud de la baie aux Oies, 53°54'N 79°07'W, Aug. 10, 1986, J. Cayouette & S. Darbyshire J86-211 (DAOM 1988); Parc des Laurentides, near Lac Malbaie, July 1, 1975, Pierre Arsenault (DAOM 150784). Northwest Territories: Baffin I.: Frobischer Bay, June 17 and July 2, 1948, H.A. Senn & J.A. Calder 3702, 3760 (DAOM 21536, 21525), Inugsuin Fiord, July 26, 1967, J.A. Parmelee & J.R. Seaborn (DAOM 117612, 117597); Horn Plateau, Lake-on-the-Mountain, 62°08'N, 118°07'W, July 30-Aug. 2, 1959, J.W. Thieret & R.J. Reich 6283 (DAOM 64106). Yukon Territory: Ogilvie-Wernecke Mts.: 5500', 64°22'N, 137°59'W, July 2, 1984, J. Ginns 8034, 8038 & W.J. Cody (DAOM 190504, 190506), Mt. Patterson, 5900', 64°04'N, 134°38'W, July 3, 1984, J.G. 8058 & W.J.C. (DAOM 190507), 5500', 64°18'N, 138°00'W, July 4, 1984, J.G. 8080 & W.J.C. (DAOM 190508), 16 mi SW Chapman L.,64°40'N, 138°28'W, July 5, 1984, J.G. 8096 & W.J.C. (DAOM 190510), Cassair Dome, 4300', 64°15.5'N, 140°22'W, July 7, 1984, J.G. 8121 & W.J.C. (DAOM 190512), 15 mi NNW of Mt. Gibbon, 5300', 64°55'N, 139°19'W, July 12, 1984, J.G. 8202 & W.J.C. (DAOM 190514), 3700', 65°11'N, 138°57'W, July 13, 1984, J.G. 8214 & W.J.C. (DAOM 190515),

5500', 64°19'N, 137°34'W, July 14, 1984, J.G. 8233 & W.J.C. (DAOM 190516), Mt. Tyrrell, 4700', 63°42'N, 140°03'W, July 20, 1984, J.G. 8528 & W.J.C. (DAOM 190523); Reindeer Mt., SW of Dawson, (DAOM 190520); Richardson Mts.: 66°16'N, 135°48'W, July 6, 1982, J.G. 6695 & W.J.C. (DAOM 190524, 190545), 66°48'N, 136°13'W, July 7, 1982, J.G. 6780 & W.J.C. (DAOM 189781), 67°10'N, 136°17'W, July 9, 1982, J.G. 6794a & W.J.C. (DAOM 190525), 67°44.5'N, 137°28'W, July 15, 1982, J.G. 6903 & W.J.C. (DAOM 190526), Mt. Russell, 2500', 67°44'N, 136°29'W, July 3, 1984, W.J.C. 29891 & J.G. (DAOM 190527); Shingle Pt., 68°56'N, 137°12'W, July 12, 1963, J.A. Parmelee (DAOM 96376). Great Britain: Scotland, Perthshire, Cairnwell, July 3, 1976, Bas 6769 (L 976.58.289) and Sept. 27, 1983, Kuyper 2449 (L 983.236.61). Finland: Kevo, Aug. 11, 1978, Noordeloos 639 (L 979.256.658). Norway: Finnmarken, Varanger Fjord, Aug. 29, 1973, Bas 6082 (L 974.225.994) and Aug. 12, 1978, Noordeloos 661 (L 978.18.022). Switzerland: Alp Trida, Aug. 28, 1984, M. Lange (DAOM 191220); Wengen, Saustal, July 22, 1984, van Crevel (L 983.236.141).

CONTRIBUTORS

Ammirati, J. F. (see Laursen, Ammirati & Farr)
Borgen, T. (see Knudsen & Borgen)
Debaud, J. C. Ecophysiological Studies on Alpine Macromycetes: Saprophytic *Clitocybe* and Mycorrhizal *Hebeloma* associated with *Dryas octopetala*.
Dissing, H. Three 4-spored *Saccobolus* species from North East Greenland.
Döbbeler, P. Ascomycetes Growing on *Polytrichum sexangulare*.
Farr, D. F. (see Laursen, Ammirati & Farr)
Gulden, G. The Genus *Galerina* on Svalbard.
Holm, K. (see Holm & Holm)
Holm, L. and K. Holm. Nordic Juncicolous Mycospaerellae.
Horak, E. *Astrosporina* in the Alpine Zone of the Swiss National Park (SNP) and Adjacent Regions.
Huhtinen, S. New Svalbard Fungi.
Knudsen, H. and T. Borgen. Agaricaceae, Amanitaceae, Boletaceae, Gomphidiaceae, Paxillaceae and Pluteaceae in Greenland.
Kuyper, Th. W. (see Redhead & Kuyper)
Lamoure, D. Agaricales de la Zone Alpine. Genus *Cortinarius* Fr., Subgenus *Telamonia* (Fr.) Loud. Part III.
Lange, M. Arctic Gasteromycetes. The Genus *Bovista* in Greenland and Svalbard.
Laursen, G. A., J. F. Ammirati and D. F. Farr. Hygrophoraceae from Arctic and Alpine Tundra in Alaska.
Leuchtmann, A. *Phaeosphaeria* in the Arctic and Alpine Zones.
Magnuson, J. A. (see Müller & Magnuson)
McKnight, K. H. (see Moser & McKnight)
Metsänheimo, K. Sociology and Ecology of Larger Fungi in the Subarctic and Oroarctic Zones in Northwest Finnish Lapland.
Miller, O. K., Jr. Higher Fungi in Tundra and Subalpine Tundra from the Yukon Territory and Alaska.
Moser, M. and K. H. McKnight. Fungi (Agaricales, Russulales) from the Alpine Zone of Yellowstone National Park and the Beartooth Mountains with Special Emphasis on *Cortinarius*.
Müller, E. and J. A. Magnuson. On the Origin and Ecology of Alpine Plant Parasitic Fungi.
Petrini, O. Endophytic Fungi of Alpine Ericaceae. The Endophytes of *Loiseleuria procumbens*.
Redhead, S. A. and Th. W. Kuyper. Lichenized Agarics: Taxonomic and Nomenclatural Riddles.
Schumacher, T. and S. Sivertsen. *Sarcoleotia globosa* (Sommerf.: Fr.) Korf, Taxonomy Ecology and Distribution.
Sivertsen, S. (see Schumacher & Sivertsen)
Watling, R. Larger Arctic-Alpine Fungi in Scotland.

Fig. 1. ISAM-II Participants

Row 4		1	2	3	4	5	6	7	8
Row 3	9	10	11	12	13	14	15	16	17
Row 2	18	19	20	21	22	23	24	25	
Row 1				26	27	28			

1. Seppo Huhtinen
2. Jean Debaud
3. Gary Laursen
4. Orson Miller
5. Henning Knudsen
6. Scott Redhead
7. Roy Watling
8. Peter Döbbeler
9. Yosio Kobayasi
10. Lennart Holm
11. Morten Lange
12. Orlando Petrini
13. Egon Horak
14. Emil Müller
15. Meinhard Moser
16. Henry Dissing
17. Sigmund Sivertsen
18. Beatrice Irlet
19. Kerstin Holm
20. Hope Miller
21. Gro Gulden
22. Katriina Metsänheimo
23. Denise Lamoure
24. Elizabeth Watling
25. Julie Magnuson
26. Adrian Leuchtmann
27. Trond Schumacher
28. Ivano Brunner

GENERAL INDEX TO GENERIC AND SUBGENERIC NAMES

INDEX TO NEW NAMES